T035036.3

COMING OF AGE WITH QUANTUM INFORMATION
Notes on a Paulian Idea

A passionate and personal account of the early days of quantum information and quantum computing, this unique book is a collection of more than 500 letters between the author and many of the founders of these intriguing fields.

Christopher A. Fuchs is one of the most penetrating modern thinkers on the philosophical foundations of quantum mechanics. This remarkable book follows his journey as he comes to grips with the quantum world. It contains correspondence with Charles Bennett, Gilles Brassard, Rolf Landauer, N. David Mermin, Michael Nielsen, Asher Peres, John Preskill, Abner Shimony, William Wootters, Anton Zeilinger, and many others.

Filled with diary entries, anecdotes, historical selections, and research ideas, this book will fascinate physicists, philosophers, and historians of science.

CHRISTOPHER A. FUCHS is a Senior Researcher at the Perimeter Institute for Theoretical Physics in Waterloo, Canada and an Adjunct Professor at the University of Waterloo. Previously "house philosopher" at the famed Bell Labs in Murray Hill, New Jersey, his honors include a Lee DuBridge Prize Fellowship at the California Institute of Technology, the E.T.S. Walton Award (Science Foundation Ireland), and an Albert Michelson Prize. Most recently, he is a winner of the 2010 International Quantum Communication Award and Chair Elect of the American Physical Society Topical Group on Quantum Information.

"I'm delighted that what Chris Fuchs calls his 'samizdat' – 7 years' worth of fascinating correspondence about the foundations of quantum mechanics – has finally been published in book form. I read the samizdat as a beginning graduate student, and it changed my career. Some of the specific questions Chris ponders here – why, for example, does quantum mechanics involve complex numbers, rather than real numbers or quarternions? – have haunted me ever since. But more than that, reading Chris's letters made me feel, for the first time, like the great conversation of Bohr, Einstein, and other luminaries in the 1920s was still going on today; that their bewilderment over why God made the universe this way and not that hadn't completely given way to technicalities and jaded commentaries on commentaries. Just as importantly, Chris's writing style – brash, outrageous, hilarious, but never mean-spirited – gave me the courage to risk making a fool of myself in my own writing."
Scott Aaronson, Department of Electrical Engineering and Computer Science, Massachusetts Institute of Technology

"The discovery in the 1990s that quantum computers could outperform their classical brethren led to the birth of an entirely new field of physics and computer science: quantum information science, whose breadth now spans thousands of researchers worldwide. As a consequence of the explosive growth of this new field, those working day in and day out with quantum information have obtained an unparalleled understanding of the quantum theory of nature. This understanding has, quite unexpectedly, also led to significant advances in penetrating a subject more often thought of as fodder for philosophy departments or late night bull sessions among physicists: the foundations of quantum theory.

Christopher A. Fuchs has been at the center of the firestorm that has been the rise of quantum information, and is the quintessential example of a new breed of quantum information inspired quantum foundations researchers. Out of a personal tragedy that destroyed his personal possessions, Fuchs has assembled in this book the fragments of his life that were not incinerated during his loss – his email correspondence with friends and colleagues about quantum theory, along with the daily banter of a deep soul seeking to break a new path in the world of ideas. Anyone interested in grabbing a front seat to a new way to see quantum theory, as debated by the author and luminaries of quantum information and quantum foundations, will find this book hard to put down. As a bonus, the book brims over with an overwhelming sense of joy and wonder at our quantum universe that, when combined with a wicked sense of humor, will no doubt make this collection of correspondences a classic document of the now blossoming age of quantum information science."
Dave Bacon, Department of Computer Science and Engineering, University of Washington

"Full of humor, vivid characters, and intellectual drama, Chris Fuchs' letters are like Einstein's ideal play – 'a delicious fragment of life, scintillating with various colors according to the position of the beholder' (though the book's two guardian angels are Einstein's favorite antagonists, Bohr and Pauli!). A window into physics in the process of being untangled, this rich email tapestry is the *fin de siècle* testament to Heisenberg's dictum that 'science is rooted in conversations.'

While references ricochet from William James to Bill Wootters; Bennett & Brassard to Flatt & Scruggs; Schrödinger to Sylvia Plath, the reader starts in the middle, flips back and forth, writes in the margins and gaps – encouraged not to passively absorb, but to engage, struggle, learn, delight.

My much-annotated and coffee-stained copy sits not on my bookshelf but on my desk, and I still stumble over new delights."
Louisa Gilder, author of *The Age of Entanglement*

"Nobody today writing about quantum mechanics combines poetry and analysis to better effect than Chris Fuchs . . . The death of letters as a high literary form brought about by the telephone turns out to have been only a lengthy coma – a twentieth-century aberration . . . The thought-provoking pages that follow, which can either be read like a Nabokov novel, or dipped into from time to time, like a collection of poems or short stories, gloriously provide a twenty-first-century demonstration that the art form is again alive and well – and also, of course, that there remain profound questions to ask and to strive to answer about the real meaning of quantum mechanics."
N. David Mermin, Horace White Professor of Physics Emeritus, Cornell University (from the Foreword)

"From the beginning, quantum information promised to transform not only our technologies for communication and computation, but also the received interpretation of quantum theory itself. In this book you can watch this transformation taking place in the mind of one of the pioneers of the field. No one has done more to articulate the information-theoretical interpretation of quantum theory than Fuchs. I am unsure if this transformation will ultimately take us to a new understanding of the quantum but, as this book demonstrates, the journey is as least as important as the destination. For both physicists and philosophers alike, this book will entertain, inform, and almost certainly challenge."
Gerard Milburn, Centre for Quantum Computer Technology, The University of Queensland

"In this marvellous collection of letters we get to watch over Chris Fuchs's shoulder as he wrestles with one of the great open scientific problems of today: how should we understand quantum theory? This book reminds us forcefully just how important that problem is, and inspires and provokes us to think more deeply about the problem. It's also a pleasure to read, personable, funny, and with a stimulating turn of phrase on nearly every page."
Michael Nielsen, co-author of *Quantum Computation and Quantum Information*

"Scientists communicate their ideas in a variety of ways, through published articles, technical seminars, classroom lectures, and textbooks. Most of all we talk to one another, and increasingly we convey and solidify our thoughts in emails to our colleagues. But very few of us use email as skilfully, prolifically, passionately, or fruitfully as Chris Fuchs. Anyone fascinated by the mysteries of the quantum world is sure to be enlightened and entertained by this extraordinary collection, in which Chris wrestles with some of the deepest questions about how we describe Nature. You have never read a physics book like this one."
John Preskill, Richard P. Feynman Professor of Theoretical Physics, California Institute of Technology

COMING OF AGE
WITH QUANTUM INFORMATION

Notes on a Paulian Idea

Christopher A. Fuchs

Perimeter Institute for Theoretical Physics
Waterloo, Canada

CAMBRIDGE
UNIVERSITY PRESS

CAMBRIDGE
UNIVERSITY PRESS

Shaftesbury Road, Cambridge CB2 8EA, United Kingdom

One Liberty Plaza, 20th Floor, New York, NY 10006, USA

477 Williamstown Road, Port Melbourne, VIC 3207, Australia

314–321, 3rd Floor, Plot 3, Splendor Forum, Jasola District Centre, New Delhi – 110025, India

103 Penang Road, #05–06/07, Visioncrest Commercial, Singapore 238467

Cambridge University Press is part of Cambridge University Press & Assessment,
a department of the University of Cambridge.

We share the University's mission to contribute to society through the pursuit of
education, learning and research at the highest international levels of excellence.

www.cambridge.org
Information on this title: www.cambridge.org/9780521199261

First published 2011

A catalogue record for this publication is available from the British Library

Library of Congress Cataloging-in-Publication data
Fuchs, Christopher A.
[Correspondence. Selections]
Coming of Age With Quantum Information : Notes on a Paulian Idea : Selected Correspondence,
1995–2001 / Christopher A. Fuchs.
p.   cm
Includes bibliographical references and index.
ISBN 978-0-521-19926-1 (Hardback)
1. Fuchs, Christopher A.–Correspondence.   2. Physicists–Sweden–Correspondence.
3. Quantum theory.   I. Title.
QC16.F79A4 2010
530.12–dc22

2010042734

ISBN   978-0-521-19926-1   Hardback

Contents

Foreword

On 18 September 1996 I received email that began

> Dear Dr. Mermin,
>
> I encountered your "Ithaca Interpretation" paper this morning on the `quant-ph` archive ... and, I must say, I've been walking around with a nice feeling since. There are some things in it that I like very much!

Although I didn't really know my correspondent, it was clear from what followed that he had thought hard – probably harder than I had – about many of the matters I was trying to address. His scholarship and intelligence were so evident that I started walking around with a nice feeling myself at having received unsolicited praise from so thoughtful a source.

Little did I know that those heartwarming words of appreciation were just bait to lure me into a long critical exchange which, to my pleasure and enlightenment, has been going on ever since. A part of my subsequent education (there have also been several invariably instructive and delightful real-life meetings) can be found in Chapter 18 below.

If Chris Fuchs (rhymes with "books") did not exist then God would have been remiss in not inventing him. Foundations of quantum mechanics is a unique blend of poetry and analysis. Without the poetic vision the analysis tends to chase its own tail into more and more convoluted realms of intricate triviality. Without the analysis the poetry easily degenerates into self-indulgent doggerel. To achieve an adequate understanding of how quantum mechanics captures our efforts to express the "relations between the manifold aspects of our experience" [Bohr] requires both the poetry and the analysis. Analysis is needed to pin down the structure of the relations; poetry is required to characterize how they reflect the experience. Nobody today writing about quantum mechanics combines poetry and analysis to better effect than Chris Fuchs. By deliberate authorial choice, the 500 pages that follow are long on poetry and short on analysis, but a search on "Fuchs" will uncover in these Archives[1] many examples where the balance is more even.

The past decade has seen the growth of intense interest in applications of quantum mechanics to information processing, brought about by its deep intellectual richness, in fortuitous but sociologically significant resonance with our cultural obsession with keeping

---

[1] Explained in Introduction, see footnote 5.

secrets. Chris Fuchs is the conscience of the field. He never loses sight of the real aim of these pursuits, and if you yourself thought it had to do with secure data transmission, RSA code cracking, fast searches, fighting decoherence, concocting ever more ingenious tricks, and such, you should look up from your beautiful algorithms or candidate qubits for a few hours every now and then to browse through these pages.

The real issue is nothing less than how you and I can each construct a representation of the manifold aspects of our individual experiences (loosely known as a world), and the constraints that my representation imposes on yours, and vice-versa. By focusing explicitly on the strange information-processing capabilities inherent in the quantum mechanical description of physical reality, the new discipline of quantum information offers an opportunity to put on a sound foundation what was only hinted at in the convoluted prose of Bohr, the facile sensationalism of Heisenberg, the aphorisms of Pauli, and the poetic mysticism of Schrödinger. If it hasn't occurred to you that this is the real justification for your quantum information-theoretic pursuits, then you owe it to yourself to pause and peruse these pages.

Those who do not come at the subject from the perspective of quantum information theory may well be irritated or even outraged by some of the views expressed herein about their fondest notions. But you would have to be an exceptionally conservative Bohmian, a dry-as-dust consistent historian, a stubbornly literal-minded dynamical collapsian, a supremely surrealistic many-worldsian, or an irresponsibly post-modern correlations-without-correlatan not to be at least entertained and amused and even, every now and then, instructed by this highly readable commingling of scholarship, intellectual passion, philosophical vision, and biographical glimpses into the daily life of a young scientist.

Indeed, one of the difficulties I have encountered at a busy time in writing this Foreword with dispatch, is the distracting presence, in another window on my screen, of the text itself, constantly luring me away from my own painful efforts at expression, into its charming, and by no means fully explored, byways.

The death of letters as a high literary form brought about by the telephone turns out to have been only a lengthy coma – a 20th century aberration. Clearly the rise of email had, from the beginning, the potential to resuscitate the patient. Now we have an existence proof. The thought-provoking pages that follow, which can either be read like a Nabokov novel, or dipped into from time to time, like a collection of poems or short stories, gloriously provide an early 21st century demonstration that the art form is once again alive and well – and also, of course, that there remain profound questions to ask and to strive to answer about the real meaning of quantum mechanics.

N. David Mermin
Ithaca, New York

Introduction to the Cambridge University Press Edition,

or How to Stuff a Wild Samizdat[1]

I imagine it is hard to write a book; I am not sure I could ever do that. But since I discovered the medium of email in 1990, I have never stopped writing. I write, rewrite, and hone my ideas almost exclusively with the aid and clarifying influence of that wonderful tool. The influence may be subliminal, but it is there overpoweringly: if you really want to get an idea worked out for yourself, make it understandable to another mind through correspondence, where your face cannot be seen, but your voice must be heard – do it over and over again, gearing each presentation to each individual mind. Connect the idea to every aspect of life you can. If you stand the test, you are making progress, and often unbeknownst to yourself, transforming the initial seed of an idea into a more resilient and compelling structure. This collection of emails, graciously reprinted by Cambridge University Press with the encouragement of Simon Capelin, is predominantly a record of my honing one such idea – *The Paulian Idea* – as I came of age with the field of quantum information. If I can entertain the reader along the way, that is just as well too.

The field of quantum information did not really exist until about 1993,[2] and I was fortunate enough to be there in the middle of a crowd of great minds hammering it out from day to day. It was a time of great excitement: I could see that their concepts and methods, suitably developed, would give the tools I needed to sculpt a certain vision of the world, one that had been floating diffusely in the back of my mind for years. It was a vision I wanted more than anything else. But to tell you about that, and indeed about the origin of this volume, I should tell you the strange story of how I came to be a physicist in the first place. It had nothing to do with seeing or feeling some beautiful order in nature, like the kind of thing I would read about in the physics biographies. It was not like Einstein's childhood experience with a compass, where he came to the conviction that there had to be "something behind things, something deeply hidden."

My career, instead, is to be blamed on the 1970s television stations of San Antonio, Texas, ninety miles down the road from my little hometown Cuero. In those days, there was a steady stream of science fiction and horror films coming from the antennas every Friday, Saturday, and Sunday. And every weekday there was an episode of *Star Trek* or *Lost in*

---

[1] See http://www.imdb.com/title/tt0059287/ if the allusion is lost on you.
[2] I mark the beginning of quantum information *as a field* with the discovery of the quantum teleportation protocol.

Space waiting for me just as I got home from school. I gained a taste for two things: the image of the noisy, swirling atom that concluded each commercial break of Channel 5's late-late-night *Project Terror*,[3] and the idea that mankind knew no insurmountable bounds. I dreamed my friends and I could fly to the stars and back if we wanted. Television molded my idea of what the world ought to be: one of danger, constant adventure, and yet, with luck and hard work, one ultimately malleable to what we want to make of it.

That thought led me into junior high school, where I decided life was too short: if I wanted to see myself flying to the stars with my loved ones (successfully) waiting behind to cheer my return, I had better get to the technical details of making it happen. So I read popular physics book after popular physics book looking for the ways and means. To my genuine surprise and deep sadness, no real-world method emerged; I was left with the science fiction I had started with. I only learned from the wisdom of physics that my dream would stay a dream.

I suppose that is where my story might have ended, but it did not. Instead, it came to me slowly, and then more and more firmly, "Physics must be wrong!" I told myself that, and I was quite serious about it. I knew I had to become a physicist, not for the love of physics, but for the distrust of it.

I could not have had better luck: the year for me to go to college was 1983, and being as backwoods as I was, I never looked beyond Texas. I would go to Rice University in Houston or the University of Texas at Austin, I told myself. Rice was the prestigious school; the University of Texas was the fallback. In the spring of 1982, I visited both campuses. At Rice, I saw a beautiful lawn; at the University of Texas, I came across a beautiful idea. It was something that intrigued me like no idea ever before – John Archibald Wheeler's thought, "Nature conserves nothing; there is no constant of physics that is not transcended; or, in one word, mutability is a law of nature." *"The only law is that there is no law."* I never applied to Rice, and I never thought twice about it; Austin was exactly right for me.

Little did I know that I had arrived at the hotbed of quantum information, the field that would arrive just 10 years hence. Wheeler's Texas entourage of students, postdocs, and visitors at the time – David Deutsch, Asher Peres, Benjamin Schumacher, William Wootters, and Wojciech Zurek, among them, with Charles Bennett and Richard Feynman coming for the odd conference – would be some of the fathers of the field. Wheeler wanted to get to the bottom of the quantum, indeed to the very bottom of "Why existence?," and he felt that information theory was the key for developing his nascent notion of "law without law."

That was enough to keep me going: I endured the normal physics curriculum of classical this and classical that and hydrogen spectra and differential equations without end, not because of any love for it, but because I knew it was a rite of passage I had to survive in order to get to 1993. Though the field had no name yet, I knew exactly that I wanted to be doing quantum information. And so, two bachelor degrees followed, a slow PhD

---

[3] Look it up on YouTube.

after that – sliding from before to after the quantum-information revolution – and finally $0.3 + 1.0 + 0.5$ postdoctoral positions. I suppose it can be said I did some solid physics during that time and since, but I have never lost my penchant to deconstruct the whole structure. Where there is a rule, there should be a way to get around it.

This very volume represents a statement of that at the level of my personal life. "One simply does not publish one's email," it was once said. Let me tell the story of why I did.

If you think the 500 pages in this volume represent a lot of email, you should see the thousands more I keep in my archives that are certainly of no great consequence. Even before 1995, I had reams and reams written with the purpose of sorting out the message of the quantum. But in 1995, something changed – I came across two books that made the greatest impression on me, Wolfgang Pauli's *Writings on Physics and Philosophy* and K. V. Laurikainen's *Beyond the Atom: The Philosophical Thought of Wolfgang Pauli*. With Pauli's way of putting things, I felt I had finally latched on to the correct flavor of idea for taking Wheeler's program forward.[4] Most importantly, I felt I could entertain "law without law" without at the same time having the universe as a whole disappear in the flash of a postmodern dream. (It had always been a worry.)

As you might expect, I honed the idea in correspondence when I could, and this time with an almost religious fervor. I burdened my long established friends like Greg Comer, but I also sought out other quantum mechanicians whose papers and thoughts had a bearing on the subject. I knew that if I could get them hooked in conversation, they would become great foils to exercise against. As these pages will make clear, two of my greatest "successes" (if one should say such a thing of dear friends) were Asher Peres and David Mermin. Their engagement and probings took me to new heights. But at the same time I overreached and played with fire – I took on far too many correspondents. I could not find the time to answer afresh all the questions posed to me, even though many were simple modifications of things already well-worn in earlier correspondence. Under pressure I took an easy way out. I got into the habit of collecting the highlights of my correspondence with Peres and Mermin, as well as sprinklings of other conversations, into a single file and distributing it in place of personalized answers. The old points of conversation were neatly marked off with an "Asherism," "Merminition," or something similar to set the stage, and then followed by an often-lengthy response, an aside, some quote that intrigued me, or a pivot to take the conversation somewhere else. When I would distribute the document, I would often say something like this to the new correspondent: "I have never addressed the issue you raise more carefully than in the attached document; see page 57, just below Asherism 35." It was a behavior I was rightly chided for more than once; on the other hand, it seemed to work.

Indeed, the thing that took me by surprise was how successful the whole method turned out to be. It seemed that people were entertained at the same time as becoming more

---

[4] To get a sense of what I mean by "The Paulian Idea," please see the first three entries in Chapter 11 of this volume, "Letters to Greg Comer." For my most up-to-date statement of the idea, see Section 8 of C. A. Fuchs and R. Schack, "Quantum-Bayesian Coherence," http://arxiv.org/abs/0906.2187.

exposed to my philosophical dreams than they (or I) had bargained for. One particularly important occasion for me was when a respected colleague wrote, "Your *emails* are clearer and more carefully argued than [Professor X]'s *papers*." (Professor X was quite famous in quantum foundations.) It was joy to my ears; it meant these documents had a value-added that I had not previously imagined. I started to think to myself, "Wouldn't it be nice if I could make these writings legitimate and post them on the quant-ph archive where all my regular papers are.[5] But how?" I knew there was a serious risk that I would come to be viewed as a crackpot. I kept the idea in the back of my mind but could see no way to act on it – for the time being I suspended the dictum "the only law is that there is no law" from my professional life. I let the unwritten rules of academia govern me. My samizdat, as David Mermin started calling it, grew and grew to over 100 pages over the course of its various distributions, but as the name suggests, remained an underground publication.

That is, until 4 May 2000, when the proverbial fire started to catch up with me. At the time, I was a postdoc at Los Alamos National Laboratory in New Mexico, and on that day, the U.S. National Park Service ignited what was meant to be a controlled burn of the forest on the nearby Cerro Grande. By the end of the day it was anything but controlled, and for six days straight the people of Los Alamos watched the smoke in the distance, as the fire came closer and closer to the town. I remember a neighbor, a military man, slowly taking his family's possessions down the hill to Santa Fe day by day. He explained to me in a thick, drill-sergeant accent, "Chris, there's only one road into this town and one road out; you ought to be packing your things." Like most people in the town, my wife and I lived in denial. I thought, "This is Los Alamos National Laboratory; if the fire gets too close to town, the US Army will step in. They would not let something bad happen to *this* town." But Los Alamos apparently had no special status, and early afternoon 10 May, we awoke from a nap to the sound of sirens in the city – a sound we had not heard before. Throwing open the curtains, we could only see a blood-red sky, and we raced to get out of town. We stuffed two suitcases, our two golden retrievers, our year-old child and her car seat, and finally my computer backpack into our little 3-cylinder-engine car and evacuated like 18,000 other residents. The next day we learned we were one of the 400 families to lose their homes to the fire. Aside from the things we had placed in our car, we lost every material possession we had.[6] There were so many obvious things we had not thought to pack, so many sentimental things.[7]

We were devastated. But on the morning of 12 May, I woke up with an idea. "It was seeded by playing with fire; it shall germinate with our real fire." I knew that I would use the Cerro Grande Fire to legitimize my samizdat. I would advertise the document as a way of "backing up the hard drive," and make explicit reference to the fire when I posted it. "Perhaps too if David Mermin were to write a foreword, it would legitimize the thing still

---

[5] See http://arxiv.org/find/quant-ph.
[6] See chapter titled "Postpartum" at the end of the book.
[7] And the list is so much larger than I had grasped at the time of writing "Postpartum." For instance, we lost the beautiful quilts and tapestries my mother had sewn for us and our daughter Emma, and it only dawned on me many months later that I had left behind every letter my father had written to me.

further." I gambled that any urge one might feel to think of me as a crackpot would be counterbalanced by sympathy for our losses with the fire. "I might just pull it off."

What follows is essentially the document I posted on the Los Alamos preprint server 10 May 2001, on the first anniversary of the fire's devastation.[8,9] At the time, it was the largest posting ever on the server. For, once I decided to follow through with the plan, it seemed a shame not to include as much relevant material as I could, nor to have as many "hooks" as possible to pull the readers in – something to give them the urge to turn the page. So my little samizdat of 150 pages grew to nearly 500, and I gave it the title, *Notes on a Paulian Idea: Foundational, Historical, Anecdotal and Forward-Looking Thoughts on the Quantum*. I tried my best to make quantum mechanics come alive. Quantum mechanics is about the world, but it is about people as well. And so of this book. I hope you will enjoy reading through it – you can pick up the thread most anywhere (it was designed that way) – and learn something of our amazing quantum world.

Allow me to introduce some of the correspondents you will find herein.

David Baker: David was my best friend in high school; he remains my dearest and oldest friend. He currently manages a supermarket in San Marcos, Texas and has three children. We made the transition from childhood to manhood together, from finding our first girlfriends, to tasting our first beers, to seeing our first death (a young man drowning just off the shore of Lake Travis). When we "cruised" the streets of Cuero, Texas, we had conversations like we had with no one else: politics, religion, history, rock-song interpretation, the meaning of life. I read a line in a newspaper a few years ago from someone remembering his youth, "Half the people you knew believed that if only they could figure out what Bob Dylan was saying, the secrets of the universe would be revealed." David and I were like that, but looking for the secrets of the universe in everything we could experience. We would tell each other we were the deepest thinkers in town. We probably still think that.

Howard Baker: Howard Baker was (and maybe is, I don't know) a freelance journalist working for *New Scientist* magazine in London. At the time of our correspondence, he was writing an article on some recent quantum-channel capacity theorems. For some reason, I felt like answering him in depth.

Howard Barnum: Howard is an academic brother of mine (we were both students of Carl Caves) and is presently a research-staff member at Los Alamos National Laboratory. I have always been fascinated by the story that Paul Dirac's doctoral thesis was titled simply *Quantum Mechanics. That* is a comment on the scope of one's thesis! Well, Carl Caves had not one, but *two* students with PhD theses of the simple title *Quantum Information Theory*! Howard's was one of them; Michael Nielsen's was the other. Like me, Howard has always seen quantum information theory, in part, as a tool for better understanding

---

[8] Though the present index is much more complete, and hundreds of typos have been mended.
[9] The volume subsequently reappeared in a limited run of paper-bound editions, kindly printed by Växjö University Press with the support of Prof. Andrei Khrennikov.

the foundation of quantum mechanics itself. But unlike me, he – at least for part of his life – has had some sympathy for the Everett or "many worlds" interpretation of quantum mechanics. Our correspondence dates from a time when he would play the devil's advocate for Everett.

In the pages ahead, you will find one reference to Allen Ginsberg in my correspondence with Herb Bernstein, but Howard Barnum was actually lucky enough to meet Ginsberg himself once. It was at a book-signing in New Mexico. Ginsberg asked Howard, "What do you do?" Howard said he was studying physics. This made Ginsberg quite happy, and he asked further, "Is it true that quantum physics implies there is no reality?" I would have loved to give Allen Ginsberg a copy of this book and, at the same time, have him sign my own copy just at the point where I quote a bit of one of his poems!

Paul Benioff: Richard Feynman is usually given credit for being the first person to show that one could use the unitary time evolution of quantum mechanics to perform universal computation, but in fact it was Paul Benioff. (Rolf Landauer was quick to point this out to anyone who would listen, though most would not. In my analysis, it is an example of the social phenomenon that the rich get richer.) With that, Benioff is certainly one of the fathers of the field of quantum computation. But my interest in corresponding with him had predominantly to do with an undergraduate encounter with Rudy Rucker's book *Infinity and the Mind*, somewhere between 1983 and 1988. Rucker wrote:

Randomness is not usually an axiom explicitly assumed in quantum mechanics, but there is a strong feeling that the behavior of a hydrogen atom is, in principle, unpredictable. So we would expect most of the hydrogen atoms to be generating random real numbers. (There is one theorist, Paul Benioff, who has tried extending Quantum Mechanics by explicitly *assuming* that such hydrogen sequences would be random in the sense of answering to no finite description.)

This played well to my ears because for some time (yes, even back then) I had wondered what could possibly be the difference in character between a 50-50 probability assignment for a coin toss and a 50-50 probability assignment for the outcome of a quantum measurement. Looking at Benioff's papers on the subject[10] led me to the idea that randomness in his sense shouldn't be assumed, but maybe was already built into the quantum formalism. In other words, perhaps the combination of a quantum state and a quantum measurement doesn't generate a probability for an outcome, as we normally think, but rather quite literally generates elements of a random real number (in the sense of Chaitin, Levin, Martin-Löf, Benioff, etc.). And I wondered whether equivalence classes of such sequences might just *somehow* be isomorphic to Hilbert space.

Well, it was *that* that was my interest in Benioff. I haven't yet told the story of meeting my graduate advisor Carlton Caves (I will), but the hope that he might help me pursue this line of thought was partially responsible for my seeking him out. Little did I know that by going to Albuquerque, I would be walking into an enclave of Bayesians. (Indeed,

---

[10] Despite hardly understanding them.

little did I even know what a Bayesian was.) After arriving, I was soon discovered to have "dangerous frequentist tendencies," as Carl called it, and my fitness to the environment started to cause numerous back reactions. Like all species, I evolved in response. But each time I did so, it was my own transformation, and for broad Darwinian reasons – the real selection pressure was the consistency of my own ideas. The end product is what I am now, a rather extreme personalist Bayesian about *all* probabilities. Yet, I am also at a stage where I believe I can finally have my cake and eat it too: having subjective quantum probabilities at the same time as objective indeterminism. To be honest, the desire for the latter was the most significant motivation for going down the Benioffian path in the first place. But these are present-day developments, and if the reader wants to learn more about them, she should consult my latest posting on the quant-ph archive.[11] The correspondence with Benioff recorded here comes from a time when these ideas where just starting to become clear, and I was reviewing my roots.

Charlie Bennett: Who can describe Charles Bennett in a single paragraph? I almost dare not try. He is the most human person I have ever met, if you can imagine what that means – he is one of those rare beings who come along only every 300 years or so, kind of like an aspiring Newton or Einstein, but actually well-balanced and happy-go-lucky ... if you can imagine what that means as well. I think it is safe to say that if Charles Bennett did not exist, the field of quantum information would not exist either, or at least it would have taken a lot longer to get here. From reversible Turing machines, to Maxwell's demon, to quantum cryptography, to quantum teleportation, to quantum channel capacities, and so much more, he has left his mark – his good nature and his love of play.

I could fill a book with funny Bennett stories. Here is one about his impact on my wife Kiki (a woman who knows physicists deeply, but knows no physics). One day I was telling her that Charlie's great genius in quantum information was in asking the same question over and over: "What would change if we added a little entanglement to the problem?"[12] The shocking thing is that it is a question that anyone could ask at any time, but time and again it was only Charlie who would do it. And when he did, a great breakthrough would follow. Quantum teleportation was one example; superadditivity of quantum channel capacities was another. Examples abound. Story told; thought forgotten. Several months later, Kiki walks into my office at Caltech where Steven van Enk and I are staring at a chalkboard. She says, "What's up?" We say, "Oh, we've been stuck on this problem all afternoon." She says, "Would anything change if you added a little entanglement?" Steven and I looked at each other, and our jaws dropped! We had never thought of that and, indeed, for a couple of days thought it might break the impasse we had come to.

It is hard to resist telling one more story. This one comes from a time when I hardly knew Bennett, much less anything of his personality. He was giving me a ride from Santa Barbara, California back to Pasadena, where I lived at the time. We had been at a workshop

---

[11] C. A. Fuchs and R. Schack, "Quantum-Bayesian Coherence," posted at http://arxiv.org/abs/0906.2187.
[12] Of course, here he is speaking of quantum entanglement. If you do not know what that is, you should certainly have some feeling for it by the end of this book.

on quantum information at the University of California's Institute for Theoretical Physics, and as he drove, Charlie told me about the latest question he and his team at IBM Research were hard at work on. They were collaborating with Bill Wootters from Williams College as well. The question had to do with something called "entanglement monogamy," and Charlie illustrated the problem colorfully by telling a little story of Alice and Bob who shared some entanglement. Charlie went on, "You see, Bob is one of those kind of guys who, when he's in the locker room, likes to share his entanglement with all the other guys . . . ." He raised his eyebrow as he glanced at me quickly, and so the story went – ultimately he transformed it into a nice mathematical question, and I was enthusiastic to work on it myself. Finally, he started to finish up his presentation by saying, "Bill Wootters invented the mathematics, but I . . . " Just after saying that much, he took a significant pause. It was probably something to do with a traffic distraction, but the pause was long enough for my mind to take over and try to fill in the blank. I thought to myself, "Hmm, he's probably one of those arrogant guys who'll say something like, 'Wootters solved the mathematical problem, but I'm the one who understood the physics.'" When the pause was over, he came back to the conclusion of his tale, starting it over, but now saying it more boldly, "Bill Wootters invented the mathematics, but *I* invented the sexual innuendo." That was one of my first tastes of classic Charlie Bennett!

Herb Bernstein: If you can imagine a person at once a quantum physicist and a dyed-in-the-wool hippie, then you can imagine Herb Bernstein. Herb is a professor of physics at Hampshire College in Amherst, Massachusetts, and is in some ways my more radical soul brother. Where I merely distrust the laws of physics, Herb sometimes distrusts reality itself. I first met Herb at one of those great quantum information workshops the Institute for Scientific Interchange used to hold yearly at the Villa Gualino in Turin, Italy. It was 1994. One sleepless night, in the computer room of the Villa, Herb started to talk about "reality creation," and I knew he was my kind of man! Herb's book with Mike Fortun, *Muddling Through*, is devoted to the alternative to "reality" he hopes to propose – it is something he calls "reali*tt*y." I get something deep and lasting from every conversation with him, so much so that when my wife and I were on our honeymoon, driving around Massachusetts, New Hampshire, and Maine, I went out of our way to have breakfast with Herb one morning so we could talk quantum mechanics and do a little reality creation. [13] Herb always asks me to consider the *social implications* of this malleable reality we think quantum mechanics hints of. I hope he sees this book as a small contribution in that regard.

Doug Bilodeau: Where do I find all my correspondents? I suppose I am a good bit of the trouble: I often write to them first. (Like the time I wrote to George Will at the *Washington Post*, complaining that he misattributed a quote to Einstein, when in fact it was something John Wheeler had found on a men's room wall in Austin, Texas.) Doug Bilodeau is a correspondent I picked up in 1999, after his posting a rather nice paper, "Why Quantum

---

[13] Writing this reminds me of how I took a copy of Bernardo and Smith's book *Bayesian Theory* along for our honeymoon as well, but that is another story.

Mechanics is Hard to Understand," on the quant-ph archive. As far as I can tell, it was his only posting on the archive and he may have had only one other paper anywhere else, but write he most certainly could. By trade, he was a technician at a cyclotron facility, but looking back now, I think the subtle sway of his sparse writings might have been part of the current that opened my eyes to the vast literature on American pragmatism (particularly William James, John Dewey, and F. C. S. Schiller) – my present top inspiration for how to think about some aspects of quantum mechanics.

Gilles Brassard: All my life I have had difficulty eating fish. Not shellfish, oysters, lobsters, crab, or the like, but the kind of fish that swim propelled by their tails moving side to side. I have never been able to figure out why I have this fear; it comes out of the blue. When I am with a crowd who would like to dine on fish at a conference, say, I am forced to reveal myself, "I'm sorry, I can't do it; I have an irrational fear of fish." Inevitably someone will ask, "Did you have a bad experience as a child? Was there something traumatic?" I explain patiently, with more emphasis, "No, I said it is an *irrational* fear. That means I can find no explanation." Yet, when Gilles Brassard is present, I can eat fish, all kinds; I even enjoy it—it is the darnedest thing. There is something about Gilles that enables me.

Gilles Brassard, with Charles Bennett, is a father of quantum cryptography: The Bennett–Brassard 1984 quantum key distribution protocol started the field. But, from my perspective, he is dearest to me for all that he has enabled. Ben Schumacher once told me an anecdote worth repeating in this context. Ben says that when he first heard of the discovery of (Peter Shor's) quantum factoring algorithm, he thought it was a joke that Gilles had propagated – he didn't take it seriously. The reason he didn't is because a year earlier, at one of the "quantum information parties" Gilles regularly organized in Montréal,[14] Gilles joked around, "We're going to use quantum computers to factor numbers." This may have had no direct influence on Shor, but the tendrils of scientific discovery are sinuous, and one really never knows all the influences that come together in the making of an event. The greatest things happen not because of lone scientists in isolation, but because of communities.

Gilles and I are good friends, and call each other "malt brothers" because of our confluent fascinations with good single-malt scotch. One evening in Cambridge, England, 1999, we became a bit too fascinated, but something fantastic followed – the idea of organizing some major workshops on "Quantum Foundations in the Light of Quantum Information." (The title was a rare idea at the time, but it is something of an industry now.) Those meetings helped put a community together. Gilles enabled.

Sam Braunstein: Sam Braunstein, a professor at the University of York in England, is another academic brother of mine. Though if I were in a court of law, I might have to say "academic half-brother": for Sam's official PhD advisor was Kip Thorne at Caltech, even if Carl Caves bore the brunt of the advising. Sam's contributions to quantum information have been manifold, but I suspect the biggest news item for him has been his involvement in continuous-variable quantum information, and particularly quantum teleportation. If you

---

[14] Like the one in 1993 that led to the discovery of the quantum teleportation protocol.

want some good giggles, go to Google and search on the terms Braunstein and "beam me up." In seriousness, one thing I have learned from Sam is the sheer pleasure of calculating, and one thing I have tried my best to emulate (with mediocre results) is his amazing, rugged bravery to just plunge into a calculation. He simply says, "let's do it," and away he goes. It reminds me of the boys I grew up with in Texas who would just jump into the muddy river, not having a clue of how deep or how shallow or how swift or how rocky the waters were; some boys would say that's the way you should learn to swim. In my mind, I can almost hear Sam say that's the way you should learn to do physics.

Jeffrey Bub: Basil Hiley claims that I first introduced myself to him (Basil) by bounding up to a breakfast table and confronting him forthrightly, "Is it true that you're a Bohmian?!" I cannot quite recall how I introduced myself to Jeff Bub, but I suspect it had some of the same flavor: perhaps the only distinction of note was that it was a dinner table. Jeffrey Bub is a Distinguished University Professor at the University of Maryland, College Park and was a student of David Bohm's. Though not a Bohmian per se, I suspected Jeff was one of those guys (and there are many) who would say, "If there are no hidden variables underneath quantum mechanics, and yet quantum states are states of knowledge, then you have given up on the existence of reality itself." I do remember pushing him hard on this point, which I think is simply a non sequitur. A lively discussion ensued (we were with Gilles Brassard, having fish), and I eventually sent Jeff my samizdat.

Surely my little samizdat was only one factor among many, but what happened afterward still strikes me as a near miracle. Jeff had been in the quantum foundations wars for a long time; his first paper on the subject with David Bohm dates to 1966. By that point in one's career, one's point of view is often so ossified that no earthly solvent can soften it up. But Jeff was different; maybe he was ready for a change. In 2003, he wrote me this: "I began [my graduate seminar] by saying that I see the conceptual problems of quantum mechanics as encapsulated in the issues of the Bohr-Einstein debate, and that I always thought that Einstein won that debate. But recently ... I have changed my mind: I now think that Bohr was on the right track, and that Einstein ... was wrong ...." If I could only bottle that solvent!

I remember a section in Oliver Sacks' book, *An Anthropologist on Mars*, where he tells the tale of a surgeon with Tourette's syndrome, that neuropsychiatric disorder characterized by uncontrollable tics, compulsions, and sometimes swearing. The surgeon's hands would go every which way normally, but the amazing thing was that the moment he would put a knife to a patient, all his symptoms would disappear. Indeed word got out of his great surgical skills, and he was quite trusted by the townsfolk. Sacks went on to write, "Many professions, one would think, would be closed to someone with [Tourette's.] ... This would have been my own belief not so long ago. But now, improbably, I know *five* surgeons with Tourette's." Let me close my description of Jeff by saying that now, improbably, I know *three* philosophers of physics near Jeff's generation who have made a transition along the same lines, reversing from "Einstein won; Bohr on wrong track" to "Einstein off track; something like Bohr worth exploring." I am thinking of Bill Demopoulos, Richard

Healey, and Allen Stairs. My hope is that word of their thoughts will get out, and this way of thinking about quantum mechanics, too, will become trusted by the philosophical townsfolk.

Carlton Caves: Carl Caves carries the blood of an impressive set of theoretical physicists in his veins: for Franz Exner begat Friedrich Hasenöhrl, who begat Karl Herzfeld, who begat John Wheeler, who begat Kip Thorne, who begat Carlton Caves. To the extent that I have ever done anything solid in physics, it surely must have come from the genes Carl passed on to me. But nature is only a small part of all of us; nurture – I declare – is the predominant factor. And that, Carl provided to me beyond all else. He is currently a Distinguished Professor of Physics and Astronomy at the University of New Mexico in Albuquerque, but when I met him in 1991 he was an associate professor at the University of Southern California. I myself was a graduate student at the University of North Carolina, with tentative plans to do thesis work in general relativity. (Remember, quantum information as a field didn't exist in 1991.) I traveled to a meeting in Maryland, where I saw him listed as an invited speaker: I wanted to talk. I had just read a paper by Carl and Sam Braunstein, "Wringing Out Better Bell Inequalities," which gave a method for constructing *information-theoretic* Bell inequalities (being information-theoretic seemed supremely important), and further gave the best conceptual analysis I had ever seen for why quantum mechanics violates these. I was as high as a kite from the paper, and for a while it promoted Carl to hero status in my mind. I was so nervous when I finally introduced myself to him that all I could force from my mouth were the words, "Information theory!" Carl replied, "That's good stuff." We talked till late in the night. I remember him telling me that evening, "In five years, all of physics will be knocking on our door asking us to teach them information theory."

In 1992, Carl established himself at the University of New Mexico; in 1993, I followed him there. Here's the way I remember it – I wonder if he remembers the same – several mornings a week, I would make myself visible in the "interaction room" outside his office, so that I would run into him just when he arrived. Inevitably, he would say he had so much to do, there was no time to talk; but inevitably, within a few minutes (usually lured by the coffee pot), we were at the whiteboard discussing the latest that each of us knew about some problem or other. In the evenings, I would often walk with him to his car as he was trying to leave. I learned that that's how research is done.

I have in the past called Carl the Arnold Sommerfeld of quantum information. No one is more thorough in pushing the equations forward and exploring every sidelight. He would say, "you must trust the formalism," and would recount what he thought of as his greatest mistake, not trusting the formalism once, and making (an incorrect) back-of-the-envelope argument into a *Physical Review Letter*. I took him seriously! (My career has been all about the formalism.) The thing I failed to take on board was that when you're at the boys' club, at least to save appearances, you have to do what the other boys are doing: Estimate this, approximate that, understand the limits of a spherical cow. One day, I was in his office just before my PhD defense. In exasperation over something, I exclaimed, "Carl, look, you know I'm not a physicist!" Carl held his finger up to his lips and said, "Sshhh, I know that, but nobody else does. We've got to keep it that way."

But I was on about nature and nurture. Carl's research method brought out the best in me; it was the American work ethic made specific, and for a while I was able to sustain it. Since then, we, with Rüdiger Schack, have nonetheless been able to slowly build an interpretation of quantum mechanics, one all our own. And one we feel will lead to some of the deep and promising secrets our world still holds in wait. I like to say that quantum mechanics indicates a malleable world – like a child, it is worthwhile to nurture it. In the end, though, perhaps the view was already in our blood. For Franz Exner argued long ago, and long before quantum mechanics, that our world must have an open and undetermined future. He himself might well have said the words of Cormac McCarthy that Carl quotes on his webpage, "For each event is revealed to us only at the surrender of every alternate course."

Richard Cleve: I should tell a story about Richard Cleve that never ceases to entertain me. Richard is a founder of quantum communication complexity theory, and has done much work in complexity-class theory in general. The first time I ever saw him was at one of the ISI meetings at the Villa Gualino already mentioned; this one was 1995's. Richard gave a talk on some recent results to do with classifying the computational power of quantum computers. I was absolutely blown away, having never before heard of these exotic ideas from theoretical computer science. My training had been in physics where spectra, Hamiltonians, and Lagrangians were the common currency of research thought. So, when I heard Richard speak of NP, BQP, PSPACE, #P, and God knows what else, my head started to swim. I thought, "How am I ever going to learn all this stuff!? How will I ever contribute to quantum information if I don't learn these things?" I was beside myself. But as luck would have it, the very next day, I was walking through the bar area in the Villa and spied Sam Braunstein and Richard Cleve sitting on a couch discussing something about quantum computing. Just as I walked by, Richard asked Sam to pause and explain something: "What is a Hamiltonian?," he asked. I had the best laugh ever. Just as his concepts were new to me, our concepts were new to him, and yet we were all doing quantum information. It is amazing how a whole field can lift itself by its bootstraps.

Greg Comer: I have spoken of various academic brothers, but I should tell the tale of my academic *big brother*, Greg Comer. Greg is a professor of physics at Saint Louis University in St. Louis, Missouri and plays a pretty mean guitar for his honky-tonk band, Chop Shop. His main line of technical research has to do with relativistic fluids and superfluid neutron stars, but his cosmic aspirations go far deeper than that. Greg introduced me to the image of the four brothers dreaming Vishnu dreaming the four brothers. We met when I began graduate school at the University of North Carolina. Because I had expressed an interest in general relativity (I have my cosmic aspirations too), Greg was assigned to be my "big brother." What luck! Not every graduate student so loves the physics he has fallen into (I am amazed at the number of graduate lumps I have met over the years); I could sense immediately Greg was of a different stripe. He says he had never seen anything like me either: Perhaps I was a bit too confident then? We hit it off and eventually I wormed my way into sharing an office with him so that we could have nonstop "why the black hole? why the quantum? why the black hole? why the quantum?" discussions. Those were very fun days.

The confluence of events that helped bring this volume about are now pretty much on the table, but one key event remains to be revealed. Greg took his first postdoctoral position with Jacob Bekenstein at Hebrew University in Jerusalem in 1990 and, upon arrival, was immediately thrown into the turmoil of the Gulf War. I suppose for emotional comfort, Greg tried to reach back to the United States whenever he could. Most people at the time would have written a letter or made a telephone call, but Greg started to use some newfangled method for communicating that I had never heard of: it was called email. A mutual friend told me that Greg had emailed me. Apparently I had an account somewhere. "That's odd. Oh well." A week later the friend told me that Greg had emailed again. "I suppose I better find this account," I told myself. Another week passed by and another letter by Greg. Finally I logged on and wrote my first email – it was a painful thing to do. A few days later, the friend tells me that Greg had replied. "Oh, he doesn't want to just do this once and be done with it?" So, I had to log on again. And so the story goes: Eventually I was writing email on a daily basis, and the act became a large part of my very existence. If there is one reason deeper than others for this book, it is Greg Comer.

Steven van Enk: Notwithstanding his broad interests in a number of subjects, Steven van Enk is a professor of physics at the University of Oregon. We have been in some tight spaces together, first as officemates at Caltech and then as workmates at Bell Labs. Nonetheless, we still manage to talk to each other civilly, with respect even, and from time to time write a paper together. Contrasting with me, Steven is a practising Bayesian, and not merely a philosophical one: for certain subject matters, he knows his probabilities well and acts accordingly. This, of course, is due to his deep understanding of the logician Frank P. Ramsey's teachings that subjective probability (as opposed to hypothesized long-run frequency or some kind of metaphysically defined propensity) is actually a *measurable* quantity, "The old-established way of measuring a person's belief is to propose a bet, and see what are the lowest odds which he will accept. This method I regard as fundamentally sound." Over the years, Steven and I have amused ourselves with any number of silly diversions. One time we wrote a paper that we intentionally made a bit empty (long story) and never posted on the archive. It was titled, "Entanglement Is Super ... but not Superluminal!" After hearing a story of Bohr's struggle to start a paper with the word notwithstanding, we had only one structural requirement for our own: it would start with notwithstanding.

Charles Enz: Charles P. Enz is a professor emeritus at the University of Geneva and was Wolfgang Pauli's last assistant. I probably came to know his name via the Pauli book mentioned above, which he edited with Karl von Meyenn. My correspondence with him reflects the fact that I had read (or would read) many of his papers on Pauli's foundational thoughts – my files reveal 19 – and I wanted to make sure my record was complete. Very kindly, after the fire, Professor Enz replenished me with his complete corpus.

Henry Folse: Henry is a philosophy professor at Loyola University in New Orleans, but designates himself a "Bohr scholar." Indeed, if there were ever a man whose writings needed professional disentanglement, it was Niels Bohr. I have read every paper Folse ever wrote on Bohr (there are 31), and I can say without any disingenuity, I needed to!

Robert Garisto: Robert is an editor for *Physical Review Letters* and has (or at least had) a personal fondness for the Many-Worlds Interpretation (MWI) of quantum mechanics. When *Notes on a Paulian Idea* first appeared, Robert wrote me that it should be dubbed the Many Letters Interpretation (MLI). I thought that was a great compliment.

Bob Griffiths: Bob Griffiths is the Otto Stern University Professor of Physics at Carnegie Mellon University and the inventor of the consistent-histories interpretation of quantum mechanics. Bob and I were officemates at the same Santa Barbara workshop mentioned above in the Charlie Bennett story. Our conversations there led to a joint paper, "Optimal Eavesdropping in Quantum Cryptography. I," with Asher Peres, Nicolas Gisin, and Bob's student Chi-Sheng Niu. There was also a follow-on paper, "Optimal Eavesdropping in Quantum Cryptography. II," by Bob and Chi-Sheng alone, where a quantum circuit for the strategy was worked out. The absence of three authors on the second paper tells something typical of the quantum foundations debate more generally! Bob argued that the operation of the circuit could not be understood without the consistent-histories interpretation, and challenged us to find an explanation by other means. None of us, of course, believed that the challenge could not be met, but in my own case (I cannot speak for the others), it simply seemed easier to let that part of the paper go its own way. The correspondence recorded here arose in the aftermath of that, when I made a focused effort to understand what was going on in consistent histories.

Lucien Hardy: Lucien Hardy is a founding faculty member of the Perimeter Institute for Theoretical Physics and the high exemplar of what a quantum-foundations researcher ought to be. His career is nearly unique in that, since the beginning, he has engaged in open, out-of-the-closet research in quantum interpretational matters, *yet* A) he is famous, and B) he even obtained a job! (In the quantum foundational community, A, though very rare, is without doubt more common than B.) This is not the usual story of a quantum foundationalist – for those that manage to avoid taxi driving as a living nearly always do so by doing foundations on the side, maintaining the majority of their research in a more "respectable" branch of physics. The surprise I experienced upon seeing Hardy for the first time is telling. I had known of him because of "Hardy's paradox" and a few other things and imagined him a grand old man of quantum mechanics, somewhat like Asher Peres, a man in his sixties. So when this young fellow – a younger man than I – bought me a beer one evening in the Villa Gualino, I thought, "Wow! That can happen?"

The table on page lii of this book was partially inspired by Hardy. We were having a walk in Hull, England in 1997, before my own coming out of the foundational closet.

I remember prodding him, "It's hard for me to get my head around the idea of quantum foundations as a *field*. How do you know when you've made progress?" Lucien replied, "That's a good question. That's a really good question! We never ask that."

Adrian Kent: Adrian Kent is a reader in the Department of Applied Mathematics and Theoretical Physics at the University of Cambridge and a Fellow of Wolfson College. He is also an Associate with the Perimeter Institute for Theoretical Physics. In some measure, more so for this and less so for that, Adrian is a critic of *all* present approaches to the interpretation of quantum mechanics. His published criticisms thereof are some of the best in the business. For years I have hoped he would write a thorough criticism of our quantum Bayesian approach and try his best to punch an irreparable hole in the Paulian idea! We would learn so much, and "that which does not kill us makes us stronger." With any luck, this Cambridge University Press edition will hit a bit too close to home for comfort – a cowboy in the quad?! – and finally push him over the edge.

Rolf Landauer: There are some giant names in the history of quantum information; Rolf Landauer's was one of them. Rolf was the first person to realize that the only essential, nonrecoverable loss of energy in a computation is when information is erased. With Bennett's proof that all computations could be done reversibly, without erasure, and Benioff's and Feynman's proofs of principle that Bennett's reversible computers could be modeled within quantum mechanics, the stage was set for developing the notion of full-blown quantum computing – those exotic uses of quantum stuff, first envisioned by David Deutsch, for exponentially speeding up certain number-theoretic tasks.

My correspondence with Rolf had to do with his more philosophical side. In some ways, the letters recorded here were preparing the path for the *single* letter I really wanted to write to him. It was a letter I never got a chance to compose. Rolf could be an intimidating guy, especially if you crossed him. And I knew that my budding view of quantum mechanics would cross him at his core. Rolf's most famous slogan, and a kind of rallying cry for the whole field of quantum information and computing, is that "information is physical." But the phrase started to nag me more and more: I felt that on one reading, it wasn't quite right, and on another reading, it was deeply wrong. The essential point, it seemed to me, is that "information *carriers* are physical" – one should not throw away the hard-won category distinction that the word "information" represents. But how to tell him so and convince him so and not be thrown out the door? I planned the letter in my head for a long time, and one weekend while visiting Charlie Bennett at his second home in Wendell, Massachussetts, sprung the idea out loud. Charlie said, "Too late. Rolf was diagnosed with a brain tumor. Part of his brain has already been removed, and he is not expected to live for more than a few days." Since I never got to externalize it, my conversation with Rolf has been going on internally all these years, and I have no way of really telling whether I am on the inside or the outside of his door. Death remains a cruel thing.

Hideo Mabuchi: Hideo was, and is, a kind of *wunderkind* of experimental quantum information and control. (He looks to be only 25 years old even today.) He is now a professor of applied physics at Stanford University, having previously been a professor of physics and of control & dynamical systems at Caltech. When we met—at where else? the Villa Gualino in 1994 – Hideo was merely a graduate student. (One gets the sense he might have just skipped that stage and been no worse for the leap.) Hideo once flattered me many years after our first meeting by exhibiting the paper place mat I used for explaining Holevo's information bound to him; he said he thought it might be worth something when I'm famous. I suppose he is a man of far, far long-term investment.

I think of Hideo as the high prankster of our field. In 1996, when we were both at Caltech, he a student and I a postdoc, he would drop in to my office from time to time to chat. He would also drop in from time to time to check that I was *not* there – when so, trouble always ensued. At the time when I was worried about Latinizing my slogan, "quantum mechanics is a law of thought,"[15] I used to have a screen saver on my laptop that would scroll the words "mechanica quantica ex mente orta lex est" across it. One day I returned to my office to find a banana sitting on my keyboard and the screen saver now scrolling "mechanica *bananica* ex mente orta lex est." I set it back to the original phrase. A few days later I came in to find "mechanica quantica ex mente *orka* lex est" going across the screen and an individual coffee bean on each and every key of the keyboard. (Don't look for deep meaning; I just tell the stories as they happened.) After a few similar iterations, I gave up and left the screen as he wanted it.

David Mermin: I cannot honestly call David Mermin a father of quantum information, but I can call him the godfather of it – he blessed the field through its formative years and continues to watch with pride as it approaches maturity. There are many reasons I say this, but one of them is told well in his own words in, "Copenhagen Computation: How I Learned to Stop Worrying and Love Bohr," quant-ph/0305088.

Then again, David always tells things well. N. David Mermin is the Horace White Professor of Physics Emeritus at Cornell University, and for years has been the writer extraordinaire of our field. I very much knew the latter when I first wrote to him, and believe me, it did take some effort to work up the nerve to critique his papers. But I did it for the reasons I have already explained. A further aspect that I had not completely appreciated at the beginning is something David argues so beautifully in his essay, "Writing Physics":

Physicists traditionally replace talk about physics by a mathematical formalism that gets it right by producing a state of compact nonverbal comprehension. The most fascinating part of writing physics is searching for ways to go directly to the necessary modifications of ordinary language, without passing through the intermediate nonverbal mathematical structure. This is essential if you want to have any hope of explaining physics to nonspecialists. And my own view is that it's essential if you want to understand the subject yourself.[16]

I found that when I wrote to David, I wrote better – self-assessment, of course – than I did to any of my other friends. It was a question of wanting to keep up with the Joneses, I suppose, but it had this lovely unintended consequence: I felt I understood things better

---

[15] See the discussion at the beginning of Chapter 27, "Letters to Rüdiger Schack."
[16] Mild paraphrasing from David's original.

and better just by touching my keyboard and having a new David Mermin letter in mind to write. Better writing, not just any writing, made for better understanding.

I started to realize that the major part of the problem of our understanding quantum mechanics had come from bad choices of English and German words (maybe Danish too?) for various things and tasks in the theory. Once these bad choices got locked into place, they took on lives of their own. With little exaggeration, I might say that badly calibrated linguistics is the predominant reason for quantum foundations continuing to exist as a field of research. Measurement? I agree with John Bell – it is a horrible word and should be banished from quantum mechanics. But it is not because it is "unprofessionally vague and ambiguous" as Bell said of it. It is because it conveys the *wrong* image for what is being spoken of. *The* quantum state? That one is just about as bad. Who would have thought that so much mischief could be made by the use of a definite article? So many of these things came to mind as I would strive to be clear and entertaining in my writings to David.

I once had a piece of trinitite, that lovely green glass born in the heat of the Trinity test explosion, 16 July 1945. I was very proud of it and kept it displayed on a shelf in my library until the Cerro Grande Fire took it from me. It was a material reminder of Stanisław Ulam's words, "It is still an unending source of surprise for me to see how a few scribbles on a blackboard or on a sheet of paper could change the course of human affairs." But from my standpoint today, it is not those equations that changed the world. They were merely a center of attention for the layers and layers of verbal apparatus and practical action required to give them life. It is language that powers things, and in this thought David Mermin has been my leader.

David Meyer: David Meyer is a body-building professor of mathematics at the University of California, San Diego, where he heads their Project in Geometry and Physics. If that sounds strange to you, think of Joe Piscopo: one wouldn't normally think of a body-building comedian either. David is a fantastic geometer. In February 1999, I posed to him the question of whether Gleason's famous theorem for quantum mechanics might still hold in the venue of "rational vector spaces." it was a question I had been bugging a lot of people about, never really expecting an answer. But to my great pleasure, David knew the answer almost immediately: it does not. The years go by, and I am still trying to process the consequences (or nonconsequences) of that knowledge.

Jeff Nicholson: Whether or not it is acceptable to say so in an academic publication such as this, I'll just record it: Jeff Nicholson was the best beer-drinking buddy I ever had in graduate school or since, and I was proud to be the best man at his wedding. Jeff is currently a research staff member at OFS Laboratories in Somerset, New Jersey and writing papers with titles like, "Phase-Locked, Erbium-Fiber-Laser-Based Frequency Comb in the Near Infrared." We started together as officemates at the University of New Mexico in 1993, and then found ourselves together again at Bell Labs many years later. Jeff is the taller, stronger, and faster of the two of us, with over 170 publications and patents. In the days that were, we had many a good round of philosophy at the ever atmospheric Jack's Liquor and Lounge in Albuquerque – red naugahyde seats and plenty of talk of reality creation. I will well bet that J. W. Nicholson is the only person on earth to have ever built an all polarization maintaining passively modelocked, femtosecond figure-eight fiber laser with a dispersion managed cavity *and* read Richard Rorty. Certainly he was the first.

Michael Nielsen: Michael Nielsen wrote a book with Ike Chuang, a very good book, and now the world reads it. The title is *Quantum Computation and Quantum Information*, and it is the best technical introduction there is to the (few) technical terms used in this volume. I have already mentioned that Michael is an academic brother of mine via Carl Caves, but we have also crossed career paths two more times since. First we overlapped as postdocs at Caltech, where he was a Richard Chace Tolman Prize Fellow, and then again at the Perimeter Institute for Theoretical Physics where he was a faculty member briefly. In between the two, Michael was a professor of physics at the University of Queensland in Australia.

Michael has an interesting philosophy: He believes in change. And he takes it so seriously as to believe in changing careers as well, at least every seven years or so. Michael is currently a writer plotting out how the internet holds the potential to drastically change the nature of science and scientific collaboration. He is writing a book on the subject, but that only represents the paper bound side of his thinking; the play-by-play blows can be found at his widely read blog, http://michaelnielsen.org/. There are all kinds of tasty tidbits to be found there. For instance, a while back I was perusing the blog and ran across an entry ending with an "idea for an idea" he had with student Sarah Morrison, "My own favourite crazy idea for resolution of the [quantum measurement] problem is that, in fact, the projection postulate will not be derived from unitary dynamics. Instead, unitary dynamics will be derived from the projection postulate." And then I thought of John Wheeler's last published writing, a December 2000 *New York Times* article:

Many students of chemistry and physics, entering upon their study of quantum mechanics, are told that quantum mechanics shows its essence in waves, or clouds, of probability. A system such as an atom is described by a wave function. This function satisfies the equation that Erwin Schrödinger published in 1926. . . .

This picture is all right as far as it goes. It properly emphasizes the central role of probability in quantum mechanics. . . . But, to my mind, the Schrödinger wave fails to capture the true essence of quantum mechanics. That essence, as the delayed-choice experiment shows, is *measurement*.

John Wheeler begat Kip Thorne, who begat Carlton Caves, who begat Michael Nielsen.

Asher Peres: I have spoken of giant names in quantum information already, but it is said that actual giants once walked the earth. Surely it was so in the golden age of quantum physics – Einstein, Dirac, Fermi, Heisenberg, Pauli. No one in my sphere of friends was aware of this more than Asher Peres. For no one else in that sphere ever came so close to looking over the wall and seeing the other side. Asher was a physicist's physicist and made contributions to nearly every field of theoretical physics. His death at the age of 70, while he was still very active, was a great loss to the community.

This book owes a special debt to him in at least two ways. First, he was among the fathers of quantum information, and so a source of its subject. But second, he took me under his wing in June 1994 and treated me as if I were *somebody*. We already knew each other at the time (he had visited Albuquerque previously), but at a conference in honor of John Wheeler's 83rd birthday something changed. I told him that I'd like to show him my results on quantifying the information from quantum states. After letting me explain hardly a thing, he said a bit dismissively, "Yes, I have a student who already has many results on that. When I return to Israel, I will have him write you and tell you what he has done." Being new at research, I was frightened that everything I had worked on was wasted – someone else had gotten there before me. Still, I gave my talk when the time came; what else could I do? It was my first major talk. After the talk, Asher came up to me and said, "That was an excellent talk! You have done everything my student has done and far surpassed him. When I return home, I will tell him to stop work on this problem immediately and find something else." That brought me onto Asher's radar screen, and I never disappeared from it thereafter. We became very good friends.

I should only say good things about Asher, but let me say a bad thing about him that caused a good thing in its place. We never really *spoke* on anything technical or philosophical to any depth: He would not listen. Our spoken conversations consisted of small talk, or else my one-sided listening to him. Whenever I would try to show Asher something technical at the chalkboard or on a napkin, Asher would soon say, "I am very tired." Perhaps he would add, "We must think on these things," and then we would part. Initially I found it very frustrating. We nearly never had interactions like the ones I described in the Carlton Caves story.

Instead, I learned early on that if I really wanted to communicate with Asher, it had to be with email! It made all the difference in the world, and thereafter conversations went smoothly. For instance, I might write something like this to Asher (as I did on 7 November 1995):

The reason $\lambda = 0$ is true for optimal eavesdropping, isn't so difficult after all ... one just need have some patience. ... The upshot is the following. If one assumes that Eqs. (43) and (44) hold for Eve's optimal mutual information (as they must) even when $\lambda \neq 0$, then it turns out (after algebra) that the appropriate z to use there is given by

$$z^2 = 1 - \left(1 - \cos^2 2\alpha \cos^2 2\phi\right) \cos^4 \lambda.$$

Now, just looking at D in Eq. (33), one sees immediately that upon taking a derivative of the figure of merit M (defined on page 12) with respect to λ, one generates a quantity with an overall multiplicative factor of $\sin 2\lambda$. Therefore one sees that $\lambda = 0$ is an extremal point. End of story.

and actually get a reply. (In this case he found my mistake.) I would never have been able to get that much information across to him in front of a chalkboard.

The height of this peculiar relationship of ours surely must have been at the 1996 Research Program on Quantum Computers and Quantum Coherence, held at the Institute for Theoretical Physics in Santa Barbara, California. There our offices were less than a one minute walk from each other, but our main mode of communication was email. In the end, between 28 June 1994 and 31 December 2004 (the day before he died), Asher and I exchanged more than 2,350 emails.

In 2003, when Asher announced his plans for retirement in the Fall of the following year, I wrote him, "you should think of yourself as being a postdoc again." Asher replied:

Thank you for telling me that I'll be like a postdoc. It's so true. Now this will be my way of telling it to friends. Only one thing will be missing: a good advisor for my research (my first advisor in 1961 was Wheeler, although I was formally the postdoc of Misner). Will you be willing to play that role?

Of course, I accepted. Now, at the end of my own career, I will be able to look back and say Asher Peres was the best assistant I ever had.

John Preskill: Ben Schumacher says, "Anything that's worth doing, is worth overdoing." Let me tell you about John Preskill. John is the Richard P. Feynman Professor of Theoretical Physics at the California Institute of Technology, and the director of their Institute for Quantum Information. I once called John "Professor X" in print,[17] but he's not the same Professor X who appeared above – that's a wholly different Professor X. The reason I say all this is because John figured out his identity last time, and I wouldn't want him to think I was putting words into his mouth this time! Still, since I've gone this far, I might as well tell the old story again. It concerns a time just preceding my arrival at Caltech to become a postdoc in John's group:

I was talking to a friend who was a student at Caltech and would later become a professor there. "Have you met Professor X?" he asked. "You'd better watch out, he's smart." Professor X was the fellow who had invited me to Caltech, and even though I had not met him, it did not faze me. I knew plenty of people with Caltech credentials: the guy I was talking to, my PhD advisor, another collaborator and so on. I shrugged, "So what?" My friend replied in a foreboding tone, "You don't understand: He's scary smart." And so he was. The fear crept into me at that moment, and I do not think it has completely left me since.

John Preskill is a generalist of the highest order in theoretical physics – I think he wants it all, to understand *all* of fundamental physics. It is an impressive sight to see him think or lecture or to look over his shoulder and see his meticulous handwritten notes on this or that subject. If I had to put my finger on the source of the fear, I think it would be that I know full-well that I am a specialist. And specialists are always afraid of generalists. But life must go one! When I console myself, I do it by thinking of something Carl Caves told me that Ed Jaynes would say, "The really good physicist is the one who chooses a problem difficult enough that he'll solve it just at the end of his lifetime. Something not too easy and not too hard." It may be lunacy, and Darwin will decide in the end, but I feel that I have chosen a problem – understanding the nature of quantum states and quantum measurements – of just the right size *for me*. But then that keeps me the specialist (with the implication above).

---

[17] C. A. Fuchs, "Soul-Searching at Caltech," *Physics World*, November, 2003, p. 49.

When we have discussed the interpretational issues of quantum theory, I have gotten the sense that what John finds most suspicious of the quantum Bayesian approach I promote is that he *thinks* it treats observers as unphysical systems. I say, "Not at all; John, you are a physical system to me." The real issue is one of inside versus outside. Contrary to the textbook exposition of quantum mechanics, a wave function that I write down about him is descriptive not of the outside, but of the inside, namely me: it captures what I believe will happen to me if I interact with him. If he knew what I believed, he'd probably do something to surprise me.

I think John's secondary worry is that unless one takes something like the Everettian approach to quantum mechanics, one cannot do quantum cosmology – the idea is that the observer must be excised from a fundamental role in the theory if one is going to address cosmological issues. (In a phrase, what is desired is a theory that is all outside and no inside.[18]) The Everettians think they do that, but a quantum Bayesian would say, "only in a way that leaves the observer's clothes behind." I tried to impress John once by writing down $| \Psi_{universe} \rangle$ and saying, "Look, I can do that too." I don't think it worked.

Most recently, I've tried a new technique in my lectures. I draw a stick figure on the chalkboard along with a little blob near him, and say, "There's an observer with a small physical system in front of him: He can profitably use the formalism of quantum mechanics to estimate what might happen to him if he interacts with it." Then I say, "But there's nothing to keep us from talking about a larger physical system." I extend the blob, making it larger. I do this a couple of times, at first imperceptibly putting a bend in the blob, making it a little kidney-bean shaped with the observer at the center of its curvature. I make it more pronounced with each iteration and finally go for broke: The observer is completely surrounded by the physical system he is considering, and I declare in great triumph, "That's quantum cosmology!"

One of these days maybe I'll work up the nerve to show John.

Joseph Renes: When I first met Joe, I was struck by how much he looked like a member of the Red Hot Chili Peppers; he was an undergraduate at Caltech at the time. When I second met Joe, I was struck that he was driving a Saab, a Swedish car, in the middle of a desert; he was a graduate student of Carl Caves in New Mexico at the time. But when I third met Joe, he told me about a mathematical problem he was working on, a problem which has set the tone for my own research for more than four years. Joe is currently a postdoctoral researcher in Germany with a prestigious Humboldt Fellowship.

The problem mentioned is a consternating one! I fear "it will be the ruin of many a poor boy" before all is done. For one thing, it is so simple to state – already addicting just for that if nothing else – but apparently not so simple to solve. The posing of the problem goes back to Gerhard Zauner and Carl Caves, independently, in 1999, but the problem went worldwide with the Renes *et al.* paper of 2003. That paper gave hope that the problem might just be tractable, and it has received steady attention ever since. The

---

[18] On first pass, the Texan in me wanted to write "all hat and no cattle," but I repressed that.

main reason I find the problem consternating, however, is that it carries with its solution the promise of rewriting quantum mechanics in a way that might pinpoint its essence. And for that, we should spare no hardship! If you are a mathematical physicist out there, read the Renes *et al.* paper and think about how you might prove the existence of "symmetric informationally complete quantum measurements." These structures are so important that even if they do not exist, they ought to!

Mary Beth Ruskai: Beth Ruskai is a mathematical physicist par excellence. She recently retired from a long professorship of mathematics at the University of Massachusetts, Lowell and currently holds the more leisurely title of research professor at Tufts University. Beth's early fame came from proving the long-open conjecture of strong subadditivity for quantum entropy with Elliott Lieb. Since 1998, she has posted any number of delicate studies of the structure of quantum channels, entropies, and entanglement to the quant-ph archive. Part of the correspondence recorded here comes from a time when Beth was writing a comment on Bohmian mechanics *and* gender issues in mathematics – we nicely agreed on both subjects.

Rüdiger Schack: I have an impression of Rüdiger Schack that I have sustained since 1993. One day we were walking westward on Lomas Boulevard, near the Department of Physics and Astronomy at the University of New Mexico in Albuquerque, and far in the distance some mountains could be seen just peeking over the horizon. There was a moment of silence in our talk of probability theory and quantum mechanics, and we walked some distance more until Rüdiger said out of the blue, in a kind of authoritative voice, "Mount Taylor, so beautiful." I didn't respond, but I thought to myself, "You can see that? You can be moved emotionally by something so distant on the horizon?" I had an eerie feeling that he could see things that I could not, things from all aspects of life. Only years later did I learn that, from where we were, Mt. Taylor couldn't be seen at all, and Rüdiger was merely commenting on the memory of his last visit. But the feeling remains: Rüdiger can see things, deep important things, that I cannot.

Rüdiger Schack is a professor of mathematics at the University of London, Royal Holloway, and was a postdoc in New Mexico at the time I joined Carl Caves' group. We remain close collaborators to this day. If it is true that I came into Carl's group with "dangerous frequentist tendencies," it is also true that the group had only "mildly Bayesian tendencies" in the first place. Rüdiger was the leader that helped us see past the wall Edwin T. Jaynes had erected in his own Bayesianism, our model at the time. It was a question of consistency in ideas, and Rüdiger was the first in the group to dimly see that we must either make our peace with consistency, or else revert to a reified, physical notion of probability. This ultimately led us down the path of Bruno de Finetti's personalist Bayesianism, even for the probabilities of quantum mechanics – a nearly lunatic idea from the point of view of most philosophers of physics, who think that somehow the exhortation, "But quantum probabilities are given by physical law!" changes everything. It is a path we are carefully walking westward today – working out all the consequences of this radical idea is the focus of our research program. I have the feeling Rüdiger will see something on the horizon, something beautiful, that I cannot.

Robert Schumann: At the time of our correspondence, Robert was an MSc student from the University of Stellenbosch in South Africa. I helped advise his thesis work; I believe we had met when I lectured at a summer school in Italy. Robert's questions were penetrating and helped me better compose my thoughts. A search on the web shows that he is currently a PhD student at Oxford University. Congratulations Robert!

Abner Shimony: Abner Shimony is an emeritus professor of philosophy *and* physics at Boston University. In quantum-information circles, he is most famous for his work on quantum entanglement and is the S in the CHSH inequality, perhaps the most important of all Bell inequalities. But his thoughts reach much further than that, and he has had seminal contributions to several branches of philosophy – I shouldn't fail to mention Bayesian epistemology among them. Every story I know of Abner is a sweet one. Here is one from my personal files.

Some years ago, I was down and out about where I had ended up in life. I had a pure research position, it was true, but it was at Bell Labs. Bell Labs was once a prestigious institute with six physics Nobel prizes under its belt and a great many scientific discoveries beyond that – for instance, it is where Claude Shannon invented information theory – but that was its history by my time, not its promise. Within two years of my joining Bell Labs, it was very clearly crumbling beneath me. I knew that I had to leave eventually, but to where? I put out feelers in all directions and interviewed at a few universities and institutes to no avail; I was never quite what anyone wanted. That of course brought my morale quite down.

Still, I would joke that I was the highest paid philosopher in private industry. I spent my time unashamedly working on foundational things. And in time, I renewed my correspondence with Abner; I sent him some of the samizdats that came after *Notes on a Paulian Idea*. From them, Abner could see my fondness for William James, the great American pragmatist, but he recommended C. S. Peirce in his place. Abner wrote, "You love William James, and I have respect for James but reverence for Charles S. Peirce." I already had good reasons for preferring James over Peirce (to do with my quantum Bayesianism), but I thought I should know something of Peirce the man; particularly, I would see Abner at a meeting soon, and I thought it'd be nice to have a toolbox of things from Peirce's personal life to flavor our conversations. I pulled Joseph Brent's biography of Peirce off my shelf and started to read. Something unexpected happened: the book was the most depressing experience for me. The thing that stuck out in my mind was how this genius, this man whose works would live forever, could never attain a proper academic position; he died a pauper. Soon before departing to the meeting, I wrote Abner: "I have just read Brent's biography ... I couldn't help but note some of the parallels between his life and mine. Though I lack his genius, I have certainly brought some of the same troubles upon myself." Little did I know the trouble this would cause!

You see, Peirce had many vices and failings: he was a womanizer, drug user, constant dreamer, braggart, and with his flamboyant lifestyle, threw away any money that ever came to him. I had not been specific enough in my note to Abner! On the second day of the conference, Abner approached me and said with an almost shaking voice, "Are you OK? I saw you in the distance yesterday, but could not make my way to you. I have been so worried. I saw that you still have your wedding ring, so that must not be your trouble. Are your children safe?" I explained that I had only meant that no university seemed to want me. Abner said, "Is that all?! When I return home, I will talk to the chairman of my department. Perhaps something can be arranged at Boston University."

There is a reason so many people love Abner.

John Smolin: John is a quantum information theorist at IBM Research in Yorktown Heights, New York, with a generational sense of humor that works well on me. Imagine a bizarre episode of *The Simpsons* – I sometimes get the feeling John really lives in a world like that. Or at least he is the living interface between that world of the Platonic "Doh!" and ours. Consider, for instance, the shoe story.

Sometime in the late 1990s, I spent a few days with John at his parents' house in Croton-on-Hudson, New York. One morning, John's father asked if we'd like to accompany him for his morning walk. We did, walking down this and that neighborhood street and, particularly, on one that ran parallel to a small creek (maybe 10 or 15 yards away). At some point, John's father breaks away from us and picks up a pair of shoes that were sitting at the side of the road, carrying them briskly to a nearby dumpster. John calls out to his father, with a tone that's a bit shocking to my (polite) Texan ears, "What are you doing?!" John's father replies angrily, "I'm throwing out these shoes!" "You can't do that!," John says, "They're somebody's shoes." John's father, with more pent up anger, "There's always shoes around here; they're all over town. I'm sick of it! I'm going to throw them away." John says, "All over town? No one abandons their shoes! Someone probably went wading in the creek and forgot to put them back on when they came back to their car. They'll come back to get them." John's father says, "Well, I'll put the shoes back today. But if they're still here tomorrow, I'm throwing them out!" John doesn't let it go, "Shoes all over town?!"

We go back to the house, have a cup of coffee, and jump into John's car to get the workday started. John drove a black Ford Mustang in those days, and he throttled it into IBM's massive parking lot, pulling into a random parking space somewhere in the middle. Our eyes bugged out: There was a shoe sitting in it! The workday passed with lots of fun and laughs as usual, and Charlie Bennett said, "Why don't we have dinner at the Indian restaurant? I'll meet you two there." We hop into John's Mustang again and pull into the restaurant's parking lot. Right in front of us, another pair of shoes! We surmised that John's dad knew what he was talking about.

John has done important work across the board in quantum information, but as one would expect of a student of Charlie Bennett, does it all – literally all – in the spirit of fun and play. There is little discontinuity between whatever video game is his passion at the moment (apparently it is something to do with a potato gun as I write this) and any quantum information problem he is working on. John would say, in the voice of Wyle E. Coyote from the old *Roadrunner Show*, "It's because I'm a genius!"

John's friend and collaborator Graeme Smith says, "Whatever happened to Fuchs? He used to do physics!" John smiles and sometimes chimes in, but can never really hide it: he'd like to understand the reasons behind quantum mechanics just as much as we foundational folk. From time to time, he even comes to the quantum foundations meetings I organize, and has one paper on the subject, "Can Quantum Cryptography Imply Quantum Mechanics?" In rebuttal to his paper, a couple of philosophers said his construction (a counterexample for the question) wasn't fair because his physical objects had *haecceity* – apparently having that broke the rules they wanted imposed. Haecceity was a very fancy word for John. In great Bart Simpson style, he said, "Yeah, yeah. So what?"

Rob Spekkens: Rob Spekkens was a young whippersnapper when we had the correspondence recorded here. He has since developed into a very fine theoretical physicist and is a faculty member of the Perimeter Institute for Theoretical Physics. I must say that I am quite proud that, after a self-described "epiphany" in 2002, Rob became one of the most effective proponents I know of for the view that quantum states should be understood as states of knowledge rather than states of nature. This, of course, is a view that will be argued *for* over and over and from every angle in the pages to follow. As Rob wrote in his "toy model" paper, "The diversity and quality of these analogies provide compelling evidence ... to proceed with a research program wherein the quantum state being a state of knowledge is the idea upon which one never compromises." On this we agree 100 percent!

But if it is knowledge, knowledge of what? Wouldn't you know it, there's always a place to disagree! After the absolute high of learning of his epiphany, I was so disappointed to learn that we disagreed deeply on the answer to this question that I responded with a rather emotional e-rant. Rob's strategy for a further reply was genius. He sent me back an Emo Philips joke:

Once I saw this guy on a bridge about to jump. I said, "Don't do it!" He said, "Nobody loves me." I said, "God loves you. Do you believe in God?"

He said, "Yes." I said, "Are you a Christian or a Jew?" He said, "A Christian." I said, "Me, too! Protestant or Catholic?" He said, "Protestant." I said, "Me, too! What franchise?" He said, "Baptist." I said, "Me, too! Northern Baptist or Southern Baptist?" He said, "Northern Baptist." I said, "Me, too! Northern Conservative Baptist or Northern Liberal Baptist?"

He said, "Northern Conservative Baptist." I said, "Me, too! Northern Conservative Baptist Great Lakes Region, or Northern Conservative Baptist Eastern Region?" He said, "Northern Conservative Baptist Great Lakes Region." I said, "Me, too!"

"Northern Conservative Baptist Great Lakes Region Council of 1879, or Northern Conservative Baptist Great Lakes Region Council of 1912?" He said, "Northern Conservative Baptist Great Lakes Region Council of 1912." I said, "Die, heretic!" And I pushed him over.

I cannot imagine any better words to keep in mind as one enters the great quantum foundations debate. Perhaps it would have been appropriate for me to put a sign on the cover of this book, "Beware all ye who enter here."

Jon Waskan: Jonathan Waskan is a professor of philosophy at the University of Illinois, Urbana-Champaign. He is the husband of my wife Kiki's BFF (best friend since four), Laurie. When Kiki and I first met, we lived in Albuquerque, New Mexico, and Jon and Laurie lived in St. Louis, Missouri. But then by complete chance, when we moved to Pasadena, California, they moved to within 20 miles of us. That lasted some time, until they moved to Denville, New Jersey. By complete chance, when we moved to New Jersey, we moved to within 20 miles of them! Who says it is not a small world? Our time together on those two ends of the United States led to some philosophical discussions, a little bit of which is recorded here.

Bill Wootters: When the fire swept through Los Alamos, I lost over 700 books. In fact, I lost every single book I owned except one – it was William K. Wootters' PhD thesis, *The Acquisition of Information from Quantum Measurements*. It happened to be in my backpack, along with my laptop computer containing the precursors to the present book – I have said before that I am a lucky person. Bill Wootters has been a kind of hero to me since my earliest days at the University of Texas. His 1980 PhD work had been co-advised by John Wheeler, and John held it up as a kind of paragon of what might be done with regard to the "it from bit" aspect of his program. It will be noticed that Wootters, starting with W, is the last named author of many papers in quantum information. That, unfortunately, has the potential to hide from the public the depth with which he has contributed to the field. Like it is said of Roger Penrose, that he can *see* in four complex dimensions, I sometimes think that Bill can actually *see* quantum information! His formulas for the quantum subentropy, the entanglement of two qubits, and complete sets of complementary bases in all prime-power dimensions are all nearly magical.

When I obtained my PhD, I was fortunate enough to have Bill as my external examiner, and it was a great honor. As Wootters wrote in that PhD thesis still on my shelf – "[Professor Wheeler] generously gave much of his valuable time to discuss with me the problems and prospects of physics at its most fundamental level, and transferred to me his belief that the hardest problems can yet be solved." – in the years that have passed, I could say the same of Bill himself.

Anton Zeilinger: Anton Zeilinger is an experimental physicist at the University of Vienna, (ultra) famous for any number of fundamental experiments in quantum information and quantum foundations. Uncle Anton. I wouldn't say this, except for an incident that occurred three years ago, in 2006. The scene was a meeting in honor of Abner Shimony, and Abner being a professor of both philosophy and physics, I wanted to honor him with a talk that was more a mixture of the subjects than I had ever attempted. In the days preceding my talk, I worked very hard to pull something together, making many new slides to accompany some older ones. I tried to get at the core of the issue, so that perhaps Abner would stop reciting Edward Lear, "they went to sea in a sieve, they did; in a sieve they went to sea," when speaking of my "quantum Bayesian" efforts with Caves and Schack. (Abner's fear for us is a species of what I called a non sequitur in the Jeff Bub story.) What resulted was a talk that was hardly conciliatory, instead apparently very polarizing.

As I gave the talk, I felt I could see several physicists writing me off as a nutcase – Alain Aspect and Nicolas Gisin have very expressive faces. On the other hand, Wayne Myrvold, a philosophy professor from the University of Western Ontario, came up afterward and said quite excitedly, "That was the clearest talk I've ever heard from you!" Feeling the disparity, I approached David Mermin to get some understanding. He said somberly, "We need to talk. That was the worst talk you've ever given." So, we walked and talked, and I started to feel like a small child. Finally, as we returned to the conference center, we paused at the door and David said, "I feel that all I've done is hurt your feelings, but you have an important message and you should present it in a serious way."

At that moment, Anton Zeilinger bounded up from the distance and boomed, "Great talk!" Mermin replied back stubbornly, "No it wasn't!" Yes it was; no it wasn't; yes it was – I couldn't believe I was caught between these two giants of physics debating my fate that way. Finally, Anton broke the impasse by looking to me rather than Mermin, "You know what I would do when I was young, being dismissed by the old professors on the front rows of the seminar? I would not look to them as I spoke, but rather to the back rows where the young students were sitting. They were the ones ready to hear something new."

In fairness, David Mermin had many good points that day, some of which I have incorporated into my talks since. But those bolstering words of Anton Zeilinger got me through the night after a very emotional experience. Ever since, in my mind, I have called him, "Uncle Anton." (He will probably cringe if he ever reads this.) Personal anecdotes aside, Anton Zeilinger is certainly a philosophical uncle to the very core of this book, the Paulian idea, and maybe he deserves the title for that alone. Here's the way he put it in a recent interview:

I call that the two freedoms: first the freedom of the experimenter in choosing the measuring equipment – that depends on my freedom of will; and then the freedom of nature in giving me the answer it pleases. The one freedom conditions the other, so to speak. This is a very fine property. It's too bad the philosophers don't spend more time thinking about it.

Simple and clean the notion is: too bad indeed the philosophers don't spend more time thinking about it!

Onward, dear reader, to *The Paulian Idea*.

Christopher A. Fuchs
Waterloo, Ontario

Introduction to the Original ArXiv.org Posting

This document is the first installment of three in the *Cerro Grande Fire Series*. The Cerro Grande Fire left many in the Los Alamos community acutely aware of the importance of backing up the hard drive. I could think of no better instrument for the process than Los Alamos National Laboratory itself. This is a collection of letters written to various friends and colleagues (several of whom regularly circuit this archive), including Howard Barnum, Paul Benioff, Charles Bennett, Herbert Bernstein, Doug Bilodeau, Gilles Brassard, Jeffrey Bub, Carlton Caves, Gregory Comer, Robert Griffiths, Adrian Kent, Rolf Landauer, Hideo Mabuchi, David Mermin, David Meyer, Michael Nielsen, Asher Peres, John Preskill, Joseph Renes, Mary Beth Ruskai, Rüdiger Schack, Robert Schumann, Abner Shimony, William Wootters, Anton Zeilinger, and many others.

In a way, I hope this book evokes images of the kind of dusty, pipe-smoke infused gems one sometimes finds in the far corner of a used bookstore – something not unlike the copy of William James' collected letters I once owned. It takes a funny person to read such a book: one who is willing to dig in the far corner. But I would not want that imagery taken too far. For despite its nod to the happenstance that brought it about, this is not a book about the past, but about our open future. Its singular theme is the quantum. Without exception, every letter in the book is devoted to coming to grips with and extolling the virtues of our quantum world. The content ranges from the foundational, to the historical, to the anecdotal, but every piece sings or whispers of the quantum. For myself, I see some of the letters as my best efforts to date at defining a vague thought that keeps creeping into my mind – *the Paulian idea*. To the extent I have communicated its faint shadow to my correspondents and seen a head turn, it seemed worthwhile to try to give it more life. The idea pleads to be made precise. But for this, it must possess the souls of more than me.

Disclaimers

I. This document represents a unique, and hopefully entertaining, method to communicate some happy thoughts on the quantum. For precisely this reason, however, it carries a great danger to my friends. It is after all a collection of correspondence. There are two things that should not be mistaken: 1) The potential of my memory to be faulty when reporting the views of others, and 2) that the quotes taken from my correspondents were composed in anything other than a casual manner for *private* use only. With regard to the latter, I assert the right of my correspondents to deny – without apologies! – that their quotes represent accurate accounts of their thoughts. I have tried to guard against misrepresentation by keeping the number of quotes and correspondent-replies to a minimum: The ones that are used, are used mainly as springboards for *my* tendentiousness.

II. Various deletions of text have been made to the original letters. The purpose of the vast majority of these is to spare the reader from the "merely personal" in my life. A smaller fraction are for the sake of protecting the innocent, protecting the correspondents, and protecting the illusion that I am good-natured. The same holds as well for a small number of explicit changes of phrase (in my own writings, *never* the correspondents). In most cases, I have tried to make the process look as seamless as possible, with no evidence that the text may have been otherwise. In my own writings, bare ellipses should be interpreted as punctuation; bracketed ellipses indicate true editorial changes.

III. There is no claim that all the ideas presented here are coherent. The hope is instead that the incoherent ones will earn their keep by their entertainment value.

Acknowledgements

I thank Todd Brun, Jeff Bub, Carl Caves, Steven van Enk, and David Mermin for the subtle influence that gave this project life. I thank my correspondents for the use of their quotes. Most importantly, I thank my wife Kiki for allowing me to type in the middle of the night.

I want you to frame a question, as sharp and clear as possible – one to which you do not yet know the answer, but desperately want to know, and expect someday to know. Pretend to be David Hilbert. The Millennium is approaching. Issue a challenge to the quantum theorists of the 21st century. List the key questions they should seek to answer. Hard questions, but not hopelessly hard, questions whose answers could transform our understanding of how the physical world works. *I need to know what the question is. Then, perhaps, I can be more engaged in the search for the answer.*

(John Preskill)

I (CF) give it my best shot:

Quantum Mechanics
The Axioms and Our Imperative!

---

1. States correspond to density operators ρ over a Hilbert space $\mathcal{H}$. — *Give an information theoretic reason if possible!*

2. Measurements correspond to positive operator-valued measures (POVMs) $\{E_b\}$ on $\mathcal{H}$. — *Give an information theoretic reason if possible!*

3. $\mathcal{H}$ is a complex vector space, not a real vector space, not a quaternionic module. — *Give an information theoretic reason if possible!*

4. Systems combine according to the tensor product of their separate vector spaces, $\mathcal{H}_{AB} = \mathcal{H}_A \otimes \mathcal{H}_B$. — *Give an information theoretic reason if possible!*

5. Between measurements, states evolve according to trace-preserving completely positive linear maps. — *Give an information theoretic reason if possible!*

6. By way of measurement, states evolve (up to normalization) via outcome-dependent completely positive linear maps. — *Give an information theoretic reason if possible!*

7. Probabilities for the outcomes of a measurement obey the Born rule for POVMs $\mathrm{tr}(\rho E_b)$. — *Give an information theoretic reason if possible!*

The distillate that remains – the bare piece of quantum theory with no information theoretic significance – will be our *first* unadorned glimpse of "quantum reality." Far from being the end of the journey, I well believe, placing this new conception of nature in open view will be the start of the greatest adventure yet in physics.

"In the new pattern of thought we do not assume any longer the *detached observer* ... but an observer who by his indeterminable effects creates a new situation, theoretically described as a new state of the observed system. In this way every observation is a singling out of a particular factual result, here and now, from the theoretical possibilities, thereby making obvious the discontinuous aspect of the physical phenomena."

(Wolfgang Pauli)

"Like an ultimate fact without any cause, the *individual* outcome of a measurement is, however, in general not comprehended by laws. This must necessarily be the case ..."

(Wolfgang Pauli)

1
Letters to David Baker

22 July 1996, "Noodles of Nothing"

How's your academic life? Have you been reading any? Do you still subscribe to *Texas Monthly*? Or was it *The Smithsonian*, I forget which? Right now, in my spare time, I'm reading *Robert Oppenheimer: Letters and Recollections*. Mostly it's a collection of letters from when he was in college and graduate school. The fellow was an amazing writer: it would be hard to tell he was a scientist if you didn't know otherwise. Of course, my real interest is in peering into his thoughts as he was learning quantum mechanics; that's why I'm putting time into the book. But it is always good to absorb something of someone else's style.

Have you ever read anything by William James (the psychologist and philosopher – not to be confused with Henry James)? For some reason I'm just fascinated with his writing style; I would love to be able to pull the same tricks, and have it acceptable to do so. If you get a chance, take a look at the first few pages of his essay "The Dilemma of Determinism" just for the style. (You'll be able to find it in the library in any collection of his essays, probably with titles like *The Will to Believe*; the one I have at home is a Dover edition.)

There's a new phenomenon going on in my field: we've apparently reached the level where it's time for some books on the subject. While in Torino, I was approached by four people with some mention of the book they're writing on quantum computers or quantum information in general. That makes a total of five books on the drawing board that I know about personally. I wish I were in the group, but I'm lucky enough to be able to write a paper every now and then. Probably, if I'd just write a little less e-mail ... !!!

27 July 1996, "Life on Long Island"

I asked whether you had read anything by William James, and then said that he should not be confused with (his brother) Henry James. That's OK; I'm sure I can take your answer to mean that you haven't read either of the two. William was one of the great American philosophers from the end of the last century. He's also known for his work in psychology. I suppose his biggest contribution to thought was in the founding (along with Charles Sanders Peirce) of a philosophical system known as "pragmatism" – something I like very much, actually. Henry was a novelist and literary critic (and probably quite a bit more famous in general circles than his brother). I've never read anything by Henry.

I got to thinking about Mr. James again because of my last note to you. So, I was very pleased to find a book at my favorite used bookstore today titled *The Philosophy of Henry James, Sr.* I had never heard of this member of the family. As the "Sr." implies, he was William and Henry's father. Apparently he had something of a philosophical system of his own ... which, the author of this book claims, is of interest in and of itself and not chiefly because of its relation to William's ideas. Anyway, I snatched the book, ten (Canadian) bucks. Also finally bought Kierkegaard's *Either/Or* (two volumes), and three books by Piaget: *The Construction of Reality in the Child*, *The Child's Conception of the World*, and *The Child's Conception of Physical Causality*. (The philosophical mysteries of quantum theory plague me just as always; I'm not proud, I look for clues wherever I can!)

The thing that intrigues me about James and Peirce is that they both rejected the mechanical view of the world that was the rage of their Victorian time. James, in particular, was led to believe something that had quite the flavor of the "many-worlds interpretation of quantum mechanics" – something some of the foolhardies of today would say is uniquely implied by the quantum mechanical formalism. (Many-worlds quantum mechanics was introduced in 1957 by one of Wheeler's PhD students Hugh Everett; James died in 1910.)

To that view, actualities seem to float in a wider sea of possibilities from out of which they are chosen; and, somewhere, indeterminism says, such possibilities exist, and form a part of the truth.
                              (*James, 1884 (address to the Harvard Divinity Students)*)

I'm sure I spent too much money on books again, but it's a nasty itch. The store I went to today is really excellent, even though it's much smaller than our apartment here. It's a wonder what an owner with taste can do. The bookstore is also especially good for me because the owner's husband is a physicist. He does much of the work in purchasing their science and philosophy collections. I got to meet the fella today for the first time; very nice – he was curious to know what quantum cryptography is. One day I was talking to his wife, and she was telling me how she had the good fortune to see Feynman lecture once. I said that I had never seen him, but that I knew his graduate advisor. She said, "My God, he must be old." I said, "well yes," and then went on to explain that it was John Wheeler. She said, "I know of Wheeler; around our house Wheeler is God!" That's how I found out that her husband is a physicist (particle physicist in particular).

I just poured a beer, a Sleeman's Lager. I like it. Actually I like a lot of the beers up here: different ones for different moods. When you come up, I'll introduce you to the whole entourage of Québecois beers. About your coming up ... let's see, what can I say?

I think you'll have no problem finding things to do while I'm tied up. The neighborhood is full of surprises (and bars), and every museum is a metro (subway) stop away. I do have two "must see" bars for you: The Mad Hatter's Library (near McGill U) and The Yoda Den (near a smoked-meat joint). Oh yeah, there's also The City Pub ... it's got a nice Kantian/"quantum mechanics is a law of thought" feel that I like (and good cheap food to go with the beer).

03 August 1996, "Lonely Jazz"

Ms Holiday is playing in the background and I've got my first cup of the morning. I'm still pretty much exhausted. I ended up staying at IBM an extra day, and the whole affair was pretty intense from start to finish. I think we found something very, very nice ... which, of course, was expected – that's why we got together in the first place. The idea is simple and it's this. Suppose Alice needs to communicate to Bob one bit of information, either a 0 or a 1. Also let us suppose the resources available to her for carrying out this task are two (noisy) fiber optic cables and two photons, one for each cable. The bit will, in some sense, be carried in the polarization of the photons; we allow two photons in the game to give Alice a little redundancy to help her get past the noise. Note that I've said nothing about how the photons are produced; I didn't say anything about whether they came from a single source (localized in some small region of space) or whether they are generated by two coordinated but independent sources. Now the question is this: can the transmission's fidelity be helped by allowing the photons to be generated at a single source AND combining them back together before Bob performs his measurement (for gathering the bit)? That is to say, can we increase Bob's chances of guessing the bit correctly by first generating the photon at a single source – so the two are "entangled" in a strange quantum mechanical way – and then allowing Bob to make a measurement on the two together, a measurement on the whole being greater than a sum of its parts? The answer is yes, and I find that so wonderful.

Why, you ask? Because, it's wonderful ... period. And a bit unexpected too actually. I guess the thing I really like the most about so many of the questions we've been asking lately is that they really put "entanglement" to use. I just ran across a wonderful quote by Oppenheimer the other day, from a letter he wrote to Fowler soon after Hahn and Strassmann discovered uranium fission:

The U business is unbelievable. ... What do you think? It is I think exciting, not in the rare way of positrons and mesotrons, but in a good honest practical way.

The idea of entanglement has its origin in Einstein, perhaps as early as 1930 ... though I can't recall for sure right now. In any case, a pretty clear statement of it and what he didn't like about it came out in 1935 in the paper of Einstein, Podolsky, and Rosen. (It's through Nathan Rosen that I have my "Einstein Number" 3: Rosen having written many papers with Einstein, Peres having written many papers with Rosen, and I having written a paper with Peres.) Its existence was something EPR considered clear-cut evidence that quantum theory could not be a "complete theory." (I.e., by this time Einstein had already given up the idea that the theory was wrong, he just didn't think that it could be the whole story.) Since then, the notion just pretty much stayed an oddity, only thought about for the most part by the philosophically minded ... that is, until its practical resurrection in quantum information theory.

By the way, I should point out that EPR were wrong in the sense that it was finally shown in 1964 that quantum theory could not be "completed" in their sense. Any such completion (that took entanglement away) would contradict experiment – and very fine experiments on this phenomenon have been done. What we have now is that, not only is entanglement required for consistency with observation, but also that it can be exploited for something interesting (and perhaps practical).

Enough physics, right? Ms Simone is on by now, I should say. Kiki and I are about to step out for Ethiopian. It's a place with a lunchtime buffet; we haven't tried it before. I had Ethiopian in DC once and was really taken with it, but I haven't been able to recapture the experience since then. Perhaps the fourth time's a charm. Then we're gonna do a little CD shopping.

Here I'd thought that I was gonna answer all your questions about Canadian beer and such this morning, but instead I just got carried away with science. Sorry about that. I'll be back later in the day with something more on the mundane (not the derogatory meaning of the word, but rather "earthly as opposed to heavenly").

17 August 1996, "More Meaning"

A lazy Saturday afternoon. I'm listening to a new CD, *Liza Minnelli (From Radio City Music Hall)*. Having an afternoon coffee, after an afternoon beer – a strange combination. New York, New York. Kiki's lying on the bed asleep, and I'm dreaming. Today was a day stranger than most. Albert apparently had a stroke this morning. He sort of lost control of much of his left side for a while, maybe 5–7 minutes. As the day has gone he's regained more and more control. Now everything is pretty much normal except he's still a bit wobbly on his left hind leg and I find myself more sentimental than usual. (Excuse me, now we have Mel Tormé.)

The things I live and breath for. *Twilight Zone*, *Star Trek*, Quantum Mechanics, and an open future. Maybe that summarizes it all. (It's been a long time since I've lapsed into Gertrude Steinisms.) The Mars rock has really set me off lately ... so much so that I was even willing to spend \$4.00 for a *Time Magazine*. I didn't find out much more about Mars that I didn't already know, but at least I learned that Audrey Hepburn was a notch on Jack Kennedy's bedpost. Actually, that really depressed me; I have quite a crush on her ... the 1961–64 version of her, that is.

Night Thoughts of a Quantum Theorist. Tonight it's my turn to cook. (A chicken dish I suspect.) The problem's not in the seasoning, but in finding the main ingredient. Those with little imagination restrict themselves to games with seasonings. Unfortunately, that's where I stand now. It's a good thing Kiki makes me cook every Saturday; maybe I'll get the hang of it eventually.

19 August 1996, "Self Promotion"

What's with the "cheers" thing? Corporate lingo, I suppose. I picked it up from the Brits, who always sign their letters to me that way. However, it's becoming a relatively general salutation in my clique. Depending upon my mood, I rotate it with, "best regards," "best wishes," "kind regards," "very best wishes," etc. On special occasions, and, in particular, if the addressee is an old German professor, I might end with something like, "with warm good wishes." But, I tell ya, that's nothing compared to some of the curlicues I saw in the Oppenheimer letter collection!

01 September 1997, "Mo' Investigative Work"

Allow me to reach into the depths of your knowledge of history. A couple of days ago, I came across the following question in Trivial Pursuit: Who said, "The victor will never be asked if he told the truth"? The answer was "Adolf Hitler." Now I would like to pin down the actual source of that quote (i.e., from what speech or what private conversation of Hitler's, to whom and when did he say it). Do you think you might be able to tackle this task? Of course my motivation for getting this straight has to do with some silly thoughts about quantum mechanics ... as you could have guessed.

21 February 1998, "Home a Short While"

Sorry to hear about your painful taste of dharma. Try to remember instead that, in the end, Brahman = Atman. That which is without is within. I've always tended to find more solace in that anyway.

02 March 1998, "Long Finicky Flight"

Well here I am again. A long flight in front of me ... they say 10 hours 58 minutes. You may not get this note for quite a while, March 10 in particular: I doubt that I'll be connecting much, if any, while in Japan (due to the difference in network protocols, etc.). My guess is that this will turn into a long note, as it usually does when I'm on a flight and I don't feel much like working. But right now I don't have a clue as to what the subject is going to be.

What can I tell you? To a large extent I've been pretty damned brainless the last few weeks. I'm not completely sure why that is, but it is. A lot of it is "shut down" I'm sure. I've got a zillion and one things I need to be doing, more than I can possibly handle. So instead of tackling what I can, I tackle none. A corollary to that is that I watch a lot of TV now.

Only 1.5 hours into this flight, and already I have one heck of a back ache. Do you really want to travel so much as you say you do? I'm still not completely sure how this summer's travels are going to turn out. The dates for the Torino conference are now set at June 29–July 19. But it looks like I'll also be invited to a conference in Benasque, Spain (a small village in the Pyrenees) July 5–25. I'm not yet sure how I want to apportion the time. Almost certainly I will go to Spain for some of that time. However, I really can't say whether I'll be going to Torino. The travel has just gotten to be too much: the more I travel, the more shallow my thoughts become ... and I've got to stop that. Last week I found out that I had the opportunity to go to Mexico City (expense free for the American taxpayer) in August: I think it's the first time I've done this, but I turned it down.

OK, maybe it's time for a little personal philosophy. Though my tolerance for equations is becoming less and less, I am finding that my view of the world is becoming firmer and firmer. Despite my own depression, I must say that I am finding myself believing that the world is more vibrant and *alive* than I ever have before. When was the last time I sent you a compilation of my philosophical ramblings? This much I have really started to share with my friend Herb Bernstein: the notion that there is a "reality" above and beyond man, everlasting and eternal, is simply outdated. It comes from a time when science could only make progress by extricating the human element from things; it comes from a back-reaction to religion. But now, with some hindsight from the quantum revolution, it seems clear to me that the world is so much more. It's far more surprising than Baconian science would have us believe, and it's far more participatory than any of the western religions (or eastern, for that matter) ever dreamt. The world, its description, and the laws that govern it, are not simply there independent of our actions. There was a time when they were, before complex organic molecules, but now that's not the case. The world and its laws seem to me to be every bit as evolutionary as life itself. And just as the idea of radical Darwinism becomes outdated when one realizes that random natural selection fails to hold, the second one being can say to another, "I love you," so it is with the universe. The world is a big pushme-pullyou. If I could talk to the animals ... . Is that what I've been doing?

But I know you want to hear a little more philosophy. Herb describes our explorations as trying to get at a new category. He calls it "reali*tt*y" ... that is to say, reality with a little something extra thrown in. It describes the fact that the world pushes back in an unpredictable way when you push on it. And the way it pushes depends on what you do to it. And, finally, that that push is not inconsequential in the least bit. Maybe you remember John Wheeler's "game of twenty questions (surprise version)"; I guess I subscribe to it more than ever.

What is this thing called language, and how does it fit into the whole of everything I wrote you above? I always think of Linda Henderson and the introduction to post-modernism that she tried to give us way back when I ask a question like this. I don't think I would have ever believed in 1983 that I would be thinking of her words 15 years down the road. Looking back on it, I have to wonder whether she had just been making a bedtime reading of Foucault or Derrida, and had been trying to share it with us. You probably don't remember this, but John Simpson and I fought some of the things she said as silliness, tooth and nail. Linda once said, "Without language there can be no thought." We said, "That's simply ridiculous." Now I find myself saying to myself, "Without language and the collective action that it leads to (through the demagogues, the communicated scientific ethic, the body politic, the media, etc.), the world we see would scarcely be the same."

But maybe that's enough for now. I'll get back to you after something of a nap.

14 April 1998, and to Herb Bernstein, "Hawking on Evolution"

I know that you have listened to me patiently in my speculation about Darwinism becoming more and more outdated as we are ever more able to inject planning and purpose into evolution. However, in case you haven't had enough, you may be interested in listening to Stephen Hawking's side of the same story: it's a pretty good talk on the whole I would say. (I certainly don't agree with some of his Everettista views nor his "end of physics" sermons, but much of the rest of it is really good stuff.) You can find a webcast of it at the following site, http://www.sun.com/newmedia/whitehouse/stephen_hawking.html.

17 May 1998, "The Electronic Itch"

To be honest, I don't have much to say, but still I have the itch, the need to write something down. This is my essence as far as it goes.

I think I am in the last two hours of my flight to Chicago. Then an hour and a half in the airport, and finally three more hours back to LA. I arrive just at the peak of rush-hour traffic; so more than likely it'll still take me one hour from there to get home. What a life.

I come back from my second trip to the fatherland still impressed. The food in this part of Germany, very near Luxembourg, was quite different from the Bavarian style that I was getting used to. But still it was very, very good. The landscape was beautiful and full of life. Yesterday and last night, I stayed in Mainz – an old town, just on the river _____ (let's see if you remember your geography) – feeling lonely and searching through the archetypal archive in my soul. Weird things like this make me take a little stock in Jung's thought. The country seems to do something to me.

It's funny contemplating this future of ours. I keep thinking more and more about it. Some of that is spurred by my now constant watch of the exponential growth in computing power. And I am sure some of it comes about by my predisposition to think that we've yet to discover any real laws of physics – the last four hundred years being an elaborate phenomenology to fill in the gaps. I keep feeling that something really, really big is about to happen. Sixteen months ago, I bought this 133 MHz Pentium I machine with 16 megabytes of memory, a $\approx$1 gigabyte hard drive, 800×600 resolution screen, and a 6x CD-ROM, all for $3000. Right now I could buy a 266 MHz Pentium II machine, with 64 megabytes of memory, a 4 gigabyte hard drive, 1024×768 resolution screen, and a 24x CD-ROM, all for $3000. (This is much more like a factor of 4 difference, not the factor of 2 that everyone talks about in "Moore's law.") Eighteen months from now, laptop computers will be at least twice as powerful as this one. Where is this going to lead? Where can it lead? It makes me wonder; it makes me religious in a rather strange sense. I don't think it unreasonable to expect that it won't be long before we'll see a full-scale reconstruction of our species. Certainly less than two hundred years from now. But then what? This question eats at me sometimes.

14 August 1999, "Rampant Repine"

And the boredom overtook us . . .

Guess what I'm doing again? (Buzz like a bee.) I'm scheduled for an early arrival in Los Angeles in a little over an hour. What I hate about that is that I always get my hopes up only to have them dashed: inevitably an early arrival leads to sitting on the runway for extra time. It never fails that there's another plane still parked in the scheduled spot. Two seats in front of me is Hugh Grant, the actor. (For real, escorted on and everything.)

Have you ever read Borges? I bought a collection of his short stories while I was living in Canada, but I hadn't really sat down to read it until now. I read "The Garden of Forking Paths" the other night and really enjoyed it; it was quite eerie.

Lately too I've been turning my psychotic side to thinking about acid. Namely, how it must induce certain kinds of connections in the brain that aren't normally there. And similarly how it must suspend other ones that we normally rely upon. Is it a priori obvious that that sort of rearrangement of the brain would be a bad thing? We've always been told that it is surely so, that it suspends our function in society. But I've been wondering what might happen if we took a large community (whose transportation is based on the bicycle rather than the car!) like Amsterdam, and surreptitiously gave the residents a small dose of acid in their water supply for something like five or ten years. Would anything interesting and permanent crop up. The acid-eaters would in this case not be isolated in society (as they always have been in the past) but would be the complete community. What form would that community evolve into? What form would their art and literature take? And most importantly for me – i.e., the real reason I'm thinking about this – what form would their scientific investigations and insights start to take? What would they be able to see in systematic ways that we cannot see at all? Could they capture in a scientific way, whole aspects of the world that we are just blind to? Would they in their discussions ask (fruitful) questions about nature that would never have occurred to us?

There's a lot of reasons I've been thinking about this. But one certainly takes its roots in a slide that I use in some of my talks. It's a chart of the raw genetic differences percentage-wise between various species of animals. I have both man and dog marked with a yellow marker. The wonderful thing is that there is only an 11% difference between the two species!! We're so accustomed to thinking that mankind is the pinnacle of creation – and surely we are – but how this hints at a wonderful new slant on the story. This difference in intelligence and understanding that we (subjectively) suppose as almost infinite, might not be infinite at all. In reality it might itself only be 11%. How wonderful that would be!

28 July 2000, "The Role of Registered Phenomena"

I have been thinking a lot about John Wheeler's registered phenomena lately. And, I have decided that on this count John was just wrong. There simply is no real world "out there" completely independent of our interventions into it. Now that our kind is here, we are an integral component of that which constitutes nature as a whole. John had hoped that there was still something of a bedrock to the world – most likely a trapping from his training in classical physics – and it was to that role he assigned the "registered phenomenon." I am forced instead to ask myself over and over which aspects of nature can we at least treat as effectively real, if not real in any absolute sense. For certainly we see an independent world around us: we stub our toes on rocks when we least expect it.

My opinion is lately this. We tend to call something real when it is beyond our control to change it. But that is only a subjective state of affairs, one controlled largely by our lack of information or technology, and sometimes, sadly, by our lack of genuine will. As Archimedes told the king, "Give me firm support and I shall move the earth." When we know all that we can know about a physical system, modern quantum mechanics tells us that we can mold it to our purposes. And, as such, it can no longer retain an independent reality. But when we have less than maximal knowledge of it, degree by degree, it becomes every bit as real as the rocks and trees about us. By this account, the independent reality of our world comes about solely from the mystery it holds for us.

Congratulations on the birth of your daughter. She was at the moment of her birth, and will be from that time forward, a mystery with which you must reckon. She will likely be the most real thing to ever arise in your life, for you will never know her completely. Every day she will bring you a new surprise. And when, years from now, she is on her own and away from your protection, you can look back and know that you partook in creation in the most absolute of senses.

2

Letters to Howard Baker

29 September 1997, "Quantum Information Questions"

I think I'll answer your questions somewhat out of the order in which you asked them. I hope you don't mind: it may fit what I'm thinking more closely that way.

Bakercise 1: *How important do you think these papers are to Quantum Information and the quantum technologies (computing, communication, cryptography and teleportation)? Are they a major step forward? And if so, why?*

In my opinion, the Holevo–Schumacher–Westmoreland (HSW) result was the most technically difficult one in Quantum Information Theory to be proved last year. In the long run, the papers are certain to be classics. As far as importance goes, it was certainly one of the top two results (the other being a consortium of papers to do with fault-tolerant computation on quantum computers).

The HSW papers solve a long-standing problem that was on Holevo's mind as early as 1978. Holevo actually had a paper conjecturing this result in 1979, but somehow it escaped all of our attention until Richard Jozsa and I met him last September in Japan. He told us of the old paper then. (I've done a citation search since then, and, believe it or not, this paper had only been cited five times in its life. Stranger still, three of those citations were for the wrong reason – the result they were citing was in his 1973 paper!) Once Holevo knew of the earlier Hausladen, *et al.*, result – which he learned of at the Japan conference, 25–30 September 1996 – things must have fallen into place pretty quickly ... because his paper appeared on the Los Alamos e-print archive 14 November 1996.

As best I can piece it together Schumacher and Westmoreland must have found the result almost simultaneously with Holevo. They saw his paper on the archive as they were in the course of writing theirs. They were disheartened a bit, but felt that the methods of their proof were sufficiently different that they would go ahead and finish their writing and submit the paper anyway. (They related this to me when we met at the PhysComp conference in Boston around Nov 20.)

Is this a major step forward? Yes, because I think this is the first paper that shows an effective way to deal with asymptotic problems to do with quantum communication channels. In a way, this result does for classical communication on quantum channels what

Shannon did for purely classical channels in 1948. (Though it should be noted that there was no rigorous proof of Shannon's "theorem" until 1954 – the first was due to Feinstein.) It asks, given that I am building messages out of some finite alphabet of signals, what is the best use I can make of a noisy communication channel? And it answers it, though this time we had to wait 17 years for a rigorous proof. (Actually that last remark may not be all that fair: Holevo's 1979 conjecture was in no way near the stage of completion that Shannon's 1948 stuff was ... and certainly – for whatever reason – it didn't draw the same amount of attention. Also, without the whole apparatus of classical information theory, the proof of this new theorem wouldn't have been imagined, nor would it have been conjectured.)

Bakercise 2: *What information do you think Quantum Information and these papers give us about the nature of the quantum world.*

It's pretty clear that Quantum Information Theory is giving us a load of new conceptual tools with which to explore the essence of quantum phenomena. In fact, I think a fairly major attitude change has come along with Quantum Information Theory (i.e., in the last ten years or so). Previous to that, certain basic facts of quantum theory – such as quantum indeterminism and Bell inequality violations – were viewed as mostly negative principles. The Heisenberg uncertainty relations were seen as these darned things that just got in the way of our precision measurements. The Bell inequality violations were similarly these darned things that got in the way of our constructing a comfortable-feeling hidden-variable theory to go underneath quantum mechanics. Now these things are turned on their head: quantum indeterminism is something really good (quantum cryptography), quantum entanglement – i.e., the thing giving rise to Bell inequality violations – is something really good (quantum teleportation), and so on. This new attitude (and the tools it leads us to search for) gives us a way of looking at quantum theory in a way that wasn't available before ... and that can only lead to a deeper understanding of nature. Maybe, in some ways, it's teaching us to broaden or diversify our notions of the elementary essences of the world: now along with energy, etc., perhaps we should include "indeterminism" and "entanglement." (I tried to say some of these things – with respect to entanglement in particular – in a slightly poetic way to a friend the other day; perhaps I'll attach that letter to the bottom of this file ... for whatever it's worth.)

John Wheeler (a long-time professor at Princeton and later at the University of Texas) was a great advocate that information theory had something deep to say about quantum mechanics. He always exuded this air of urgency about him: "It is imperative for us to understand the meaning of quantum mechanics in the grand scheme of things! We must make as many mistakes as we can, as fast as we can, so that we can hope to obtain an understanding within our lifetime!" The last time I saw him was in 1994 at a little conference in Santa Fe – the one, in fact, where Peter Shor's factoring algorithm was announced. (Actually, come to think of it, I saw him one time later that same year ... at his 83rd birthday festschrift.) Anyway, at Santa Fe, Wheeler gave a talk (probably titled "How Come the Quantum?") that he closed with a slide depicting Planck's head (maybe etched on a coin or something). I remember, he said (roughly), "In 1900 Planck discovered the quantum. The end of the century is drawing near. We only have six years left to understand why it is that it's here. Wouldn't that be a tribute?!"

I think there is something of a feeling in the Quantum Information community that this may not be complete hogwash. In ways, there really is an exciting, sort of revolutionary feel to it all. I wish I could put my finger on it more carefully for you. In the end, of course, it's still just quantum mechanics … so there's nothing revolutionary in that sense. Instead, it's more like waking up one morning and realizing that "maybe it's not coincidental or an accident that gravitational and inertial mass are numerically the same in Newtonian mechanics." This is surely the sort of thing that hit Einstein at some point: it was then just a question of counting the time until something truly wonderful came out of it.

With all that in mind, let me focus on the specifics of the HSW result. There is one thing that is truly intriguing about the effect that powers it. When separate particles are "entangled" as they are in the Einstein–Podolsky–Rosen pairs, there is a real sense in which the "whole is not reducible to a sum of its parts." A more precise way of putting this is really something of its negative: "maximal knowledge of the whole does not imply maximal knowledge of the parts for entangled entities." The idea that I'm trying to convey with this is that, for entangled particles, even though the whole of the system may be in a completely specified quantum mechanical state, the separate parts will not have a precise state of their own. This sort of thing is useful in its own right, as I've already pointed out. The effect that powers the HSW result is something of a dual to this, though it again boils down to a statement that the "whole is not reducible to the sum of its parts." Let me try to explain.

Here one uses nonentangled particle states in a such a way that the whole is "more than the sum of its parts." Specifically, even classically correlated quantum states of separate entities are more than the sum of their parts if one is willing to allow a little entanglement to assist in the measurement. This physical fact was found by Holevo in 1979 and independently by Peres and Wootters in 1991. The HSW result carries that effect to its most extreme form in the context of a specific communication problem.

Is this result deep? I think so, specifically because of the way it stands in a dual relation with more standard entanglement. In some ways it is just as shocking as entanglement itself. The full implications of this new sort of "non-locality" have yet to be fleshed out.

Another way in which the HSW result gives rise to surprising things has to do with something I wrote about in *Physical Review Letters* last month or so. Because of the HSW theorem, it turns out that there are clear-cut situations where it is better to use a signaling alphabet composed of nonorthogonal quantum states. This in itself is rather shocking, because from the "classical perspective" (i.e., the bad intuition we get stuck with by our everyday experiences) nonorthogonal signals correspond to *noisy* signals. That is to say, it's very much like the telephone operator asking you to mumble so that you have a better chance of being understood at the receiver's end! Clearly that is nonsense within the world of classical physics, but quantum mechanics, on the other hand, has this way of making life more exciting!

Bakercise 3: *Is it possible to give some figures on the amount of information that could be carried and read from a photon or other quantum carrier?*

Let me first answer your question in generality – I'm not sure whether you're addressing specifics of the HSW theorem or whether you're asking something of a broader nature. There is a certain sense in which an arbitrarily large amount of classical information can be loaded onto a single qubit. This is because there are a continuous infinity of ways in which it can be prepared. However, it is impossible to obtain back that much information from it. For a qubit, the maximum amount of information that can be retrieved is a single bit's worth. You can find a fairly elementary exposition of these points in a paper Carl Caves and I wrote: "Quantum Information: How Much Information in a State Vector?", LANL e-print archive quant-ph/9601025.

So the real answer to the general question is that it is not the raw information that can be loaded onto a quantum object that counts. Rather it is in the novel ways that that information can be used.

Concerning the HSW result specifically, yes I can supply you with some numbers. Consider a communication channel that consists of individual two-state atoms as the qubits; that is to say, we shall use the two-dimensional Hilbert space due to an atom's ground and excited states as the spot where we place our signals. The channel's noise is due to the possibility that the atom's excited state can decay to the ground by dumping a photon into the electromagnetic field. Suppose the coupling between the atom and the field is such that the probability of this decay is 30% (as the atom makes its way from sender to receiver). Then, before the HSW result, the best rate at which one could hope to transmit classical information on this channel was 0.229 bits per transmission. With the HSW result, we know that there are ways to use this physical system to transmit 0.300 bits/transmission if we insist on using orthogonal (i.e., noiseless) signaling alphabets, and 0.311 bits/transmission if we allow nonorthogonal (i.e., noisy) signals into the game. This represents squeezing 36% more performance out of the fundamental physical laws than had been expected. (There are likely examples that make this difference much more dramatic, i.e., on the order of 100% or more, but this is the one I know off the top of my head.)

Bakercise 4: *Do you see us using quantum information and the quantum technologies in the near future?*

How do you define "near future"? I think it is inevitable that quantum information technologies will one day make an effect on the common man's life. Technology always seems to have a way of coming into existence if it's not excluded by physical law. (If it doesn't come into existence, it means we will have found something better.) It's just a question of when.

About the HSW result in particular, Hideo Mabuchi (here at Caltech and soon to be on its faculty) tells me that with a slight modification of the Turchette, *et al.*, experiment for building quantum gates in "cavity quantum electrodynamics" the rudiments of the HSW result could be seen pretty easily. (What I mean by "rudiments" here is that with a little entanglement in the measurement one could realize transmission rates that one could not without that entanglement.) However, one should not lose sight of the fact that the HSW result is an asymptotic one. Thus it establishes a fundamental limit on the rate at which a channel can be used, but to obtain that rate one needs ever larger amounts of entanglement in the measurement. So its real role is really in establishing how far an engineer can hope to push a system.

02 October 1997, "Next Round"

Below I try to answer all your questions.

Bakercise 5: *What's the next big thing to be done do you think?*

I wish I knew! The big things are usually surprises. In general – to give you something of a workaday guide – we need a deeper understanding of the similarities and differences between quantum and classical information. But from that general statement, there are several ways to go.

For instance, it would be nice to have a result similar to the HSW one but for the "(one-way) quantum capacity of a quantum channel." That would aid in the comparison. But I haven't seen anything too satisfactory in that direction yet. This is something that'll just have to build up slowly.

Another direction is to just get a better understanding of how to quantify entanglement. Maybe the best efforts in that direction so far have come from the IBM group (Charlie Bennett and company) and Bill Wootters. Once that is done, then one can hope to separate cleanly the amount of classical vs. quantum correlation between two systems.

I hope that gives you some of the flavor of things in the quantum information theory world. In our sister world, the world of quantum computation, things are about the same. Of course it would be good to find *other* new algorithms that can do wonderful things on quantum computers. But maybe the most (systematically) important question is to pinpoint in a very precise way just where it is that a quantum computer gets its power from. That is to say, in a way that is deeper than using some pat phrases like "quantum parallelism" or "quantum entanglement, of course." Then, one may have a hope of understanding the general class of problems that, in principle, can be sped up on a quantum computer.

Bakercise 6: *In what way was he* [Holevo] *thinking about these problems in 1978. Surely not in the present terms of Quantum Information?*

Holevo is an impressive soul! He posed exactly the HSW problem in ... doh(!) ... actually it was 1977 in a paper titled "Problems in the Mathematical Theory of Quantum Communication Channels" (*Reports on Mathematical Physics*, **12**(2), pp. 273–278, 1977). The paper was received 10 December 1976. In a "note added in proof" he writes something that essentially says that he has an *example* that shows that the classical capacity (of a quantum channel) can be increased from what had been previously thought by allowing the receiver to perform entangled measurements. (In the body of the paper, he had already said it was an open problem.) By the 1979 paper, he had *conjectured* what is now the HSW formula.

Now let me try to clarify what sorts of things were in vogue in the early to mid 1970s (especially in Russia) along these lines. You are correct in guessing that things were not stated in terms of the present lingo of Quantum Information Theory. However, the question of how best to use a quantum mechanical system to transmit *classical information* goes back at least to a 1962 Masters Thesis at MIT by G. D. Forney; the earliest published paper on the question may be one due J. P. Gordon in 1964. That's as far as I've been able to trace it back. Now you do understand what is meant by this: using a quantum channel to send "classical information" is to use a quantum mechanical system to carry distinguishable symbols that can be decoded at a receiver. Examples of sending classical information are putting ink on a newspaper (the quantum carrier is the paper), talking on the telephone (the quantum carriers are the electrical currents in the wire), or broadcasting a television show (the quantum carrier is the electromagnetic field). This is to be held distinct from the uses of a quantum channel that came up in the 1980s and 1990s: that is, in using quantum mechanical systems to carry incompletely distinguishable (i.e., nonorthogonal) states and/or entanglement from the sender to receiver. These applications are new: Holevo certainly never thought of them – he's told me this, in fact. Nevertheless the concept of entanglement has been around for a long time, at least since Schrödinger coined (a German version of) the term in 1935. What Holevo proposed was, essentially, to use entanglement in the aid of the measurement or decoding step in the classical information problem.

Aside: The English version of the word "entanglement" is also due to Schrödinger: it too came out in 1935. Maybe it'll also be useful for you to know that, in a rudimentary form, Einstein was aware of the concept as early as 9 July 1931. This can be seen from a letter of that date from Paul Ehrenfest to Niels Bohr that describes a conversation with Einstein. Einstein gave a rudimentary version of the Einstein–Podolsky–Rosen argument. Max Jammer wrote a nice historical article on this; if you'd like I can try to dig up the exact reference.

Bakercise 7:

"I had been working on the problem myself somewhat, but was never able to pinpoint a measurement good enough for the decoding. That's where the real genius in their proof lay."

Can you explain this?

Yeah, sure. There was a paper by Hausladen, Jozsa, Schumacher, Westmoreland, and Wootters that proved a very limited case of the HSW result. Some versions of the paper were floating around in 1995. The paper was a little muddled as to the purpose or goal of the problem, but it was all essentially there. Anyway, at the end of the paper, they conjectured the full HSW result. But then upon meeting and talking to Holevo and reading his 1979 paper, I said, "Aha, that's what it's all about." Plus Holevo, in that paper, had also proven that the conjecture was at least an upper bound on the true answer. So I talked to Richard Jozsa about working on the thing, and he said, "Sure, jump in" (or something to that effect).

So here's where it stood. To use a (noisy) quantum channel to send classical information, here's what you have to do. The sender and receiver have to decide on the set of possible messages that may be sent. (For instance, you and I have tacitly agreed to speak in English rather than in French.) Then they have to decide on a good coding of the messages so that they will become resilient to the noise. (As an example here, if we knew that there was an evil "third letter thief" listening in to our conversation we might start douubling thee thiird lettter of eveery worrd thaat we wriite to eacch othher.) Finally, the sender has to decide how she will actually implement this code in terms of preparations of a quantum system (the ones to be sent down the channel), and the receiver has to decide upon which measurements he will perform to help reveal the message. (You see, when Alice and Bob communicate via written letters, it is clear what Bob has to do to retrieve Alice's messages in the most efficient manner: he has to "look" at the words on the paper. However, in general, there are other things he could have done. For instance – I know this is silly – he could have felt out the indentations on the paper instead of looking.) With regard to quantum systems, it is generally not obvious as to which of the many possibilities the optimal measurement will turn out to be. "Should I look at spin in the x-direction, or spin in the z-direction? Quantum mechanics won't let me do both, so I'd better be careful in picking my protocol."

Of course all these steps in building a good use of the channel are intertwined. Luckily the coding part of the scenario above can be taken out pretty easily by a minor extension of classical information theory – this Holevo did in his 1979 paper. After that, one can simplify the problem by holding fixed everything but the decoding/measurement part at the receiver's end. Even with that, though, the problem becomes a bear. And that's what I was talking about. Holevo and Schumacher/Westmoreland both finally stumbled onto a measurement that was good enough to achieve this upper bound that Holevo had already proven. That's where all the genius in these last articles lay.

Bakercise 8:

"Is this result deep? I think so, specifically because of the way it stands in a dual relation with more standard entanglement. In some ways it is just as shocking as entanglement itself. The full implications of this new sort of "non-locality" have yet to be fleshed out."

Are you suggesting there is a sort of residual entanglement still available to classically correlated states?

Not really. I think putting it the way you did in this question is probably dangerous – you should steer clear of that. The problem is that I don't understand the effect so much yet that I have a good everyday-language way of saying it. (You see, the reason we use so much mathematics in physics is that we don't have to think too hard then!) Maybe the best way to put it is simply that these classical correlations between quantum states have more visibility than meets the classical eye. That is, if you allow entanglement to be used in the measuring process, then more about those correlations can be ferreted out than would have been the case otherwise. However, there is no residual entanglement in the states themselves: the acid test for this is that there is no way to make these classically correlated states violate any Bell inequalities.

Peres and Wootters put it like this in their 1991 article (*Phys. Rev. Lett.* **66** p. 1119, 1991):

It is well known that composite quantum systems, consisting of noninteracting parts, can possess nonlocal properties. In particular, a composite system can exhibit correlations which cannot be reproduced by any theoretical model that involves only variables belonging to each subsystem separately. ... In this Letter, we consider a different kind of composite system. Its parts never interacted in the past. They may have been prepared in different laboratories. However, they are prepared according to the same set of instructions. Therefore, these subsystems are in the same quantum state – insofar as their internal variables are concerned. ... Our work suggests that one can indeed obtain more information by measuring the two particles together ...

I hope that helps.

Bakercise 9:

"There is a certain sense in which an arbitrarily large amount of classical information can be loaded onto a single qubit. ... For a qubit, the maximum amount of information that can be retrieved is a single bit's worth."

I need to make this clear before I start writing the feature so forgive me for repeating what might be the very obvious.

Is this statement correct?

Yes it is. Though I think you will be able to understand the flavor of it better if you read the first four pages of the thing I wrote with Caves.

The remark you quote above, as I say, is true. But it generally addresses a different sort of scene than the one addressed in the HSW problem.

Bakercise 10: *If you have classical information on a quantum channel (the HSW system), you can load a large amount of information on but you can only take one bit off if you use orthogonal states. If you use non-orthogonal states you are limited to the von Neumann entropy which will be less than one bit.*

Is this statement correct?

Somewhat. Regardless of whether your encoding takes (classical) bits to orthogonal or non-orthogonal states, the most information an observer can hope to retrieve about the encoding is the von Neumann entropy of the density operator associated with the states in that encoding. The von Neumann entropy of any density operator is bounded above by the logarithm of the dimension of the Hilbert space. For a single qubit, the von Neumann entropy is bounded above by $\log 2 = 1$ bit.

But that issue is not so important for the HSW question. Their question is more to do with: what is the best way for me to stuff information into a qubit so that I have the greatest chance of retrieving it after the qubit has encountered noise. In particular, it can be shown that it is never a good idea to try to stuff more than 4 classical bits into a qubit if your ultimate goal is to retrieve the most information at the (noisy) channel's output. (If there is no noise at all, it is never a good idea to try to stuff more than a single classical bit into it.)

Bakercise 11: *If you have quantum information on a quantum channel the main problem is to keep the quantum system intact (quantum fidelity) until it reaches the other end. There it will be used by a system needing quantum information input e.g., a quantum computer.*

Yes this is correct. There are at least two tasks to which one might submit a quantum communication channel. The first, as we've already discussed at great length, is to reliably send classical bits down it. The second, which we haven't touched on too much, is to use it to transmit quantum information. This breaks into two categories: A) to use it to send arbitrary quantum states drawn from some Hilbert space, and B) to transmit entanglement over it, so that the sender and receiver become entangled to their hearts' content. In category A, the criterion of a good transmission is that the outputs be as much like the inputs as possible. I.e., the fidelity between input and output be good. In category B, it's that the entanglement be as high as possible. Now the way to ensure that these things happen is to spread the real thing you're trying to send over many transmissions through the channel. That's what Peter Shor's insight into quantum "error-correction" is about. Once you know that such a thing will help, then the next question is: how many transmissions do I need per qubit to make the signal fidelity or final entanglement good? That's the question addressed by the issue of the "quantum capacity of a quantum channel." (But we are quite some distance from that answer.)

Bakercise 12: *Can you answer these questions?*
Is there a way this quantum information can be used by a classical system, e.g., a television, normal computer? And if so, how much information can be taken off? From what I have read so far it should still be one bit, but there are hints it might be more, much more. What is the correct answer?

Nope, sorry, it doesn't really look like that's the case. What I can tell you is what I've already said. I guess the thing I'd like you to take away from this conversation is that the thing we're learning from Quantum Information is a lot of new ways to think: we're learning to view quantum mechanics in a really positive light now. Entanglement, for instance, is a new resource that should be taken into account when approaching a problem. Sometimes it's helpful and sometimes it's even a gold mine (like when used for factoring large numbers, Shor's factoring algorithm). But it's not a panacea. Learning all the things we can about this stuff, including the limitations, is our great challenge. Good and even amazing things will come – in my opinion – but they'll likely be concerned with things we don't have the imagination to think of today. Quantum Information just establishes a good direction in which to think.

06 October 1997, "49th Parallel"

Bakercise 13: *I have another question prompted by your remarks on entanglement and energy. How close is the parallel? Do you think there is an $E = mc^2$ for entanglement or a $q = DU - w$ or $G = H - TS$?*

Wouldn't that be great?! But, unfortunately, I don't think anyone can say presently how close the concepts might be. Some of the questions raised concerning entanglement certainly have the flavor of thermodynamics, but I think it's just too premature to draw any rigorous parallel. Sorry.

3

Letters to Howard Barnum

02 June 1999, "Please Do"

You know what I'm really trying to get at here: I want the structure of quantum mechanics to be extremal for some kind of more abstract (i.e., QM-independent) information-disturbance problem. But the first thing that needs to be done is to pin down what quantum mechanics itself has to say.

Another thing that keeps coming to mind is that the assumptions behind Gleason's theorem might have something to do with the allegory I presented you with at Charlie and Theo's house. God said, "Ok I will make information gathering invasive, but not so invasive that you will be forever science-less." The connection is this. I think one can split the assumption behind Gleason up into at least two parts: (1) The questions that can be asked of a quantum system *only* correspond to orthonormal bases over some vector space, with there being no good notion of measuring two distinct questions simultaneously (there being no good notion of an AND operation). And (2) It is the task of physical theory to give probabilities for the outcomes of those questions, but we can say at least this much about the probabilities. They are context-independent in the sense that, for a given outcome, it does not matter which physical question (which basis) we've associated it with. I.e., the probabilities can be assumed to be of the form of a frame function.

It seems to me that the first assumption captures to some extent the idea that information gathering is invasive. If you gather some information and I gather other information, there is no guarantee that we can put the two pieces of information into a consistent picture. The second assumption, though, appears to be more of the flavor that, nevertheless, that information gathering is not *too* very invasive. For otherwise I might imagine the probabilities to depend upon the full context.

17 June 1999, "So Slow"

Kochen-Specker, no it's not too important. It just immediately throws out a certain naive class of hidden variable theories. That's good for illustrative value, but not much more. The thing that intrigues me about David Meyer's result is that if we work over the rational

20

field, then we can no longer get that illustrative example. Because of that, I keep wondering whether there is something fishy here ... it's just a vague feeling.

Wallach's result has to do with trimming down the assumptions behind Gleason's theorem on tensor-product spaces. It's not related at all to the stuff about Meyer. About extra assumptions for getting Gleason's result on rational Hilbert spaces, I don't know. I was just speculating: for instance, linearity would probably give it. But then that already throws away all the beauty of it.

All this does cause me to wonder about on which kinds of discrete structure we might hope to prove an analogue of Gleason's theorem. For instance, let us not fixate on the field over which the Hilbert space is defined. But perhaps instead let us consider as valid quantum mechanical "questions" (i.e., measurements, but I no longer like that term), any orthonormal basis that is rotated away from a standard basis by angles which are rational fractions of π. Can we prove Gleason's theorem on such a structure? I suppose I should ask either Wallach or Pitowsky this question.

04 July 1999, "No Microphysical Reality?"

Oh the accusations you make about my poor philosophy! It's the middle of the night, and there's a wonderful thunderstorm going on right now. I've been up the last hour or so rereading a lot of your emails from last week.

It seems you think I believe there is NO reality? Anyway, I think the notes below best express my present views. Have patience with them: read all the way to the bottom, and then come back and read the top.

[NOTE: In place of the notes I originally recommended to Howard, see this whole book!]

18 August 1999, "Modalities"

It dawned on me on my flight home that we've never discussed the so-called "modal interpretations" of quantum mechanics. Take for instance the Spekkens and Sipe poster we saw in Baltimore as a good example. The essence of these interpretations seems to be the following. One starts with a Hilbert space for the universe and assumes that it has a fixed tensor-product structure. The state vector of the universe is not identified with reality immediately but instead gives a catalog of what is *possible*. The properties that are *real* in this theory are the state vectors associated with the factors of the big Hilbert space – but only one of those product states "actually" obtains. For instance, in the Spekkens–Sipe proposal one takes the universal wavefunction, decomposes it into a superposition of orthogonal product states (with a certain extra entropy-minimizing criterion) and that defines what is possible. The possible physical states of the universe are those product states. Their (objective) probability for being real is given by the absolute-square of the associated coefficient in the decomposition. But the real world only actually has the properties associated with one of those components. At every moment in time one has this, and thus the world has its properties stochastically and nonlocally determined.

It seems that the main difference between this point of view and the many-worlds view is that in this set-up only one world actually obtains. Also I suppose that it is crucial that there is a fixed tensor-product structure and at least some adherents of the many-worlds view would say that that's not so crucial.

What's your opinion of all this?

30 August 1999, "It's All About Schmoz"

I read your paper ["Dieks' Realistic Interpretation of Quantum Mechanics: A Comment"] on the plane the other day: essentially I agree with you that the modal interpretation is just a hidden variable theory. Jeffrey Bub in his new book even classifies Bohmian mechanics as a modal interpretation. For some reason the modal guys just don't seem to like the label "hidden-variableists."

Barnumism 1: *the only thing that can tell us the other branches are there is a "disentangling" interaction ... something like the reversal of a measurement. This is what I think is usually meant when e.g., relative state'ers tell you that the reason for having the other branches around is "interference": we might, in principle, be able to set up an experiment involving us* and *the rest of the universe (or whatever portion of it is entangled with us in respect of the measurement we're trying to reverse), and arrange to interfere them in such a way as to disentangle the state, erasing our observation record. Sure this is incredibly difficult. But, do we have a rigorous theory of its* impossibility?

This passage lays bare the very core of our difference! In the Everett interpretation, we can never ARRANGE to "disentangle the state." There is no sense in which this phrase or the gist of your passage is even meaningful in that interpretation. The Everett world just IS. Experimenters have NO choice to arrange anything in an Everett multiverse; their actions are at best a grand conspiracy set by the initial condition of the universe. They have no will, no possibility to do anything that wasn't preordained. If an experimenter in the Everett theory perceives himself reversing the entanglement of two systems, it can be but ILLUSION that he had any choice in the matter.

Barnumism 2: *no-one [...] has ever reversed a measurement whose result they were conscious of [...]. So, as a state-vector-isn't-real-but-is-just-a-representation-of-our-knowledgeist, you might say these interference effects haven't been observed, and you needn't worry about them as you blithely reduce the wavefunction, and say you just got new knowledge, changing what you can say about the probability of future experimental outcomes. Fine, but now your theory differs from no-collapse theories on the outcomes of at least some* in-principle *possible experiments.*

You just don't get it! It is precisely for the state-vector-is-knowledgeist, that there is any substantial sense to being able to reverse the evolution of a quantum system. There is nothing extra to standard UNITARY quantum mechanics in this picture: one simply must be sober enough in thought to recognize that the state vector is NOT reality and that is that.

Suppose I know everything that I can validly know about an electron, and so represent that knowledge of it by some pure state $|\psi\rangle$. Suppose too that you believe the same thing of the electron that I believe, and hence you yourself use $|\psi\rangle$. Finally, suppose that just as with the electron, I know everything I can validly know of you including your intent to measure the electron's spin. I describe you via the quantum state $|\eta\rangle$. Time passes and true to form you measure the electron. You learn something about the electron and thus update your state of knowledge (perhaps discontinuously as we usually presume); you change the electron's state assignment to something new, perhaps $|\tau\rangle$. I, not interacting with you or the electron any further, describe the state of affairs via some unitary evolution that entangles you and the electron. My state of knowledge about you AND the electron is now some entangled state. The difference in our two DESCRIPTIONS is one due to an outside point of view as compared to an inside point of view. In that sense, it is the same disparity as one encounters in the old Maxwell demon problem.

Now, what about reversibility? It is precisely this: my state of knowledge of you and the electron is still maximal. The fact that from the inside point of view you have learned something and thus made some discontinuous change in your description of the electron makes not one iota of difference. My state of maximal knowledge allows me certain privileges. For instance, since I know both you and the electron so well, I can in principle construct some complicated apparatus to control you in any way I wish. In particular, I can, IF I SO CHOOSE, reverse your measurement, wiping out all memory in you of ever having gone through the motions of the measurement. There is no inconsistency here; there is nothing external to unitary quantum dynamics. The only thing that one must do to see the logic is HUMBLE oneself to the point of view that quantum mechanics is about what we know of systems and how we can manipulate them in the light of that knowledge. In this point of view, the ontological content of quantum mechanics is not in the state vector ... it lies somewhere else in the theory.

You used the phrase, "no one has ever reversed a measurement whose result they were conscious of." This I agree with wholly. When one learns information, that process is by its very nature irreversible: within the Shannon theory, information gain cannot even be defined without the notion of prior ignorance. Probabilities must change irreversibly for information to have been gained. But REALIZE that is from the inside point of view. IF there is an outside point of view (as there was in the example above), then there is nothing sacred about irreversibility ... but that is purely with respect to that outside point of view.

In general what I dislike about the Everett extravagance is that it makes a concerted attempt to ignore the agent's – be it a man, a scientist, a chimpanzee, or a HAL-9000 – place in science and place in the universe. It is a mysticism not unlike Kepler's theory of planetary motion.

If you do one favor for me this year, please read and think about all that I said above. And please read and think thoroughly about the old Fierz paper that I'll place below. [The paper is reprinted in full in a note to Rüdiger Schack, titled "Penrose Tiles," dated 29 August 1999.] Please try to put yourself in my mindset when reading all these words ... and I think you'll find it more reasonable than you ever thought you could.

05 September 1999, "New Schmoz Cola"

You know, I probably shouldn't even be writing you, if for no other reason out of punishment: I still haven't seen your crypto paper on the archive (or in my mailbox)!!!

Barnumism 3: *Ah! I have come to understand (or remember, perhaps) the importance of leaving a place for free will in your views on the foundations of QM. I'm sure you realize that a Newtonian, or perhaps better, Laplacian universe would have exactly the same problems? This didn't cause physicists to abandon Newtonian mechanics, though ...*

I agree with your second sentence pretty much, though I think Everett is still worse for other reasons. As I told Herb a few weeks ago:

I think many-worlds empties the world of content in a way that's even worse than classical determinism. Let me explain. In my mind, both completely deterministic ontologies and completely indeterministic ones are equally unpalatable. This is because, in both, all our consciousnesses, all our great works of literature, everything that we know, even the coffee maker in my kitchen, are but dangling appendages, illusions. In the first case, the only truth is the Great Initial Condition. In the second, it is the great "I Am That I Am." But many-worlds compounds that trouble in a far worse fashion by stripping away even those small corners of mystery. It is a world in which anything goes, and everything does. What could be more empty than that?

The Laplacian universe did indeed have the problem you mentioned, and NO that didn't cause physicists to abandon Newtonian mechanics. But, as you saw from my last note, I have no intention of abandoning unitary quantum mechanics: your reprimand slides off me like water. The solution is in taking the Fierz article to heart. Physical theories are here to help us make our way through the world; they should, for the most part, be viewed as predictional meat grinders and not ontological statements. In the classical case, the input is the initial Liouville distribution, the output is the final Liouville distribution. Both are states of knowledge: the inputs and outputs should not be viewed as states of the world.

Barnumism 4: *From the inside, the experimenter has a choice; from the outside, no. Here's a caricature, so feel free to object: Bell's worry about the foundations of QM has been: that we have "measurement" as an "unanalyzed primitive" of the theory. Everett shows us how to get around that. You don't like Everett's resolution because you want to have an unanalyzed primitive around so it can be the locus of free will.*

Not exactly. From the inside, the experimenter has a choice; from the outside, the experimenter has a choice. Quantum mechanical statements are statements about the extent to which things can be predicted. Get out of the ontological mind-set! Concerning "Everett showing us how to get around that," as much effort as I have given it, I have never been able to see it as more than a statement of faith. It, like most religions in this world – it seems to me – leads to not much more than comfort. About my "wanting a locus for free will," there is little doubt about that. The thing about quantum mechanics is that it helps me believe we're closer to that situation than I once thought possible.

I find it hard to gulp that my whole life is an illusion. I find it immensely improbable that the design of this Dell computer was written in some partial way in an ontological initial condition for the universe. It strikes me as almost laughable that though we are essentially automata – even if splitting into minor variations higgledy-piggledy – we are nevertheless automata lucky enough to have the built-in illusion that we are understanding and discovering things anew with each day.

This will be hard for you to believe, but if you ask me, I think my view is far more Copernican than yours. You're willing to risk the guess – when you're playing devil's advocate for Everett, that is – that the Everettista model can be placed in an encompassing correspondence with all that is. "This little thing that OUR minds dreamed up captures reality in its very essence." That is an extreme anti-Copernican chutzpah, it seems to me. In contrast, in order to get my view off the ground all I have to do is acknowledge that our knowledge is limited, that I am perhaps only 11% better off in my understanding of the world than my dogs. I do acknowledge that our actions can change the world in a very limited way (at present at least), but I do NOT acknowledge that they can change things "before we were here." It is a humble point of view, but a point of view with hope (for in it there are aspects of the world that are not set in immutable stone).

Barnumism 5: *Your discussion of the experimenter (Dr. Science) who does the reversal, is fascinating. I am glad you (and also Herb) agree that Dr. Science could wipe out your memory of the measurement. The inside/outside business (also emphasized by Herb at lunch the other day) is nice. But, if I (who will [undertake] the experiment), know what Dr. Science is going to do, won't I want to make predictions that agree with Dr. Science's?*

But there is no sense in which the two descriptions disagree, if you take the appropriate view of what it is the state vector symbolizes. When one has a pure-state description of a system, one has *maximal* knowledge, but maximal knowledge is not complete knowledge in quantum mechanics – there is nothing in quantum mechanics that plays the role of complete knowledge in classical physics, i.e., the phase space point. I assert these sentences with great confidence: it would however be an excellent research project for my point of view to find a formal way to show there is enough leeway in the alternative quantum descriptions that one never gets into trouble.

What is it that the knowledge is of? I'll bet your gut reaction is to think I'm thinking the knowledge is of some kind of underlying "reality." Well, I'm not. Lately I've been thinking the best language to describe the situation is the same as in L. J. Savage's treatment of classical decision theory. Quantum "measurement" is more rightly viewed as being about decisions, acts, and consequences. When confronted with a quantum system, we can *decide* to act upon it in one of any of a number of nonequivalent ways. The set of acts available to us is the set of positive operator valued measures: each POVM corresponds to an *act* we might perform on the system. The set of possible *consequences* associated with an act corresponds to the separate operators within the act's affiliated POVM. The quantum system is the conduit from the acts to the consequences; it takes the place of the "states of the world" in the Savage system, but it is a black box that cannot be cracked. The quantum state is only an "if-then." It stands for a compendium of probabilities that summarize our beliefs about the consequences of our acts. That is its sole content. [As an aside, you might notice that I associated the POVM elements rather than the associated quantum operations with the consequences of an act. That was intentional; in my view the operations have to do solely with the updating of subjective beliefs in analogy to Bayes' rule.]

Barnumism 6: *Herb at lunch kept saying (I paraphrase severely) "sure these other terms exist, but only in the wavefunction from the outside point of view ... not in the wavefunction you can use." But, that's exactly what's being argued about: if an experiment is going to happen, and I, while "inside" the experiment, can understand what's going to happen by taking the viewpoint you label "outside", who are you to tell me I can't take that viewpoint? One could then see the Everett interpretation as a systematic attempt to be as "outside" as one can in one's model of the world (viewed as fundamentally quantum mechanical).*

I've said it before, I'll say it again: can a dog collapse a state vector? Dogs don't use state vectors. I myself didn't collapse a state vector until I was 20 years old. More to the point,

QUANTUM STATES DO NOT EXIST.

If you take the Savage language presented above seriously, then it is an immediate consequence that you on the inside cannot see things as if you were on the outside. (Knowing of an outside viewpoint's POSSIBLE existence is a different thing than taking it.) You cannot take a view that includes yourself within it because you can act upon what you can act upon and nothing more. How can you get outside yourself to act upon yourself?

It does appear to be true that I cannot stop you from attempting to take the most outside view (naively) imaginable, the one that Everett wishes. But I would only be a broken record if I were to say that I don't see how that point of view can be but stale and counterproductive. When you ontologize the state vector, you force the world to be an empty shell – that's the way I see it.

Barnumism 7: *Not ignore, but to allow the possibility of representing the agent as made of the same stuff as the rest of the world ... a very monistic kind of theory ... and to allow the possibility of taking a point of view "outside" the agent as well as the laboratory apparatus and quantum systems under study, or whatever. I'm just afraid you dislike this approach so intensely that you want to claim it's impossible, illogical, etc ... whereas my view of the Everett interpretation is that it shows there's nothing special about quantum mechanics that makes taking the outside view impossible, although perhaps it makes it weirder, and less appealing, to take that view.*

In my view, I explicitly do NOT see the agent as being made of different stuff than the rest of the universe. If I did, I would not be able to trust modern medical science to develop new drugs to cure old diseases; I would not be able to imagine a Dr. Science who might reverse my measurements. That is just the point. There is the world, and then there is what we can say about how it reacts to our experiments, our punches and our pushes. Quantum mechanics is about what we know of the right-hand part of that sentence.

I do dislike the Everettista movement fairly intensely, but I don't feel that I have gone so far as to claim that it is "impossible, illogical, etc." My main points are these. 1) I have not been convinced that Everettism is less of a religion than my own personal point of view. To me, in fact, it seems the opposite: I see almost no possibility of extracting the world of our common experience from an imagined universal wave function. This could be a weakness on my part; I am not all knowing. But if that is true, I still have: 2) Everettism simply seems to be a dead end for the furtherance of our understanding this wonderful world. I could be following a false prophet, but if my view succeeds then we will know that there is room for something new under the sun. How could we not want that?

4

Letters to Paul Benioff

09 April 1996, "Randomness"

I don't know if you remember me, but we met at the Santa Fe Institute Workshop on the Physics of Information two summers ago. I spoke to you about your old papers on randomness and quantum mechanics.

In any case, I am very much interested in them again. (Presently, in particular, the ones titled "Finite and Infinite Measurement Sequences in Quantum Mechanics and Randomness: The Everett Interpretation" *JMP* **18** (1977) 2289, and "A Note on the Everett Interpretation of Quantum Mechanics" *Found. Phys.* **8** (1978) 709.) Carl Caves and I are writing a book on Quantum Information and are trying to get a feel for how far these sorts of considerations really go toward deriving the standard probability expression of quantum theory.

Last night I reread the latter of the two papers mentioned above. The main question on my mind now is about Section 4 "The Asymptotic Description." At the end you state, "Thus it is an open question whether or not there is any possible state description of the asymptotic situation that includes component states each corresponding to a possible universe as perceived by an observer with memory trace ν." Has there been any progress in this direction since then?? If there has, could you please send me a reference or outline the solution?

Paul's Reply

I think I do remember you from the Santa Fe conference, which I enjoyed very much. I am sorry for the delay in responding to your letter. I dug out my paper and reread it as I had forgotten about it. Yes the problem still exists, as one cannot discuss in quantum mechanics within Hilbert space infinite sequences without using some restrictive tail conditions to limit consideration to a countable infinity of sequences described in a separable Hilbert space. From a measure theoretic viewpoint, any countable set is a set of measure zero.

One way around this may be the use of quantum field theoretic techniques such as lattice gauge theory. In this manner systems with an infinite number of degrees of freedom can be treated, but I have not pursued this approach.

I have been much interested in randomness especially in relation to quantum mechanics. As you may know, there are an infinite number of different definitions of randomness each depending on the definition of statistical tests for randomness. In 1976 [*JMP* vol. 17, pp. 618, 629 (1976)] I showed that for sufficiently strong definitions, physics had something to say about the foundations of mathematics in that there were mathematical universes as models of ZF Set Theory which had no random sequences and thus could not be mathematical universes for physics and especially quantum mechanics. The problem was that I could not prove and still cannot prove that these very strong definitions are necessary. Weaker ones, e.g., those of Chaitin and others, seem sufficient. It appears that this approach is not productive.

In any case I support your efforts in this direction. Please keep me posted on your progress and give my greetings to Professor Caves. I hope we meet again at some conference.

15 April 1996, "More Randomness"

Thanks for clarifying things for me: I was mainly interested in knowing whether there are other mathematical tools lying around for the asymptotic situation that might make my life easier. If you don't know of any, that's OK; in the end, I haven't been convinced that the full asymptotic situation is important anyway.

Mostly I'm interested in recasting these old results by you and by Hartle (*AJP*, vol. 36, p. 704, 1968) in a Bayesian (or subjectivistic) light anyway. For that I don't think I need to get a handle on the full asymptotic case.

Thanks also for your comments on your 1976 *JMP* papers. These are two that I was not aware of! (Though I should have been because they are referenced in your others.) Let me ask you one question right off. Is your result true for finite dimensional quantum systems? Or rather must you consider repeated measurements on identical preparations of *infinite* dimensional systems to make it go? If the first option is the case, then I'll certainly try to dig them up at the library here. (Unfortunately I have to make a special request to get hold of older papers around here.)

There was a time when I was a strong believer that quantum mechanical probabilities are more closely allied with the frequentist notion of probability than with the subjective or Bayesian versions. However, I had hoped that that would somehow make itself known through the already-given structure of standard quantum mechanics ... that is to say, rather than introducing "randomness" explicitly as you did. But, of course, I had no clear idea about how to check that such a thing might actually be true.

Presently I'm more taken with Bayesian notions. (Evidence that Carl Caves has had some influence on me!) In case you're interested in what I mean by this, I'll attach a letter I wrote to Michiel van Lambalgen a few days ago. van Lambalgen is also interested in "randomness," though he does not believe that it is properly defined within the framework of classical, Platonistic mathematical logic. (He's told me that he's corresponded with you before.)

Paul's Reply

First to answer your question re my 1976 papers in *JMP*. The results are true for finite dimensional systems and their limits. That is one infinite repetition of a measurement process in which one prepares a system makes a measurement and discards the system is considered. One does not have to consider finite or infinite repetitions of infinite repetitions of measurements.

My 76 papers are based on the existence of many different definitions of randomness. (This is briefly discussed in the last section of the paper.) In fact there are an infinite number of possible definitions. The basic problem is that I know of no way to use physics or mathematics to place a floor under the definitions other than the relatively weak floor, namely that each statistical test must be at least effectively enumerable (Martin-Löf or Chaitin). The problem is very intriguing though because randomness is so much more an essential part of quantum mechanics than of classical mechanics. This is what motivated my work in the area.

I remember a brief conversation I had with Leonid Levin many years ago on a bus at the end of a meeting. To him it was obvious that the proper definitions of randomness should be very strong so that the results of my 76 paper would apply. However, I did not understand his arguments at the time and did not pursue the matter further. Dr. Levin was at Boston University the last I knew. Keep me posted if you pursue this further.

30 April 1998, "Randomness Again"

Once again I am thinking about randomness and quantum mechanical indeterminism in a relatively serious way. In particular, I just read your article:

P. Benioff, "On the Correct Definition of Randomness," in *PSA 1978: Proceedings of the 1978 Biennial Meeting of the Philosophy of Science Association, Vol. 2*, edited by P. D. Asquith and I. Hacking (Philosophy of Science Association, East Lansing, Michigan, 1981), pp. 63–78.

and the follow-up article by Hellman:

G. Hellman, "Randomness and Reality," in *PSA 1978: Proceedings of the 1978 Biennial Meeting of the Philosophy of Science Association, Vol. 2*, edited by P. D. Asquith and I. Hacking (Philosophy of Science Association, East Lansing, Michigan, 1981), pp. 79–97.

I must say, I find many of Hellman's criticisms pretty convincing. In particular, his remark at the top of page 86 and his Note #4 resonates well with me. Did you ever write anything countering his arguments? Or do you have any thoughts on the matter that you never wrote down? If you're interested in recording them, I'd love to hear!

Paul's Reply

Randomness is a fascinating subject that won't go away. I had forgotten about the papers you referred to but found them. I have not written a critique or rebuttal. But the following expresses my views.

I have much criticism with Hellman's critique. One main one is that he does not distinguish carefully between classical mechanics and quantum mechanics. In CM randomness is epistemic as it is due to an observer's lack of knowledge about the properties of a system. In QM it is present in principle and has nothing to do with what an observer happens to know or not know or can know about a system.

I do not understand what Hellman is driving at in his p. 86 comments. I do not use the terms as nondeterministic and acausal. A process such as an infinite repetition of a preparation of a QM system and measurement of an observable that does not commute with the prepared state of the system generates a random sequence of outcomes. (For simplicity assume a new system is prepared for each measurement.) A product probability measure describes this ordered ensemble of repeated measurements. The randomness of this outcome sequence is a physical property of the ensemble just as position or momentum is a property of the individual systems in the ensemble.

For me the main question is if the outcome sequence is random, how random is it or random according to what definition? I do not know the answer to this question. The main point about my *J. Math Phys.* papers was to show that if all such sequences are random according to sufficiently strong definitions, e.g., Solovay random, then physics has something to say about the foundations of mathematics in that not all ZF models can be used as the carriers of the mathematics of physics. Hellman does not find this of interest because the usual model of ZF is much bigger than the minimal model. I do not agree. Thus for Hellman to emphasize, p. 81 bottom, "if we believe in ZF there is nothing for physics to say" is not right. If a sufficiently strong definition applies, then physics would have shown that we must go outside ZF, at least for a mathematical description of randomness of QM outcome sequences.

The requirement I proposed in the paper that the weakest definition of randomness be such that no contradiction can be derived from an empirical outcome sequence seems reasonable. However I have to take my own medicine. This may or may not be relevant to the strength of randomness associated with a QM outcome sequence. In this sense I agree with Hellman's criticisms.

However, the last part of Hellman's note 4 on p. 91 is simply wrong in my view. The invariance of limit means under subsequence selections is intuitively assumed by every physics experimenter. If the even numbered elements of a sequence showed evidence of converging to a different limit mean than the odd elements the experimenter would conclude something was wrong with the experimental setup. This holds also for the sequential tests of randomness even in the widely accepted and weak definitions of randomness, e.g., Martin Lof or Chaitin randomness. This also applies to every outcome sequence generated by experiment, not "almost every one" whatever that means physically.

I do not know what definition of randomness applies to the sequences of outcomes in QM. So far no one has shown that there are any problems with the weak definitions, however, I do not know how hard people have looked. I stopped working on the problem because the approach did not seem fruitful. But the problem has not gone away for me.

As you might guess I am extremely interested in the connections between the foundations of physics and mathematics. This is a main motivation of my work on quantum computers and recent work on quantum robots and environments.

I am quite interested in your thoughts on this. Keep me posted on what you come up with.

10 June 1999, "Definitive List?"

I just finished compiling what I think may be a definitive list of all your writings to do with quantum mechanical randomness. It's pasted below. Would you mind having a look over it to see if I've missed anything? If I have missed something, could you email me the appropriate references? Thanks a million.

Just recently I read your paper "Some Foundational Aspects of Quantum Computers and Quantum Robots" and enjoyed it very much! Indeed, you point out an issue that is in the greatest need of exploration: how do the theory-making entities in this quantum world bootstrap their way into a theory of it? In turn how do they validate the theory they've come across? I think if we understood that, we would have a much better grasp on the "why" of quantum mechanics than we do today. (Today, I think we have essentially *no* grasp on why the theory takes the precise structure that it does. The contrast of this situation to the one in special relativity, for instance, is stark. There one can see in almost an immediate way the physical assumptions behind the apparatus of the theory. It is almost a crying shame that we are nowhere close to that with quantum mechanics, given that it is over 70 years old now.)

I expressed my own ideas in this direction to an old friend in a rather poetic way a couple of months ago. I'll attach that text too, in case it might pique your interest. [See note to Greg Comer dated 22 April 1999.] (Don't take the religious imagery as a serious reflection of my views; I used it only to help drive a point home.) Too bad we couldn't get together at the Gordon conference to discuss these things. Will you be going to the Newton Institute meeting this summer? I'll be there June 16 through July 17.

1. P. A. Benioff, "Some Properties of the Contact between Theory and Experiment," *J. Math. Phys.* **9**, 514–524 (1968).
2. P. A. Benioff, "The Imposition of Empirical Acceptability Conditions on Sequences of Single Measurements," *Z. Naturforsch.* **24a**, 86–96 (1969).
3. P. A. Benioff, "Some Aspects of the Relationship between Mathematical Logic and Physics. I" *J. Math. Phys.* **11**, 2553–2569 (1970).
4. P. A. Benioff, "Some Aspects of the Relationship between Mathematical Logic and Physics. II" *J. Math. Phys.* **12**, 360–376 (1971).
5. P. Benioff, "A Theorem on Undefinability," *J. Symb. Logic* **36**, 377 (1971).
6. P. A. Benioff, "Decision Procedures in Quantum Mechanics," *J. Math. Phys.* **13**, 908–915 (1972).
7. P. Benioff, "Possible Strengthening of the Interpretative Rules of Quantum Mechanics," *Phys. Rev.* D **7**, 3603–3609 (1973).
8. P. Benioff, "On the Relationship Between Mathematical Logic and Quantum Mechanics," *J. Symb. Logic* **38**, 547–548 (1973).
9. P. Benioff, "On Definitions of Validity Applied to Quantum Theories," *Found. Phys.* **3**, 359–379 (1973).
10. P. Benioff, "Some Consequences of the Strengthened Interpretative Rules of Quantum Mechanics," *J. Math. Phys.* **15**, 552–559 (1974).
11. P. Benioff, "On Procedures for the Measurement of Questions in Quantum Mechanics," *Found. Phys.* **5**, 251–255 (1975).
12. P. A. Benioff, "Models of Zermelo Frankel Set Theory as Carriers for the Mathematics of Physics. I" *J. Math. Phys.* **17**, 618–628 (1976).
13. P. A. Benioff, "Models of Zermelo Frankel Set Theory as Carriers for the Mathematics of Physics. II" *J. Math. Phys.* **17**, 629–640 (1976).

14. P. A. Benioff and H. Ekstein, "States and State-Preparing Procedures in Quantum Mechanics," *Il Nuov. Cim.* **40B**, 9–26 (1977).
15. P. A. Benioff, "Finite and Infinite Measurement Sequences in Quantum Mechanics and Randomness: The Everett Interpretation," *J. Math. Phys.* **18**, 2289–2295 (1977).
16. P. Benioff, "A Note on the Everett Interpretation of Quantum Mechanics," *Found. Phys.* **8**, 709–720 (1978).
17. P. Benioff, "On the Correct Definition of Randomness," in *PSA 1978: Proceedings of the 1978 Biennial Meeting of the Philosophy of Science Association*, Vol. 2, edited by P. D. Asquith and I. Hacking (Philosophy of Science Association, East Lansing, Michigan, 1981), pp. 63–78.
18. P. Benioff, "Randomness Yet Again," personal communication to C. A. Fuchs, 2 May 1998.

Paul's Reply

It is very pleasing to see your research program direction as your views have many points in common with mine. In essence I believe in the need for a coherent theory of mathematics and physics (as quantum mechanics) that (1) refers to its own validity and completeness to the maximum extent possible (TMEP), and (2) is maximally complete and valid. Such a theory is also a theory of everything. The theory will also include a description of quantum systems (e.g., us) constructing and validating the theory. I have made these points in a recent paper quant-ph/9811055 which will appear in *Phys. Rev.* A June 99.

Probably the most radical aspect of this viewpoint is that one does not start with the basic aspects of physical reality as given. Rather they are consequences of the requirement that a coherent theory of math and physics refer to its own validity TMEP and be maximally valid and complete. (An equivalent way to state this requirement is that the theory be maximally internally self-consistent.) In this view the basic aspects of physical reality, such as why there are 3 space and 1 time dimension and why the basic forces and particles have the properties they do, is a consequence of these requirements. Physical reality in its basic aspects is thus an emergent property. It emerges from the more basic requirement of maximal internal self-consistency that a coherent theory of physics and mathematics must satisfy.

More details on this viewpoint are in the material appended to this letter. It represents a section I wrote for inclusion in the paper you just read on "Some Foundational Aspects–" but had to delete because of length restrictions imposed by the editor. It expresses with minor differences my views today. As always comments are appreciated.

19 March 2000, "Probability Validation"

Thanks for your encouraging remarks about the *Physics Today* article Asher and I wrote. You write,

Benioffer 1: *Regarding your example of the weather prediction of 35% rain, it is true that there is only one unique time series of tomorrows. But this series is used by the observer and the weather predictors to determine if the percentage predicted is valid or not. I won't take an umbrella if the predictor has no credibility with his prediction. You did not discuss in your article how one determines validity for probability predictions.*

In an earlier draft, I wrote the following passage:

Here it is essential to understand that the statistical nature of quantum theory does not restrict its validity to situations where there is a large number of similar systems. Statistical predictions do apply to single events. When we are told that the probability of precipitation tomorrow is 35%, there is only one tomorrow. This tells us that it is advisable to carry an umbrella. Probability theory is the formal quantification of how to make rational decisions in the face of uncertainty; this carries over as much to quantum phenomena as it does to anything else. When one makes a probability statement concerning a quantum measurement outcome, one is essentially making a *bet* about what will be seen. As long as the probability calculus is used, an adversarial gambler can never force the bettor to a sure loss.

Of course this won't clarify too much for you, but it does push a little further the direction in which I am thinking. For the Bayesian, there is no such thing as a valid or invalid probability prediction … as far as a particular numerical value is concerned. The only empirically enforced criterion is that the probability *calculus* be obeyed. What I am talking about above is something called the Dutch book argument. You can find it all nicely laid out in the following two books:

1. H. E. Kyburg, Jr. and H. E. Smokler, eds., *Studies in Subjective Probability*, Second Edition, (Robert E. Krieger Publishing, Huntington, NY, 1980).
2. J. M. Bernardo and A. F. M. Smith, *Bayesian Theory*, (Wiley, New York, 1994).

Below I'll place a piece of an exchange I had with Asher when I was reading through a draft of one of his other papers. They make a further more detailed connection to your comment. [See note to Asher Peres, dated 1 December 1998, titled "Here Comes the Judge."]

20 March 2000, "Small Addendum"

I was thinking about something you wrote me a couple of years ago, and I thought I should say something about it before you readdress the point in the context of yesterday's email.

Benioffer 2: *I have much criticism with Hellman's critique. One main one is that he does not distinguish carefully between classical mechanics and quantum mechanics. In CM randomness is epistemic as it is due to an observer's lack of knowledge about the properties of a system. In QM it is present in principle and has nothing to do with what an observer happens to know or not know or can know about a system.*

I agree with you wholeheartedly that in quantum mechanics the "randomness" of measurement outcomes is NOT epistemic. However, that does not preclude my view that all probabilities (including quantum mechanical ones) are epistemic in nature. They quantify how much we can say about a phenomenon based upon what we know. It so happens in the quantum world that we cannot tighten up our knowledge to the point of removing all ignorance (about the consequences of our interventions), and in that sense the randomness is ontological – it is a property of the world that was here long before we ever showed up on the scene. But it takes epistemic tools to describe that property. That's the direction I'm coming from.

20 October 2000, "Not Instructions, but Information"

Let me try to answer your questions, which I think are good ones. (I'm on a flight now heading for Texas and should have plenty of time to answer you.)

Benioffer 3: *To me that is an interpretation of QM. Interpretations are what give otherwise empty theories their meaning.*

You're quite right about that. What Asher and I wrote about is indeed a kind of interpretation of the quantum mechanical formalism. The title and the ending words of the article were more for attention-getting than anything else. Also, though, the words were meant to be a small slap in the face to some of the extremes people have gone to (like Everett worlds, Bohm trajectories, and Ghirardi–Rimini–Weber stochastic collapses) just to hold on to a philosophic view that came around long before quantum mechanics was ever heard of. (Talk about people being set in their ways!)

Benioffer 4: *If I recall correctly you wrote a paper with Peres on QM without interpretation but stated that a quantum state corresponds to an algorithm for preparing it.*

In this, though, you're confusing two things. Asher wrote a paper in the early 1980s for *AJP* titled "What is a State Vector?" and in that paper he took the point of view you mention. That is, that a quantum state corresponds to nothing more than the (equivalence class of) instructions for preparing a system one way or the other. I've never felt completely comfortable with that point of view, so in the paper that Asher and I co-wrote for *Physics Today* last year we worked around that. I think also that I may have even swayed Asher's opinion on this issue, but you would have to check with him directly. (Perhaps I'll just carbon copy this letter to Asher.) Here's how I put it in a note to Asher 15 November 1999: [...]

I believe this point of view is adequately expressed in the article Asher and I wrote for *Physics Today*: a quantum state is nothing more and nothing less than one's best (probabilistic) information on how a system will react to our experimental interactions with it. How we may have come by that information – be it through a preparation, through a sheer guess based on all available evidence, or the principle of maximum entropy – is something I view as largely outside quantum theory proper. The structure of quantum theory instead codifies how we should manipulate our information (this is what time evolution and the collapse rule is about) and enumerates the varied ways with which we may gather new information (this is what the structure of observables or POVMs is about).

In this sense, I would call what we are talking about an "information interpretation" or "Bayesian interpretation" of quantum theory, rather than an "algorithmic interpretation." For the most part, however, I would like to avoid the wording of an interpretation.

I will go further – and this is one point where I may diverge from Asher – and say that I do suspect that we will one day be able to point to some ontological content within quantum theory. But that ontological statement will have more to do with our interface with the world – namely that in learning about it, we change it – than with the world itself (whatever that might mean).

Benioffer 5: *To my main question, which is related to your algorithmic belief. In essence exactly what is meant by saying a state is physically preparable, an observable is physically measureable, or a Hamiltonian physically realizable? If one accepts the belief that a state is preparable if and only if there exists a set of instructions for preparing it, then, by cardinality arguments, most states are not preparable. The same holds for observable measurements and Hamiltonian realizability.*

There is a lovely line in John Wheeler's paper

J. A. Wheeler, "World as System Self-Synthesized by Quantum Networking," *IBM J. Res. Develop.* **32**, 4–15 (1988)

that he took from W. V. Quine about the real numbers being a sort of convenient fiction. I wish I could repeat that line now, because I think it summarizes my point of view about this question. In fact, it's only the tip of an iceberg for me.

Let me give an example, suppose the *only* thing I know about a quantum system (beside the dimensionality of its Hilbert space) is the expectation value of some observable. Suppose even that that expectation value happens to be a rational number. The only question that's relevant to me is how do I use the information I have to make an estimate of the probabilities of outcomes for any other measurement I might make. Someone who believed in an ontological content for the quantum state itself might just stop and give up at this point: for to him the probabilities I am speaking of are actual physical properties pertaining within the system. But for me that makes no never-mind: Bayesian probability theory gives a host of things I can do in situations of incomplete knowledge. In particular, Jaynes' principle of maximum entropy is perfect for this situation (where an expectation value is absolutely the only information I possess). Through the principle of maximum entropy I am able to make a quantum state assignment, and when I have done that, I have done the most I can do.

Now, that quantum state assignment allows me to say things about all kinds of observable. For some of these observables, it will predict irrational even uncomputable probabilities. In fact, it will even let me predict probabilities for the outcomes of uncomputable observables (I'll presume you understand what I mean by this phrase). Does that bother me? No. That is because these numbers are derived from a quantum state which is in itself, as far as I am concerned, a kind of convenient fiction. These numbers and that state simply stand for the best I can say about any potential measurement (whether that measurement has a finite description itself or not).

Benioffer 6: *The main problem with this view, which I think is probably right, is what is the exact meaning of a set of instructions? In essence the problem is that, unlike the case for computable functions, there is no Church–Turing thesis equivalent for preparable states, measurable observables, or physically realizable Hamiltonians. There does not seem to be some precise concept which is generally accepted as equivalent to the informal notions of preparable states, etc.*

What are your views on this? Can you shed any light on this problem?

Above, I said that seeing the real numbers as a convenient fiction is just the tip of an iceberg for me. This is because I suspect the rational numbers themselves share no less in that sin. Indeed I sometimes think that the mathematization of reality (in any form or fashion) is nothing but a convenient fiction, one that helps us coordinate our activities and communicate with each other more easily.

Should there be some kind of kind of Church–Turing thesis for quantum observables? For all the reasons above, I suspect I'm less moved by this question than you are. But still I think one can make some sense of it, if one insists. If you are interested in my detailed views on this point, I will try to articulate them in a later note.

5

Letters to Charlie Bennett

25 December 1997, "Bennett Festschrift"

The Day I Was Worth 30 Pounds

Charles Bennett is so much more than a great scientist: his personality is the source of great science. In some ways, I never feel more comfortable or secure than when I'm in his presence. Last year, while visiting England, I decided on a whim to attend a small quantum computing conference with a friend. The friend had no official travel funds, so we were trying to spend as little as possible. It turned out that Charlie was at the same conference.

When time rolled around for the conference banquet, Charlie happened to catch me in the hotel lobby and asked, "Where is this banquet? You're going, aren't you?"

"No! It's too expensive for my blood; the price is 30 pounds." You see, I thought I would sound a little noble that way. I really did want Charlie's company, but there was my absent friend to think of, and he couldn't afford it.

Charlie said, "Well look, I make a lot of money now; I'll pay your way."

I was in a pinch! Bad choice for an excuse: everyone knows what a sponge I can be! But, just then, my friend walked in, and I had an out. "Actually we're going to try to find a pub or something. He feels the same as I do about this ridiculous price."

"Oh come to the banquet. I'll contribute 15 pounds toward each of you … for good conversation."

"No, we can't really. That's too much of an imposition. A pub would be so much better than a stodgy old banquet anyway." Charlie yielded: he would join us at the pub! (But you must realize: there was no arm twisting. It was his decision … though with a little reluctance. After all, he *was* an invited speaker. The conference organizers *had* flown him all the way to England. But no one had told him that he was expected at the banquet … or so he reasoned.)

We turned to the concierge to get directions and, in the process, picked up another companion for the trek. This one was a stray; none of us had ever met him before. But he was a nice fellow. And he was enthusiastic, wanting to learn everything he could about entanglement and the analogy between it and a shared secret key. Off we went.

One block. Two blocks. Almost across the campus, when we run into a nicely dressed man – jacket, tie, shiny shoes – who seemed to know Charlie like a friend. "Where are you going," he asked. "Taxis for the banquet are this way."

Charlie's "little reluctance" got the better of him. He looked to the slick fellow, "Do you think I'm expected to be there?"

"Oh, why yes. Of course." And he threw in a couple of proper English "hurrummphs" for good measure.

Charlie's face turned a little red with embarrassment. "Look guys we have to huddle. I really should go to this banquet. I think that's probably expected of me; these guys did spring for my airfare and all. I tell you what, I know this banquet is expensive; I'll contribute 10 pounds toward each of your meals. Come on, we'll have a good time."

But there was really no changing the plan now: we held fast, thanked him for his kindness. Charlie went his way; we three went ours.

What does this have to do with science? Nothing really. But it does have everything to do with the atmosphere of warm comfort that Charlie has built around his group at IBM. I think it goes some way toward explaining how so many wonderful results have come from that little area of New York in the last few years. At the end of the evening, after a little contemplation, I knew I was worth 30 pounds. And that may be one of the nicest compliments I have ever received.

05 May 1998, and to many others, "Probability Does Not Exist"

To all of my friends who may have been influenced by Ed Jaynes as much as I was, I thought I should spread the word. Professor Jaynes left this world last Thursday, April 30. Beside the human loss, we as scientists will lose so much from the fact that he was not able to finish the magnificent book he was writing *Probability Theory: The Logic of Science*. His guidance will be missed.

07 December 1998, and to the other teleporters, "Swedish Bikini Team"

The Swedish television crew came and went yesterday with little incident. No questions at all about the soul this time: I was spared (and maybe so were you). But I didn't pass up the chance to try to get an unsuspecting audience to think deeply about the wave function. (This audience, in fact, really will be unsuspecting: the footage was for a Swedish children's show.)

The question came, "What material is teleported? What is it that is transported instantaneously from Alice to Bob?"

I said, "The only thing that is teleported in quantum teleportation is what the preparer HAS THE RIGHT TO SAY about Bob's system. It is his description, his predictions, that jump instantaneously from one system to the other."

A blank stare, "But what MATERIAL is teleported?"

"That is the material."

He seemed pleased and left it at that. When he left, I had a look at my atlas to refresh the old memory: Sweden was indeed not so very far from Copenhagen.

10 December 1998, "More Swedish Bikini Team"

Bennettism 1: *Why bikini? That is even harder to understand than meat slicer.*

Don't you remember the old beer commercials (I think it was Miller Lite), that flashed their viewers with poses of the "Swedish Bikini Team"? The team consisted of about 12 buxom, snow-white haired, darkly tanned women in blue bikinis. Why did I title my note after them? Because I wanted you guys to read it; I figured the SBT was just as relevant to my note as it was to drinking beer.

Bennettism 2: *By "material" I think they [meant] physical matter, with mass, etc., not material as the word is sometimes used by wordsmiths to mean information, as in the background material behind a newspaper story.*

Oh, I'm sure of that. That was meant to be implied by what I wrote; did you not get it? But the point I was slyly trying to make – I'm sure it made no impression on them – is that not only is information physical, but it's just about all there is.

Bennettism 3: *If they asked me what material is transported I would say none. Only information is transported, indeed a subtle kind of information that would be spoiled by observing it.*

Maybe there's hope for you yet, Charles Bennett. Maybe one day you actually will be induced to leave the tranquilizing comfort of that silly church you attend (the one of the larger Hilbert space). I wonder, and I wonder all the time, if the structure of quantum mechanics isn't just a result of a more fundamental fact: that information gathering and disturbance (to others' information) go hand in hand in our world. In such a world, we construct the best scientific theory we can subject to that constraint and … (poof!) it's quantum mechanics.

By the way, since it's after 1:00 am and I'm sleeping and dreaming anyway, tell me this. What could the phrase "information is physical" possibly mean in an inanimate, mechanistic world? Imagine a world so unlucky that information processing units like ourselves or my two dogs or Deep Blue never arose. Indeed what would information alone mean in such a world? My only knowledge of information – through the Shannon theory – always requires a probability distribution around in order to define information. One needs "ignorance" before one can have "information." Do you really feel comfortable with ignorance as a concept defined within the church? If you start talking about partial traces and von Neumann entropies right off the bat, I probably won't be satisfied. What gives the right to associate such things with ignorance? Our main reason now is that it matches the Shannon theory in all the appropriate cases. And if you start talking about algorithmic information right off the bat, I probably won't be satisfied either. (For other religious reasons.) But still, if you have time, you should give it your best shot.

10 December 1998, "More Bikinis"

By the way, let me ask you a more serious question. You said:

Bennettism 4: *If they asked me what material is transported I would say none. Only information is transported, indeed a subtle kind of information that would be spoiled by observing it.*

On the other hand, I had said:

I said, "The only thing that is teleported in quantum teleportation is what the preparer HAS THE RIGHT TO SAY about Bob's system. It is his description, his predictions, that jump instantaneously from one system to the other."

Do you see our answers as different? From my point of view, they look about the same (except I didn't mention the disturbance business here). If you see a difference, what is it?

Charlie's Reply

I have a couple of serious (for me) answers. My first is that my idea feels right, while yours is hard to get my head around. The capital letters make me think that even you may be using them to overcompensate for an otherwise shaky idea. Also, rights are the province of lawyers and ethicists, a very ugly province indeed compared to physics. But I am probably wrong about both these things. Probably to one uncorrupted by many worlds, your idea seems perfectly sensible and beautiful. My second thought is the realization is that you encountered Jaynes in your formative years, at around the mental age I was when I encountered whatever makes me loyal to my Church. In other words, in both cases, we encountered what might be described as a pedophilic idea, an idea so seductive as to be dangerous to our youthful selves and make us dangerous to others when we grew up, or were thought by others to have grown up.

10 December 1998, "500 Swedish Irregular Verbs"

You made my Thursday morning! Now I'll always keep our pedophilic tendencies in mind when discussing foundational issues with you! So you're trying to make me feel perverse: is that because I shunned a date with your photos? Thanks for the commentary: it's quite useful to see what phraseology causes trouble. Discarding the "right," however, I still don't see the difference between the words "information" and the "sum total of what one can say." They're both just euphemisms for the quantum state, no?

By the way, it's interesting that you wrote the title you did. Just yesterday I wrote two notes, one titled "177 words of curly hair," and the other "175 words that touch upon truth." The first contained an abstract that Wojciech Zurek wrote for a paper about his existential interpretation of QM (whatever that is). The second was as an abstract I wrote for a paper that should capture the essence of my pedophilia. I used this trick to try help convince my coauthors that my abstract wasn't too long – probably won't work, I use too many tricks.

13 January 1999, and to many others, "Emma Jane Fuchs"

Dear Family, Friends, and Colleagues,

Emma Jane Fuchs was born into the world healthy (and by all indications happy) yesterday 12 January 1999 at 9:50 pm. She is beautiful like her mother and already striving hard to understand her place in the world like her father. The labor near the end was a tough one, though looking back on it, it is no wonder: 8 lbs, 8.8 ounces is a lot of little girl to get through a birth canal. Kiki is doing just wonderfully now, very proud of her daughter and the cleft chins the two of them share.

On Emma's behalf, I send greetings to all.

10 February 1999, "Computational Power of N w/o E??"

I'm going to be travelling and out of email contact for a few days, but I wanted to record the following vague thought before it slips out of my head ... and see if it elicits any reaction.

Have any of you thought about the "computational power" of nonlocality without entanglement? By this I mean the following. What tasks can be performed efficiently by a quantum computer or Turing machine whose discrete time steps are always restricted to take it from one separable state to another? This of course means that the state space available to such a computer is a Cartesian-product space and not a tensor-product space – Ekert and Jozsa make a big to do about the difference of the two products in `quant-ph/9803072` where they argue that entanglement is the essential feature of quantum computation – but it's a heck of a lot of a larger space than the Cartesian-product space of classical states (all of which would be orthogonal presumably). Don't you think that the 9-state example hints that there just might be something here? The key would be in making use of all the nonorthogonal possibilities in the separate spaces. Another hint comes from the original Deutsch problem (or was it D-J?), where one need not entangle the target and the register to get something resembling a true blue quantum computation.

Could we invent a *computational* example where something good (something distinctly nonclassical) comes out of nonlocality without entanglement?

05 June 1999, "Emma Jane"

Bennettism 5: *PS What is Emma Jane up to these days?*

- smiling at people when they smile at her
- smiling when one parent or the other rescues her from her lonely bed first thing in the morning (sometimes there is a race to get this lucky prize)

- laughing when her arm pits are tickled
- laughing when her dad mimics the dogs barking outside the window
- sleeping through most nights
- getting cranky most evenings just before bedtime
- bouncing with delight in her bouncing contraption that hangs from the ceiling
- struggling to crawl (great sounds of exasperation when her legs and arms move, but that belly on the floor just has too much friction)
- rolling over rather easily, but clockwise only (this is her best form of locomotion presently – on one occasion she's moved at least four feet like this)
- taking in all the sights in her daily walks hanging from mom or dad's belly – I think this is her most content part of the day; she never complains and stays awake as long as she can
- teething: gnawing on her mom's fingers, gnawing on her little green beany frog, gnawing on her teething blanket, gnawing on the edge of her high chair, gnawing, gnawing, gnawing
- learning to voice her opinion when she wants to be played with ("Damn it, what am *I* getting out of your cooking dinner? Come play with me!")
- eating solid – what a euphemism! – foods surprisingly well ("If any falls on the edge of the table, not to worry: I can suck it right up. You may need a spoon; I don't!")
- getting familiar with the numbers 1 through 10: one kiss (smooch), two kisses (smooch, smoooch), four kisses (smooch, smoooch, smooooch, smooch), three fingers, . . .
- having the attention span for about one book per sitting
- loving her bath time, playing in the water, acting like an Olympic swimmer
- inspiring her parents to great happiness
- giving her dad more reason to suspect that the world is so much more than a mechanical contraption clinking along

27 November 2000, "An Emoticon"

I like the title of your talk:

Charles Bennett is Technion's guest of honor in the program "Israel Pollak Distinguished Lecture Series". He will give two more lectures:

Wednesday 6 December, 15:00, Physics Building, Auditorium 323 Lecture of general interest:

QUANTUM INFORMATION PROCESSING: Uses for a kind of information so fragile that it cannot be observed without disturbing it

Now that sounds like a talk that even I would want to attend! Are you going to mention as one of your "uses" that it also gives us a clue as to what quantum mechanics is about in the first place? (Grin.)

6

Letters to Herb Bernstein

25 December 1996, "Reality Steaks"

Holiday cheers! I'm sitting in Geneva, connected to my machine at Caltech, thinking about an old fart in Massachusetts. I was just reading an article (in the *New Yorker*) about Woody Allen and came across the most wonderful quote: "I hate reality, but, you know, where else can you get a good steak dinner?" Like it? Quantum mechanics everywhere you turn.

17 February 1997, "Prophetic Herberts"

This is the second time you've intrigued me with a phrase or two: I'm not letting you off the hook this time. Please explain in more detail what you mean by the following. The first quote comes from 25 December 1996:

Herbal Treatment 1: *Actually that Austrian reaction to enhancing classical communication was probably part of a reality-loving or at least a quantum-preferring inclination which is some-thing beneficial to the reality-seekers amongst us – if everyone were so crazed about measurement and what-all, Charlie would never have figured out that it wasn't the knowing of an answer which introduced the irreversibility; it was the erasure of the "garbage" produced in the calculation. And without all the fuss over reversible classical computation, we wouldn't have had so much fun with quantum comp.*

The second quote comes from 12 February 1997:

Herbal Treatment 2: *I was impressed similarly by Charlie's hint that thinking beyond the "big deal" everyone since Szilard seemed to make of measurements was crucial to his realization that the entropy generated by Maxwell's demon for a 1-molecule gas came from forgetting which side of the door it was on. The entropy generation didn't come in measuring which side of the barrier contained the molecule, and memory of the side would make the process reversible, etc. etc. etc.*

Are you saying that Charlie's solution of the "Maxwell Demon Paradox" might have something to say about the solution to the "Quantum Measurement Paradox"? If so, let me know what you're thinking – I'm intrigued.

Herb's Reply

I am saying that I suspect Charlie was UNimpressed by the notion that something very special occurs in the act of measurement *per se* – and I do mean UN. He was able to see past where Szilard went unlike many other people, because he didn't believe the measurement automatically meant irreversibility. And he analyzed what it would take to make the "measurement" i.e., determination of which side of the barrier the molecule was on recoverable; indeed he pushed the analysis until he could say for sure the state was unrecoverable, and discovered that the issue was getting rid of the garbage that accumulates in a series of measurements, like the garbage that accumulates during a long calculation.

So the moral of this long-winded story is that we who are impressed with the quantum requirement to make a decision somewhere about what it means to do certain interactions, i.e., those of us who remain Bohrian enough to believe a phenomenon is not a piece of reality until it is a registered phenomenon, must be cautious. We admit it is an assumption that this implies reality creation IS a part of scientific inquiry and we examine this – and others – of our assumptions always; by doing so we will also clarify when they are wrong or misleading and when they just slightly misdirect our gaze.

In short there is a rather complex and muddled connection between C. Bennett's "Maxwell's Demon" work and our continuing investigation of the need and privilege to CREATE the reality we study in quantum mechanical situations of all kinds.

23 February 1997, "Herbertic Realities"

Thanks for your comments on Charlie and demons; I understand you better now. I especially liked your comment: "... indeed he pushed the analysis until he could say for sure the state was unrecoverable, and discovered that the issue was getting rid of the garbage ..." I'm not sure what I want to say about that, but I want to say something – I just love the intent there. What is the difference between getting rid of information by dumping it into the unknown and getting rid of it by spreading the information to an uncontrollable group of free-willed communicators. Give me a few days to try to formalize it all.

27 March 1997, "Consistent Herb"

In the next two mailings, I'll send you the correspondence I had with Griffiths on "consistent histories." (The first is my writing, the second is his.) I would still like to construct a detailed reply concerning this analogy he likes to draw between "frames of reference" in special relativity and "Boolean frameworks" within his version of quantum mechanics. I don't like the analogy and I think it is misleading precisely because quantum mechanical probabilities are not about real existent things that are or are not the case – as best I can tell, they are solely about contingencies. Alas though, my life and time are not infinite. (I'll get to it eventually.)

I had a very nice conversation with Charlie on the drive back to Wendell from your house, that, I think, has allowed me to sharpen what I think about things in quantum mechanics. In particular, I would like to work on a point of view that substitutes in place of the "Church of the Larger Hilbert Space," something along the lines of "Church of the Big-Enough Hilbert Space." However, it'll require some writing for me to get it down coherently. I'll try to combine it with my thoughts on Mermin.

Anyway, I had a really great time last week at your place. This quantum mechanics business is so wonderful, it's almost no surprise we can entertain ourselves with that, a few fried potatoes, and not much more! Greetings to Mary and the family.

29 March 1997, "More Reality"

Did you, by chance, read the article titled "Get Unreal" in the March 17th *New Yorker*? In case you didn't, I'll share with you my favorite quotes from it (that I want to tabulate anyway . . . I love to use friends as good excuses).

"People experimenting with high-definition television, or HDTV . . . report that the resolution of the image is such that details invisible on ordinary television screens, like scuff marks on the anchorperson's desk . . . leap out at you on HDTV. The Times predicted the other day that television production designers, formerly accustomed to tacking a studio set together out of plywood and duct tape, will have to start building the real thing. On standard television, mahogany stain looks like mahogany. On HDTV, it looks like stain. To maintain a realistic image, in other words, television will have to construct a better reality. The simulation is getting ahead of the simulated."

"In the case of the music industry, the rationale for persuading people to ditch their phonographs and LPs and replace them with CD equipment was that digital CD sound was more lifelike than the analog sound of the old LPs. Actually there was nothing wrong with analog sound. The ear learned to block out the occasional hiss and pop, and found the result adequately lifelike. You do not, after all, listen to live music in a soundproof chamber. . . . Once CDs became the standard, though, *they* became 'lifelike,' and analog recordings started sounding wobbly and staticky."

"Human kind cannot bear very much reality," says the little bird in T. S. Eliot's "Four Quartets."

"The craving for more and more realistic representations of reality is at bottom an aesthetic craving, and one that people are perfectly capable of indulging purely for its own sake – an indulgence beautifully illustrated by the tremendous expense and effort being invested in getting a computer image to look and function exactly like a piece of paper. But the effort to satisfy such cravings kind of misses the point of what the true pleasure of representation is. The true pleasure of a representation does not come from its indistinguishability from the real thing. It comes from its distinguishability. An Elvis impersonator gives pleasure precisely because he's not Elvis, and it is crucial to the effect that we never forget he's not."

07 April 1997, "Howling Quanta"

"The universe is a thing of dream
substance naught & Keystone void
vibrations of symmetry Yes No
Foundation of Gold Element Atom
all the way down to the first Wave
making opposite Nothing a mirror
which begat a wave of Ladies marrying
waves of Gentlemen till I was born in 1926
in Newark, New Jersey under the sign of
sweet Gemini"

(Allen Ginsberg)

I guess you've heard that Mr. Ginsberg passed on? Maybe that makes today the appropriate day for writing this note. You ask, why are we (you and I) so in sync on this issue of reality creation? I suspect it's most likely a chance fluctuation ... or a "selection effect" – why else would *we* end up working on quantum information?

Of course, the sixties did play a part, apparently for both of us. But for me, I suspect, in a very different way than for you. I was a kid in south Texas in the late sixties, very far from hippiedom. My sixties were filled with conservative democrats, cowboys, oil fields, and an occasional trip to the big city of Houston – there was no great connection to a world of ideals. I think surreality slipped into me from a much smaller world of turmoil. Somehow in the early years I gained the habit of making things look magical to myself. In my head, the most mundane objects became spaceships. For years I imagined that an imaginary friend and I were in search of the "end of space." The fantasy mostly consisted of us travelling outward, and building ever faster vehicles as we were doing so. There was no time to waste!

Did this shape my thoughts on quantum mechanics and, in particular, make me more receptive to it as an almost mystical structure? I think so. I'll place a couple of passages below that build some imagery about it. The first I wrote sometime in 1993; the second is a little poem that came out in the DFW airport about a month or two ago. [For the poem, see the beginning of the chapter on Letters to Greg Comer.]

You asked me if I have ever thought about the immense responsibility entailed by the possibility of reality creation. Yes I have, but never in any very organized fashion. In relation to this, my thoughts keep coming back mostly to the Holocaust, but also to instances of self-devastation like with the Jim Jones gang in Guiana. I think it was just this sort of thing that Einstein feared most about quantum mechanics: that reality ... and consequently right and wrong(!) ... might even be creations of the participants. (By the way, you can find an expression of this in the bad "free form" poetry below.) I understand the fear of E ... but, also, I don't know how to get around it.

Everyday Albuquerque reveals. Hollywood knew the quantum; giant scorpions in a desert. Fluid reality. The shape of entertainment. Notorious youth making out in cars; paying no attention to surreality itself. Paying no attention to the mystery and depth. There is a faint image that lives within me; sometimes it speaks. And sometimes I know it contains the room across the hall. Sometimes I know the room across the hall was foreseen somewhere. Something is built in. Somehow it registers; it picks out a unique reality. The tome of this world. Einstein must have foreseen the evil in our dear quantum. I don't envy the pain; the blood is so very frightening. No evil in A-bomb, but the evil of no image; the evil of no substrate; the burning people in Waco; my father tied to a bed. The poor roaming wisdom of E. On a movie set; money to be gained. Sexuality in images for money. This I see in THE Friday night. And this gnaws; frightens; feeds; asks for stability. An image that asks for expression. But with mysticism that can't be controlled. In paint delicately placed on a manifold. Spacetime is there somewhere in all the projects asking for flight. I see jeeps with water canisters on the back. I see a snake near a water faucet. I see a child giving up. Emptiness in a world nonexistent. Emptiness as definition; and emptiness for all its sake. There was an old stump with a few nails of differing size driven into it; it was a console and control. There were mysteries at the grass and at the curb. Tall grass; dreams of soldiers; boys dying across the world. The fence near the control protected; gave barrier. The nurturer of surreality was somewhere there too. Orwell came to visit somewhere in those dreams, at the control and at the creek. He gave me a gun from afar, and a yellow bicycle. And with the gun I learned to count. And with the bicycle I learned fear and dread. Dreams of reason in this reasoned state. Orwell saw my youth from afar; and then gave his mind away. His novel is my world; the scaffold on which to build what can be built. Orwell shaved his head and came to visit when I was sick. The tears came when they said "love." Albuquerque made its presence known somewhere in Texas long ago. The Germanic blood in my veins saw it then. The world was portended.

(CAF in a weird mood, 1993)

05 June 1997, "Dreams of a More Ethereal Quantum"

I thought I'd just say hi and let you know that I've been visiting "reality town" for the last couple days. I've finally (after all this time!) really started thinking about Mermin's Ithaca Interpretation. (He's now jokingly calling it the "75 Hickory Road Interpretation" since not everyone else in Ithaca is so enthused about it. Charlie B. asked him if everyone in his family was actually in agreement!) Much of this was spurred by my meeting him last week in Montréal.

I'm not sure what to make of it. I guess I think it's all too vague for me to have any strong opinion one way or the other yet. This has been helped out by my impression that Mermin himself doesn't really know what he's trying to get after yet. The key ideas seem to be:

1) The density operator for a "system" is a convenient icon/shorthand for all possible correlations between all possible "subsystems." However, the quotation marks in the preceding sentence are absolutely necessary; for it is the correlations themselves and not the individual properties of the systems that form the "elements of reality."

2) The real interpretational problems in quantum theory are really in understanding a proper notion of "objective probability." The only hint we have for understanding such a notion presently is the quantum theory itself – the very thing we were working hard to interpret in the first place. [This attitude, by the way, traces back to at least his 1983 essay "The Great Quantum Muddle"

critiquing Popper's books, where he writes, "... what is most marvelously intricate and subtle in the behavior is just a mystery and a horror to be dispelled by some clear thinking about [[objective]] probability."]

You can see from Appendix B in the paper (on the Hardy Paradox) that he wants to tie these two ideas together very closely. Namely that the flavor of objective probability is that some conditional probabilities cannot be properly formed. However, I don't see really how that is so distinct from Copenhagen: "unperformed measurements have no outcome." What is it that's really new here? (... beside the over-use of the phrase "objective probability"?) What am I missing? I guess the idea is to use the standard interpretation to figure out which (conditional) probability statements are meaningless and then abstract/bootstrap these properties into the new interpretation.

I just looked back at the last note you wrote me on this subject. I'm not sure I can answer your questions. Have you thought about them more? The best I can say is that I don't think he intended the density matrix as the end-all and be-all of everything ... only of the correlations between the subsystems which it encompasses. The density operator itself says nothing of how the overall system is correlated to the rest of the world (which I think is what your example was getting at). Apparently the only thing that one can glean ... or, at least this is what I think he'd say ... is that if the state is not pure then the system of interest is definitely correlated to something without (i.e., external to it).

I'll keep thinking about these things. Within a day or two I'll be composing a note to Mermin on "objective probabilities"; I'll cc: you something when it's written.

In the mean time, I place below – for your amusement – a note I just wrote Asher. I think it contains the most succinct statement I've written yet concerning the program of "Mechanica quantica ex mente orta lex est."

04 August 1997, "The Nielsbohricon"

Anyway, I was sitting here feeling low about not having yet finished the long note on quantum probabilities that I am in the process of writing you. And I was thinking about your Niels Bohr "roundtable" plans. You guys at the liberal artsy places get to do such fun things! In case it's of any use to you, below I give a small offering of a "starting point" for your fun and games: these are some of the things that I myself have perused and found useful. (The list has no pretension of being complete.)

I think these things are useful to read ... not because they'll paint a consistent picture of what Bohr thought (I doubt there's one to be painted) ... but because they give a good hint of all the issues that our own project must come up against. The reality project, that is. I agree with Beller in that, "My aim is not to cure this 'schizophrenia' by eliminating the inconsistencies, but to analyze the sources, uses, and aims of such shifting philosophical positions ..."

I think the best things in the list are Honner, Holton, and Faye's three books. Kalckar is good because it has a lot of previously unpublished personal letters to and from Bohr on meaty reality issues. If you're looking for the most on Møller's book, take a look at Feuer, Faye, MacKinnon, and Holton. The collection edited by Faye and Folse is really good too. I'd especially recommend "Description and Deconstruction: Niels Bohr and Modern Philosophy," by John Honner, in it. The book by Plotnitsky was strongly recommended by David Mermin, but I haven't made much headway into it – the subject is deconstructionism, Derrida, and the connection of all that to Bohr and Gödel.

I have still several more things to put in the list for you, but you'll have to give me some time to dig them up.

1. Mara Beller, "The Rhetoric of Antirealism and the Copenhagen Spirit," *Philosophy of Science* **63**, 183–204 (1996).
2. Jan Faye and Henry J. Folse, eds., *Niels Bohr and Contemporary Philosophy*, (Kluwer, Dordrecht, 1994).
3. Jan Faye, *Niels Bohr: His Heritage and Legacy (An Anti-Realist View of Quantum Mechanics)*, (Kluwer, Dordrecht, 1991).
4. Lewis S. Feuer, *Einstein and the Generations of Science*, (Basic Books, New York, 1974).
5. Henry J. Folse, *The Philosophy of Niels Bohr: The Framework of Complementarity*, (North Holland, Amsterdam, 1985).
6. A. P. French and P. J. Kennedy, *Niels Bohr: A Centenary Volume*, (Harvard University Press, Cambridge, MA, 1985).
7. Gerald Holton, *Thematic Origins of Scientific Thought: Kepler to Einstein*, Revised Edition, (Harvard University Press, Cambridge, MA, 1988).
8. John Honner, *The Description of Nature: Niels Bohr and the Philosophy of Quantum Physics*, (Oxford University Press, Oxford, 1987).
9. Jørgen Kalckar, *Niels Bohr, Collected Works: Volume 7, Foundations of Quantum Physics II (1938–1958)*, (Elsevier, Amsterdam, 1996).
10. Edward M. MacKinnon, *Scientific Explanation and Atomic Physics*, (University of Chicago Press, Chicago, 1982).
11. Dugald Murdoch, *Niels Bohr's Philosophy of Physics*, (Cambridge University Press, Cambridge, 1987).
12. Abraham Pais, *Niels Bohr's Times, in Physics, Philosophy, and Polity*, (Clarendon Press, Oxford, 1991).
13. Arkady Plotnitsky, *Complementarity: Anti-Epistemology after Bohr and Derrida*, (Duke University Press, Durham, NC, 1994).

12 October 1997, "Reality as Realty"

I'm just thinking of you, as I often do on my leisurely Sundays. Today's treat comes from the LA Times Book Review: an article by Martin Gardner, titled "Mathematical Realism and Its Discontents."

Consider $2^{1,398,269} - 1$. Not until 1996 was this giant number of 420,921 digits proved to be prime . . . . A realist does not hesitate to say that this number was prime before humans were around to call it prime and that it will continue to be prime if human culture vanishes. It would be found prime by any extraterrestrial culture with sufficiently powerful computers.

Social constructivists prefer a different language. Primality has no meaning apart from minds. Not until humans invented counting numbers, based on how units in the external world behave was it possible for them to assert that all integers are either prime or composite ... . In a sense, therefore, a computer did discover that $2^{1,398,269} - 1$ is prime, even though it is a number that wasn't "real" until it was socially constructed. All this is true, of course, but how much simpler to say it in the language of realism!

The last sentence just pegs it for me. Take reality when you can, I say! But when it's simpler to let it go, you've got to do that too. As I see it, that's the place quantum mechanics leaves us. For instance, Bohm's realistic version of QM seems to be perfectly consistent and to give the same observable results. All this is true, but how much simpler to say it in the language of anti-realism!

Oh well, enough of that.

13 November 1997, "Tria Juncta In Uno"

I'm starting to think that writing proposals might be more useful than I could have imagined. Look at what I just found in my electronic thesaurus for the word "REAL". In particular notice "well-documented"!

REAL: real, essential, substantive, substantial, not imagined, uninvented, actual, positive, factual, genuine, well-documented, historical, grounded, well-grounded, true, natural, of nature, physical, flesh and blood, material, concrete, solid, tangible, dense

23 November 1997, "Sunday Baudrillard"

I'm reading Jean Baudrillard and eating Kiki's famous sourdough raisin-pecan biscuits this morning. Ever heard of the guy? Baudrillard, that is. Anyway, I ran across the following little quote:

The transition from signs which dissimulate something to signs which dissimulate that there is nothing, marks the decisive turning point.

When do you think that was? 1926–27, maybe?

28 November 1997, "Muddling Through Reality"

I look forward to reading your book: just send it when you can. (If it arrives here before December 20, I can take it to Texas with me ... for reading on my vacation, Dec. 20 through Jan. 2.)

Yes, Baudrillard is contemporary. The reference to 1926–1927 in my first mention of him was a joke! It was meant to signify the birth of quantum mechanics.

10 December 1997, "Reality Alert I"

I've got a tip for you! Yesterday, I ran across a wonderful article that I know will please you immensely. It's by S. S. Schweber: "The Metaphysics of Science at the End of a Heroic Age," in *Experimental Metaphysics*, edited by R. S. Cohen, M. Horne, and J. Stachel (Kluwer, Dordrecht, 1997), pp. 171–198.

I think one of the Ss in Schweber's name stands for Sam ... is that true? Anyway, the subject is roughly speaking "law without law" ... the mutability of physical laws ... the creation of reality ... and an evolutionary universe. Roughly all the stuff of which Wheeler used to speak. The twist here is that he spends a little time exploring ethics and morals in such a kind of world. Let me give you an excerpt:

Since ... the scientific enterprise is presently involved in the creation of novelty – in the design of objects that never existed before in the universe and in the creation of conceptual frameworks to understand the complexity and novelty that can emerge from the *known* foundations and ontologies – they must assume moral responsibility for these objects and representations. My emphasis on the act of creation ...

and, quoting C. S. Peirce,

Under this conception, the ideal of conduct will be to execute our little function in the operation of the creation by giving a little hand toward rendering the world more reasonable whenever, as the slang is, it is 'up to us' to do so.

Besides, the article also has a lot of wonderful references to stuff on "law without law" that I hadn't known about. You should take a look. (There is, however, an annoying thing in it: he consistently misspells Peirce's middle name – it is Sanders not Saunders.)

David Mermin spent four days here last week, and we had a lot of time to discuss the "Ithaca Interpretation of QM." I would say that, to a large extent, he's on our side of the game. I.e., the anti-realism side. That, somehow, though hasn't come out in this particular twist of his thinking. I showed him something you once wrote me complaining that he's slipping away from our side, and he had a good laugh. He's a very good guy; I like him a lot.

10 December 1997, "List of Sins"

Herbert, Herbert, Herbert,

I am so disappointed in you. What's it take for a guy to be treated with respect by you ... perhaps a *faculty* appointment at Caltech??? Let me list your sins:

1) Not only did you send Hideo's copy of your book by priority mail ($4.00 postage), while only sending mine by "special standard mail,"

but

2) You placed $1.28 of postage on it when it apparently required $2.24 worth. Caltech had to pay 96 cents for me to have the privilege of your words!

I just hope you know that I'm hurt. Please return the e-mail I wrote you this morning about Schweber's article: I can no longer bear knowing that I'm sharing my thoughts with you. (This time though, please use the correct postage!)

Christopher

PS. The zip code at Caltech is 91125, not 91109.

27 December 1997, "Reality Alert II"

Thanks for your comments about Schweber, etc. This "reality" issue is becoming more acute, isn't it? I've very enthusiastic about it all again lately. I read the preamble and the amble of your book the other day on the plane. So far at least, I've been enjoying it immensely.

Mermin really is a good guy. His take on reality is, I think, somewhere between the Bernstein–Fuchses of the world and the hardcore realists. In general, he gives a thumb's down to reality ... so I guess that makes him much closer to our side of the mark. He just doesn't seem to think that quantum mechanics itself is about reality creation. That takes place – according to him, that is ... or at least I think so – at a somewhat higher level.

I wrote him an extensive critique of the Ithaca Interpretation (in its latest incarnation). After he puts his long paper on the archive, I may make my note available for public perusal (but it's not something I would put on the net). In general, there is a lot that I like about his way of viewing things – but I think it points much more strongly than he believes to the idea that quantum mechanics is about *information* and nothing more. By studying him, I've been able to come to much better grips concerning what I think quantum mechanics is about. I've been summarizing my latest understanding with the pithy phrase, "Bohr was Bayesian!"

29 December 1997, "Cradle of Reality"

Guess where Kiki and I are heading off to this morning? Austin, Texas: for me, the cradle of reality-creation. It was there that I first became acquainted with John Wheeler's ideas, and somehow the town has a mystical hold on me. Wish me luck for a little more insight.

04 January 1998, "Bohr, Bayesians, and Bernsteins"

I'm back in California now. On the ride home, I managed to get another chapter read from your book with Mike Fortun. (Yes, I would like to meet him.) I'm sorry that my reading is going slower than I had imagined it would. I just found that I was in hardly any mood to be intellectual over the holidays. So I still can't give you much feedback on it other than that I am wholly in agreement with the point of view espoused in Chapter 1. Most important to me, in particular, is the issue of articulating what exactly this new category "realit*t*y" consists of. What concept or category can we take to replace "reality" in a world where our actions actually make a difference? I think that is a lovely question!! I don't think I had ever seen it posed so pointedly before reading this Chapter. So thanks for giving me the opportunity. (If I can dig them up, I'll send you some of my old meager attempts at *even* formulating this question – I didn't come quite so close to the mark as you guys have!)

Thanks for commenting on my phrase "Bohr was a Bayesian" and the idea that quantum mechanics is about information. Actually I didn't mean by it anything so detailed as Bohr's particular take on the "frequency interpretation of probability," etc. I think what I really mean by the phrase is encapsulated in the following two-sentence explanation. Bohr was willing to own up to the idea that quantum theory is a theory of "what we have the right to say" in a world where the observer cannot be detached from what he observes. He was willing to own up to the idea that it is that and nothing more. (Making these two sentences rigorous and useful is roughly the big picture of my present research program. Did you, by chance, read the research proposal that I sent you? If you have any feedback, I *would* like to hear it.) My most up-to-date attempt to say this clearly is in a couple of notes – one to Comer and one to Mermin – I'll go ahead and forward them on to you in case you're interested. (Read the one to Comer first, for the overview.)

Oh yeah, and on a related note, I did want to comment on something you said,

I got the idea that "Ithaca" was about the possibility that q-mech somehow was only connected to information. To my mind this is a bit close to the folks in our field who think the universe is only information. That always seemed to trivialize what a marvelous theory quantum mechanics is – both philosophically and practically. Just misses the whole point, like the related stance of many-worldism.

To say something like that is to "ontologize" information. When I say, "Quantum mechanics is almost totally about information," that is not at all the sort of thing that I am trying to imply. Instead, I am trying to relate that almost all the formal structure of quantum theory (i.e., Hilbert spaces, state vectors, unitary evolution, etc.) is not really about *physics* at all. There is almost no physics in that formal structure: it is rather something of a more Bayesian-like character, it is about the formal tools for describing what we know. The *physics* behind quantum theory is that: "reality" *must* be replaced by "realit*r*y" (to borrow a little from your terminology).

14 January 1998, "Quote-A-Day"

New quote. (My computer gives me one at random every day.) I'm not sure why I like it, but somehow it strikes me as having something to do with the "muddled middle" you speak of in your book.

The very purpose of existence is to reconcile the glowing opinion we have of ourselves with the appalling things that other people think about us.

It's by Quentin Crisp (b. 1908), British author, from *How to Become a Virgin*, ch. 2 (1981).

25 January 1998, "Shame Dance"

I'm ashamed to say, but only now I've just finished reading Chapter 2 ("Articulating Experiments") of your book with Fortun. Time (and much of my sanity) has just slipped away from me this semester: true to your analysis, my life is very much a kludge job. All I can say still is that I very much like this idea of "reali*tt*y." It shines through to me as the most important of ideas. But with all the silly travel that I have to do this coming month (and the preparation of talks for it), it'll probably be some time before I can really get into the book again. Let me apologize for that right now: I know that I'm not being of very much use to you in your publishing process. But there certainly is enough in the book to keep me wanting to come back to it.

02 March 1998, "Early Morning Japan"

It's early morning in Japan right now, and I'm having trouble sleeping. My clock's still running on California time. I was thinking of you, wondering how you're doing. I've spent the last few listless hours in bed trying to compose a little reali*tt*y, trying to give the world a little form in exchange for the form it gave me. You should see this joint I'm staying at! It's the Tamagawa University guest house, and I'm sure by Japanese standards it is a palace. Very comfortable. Alexander Holevo and I are sharing it for seven days. Then I move on to visit the Communications Research Laboratory of the Ministry of Posts & Telecommunications for three days. I've been warned that my living arrangements there will not be so nice.

08 March 1998, "The Soul of Daibutsu"

This is just a short note to wish you well before I leave Japan. A few days ago I saw the most moving sight: The Great Buddha of Kamakura, Daibutsu. It's a 44 foot tall bronze buddha; I was even able to walk inside its belly. Dreams of philosophy made my heart pound. At a nearby shrine, I placed a 500 yen wish on their "wish board" (with about 10,000 previous wishes): Wish for Reali*tt*y. I hope you won't mind my pirating your phrase, but there was just nothing else to better describe what I was truly wishing for at just that moment.

Most honorable regards. Your faithful bhikku,

Christopher

29 March 1998, "Philosophical Purgation"

I'm doing my usual thing of being philosophical on Sunday. I've come across a little piece of an interview with David Bohm that I sort of like the flavor of ... so I'm entering it into my usual Sunday database: my friends. I'll pick on you because I know that your wife (and maybe you) resonates with some of Mr. Bohm's ideas.

Q: You said that there was difficulty in understanding quantum mechanics.

A: Yes. I think that the difficulty is that we have no way of understanding what is actually happening, or what I call the *actual fact*. If I may paraphrase Bohr, we have only the phenomena, i.e., the observed phenomena, which are essentially classical in their description. Ordinary classical phenomena – the observation of a dot or a click – were previously understood to signify information about particles, and the particles were independent of these phenomena. Now, if you analyse the Heisenberg microscope experiment, you come to the conclusion that the experiment cannot give you unambiguous information about the structures you are supposed to be observing. Therefore, there is no clear way of considering the unknown reality which is responsible for the experimental results.

Q: Wouldn't Bohr have said that this is a fundamental property of the world?

A: In effect he did say that. I don't think he ever said it directly, but it was implied. But if he said that it is fundamental, then I ask: how does he know it's fundamental? It's only fundamental as long as the present theory works, and there are many ways in which it doesn't work, as we know. We certainly just can't accept it on authority that it is fundamental. . . .

I think the reason I like this little passage is that Bohm shows that he did indeed have a good understanding of the Copenhagen attitude. He simply chose to reject it. There's nothing in my mind better than an honest assessment of one's motives. I myself keep playing with the idea that all of quantum mechanics is about one fundamental ontological statement: Wheeler's game of twenty questions. Almost all of the detailed structure of the theory is about reasoning, betting, assessing, and dealing with a world possessing exactly that fundamental property. Once we clear up that that is what quantum mechanics is really about, then we will really be in a position to find something truly new and wonderful. So the point that Bohm wishes to question is just the point that I want to take as basic, the one I wish to see if it can be built upon.

That's all. Happy philosophy.

02 January 1999, and to Greg Comer, "A New Year's Toast"

Happy New Year! Please allow me to toast it in with a small gift, one that I think concerns all three of us. It's a beautiful little article from *Science* magazine by John Banville (a reporter at the *Irish Times* in Dublin). I scanned the article in in its entirety because the whole thing seemed so relevant. (Only partially reprinted here.) It contains a truth that I think we three in particular know all too well . . . a truth that we three in particular want to see become a *productive* truth.

Beauty, Charm, and Strangeness: Science as Metaphor

(From Science Magazine Online)

by John Banville

I wish to advance a thesis which, were they to take note of it, the academies would decry as scandalous. My thesis is that modern science, particularly physics, is being forced, under pressure of its own advances, to acknowledge that the truths it offers are true not in an absolute but in a poetic sense, that its laws are contingent, that its facts are a kind of metaphor. Of course, art and science are fundamentally different in their methods, and in their ends. The doing of science involves a level of rigor unattainable to art. A scientific hypothesis can be proven – or, perhaps more importantly, *disproven* – but a poem, a picture, or a piece of music, cannot. Yet in their *origins* art and science are remarkably similar. It was a scientist, Niels Bohr, who declared that a great truth is a statement whose opposite is also a great truth. Oscar Wilde would have agreed.

Since the Enlightenment, the chasm between art and science has yawned ever wider with each new stage in the campaign to subdue nature to man's will. The human race cannot abide nature's indifference, and uses the physical sciences to attempt to wring from it a word of acknowledgment. Yet what we today think of as science is for the most part not science at all, but *applied* science, that is, technology. The machinery of modern science is so elaborate, and the building of it requires so much ingenuity – requires, indeed, so much *science* – that we naturally confuse the thinking with the doing. The great particle accelerator at CERN, for example, is for us the very image of modern science: a vast and inconceivably expensive machine built to perform minute and unimaginably complex operations whose results can be interpreted only by a handful of physicists. But we are willing to pay the cost of building these machines, are willing to allow the physicists their arcane rules and specialized language, because we believe that they are getting their hands into the very bowels, or, rather, the very synapses, of nature. And at some point, we believe, they will bring forth news of another advance, another boiled-down version of the world's variousness, another $E = mc^2$, only bigger and better. Perhaps this time they may even discover the final equation, the Grand Theory of Everything. Then, as Stephen Hawking puts it, "we shall all, philosophers, scientists, and just ordinary people [I am struck by that distinction, by the way], be able to take part in the discussion of the question of why it is that we and the universe exist. If we find the answer to that, it would be the ultimate triumph of human reason – for then we would know the mind of God."

Such foolhardy talk, from such an eminent source, misleads us into the notion that the aim of science is to find the "meaning" of the world. That there must be a meaning seems certain, otherwise how is it that there is such a thing as progress? Science keeps uncovering more and more secrets, keeps getting closer and closer to ... well, to *something*, in the same way that computations in the infinitesimal calculus keep approaching nearer and nearer to infinity without ever getting there. Progress must be progress *toward* something, surely, some final end to the quest for knowledge? But to my mind the world has no meaning. It simply *is*. Leibniz's thrilling question, "Why is there something rather than nothing?" is significant not because an answer to it is possible, but because out of the blind, boiling chaos that is the world, a species should have emerged that is capable of posing such a question.

Science and art are different ways of looking at the same thing, namely, the world. Let us take the case of Goethe. In his role as amateur scientist, he was vehemently opposed to Newton's mechanistic model of reality. He was mistaken – that is to say, his science was bad science, although his scientific writings are not bad philosophy, and still less are they bad poetry. Goethe demanded that science should always hold to the human scale. He opposed the use of the microscope, since he believed that what cannot be seen with the naked eye should not be seen, and that what is hidden from us is hidden for a purpose. In this, Goethe was a scandal among scientists, whose first, firm, and necessary principle is that if something *can* be done, then it *should* be done. Yet his furious denial of Newton was more than merely the bloodshot jealousy of one great mind drawing a bead on another. Goethe's theory of light is wrong insofar as the science of optics is concerned, yet in the expression of his theory Goethe achieves a pitch of poetic intensity that is as persuasive, in its way, as anything Newton did. But persuasive at what level?

There is a world beyond politics, says the poet Wallace Stevens, and we might adapt that to say that there is a world beyond science, or, at least, there is a world beyond *the current state* of science. At the end of the 19th century professors of physics in the great European universities were steering students away from the discipline because they believed that there was very little of interest left to be discovered about the nature of physical reality. Then came Einstein. As we approach the end of the 20th century, we are still guilty of hubris, as evidenced by Stephen Hawking's statements quoted above. Probably a Unified Field Theory will be achieved, and will seem for a time, perhaps even as long as the period between Newton's *Principia* and Einstein's first paper on the theory of relativity, to explain everything; then a Heisenberg or a Gödel will come forward and point to a loose end which, when pulled, will unravel the entire structure.

This is a truth that both clear-sighted artists and scientists – that is, those not blinded by hubris, or a cramped imagination, or both – have always acknowledged: There is no end to the venture. The difference between the two, however, is that while the artist acknowledges that in art there is nothing new to be said, only new ways of saying the old things, new combinations of old materials – a process, paradoxically, that *makes* a new thing, namely, the work of art – science seems always to be pressing on into hitherto uncharted territory. Yet the fact is, science is not *making* this new landscape, but *discovering* it. Einstein remarked more than once how strange it is that reality, as we know it, keeps proving itself amenable to the rules of man-made science. It certainly is strange; indeed, so strange, that perhaps it should make us a little suspicious. More than one philosopher has conjectured that our thought extends only as far as our capacity to express it. So too it is possible that what we consider reality is only that stratum of the world that we have the faculties to comprehend. For instance, I am convinced that quantum theory flouts commonsense logic only because commonsense logic has not yet been sufficiently expanded.

03 January 1999, "Mary and Banville"

I forgot to tell you: also please share the Banville article with Mary. I would be interested in knowing her opinion on it. Does the artist think this is as much hogwash as the average physicist probably does? (You and I aren't so average.)

The point I find most eloquently stated in that article is where Banville says:

Einstein remarked more than once how strange it is that reality, as we know it, keeps proving itself amenable to the rules of man-made science. It certainly is strange; indeed, so strange, that perhaps it should make us a little suspicious. More than one philosopher has conjectured that our thought extends only as far as our capacity to express it. So too it is possible that what we consider reality is only that stratum of the world that we have the faculties to comprehend.

To bring a similar point home in a talk I give on the "real meaning of quantum information," I use three slides that I'm particularly proud of. This first is one I stole from Seth Lloyd from his talk at the QUIC Kickoff meeting. It shows a tall mountain labelled with "novel quantum states" at the bottom and "factoring" at the top, and some clouds in between. I say, "It has become popular to show slides like this … to depict graphically the mountain we must climb in quantum computing. It gives the idea that the thing we really want is that peak. But if you ask me, the real reason to climb from the base to the peak, is to gauge the distance between the two." Then I change slides to one of an iceberg that Kiki drew for me. (There's a little boat near the iceberg, and a penguin sitting on its peak.) "If we can gauge that distance, then we'll have a better feel for the other 8/9 that's still below the surface!" Then I say that the source of that belief has no logical justification, but it seems that physicists have lately gotten in a habit of ignoring a pretty basic fact of the world. Then I show a slide copied from a biology book, "Amino Acid Difference Matrix for 26 Species of Cytochrome c." I've got *man* and *dog* highlighted in yellow on it; the difference is only 11 percent. Then I say, "My dog isn't even close to a Grand Unified Theory of the universe. There are some things he just can't see, no matter how hard I try to train him. Why should we be much more than 11% ahead of him in the game?"

24 April 1999, "Genetic Genesis"

OK, I still haven't read through Fleck again. But I keep thinking I'm going to. Does that count for anything?

Anyway I really would like you to write up a report on that wonderful conversation we had with Charlie the other night. It would be sad if it faded from our collective memory. Of the three of us, you seemed to have the deepest understanding of all that was being said and, on top of that, to have a glimpse of how to conciliate all three points of view. So please, please do it for me.

As a bit of a peace offering in the meantime, let me send you two things. The first is a small essay I wrote a couple of days ago to firm up how I might present my research program in a sort of evocative way. It's based on the Adam and Eve slide I showed you. I think you'll enjoy it; it's placed below. [See note to Greg Comer, titled "Fuchsian Genesis," dated 22 April 1999.] The other thing, coming in the next email, is a passage from a paper by Doug Bilodeau. [See letter to Asher Peres, titled "The Deep Intervention," dated 26 December 1998.] It expresses very clearly what I was trying to convey to Charlie about the scientist as the setter of initial conditions. He thinks that is absolutely incidental and unimportant for science (an epiphenomenon?), but I just don't think so. I guess the essay below and Bilodeau's passage both express this. I'd love to hear your opinions (along with the stuff I begged for above).

08 May 1999, "A Fleck of Quantization"

The last time I wrote you about Ludwik Fleck's book *Genesis and Development of a Scientific Fact* – 1 October 1995, can you believe it!? – I said the following:

Another piece of news is that I've finally finished reading Fleck's book. It was indeed worth taking a look at, though I'm not yet sure what to make of it. In this "reality generation" business, I see the quantum as a crucial ingredient. But Kuhn and Fleck seem to argue otherwise. It's amazing how deeply ingrained I find the "realist" tendency. Because quantum theory apparently forces me to relinquish some of my intuitions, I'm willing to give these spookier ideas a shot … but always some resistance remains. In any case, I was intrigued that Fleck and I have come to similar ideas about the "stability of reality." (Cf. one of my old notes titled "Insights from Cuero" or something like that – it's the one that contains stuff about the burnings in Waco.)

These are the little notes I compiled; maybe you'll find them useful too.

1. crossing thought style – p. 2
2. reality – p. 10, 28 (bottom), 127 "Our reality did not exist for them", 156
3. resistance – p. 27
4. individual cognition dependent upon community – p. 38
5. contribution of the individual – p. 40
6. slacker reference – p. 44
7. stability of reality – p. 47, 99, 102
8. thought collective being cultural and not across the board for humankind – p. 174 (note p. 49)
9. French bread – p. 50
10. existence is lawless – p. 51
11. truth – p. 100, 116, 125
12. fact as thought-collective resistance – p. 101
13. objects created by thought – p. 181
14. how to use philosophical principles – p. 181
15. symbols – p. 125
16. name as property – p. 136
17. meaning as a property of the object – p. 137

———————————

Apparently this made little impression on you: you told me you didn't even remember my sending it! So let me bless you with something of an expanded version.

In what follows I revisit those marked sections and copy down what it was that I found interesting. In a small number of cases, I couldn't figure out what I found interesting last time; so those citations got skipped this time around. Also, a few of the subjects got renamed.

Reading over it all again, I guess I find that I'm significantly more impressed than I was last time. Perhaps the years have made me wiser. Or perhaps they've caused me to return to my youth: you'll see what I mean shortly. I think your forcing me to look through Fleck again was particularly timely. This is because of the little tale "Fuchsian Genesis" I sent you the other day. I continue to think quantum mechanics supplies a crucial piece that Mr. Fleck couldn't have foreseen.

Throughout I'll make commentary here and there, and even annotate with some of my old writings. At the end, I'll sum up my present thoughts on the *Denkkollektiv*. And then you must fulfill your promise!

A Few of Fleck's Thoughts (with Annotation)

Resistance – p. 27:

Once a structurally complete and closed system of opinions consisting of many details and relations has been formed, it offers enduring resistance to anything that contradicts it.

A striking example of this tendency is given by our history of the concept of "carnal scourge" in its prolonged endurance against every new notion. What we are faced with here is not so much simple passivity or mistrust of new ideas as an active approach which can be divided into several stages. (1) A contradiction to the system appears unthinkable. (2) What does not fit into the system remains unseen; (3) alternatively, if it is noticed, either it is kept secret, or (4) laborious efforts are made to explain an exception in terms that do not contradict the system. (5) Despite the legitimate claims of contradictory views, one tends to see, describe, or even illustrate those circumstances which corroborate current views and thereby give them substance.

Individual cognition dependent upon the community – pp. 38–39:

In comparative epistemology, cognition must not be construed as only a dual relationship between the knowing subject and the object to be known. The existing fund of knowledge must be a third partner in this relation as a basic factor of all new knowledge. It would otherwise remain beyond our understanding how a closed and style-permeated system of opinions could arise, and why we find, in the past, rudiments of current knowledge which at the time could not be legitimized by any "objective" reasons and which remained only pre-ideas.

Such historical and stylized relations within knowledge show that an interaction exists between that which is known and the act of cognition. What is already known influences the particular method of cognition; and cognition, in turn, enlarges, renews, and gives fresh meaning to what is already known.

Cognition is therefore not an individual process of any theoretical "particular consciousness." Rather it is the result of a social activity, since the existing stock of knowledge exceeds the range available to any one individual.

The statement, "Someone recognizes something," whether it be a relation, a fact, or an object, is therefore incomplete. It is no more meaningful as it stands than the statements, "This book is larger," or "Town A is situated to the left of town B." Something is still missing, namely the addition, "than that book," to the second statement, and either, "to someone standing on the road between towns A and B while facing north," or "to someone walking on the road from town C to town B," to the third statement. The relative terms "larger" and "left" acquire a definite meaning only in conjunction with their appropriate components.

Analogously, the statement, "Someone recognizes something," demands some such supplement as, "on the basis of a certain fund of knowledge," or, better, "as a member of a certain cultural environment," and, best, "in a particular thought style, in a particular thought collective."

If we define "thought collective" as *a community of persons mutually exchanging ideas or maintaining intellectual interaction, we will find by implication that it also provides the special "carrier" for the historical development of any field of thought, as well as for the given stock of knowledge and level of culture. This we have designated thought style.* The thought collective thus supplies the missing component.

Contribution of the individual – pp. 40–41:

Cognition therefore means, primarily, to ascertain those results which must follow, given certain preconditions. The preconditions correspond to active linkages and constitute that portion of cognition belonging to the collective. The constrained results correspond to passive linkages and constitute that which is experienced as objective reality. The act of ascertaining is the contribution of the individual.

The three factors involved in cognition – the individual, the collective, and objective reality (that which is to be known) – do not signify metaphysical entities; they too can be investigated, for they have further relations with respect to one another.

These further relations consist in the facts that, on the one hand, the collective is composed of individuals and that, on the other, objective reality can be resolved into historical sequences of ideas belonging to the collective. It is therefore possible from the viewpoint of comparative epistemology to eliminate one or perhaps even two factors.

Although the thought collective consists of individuals, it is not simply the aggregate sum of them. The individual within the collective is never, or hardly ever, conscious of the prevailing thought style, which almost always exerts an absolutely compulsive force upon his thinking and with which it is not possible to be at variance.

Minimum thought collective – pp. 43–44:

A kind of superstitious fear prevents us from attributing that which is the most intimate part of human personality, namely the thought process, also to a collective. A thought collective exists wherever two or more people are actually exchanging thoughts. He is a poor observer who does not notice that a stimulating conversation between two persons soon creates a condition in which each utters thoughts he would not have been able to produce either by himself or in different company. A special mood arises, which would not otherwise affect either partner of the conversation but almost always returns whenever these persons meet again. Prolonged duration of this state produces, from common understanding and mutual misunderstanding, a thought structure that belongs to neither of them alone but nevertheless is not at all without meaning. Who is its carrier and who its originator? It is neither more nor less than the small collective of two persons. If a third person joins in, a new collective arises. The previous mood will dissolve and with it the special creative force of the former small collective.

We could agree with anybody who calls the thought collective fictitious and the personification of a common result produced by interaction. But what is any personality if not the personification of many different momentary personalities and their common psychological Gestalt? A thought collective, by analogy, is composed of different individuals and also has its special rules of behavior and its special psychological form. As an entity it is even more stable and consistent than the so-called individual, who always consists of contradictory drives.

A moral? Even as little as two students at Columbine High School in Littleton, Colorado can form a thought collective. And indeed they did: look at the drastic reality they were able to create (destroy).

Stability of reality – pp. 46–47:

Gumplowicz expressed himself very poignantly on the importance of the collective. "The greatest error of individualistic psychology is the assumption that a *person* thinks. This leads to a continual search for the source of thought within the individual himself and for the reasons why he thinks in a particular way and not in any other. Theologians and philosophers contemplate this problem, even offer advice on how one ought to think. But this is a chain of errors. What actually thinks within a person is not the individual himself but his social community. The source of his thinking is not within himself but is to be found in his social environment and in the very social atmosphere he 'breathes.' His mind is structured, and necessarily so, under the influence of this ever-present social environment, and *he cannot think in any other way*."

Jerusalem dealt with this problem in a number of essays, the last of them bearing the apposite title "Social Conditioning of Thinking and of Thought Patterns." "Kant's firm belief in a timeless, completely immutable logical structure of our reason, a belief that has since become the common heritage of all who adopt an a priori point of view and is maintained with great tenacity also by the latest representatives of this direction of thinking, has not only failed to be confirmed by the results of modern ethnology but proved to be definitely erroneous." "The primitive individual feels himself only a member of his tribe and clings to its traditional way of interpreting sensory perceptions with absolutely incredible tenacity." "I have no doubt, and it is confirmed through the diverse institutions found in primitive societies, that tribesmen reinforce each other's belief in the ubiquity of spirits and demons, which is already sufficient to give these figments of the imagination some degree of reality and stability. This process of mutual corroboration is by no means confined exclusively to primitive societies. It is rather prevalent today, fully effective in our everyday lives. I wish to designate this process and any structure of belief formed and fortified by it *social consolidation*." "Even particular and objective observations ... require confirmation by the observation of others. Only then will they become common property and thus suitable for practical utilization. Social consolidation functions actively even in science. This is seen particularly clearly in the resistance which as a rule is encountered by new directions of thought."

Stability of reality – p. 99:

Because it belongs to a community, the thought style of the collective undergoes social reinforcement, as will shortly be discussed. Such reinforcement is a feature of all social structures. The thought style is subject to independent development for generations. It constrains the individual by determining "what can be thought in no other way." Whole eras will then be ruled by this thought constraint. Heretics who do not share this collective mood and are rated as criminals by the collective will be burned at the stake until a different mood creates a different thought style and different valuation.

Stability of reality – p. 102:

The fact thus defined as a "signal of resistance by the thought collective" contains the entire scale of possible kinds of ascertainment, from a child's cry of pain after he has bumped into something hard, to a sick person's hallucinations, to the complex system of science.

Facts are never completely independent of each other. They occur either as more or less connected mixtures of separate signals, or as a system of knowledge obeying its own laws. As a result, every fact reacts upon many others. Every change and every discovery has an effect on a terrain that is virtually limitless. It is characteristic of advanced knowledge, matured into a coherent system, that each new fact harmoniously – though ever so slightly – changes all earlier facts. Here every discovery is actually a re-creation of the whole world as construed by a thought collective.

A universally interconnected system of facts is thus formed, maintaining its balance through continuous interaction. This interwoven texture bestows solidity and tenacity upon the "world of facts" and creates a feeling both of fixed reality and of the independent existence of the universe. The less interconnected the system of knowledge, the more magical it appears and the less stable and more miracle-prone is its reality, always in accordance with the thought style of the collective.

This issue about community-created realities – from whence comes their stability – used to be on my mind a lot. Let me throw in at this point a little essay I wrote my friend Greg Comer way back (28 March 1993). I think it is quite in line with much of what Fleck is saying:

Insights from Cuero

I login to the computer only to find an "empty tray" representing the status of my new electronic mail. Woe is me. You know it is the mail that makes me live and thrive.

I wonder how close I really am to getting this idea of the quantum straight. Sometimes I think I'm very close and then at others (like today) I feel completely lost. And it's you that gets to hear about all this over and over.

I wish I could fill your ears with the technical details of how to "derive" the quantum from a few simple desiderata for LWL. But I can't and that's nothing new. So to fill my time and make my fingers tired I plan to fill this note with pure unadulterated philosophy ... Insights from Cuero. There'll be no hint of real science here; I make no apologies.

From where and of what utility comes this notion of an "objective world" independent of man, woman, animal, and plant? My opinion is that (in the end) "objective reality" is posited for nothing more than to have a device for coordinating the various experiences common to all those who communicate. Mark Twain once wrote, "If you tell the truth, you don't have to remember anything." It seems to me that that pretty much sums it up. The objective truth saves us from having to make sure that the stories we tell are consistent; it saves us from having to remember all aspects of the past. It gives us a means for determining whether someone's behavior is insane. It gives us a means by which to determine the guilt or innocence of an accused murderer. Is there any other real motivation for positing an objective reality? I can think of none ... but why should I, the reasons listed above should be powerful enough argumentation for anyone involved in the sciences.

Nevertheless, just how essential is this notion of an objective reality? The "idealists" have been trying to do away with it since the days of Bishop George Berkeley. "Esse est percipi." Why be fixated on a "material reality" that can never be confirmed; all that is truly available to our discerning is our thoughts and our sense impressions. As far as thinking beings are concerned, material/objective reality is a mental construct ... whether it is indeed ultimately "out there" or not. So just do away with it; superfluousness. All that is really needed to make sense of this world is "mind."

Cute idea, I say. But what a strain when the so much simpler picture of "objective reality" will do. This is the side of the razor I'd put Ockham on if I could. Mais, c'est un monde que je ne vois pas.

The problem we face in modern times is that the existence of quantum phenomena seems to cry out for the relinquishing of this objective reality notion. How else can we consolidate the experimentally confirmed violation of the Bell inequalities with the equally well experimentally confirmed Lorentz character of spacetime? (Well, of course, there are ways ... if one is of such a mind. Bohm's 1952 nonlocal hidden variable theory, for instance, is a step in that direction. Or at worst, one could say that Allah wills each and every quantum mechanical measurement outcome. But what a price these both are to pay. In the first case, one relinquishes simplicity in the equations. In the second, one relinquishes science and reasoned philosophy.) The case in the end seems to be that quantum mechanical systems have properties only insofar as those properties are *created* by (freely-chosen) human "measurements."

So what are we left with? Either we decide it's simpler to work on some muck like the Bohm hidden variable theory to save objective reality ... risking the simplicity of the best physical theory we've yet constructed, not to mention the fact that so far it has seemed to lead to no new physics. Or we decide it's simpler to work out how our experiences can be coordinated without an objective reality. You know which way I lean.

But how can we easily get by with simply "consistency" and "coordination" for our "community of communicators" instead of "reality?" From where now can we pull a man-independent notion of consistency? The answer must be "nowhere." We are left with pulling it from the community itself. And it is of course this very fact that worries us – the individual and community opinions being demonstrably (through standard interpersonal relations and history) so *damned* fluid. You asked a few days ago:

> People are different, so surely the rules must constantly change, even if slightly. This is
> what troubles me these days: the rules must constantly change. ... If there are no laws,
> then how can we count on the permanence of anything, especially the requirement of
> "consistent"? ... But if people are truly different how can nature be consistent?

No answers today, of course, but maybe some hints: force and diplomacy.

Sometime around the year 1985 or 1986, I was walking across the University of Texas campus ... from the Student Union (I think) to the Perry-Casteñada Library. On the South Mall between the UT Tower and the TX State Capital, I approached a man talking to himself fairly audibly walking in the same direction as myself. I slid up behind him as closely as I could without being detected and tried to listen in on the conversation. (He did, by the way, later notice me ... and I became scared stiff as he subsequently followed me all the way into the library!) Sure enough, the conversation seemed, as far as I could tell, like any other conversation one might hear on a Sunday dusk ... only one could not hear or see the second participant. (The day, by the way, was indeed Sunday.) This caused pause for a lot of reflection in that short time ... and consequently over the years.

The first thought, of course, was that the man was crazy. There was no one else there for him to talk to. But then I started thinking about what might happen if I were to confront him with this observation. One possibility that occurred to me was that he would look at me, tell me I was "crazy", and introduce his friend Joe (for instance). At that point, given exactly that situation, how on earth would I really know that it was he and not I that was crazy? Of course, by nature, I would have to believe that it was he that was insane, but how could I really prove to myself that that was the case? It dawned on me that I needed only to find a passerby to confirm my opinion. But what if our man of interest said, "You're both crazy. Can't you see my friend? This is the strangest thing I've ever encountered; two crazies at one time."??? How could we know that we weren't both crazy? Well, in the same vein, we could just ask for confirmation of the man's insanity from a third person; then we would feel quite confident, wouldn't we? Yeah, we probably would, but this is now a question of principle. What if the man said, "All three of you are crazy as birds. My friend Joe is here as plain as day. Joe, we need to call the authorities about these people. You heard what I heard didn't you? I thought you did. Yeah, that's a good idea; you run off to the library. There's a phone there." Just what if? Then we would have to pull in a fourth person for that extra bit of confirmation to assure ourselves of our own sanity. And so on the story could go.

What is the point of all this? Without an objective reality underlying our experiences, even one bad seed can destroy the possibility of a comfortable background for their coordination. We can never be sure that that one "crazy" man is not in fact sane and we, on the other hand, are all insane. This tells us that questions of principle – to some extent – go down the tubes when the notion of objective reality is wiped from our repertoire of physical concepts. That said, though, what would happen *in practice* in a scenario like that described above? Easy. The singular man would be locked away in some asylum in San Antonio, TX. Majority rules ... or those with power rule ... or the ruling class rules ... or .... The point is some faction makes the rules, and those rules are enforced. It is this enforcing that gives a certain uniformity to our existence. Any unbearable non-conformity is either isolated or destroyed.

Before we run with this idea a little further, let's take a look at a slightly more difficult case. Consider the events that have been taking place in Waco, TX the last few weeks. There a cult leader, David Koresh, has declared himself Jesus Christ. He and his following (originally over 100 strong) have blockaded themselves along with quite a bit of firepower within some compound either near or in the city. At least six or so people have been killed in the ensuing struggle. Can we consider these hundred people sane? I would say, "no!" One would think they would surely sense the consequences of their actions ... if they were sane. They will eventually lose; they will all eventually be arrested or killed. There is no doubt about this ... *to all of us who observe from the outside*. From the "inside", though, it must be a completely different story: David Koresh *is* Jesus Christ; God *will* save them; killing others *is* justified. By any standard criterion of insanity, this must be a "mass insanity." (In perhaps different words, many of the convicted will probably plead this when trial comes.) No longer here are we isolated to a *single* man doing crazy things; now it is a mass of 100 people. [Of special interest, I find it particularly enlightening that Koresh's right-hand man is a "Harvard educated lawyer." !!! This is just a hint of what we know must eventually be the case in this observer created reality we are trying to investigate. "Rational powers of thought" (as exhibited by the LSAT score required for Harvard) must have nothing to do with anything in the end. Reality is defined by the community. Period.] The point now: be it the insanity of one or a hundred, the faction making the rules will snuff that insanity out in one of the ways listed above.

Now let's carry this to the extreme. What of Germany in 1941, perhaps at the peak of its Nazi prowess? It very well could have become the "faction making the rules." And then it would have enforced those rules. Surely an insane faction by present-day standards! But if it had become the ruling faction, it is our present-day notions that would be insane. This is what we have to gulp in our picture of a reality created by the community of communicators – not only for social interactions but for the "physical world."

Just in summary let me reiterate this small insight or missing link: Not only is reality created by the community, but also *enforced* by the community. It is this enforcing that gives a certain uniformity to our existence and ultimately circumscribes any good notion of "consistency."

How formal this notion can be made, I'm not sure. I don't even want to contemplate it now. What I'd like to do in the remainder of this note is briefly sketch the importance of Wheeler's phrase: "The past exists only insofar as it is recorded in the present." Lately here in Albuquerque we have been a little worried about the problem of the arrow of time and its relation to the Second Law of Thermodynamics. I've told you this before. David Wolpert asks, "Why can we only remember the past and not the future given that all physical laws are time symmetric?" If we take Wheeler's standpoint, this is not a problem at all but a tautology, a triviality. For what is in memory is, by definition, "the past." The past has no existence otherwise. It too is not simply something "out there" independent of man, woman, animal and plant. This point we can use to tie together a loose string in our discussion.

If a ruling faction enforces a certain uniformity and consistency in the present, then by Wheeler's maxim, it also enforces that same uniformity and consistency in the past. You ask how can "consistency" be stable? Well here's the direction of an answer ... if we take all this seriously. The present makes the past and thus the continuity of all its policies with those of the past ... including the continuity of physical law. Perhaps there is a bit more than social commentary to be learned from George Orwell's novel *1984*! Recall how history was constantly being rewritten in that novel by an official government agency. Any dissenters from the official account were snuffed out. No less can we expect in our account of physical law.

This ends our tale so far. Please, please recall though that I warned you that this note would contain only pure unadulterated philosophy. Not a bit of science. So I expect no less than a million loopholes in these thoughts. Nevertheless I hope there is at least something worthwhile in here.

Does this tell you that I have been inclined to the idea of the *Denkkollektiv* for a long, long time? Notice again, though, that my motivation came largely from quantum mechanics.

Requirement that there be a community – p. 174 (note from p. 49):

But we soon read: "Not every observation by an individual must in itself be valued as an experience. Only after a stock of general and well-confirmed knowledge has formed as a result of mutual agreement and reinforcement in the course of continued cooperation of the intellects involved should we speak of experience. General and well-confirmed experience, however, must be considered the sole criterion of truth." Confrontation of these contradictions does not constitute a criticism of Jerusalem. It merely exemplifies that when new thought styles are evolving, contradiction sets in as an expression of the intellectual "contest of the fields of view."

French bread – p. 50:

What is the reason for this special position of current scientific statements as required by the philosophers just quoted?

They believe that our present-day scientific opinions are in complete contrast with all other ways of thinking. As if we had become wise and our eyes had been opened, they believe that we have simply discarded the naive self-consciousness of thought processes which are primitive or archaic. We are supposedly in possession of "correct thinking" and "correct observation," and therefore what we declare to be true *is* ipso facto *true*. What those others such as the primitives, the old people, the mentally ill, or the children declare to be true *seems to be true only to them*. This arch-naive view, which prevents the building up of a scientific epistemology, reminds us very much of the theory of a French philologist of the eighteenth century who claimed that *pain, sitos, bread, Brot, panis* were arbitrary, different descriptions of the same thing. The difference between French and other languages, according to this theory, consisted in the fact that what is called bread in French really was bread.

Connection to Wheeler's "Law without Law" – p. 51:

The views outlined here should not be construed as scepticism. We are certainly capable of knowing a great deal. If we cannot know "everything," according to the traditional position, it is simply because we cannot do much with the term "everything," for every new finding raises at least *one* new problem: namely an investigation of what has just been found. The number of problems to be solved thus becomes infinite and the term "everything" meaningless.

An "ultimate" or set of fundamental first principles from which such findings could be logically constructed is just as nonexistent as this "everything." Knowledge, after all, does not repose upon some substratum. Only through continual movement and interaction can that drive be maintained which yields ideas and truths.

I find this passage greatly reminiscent of David Deutsch's description of John Wheeler's idea of "law without law" [D. Deutsch, "On Wheeler's notion of 'Law without Law' in Physics," *Found. Phys.* **16**, 565–572 (1986)]. Deutsch wrote:

Is it possible that there is an ultimate law of physics, a principle $\mathcal{P}$ from which follows everything that is knowable about the material world? If so, what can we already infer or postulate about the form that this principle takes? Wheeler's idea of "law without law" is an attempt to begin to answer these questions.

If there were no all-explanatory physical principle $\mathcal{P}$ approachable by the methods of science, this would presumably mean that there exist aspects of the natural world that are fundamentally inaccessible to science. This would run directly counter to rationalism and to our view of physics as the universal science, which have hitherto been the driving forces behind progress in the subject and which we should be extremely reluctant to abandon.

But if $\mathcal{P}$ were itself a law of physics, then the problem of $\mathcal{P}$'s own origin – why that particular principle holds in nature rather than some other – would be forever insoluble. And hence $\mathcal{P}$ would not be all-explanatory within physics. So, paradoxically $\mathcal{P}$, the ultimate principle of physics, cannot be a "law" (of physics). Hence the expression "law without law."

Honestly, though, I'm not sure to what extent Wheeler really saw this argument as his own motivation: it might just be a case of Deutsch looking at Wheeler through Oxford-colored glasses. I don't recall ever seeing Wheeler write the argument himself. Instead, it seems that his motivation for "law without law" had been that every law he had ever seen formulated had ultimately been "transcended." This itself built a bit of scepticism. But then on top of that was quantum mechanics with its random measurement outcomes, "each individual one shunning all law."

What is truth? – p. 100:

A historical connection thus arises between thought styles. In the development of ideas, primitive pre-ideas often lead continuously to modern scientific concepts. Because such ideational developments form multiple ties with one another and are always related to the entire fund of knowledge of the thought collective, their actual expression in each particular case receives the imprint of uniqueness characteristic of a historic event. It is, for instance, possible to trace the development of the idea of an infectious disease from a primitive belief in demons, through the idea of a disease miasma, to the theory of the pathogenic agent. As we have already hinted, even this latter theory is already close to extinction. But while it lasted, only one solution to any given problem conformed to that style. ... *Such a stylized solution, and there is always only one, is called truth.* Truth is not "relative" and certainly not "subjective" in the popular sense of the word. It is always, or almost always, completely determined within a thought style. One can never say that the same thought is true for A and false for B. If A and B belong to the same thought collective, the thought will be either true or false for both. But if they belong to different thought collectives, it will just *not* be *the same* thought! It must either be unclear to, or be understood differently by, one of them. Truth is not a convention, *but rather* (1) *in historical perspective, an event in the history of thought,* (2) *in its contemporary context, stylized thought constraint.*

This is a wild idea, but notice how the sum total of thought collectives almost hints of a "partial Boolean algebra" structure. That is to say, the set of thought collectives looks a little like a Hilbert space: the set of propositions concerning a quantum system can be viewed as a collection of "local" Boolean algebras with a certain "pasting" condition for connecting them together – in total one gets a so-called Boolean manifold. So too, it might be the case with thought collectives.

Fact as thought-collective resistance – pp. 101–102:

In the field of cognition, *the signal of resistance* opposing free, arbitrary thinking is called a *fact*. This notice of resistance merits the adjective "thought-collective," because every fact bears three different relations to a thought collective: (1) *Every fact must be in line with the intellectual interests of its thought collective,* since resistance is possible only where there is striving towards a goal. Facts in aesthetics or jurisprudence are thus rarely facts for science. (2) *The resistance must be effective within the thought collective. It must be brought home to each member as both a thought constraint and a form to be directly experienced.* In cognition this appears as the connection between phenomena which can never be severed within the collective. This linkage seems to be truth and conditioned only by logic and content. Only an investigation in comparative epistemology, or a simple comparison after a change has occurred in the thought style, can make these inevitable connections accessible to scientific treatment. The principle of immutability of species characteristics was valid for classical bacteriology, according to the interpretation of the time. If a scientist of that time had been asked why the principle was accepted or why the characteristics of species were conceived in this way, he could only have answered, "Because it is true." Only after a change in thought style did we learn that the opinion was constrained mainly by the methods applied. The passive linkage between these principles was transformed into an active one. (3) *The fact must be expressed in the style of the thought collective.*

Objects created by thought – p. 181:

The boundary line between that which is thought and that which is taken to exist is too narrowly drawn. Thinking must be accorded a certain power to create objects, and objects must be construed as originating in thinking; but, of course, only if it is the style-permeated thinking of a collective.

How to use philosophical principles – p. 181:

... philosophical principles are like money. They are very good servants but very bad masters. Principles should be made use of, but not blindly accepted as guides.

Symbols – p. 125:

This is how chemistry was described before it entered the modern age. Such mystical allegories and comparisons and the strongly emotional images exhale an atmosphere that is completely alien to our scientific thinking. The comparison of gold with the sun and of silver with the moon survives only in popular imagination. Associating lead with Saturn and tin with the devil has lost all meaning even in popular thinking. It is a special, self-contained style, consistent from its point of view. Those people thought and saw differently than we do. They accepted certain symbols that to us appear fanciful and contrived. What if we could present our symbols – the potential, or physical constants, or the gene of heredity, etc. – to thinkers of the Middle Ages? Could we expect them to be delighted with the "correctness" of the symbols and instantly listen to reason? Or, conversely, would they find our symbolism just as fanciful, contrived, and arbitrarily devised as we find theirs?

Name as a property – p. 136:

A name here has a completely different significance from what it has today. It is not an arbitrary, conventional designation or one that arose by historical accident. The meaning is inherent in the name, and its investigation constitutes an integral part of acquiring knowledge about what it names. The name ranks as a property of its object of reference.

Meaning as a property of the object – p. 137:

We are thus confronted with ideograms, or graphic representations of certain ideas and certain meanings. It involves a kind of comprehending where the meaning is represented as a property of the object illustrated.

The Denkkollektiv and the Quantum

The Denkkollektiv, what do I make of it? Something about it sounds so nice and right, but still I worry. Let me try to make much clearer what I already tried to express to you in the old letter of October 1995.

The essential point of Fleck, as I see it, is that science never grasps the *actual* "thing in itself." The most science can do is toy with the "things in themselves" that it itself constructs – that is, as part of a community-oriented project to grasp and codify the world. With this, I am inclined to believe. But still one can ask, is there an actual thing in itself, the measly idea of which can at least be abstracted from the phenomena we observe? Could it be that science, despite its socially constructed character, is steadily (or even jerkily) moving toward being a better and better reflection of *what is* (i.e., that which was there prior to our attempts to understand it)? As far as I can recall, Fleck gave no sound argument for why that *could not* be the case. As far as I can tell, nothing about Fleck's system would crumble if underneath it all there really were a "real world" that science might or might not grasp. Of course, the scientist may never know that he is there, but that doesn't preclude the existence of such a substrate.

This contrasts with the world indicated by quantum mechanics. It is a world so sensitive to the touch that the most natural understanding of it may **only** come about via Fleckian lines of thought. This is my present feeling. And this is what I was trying to express in the note I forwarded to you titled "Fuchsian Genesis" (it was attached to the bottom of a note titled "Genetic Genesis").

Charlie Bennett might say, "Of course, there's a reality out there. That's what gives the scientist a reason to be." But maybe there isn't: the quantum points in just that direction. Quantum mechanics teaches us that the questions we can ask of the world have this wonderful non-Boolean property: I can ask a question X or a question Y, but there's no good sense in which I can ask the question $X \wedge Y$. Thus my free will (and your free will) play a fundamental role in the evidence we amass for our world picture, and necessarily so. This in turn also makes the Denkkollektiv more fundamental than one might have thought. It is not the case that if we as scientists cleaned up our act sufficiently well, we might find ourselves sitting atop that hallowed edifice called reality.

The reality we have is in nonnegligible part the one we create. And, it seems to me, there is no reasonable way of getting around that when – but only when – quantum phenomena are recognized as part of the world.

15 May 1999, "Tsk, Tsk"

Almost like I was 28 years younger, I bounded out of bed this morning with this greatest of glee! But this time it wasn't for the Saturday morning cartoons, no. Somehow I was sure that today you would surprise me with some meaty philosophical speculation about the realities we create – a report on the connections between your thought, my thought, and Mr. Bennett's. I opened my mailbox in childlike anticipation ... ... and then ... slowly ... ... sulked back to bed. There was no more comfort than the protection of the sheets over my head. The world was left better unfaced.

22 May 1999, "Reality in Lottery"

Kiki told me you called. Sorry I haven't gotten back in touch with you; lots of things have been going on this week, and I've hardly had a chance to get any email done. If you'd still like to talk, give me a call Saturday. Or if you want, I can call you Sunday (when rates are significantly cheaper for us). Let me know (when I should be listening for the phone or commanding its dial).

In the meantime, have a look at what I swiped from the New York Times this morning. It's from an article titled "Living Off the Daily Dream of Winning a Lottery Prize." Would you call that reality creation? Would you say Newsome lives in one Denkkollektiv, while the writer of the article lives in another?

Newsome, who works at a nearby powder-coating plant, applying protective coatings to metal products, has developed his own system of tracking past winning numbers to determine patterns. In a ritual that is performed in the family's house each morning, Newsome rises at 4:30 a.m., puts on his robe and walks downstairs to the kitchen.

After turning on a pot of water for his instant coffee, he sits at the table and in front of him he places several memo pads and worn file cards that have hundreds of three-digit numbers written on them.

Painstakingly, Newsome creates charts based on factors like which lottery hostess on television picks the numbers on which day of the week. (A lottery drawing is held every day except Christmas in New Jersey.) Although the drawing is totally random, Newsome is convinced that there is a pattern.

"I see here that 994 has come out three times on Sunday," he said as he sat down to make his selections one recent Sunday morning. Lighting up a Marlboro 100 and scooping four heaping tablespoons of sugar into a large mug of coffee, he predicted, "I think that 994 is going to come back today." (It did not.) Based on the information Newsome compiles each morning, he writes down his 20 favorite numbers for the day, and leaves a copy on the table for his wife. She decides which ones she will play.

It is hard to tally how much the Newsomes spend or win, because in addition to the game itself, there is a little gamesmanship between the two. A couple of months ago, Mrs. Newsome played one of the numbers her husband had suggested and she won more than $300. When Newsome asked her whether she had played it, she told him she had not. "I know he's hitting a lot of times and ain't saying nothing to me," she said.

Last June, according to Newsome's day planner, the couple won on four consecutive days, for a total of more than $1,000. "That was a good week," Newsome said. But throughout most of two months early this year, the Newsomes were playing every day and were not having any luck.

By the middle of one month, Newsome was growing anxious. "We could sure use the money," he said one evening as he walked up to the counter at Home Dairy and played 172, 578, 198, and 574. (They were all losers; the winning three-digit number that evening was 416.)

02 August 1999, "Epiphenomena Chez Dyer"

Thanks so much for the long note about Charlie, Howard, and the other many-worlders. I enjoyed it very much and think we're finally getting somewhere: the conflicting desires for many-worlds and Copenhagen are rooted in different ideas about the goals of science. That does have a ring of truth, doesn't it?

I'm sorry to be writing back at such a late date over this. My time in Europe turned out to be much more hectic than I imagined it could be. Actually I started to write you a note on the conversation Chez Dyer over two months ago, just after I got your first long note about it. But for one reason or other I just never finished it. I've decided to paste below what *was* written and simply not bother with finishing it. I think the language I was using in that note was a little overblown anyway. Included in that note is a quote of Henry Stapp that I never got to the point of explaining; it's from his article "Attention, Intention and Will in Quantum Physics" quant-ph/9905054. It is from that article that I snagged the phrase "dangling appendage" as you'll see momentarily. In general I disagree with almost everything Stapp says, but I thought this quote was noteworthy and quite relevant to our conversation Chez Dyer.

Essentially the thing that struck me at Chez Dyer was the uncanny feeling that all three of us were saying precisely the same thing. But for some reason, Charlie felt that he was saying something different. That confused me greatly and, in fact, still confuses me.

In Charlie's mind, somehow you and I have crossed the bounds of what is real science; somehow he views his language as more neutral (less anthropocentric) than ours ... and therefore better.

The greatest danger I see in the many-worlds/one-Hilbert-space point of view (beside the ridiculous silliness of it all) is the degree to which it is a dead end. The degree to which it is morally bankrupt. Charlie, by thinking that he has taken some of the anthropocentrism out of the picture, has actually emptied the world of all content.

Beyond that though, I think, many-worlds empties the world of content in a way that's even worse than classical determinism. Let me explain. In my mind, both completely deterministic ontologies and completely indeterministic ones are equally unpalatable. This is because, in both, all our consciousnesses, all our great works of literature, everything that we know, even the coffee maker in my kitchen, are but dangling appendages, illusions. In the first case, the only truth is the Great Initial Condition. In the second, it is the great "I Am That I Am." But many-worlds compounds that trouble in a far worse fashion by stripping away even those small corners of mystery. It is a world in which anything goes, and everything does. What could be more empty than that?

My most technical criticism though, is that I don't see anything in the quantum formalism that compels the many-world point of view. One could have constructed such a free-for-all world in 1884. William James came close:

Indeterminism, on the contrary, says that the parts have a certain amount of loose play on one another, so that the laying down of one of them does not necessarily determine what the others shall be. It admits that possibilities may be in excess of actualities, and that things not yet revealed to our knowledge may really in themselves be ambiguous. Of two alternative futures which we conceive, both may now be really possible; and the one become impossible only at the very moment when the other excludes it by becoming real itself. Indeterminism thus denies the world to be one unbending unit of fact. It says there is a certain ultimate pluralism in it; and, so saying, it corroborates our ordinary unsophisticated view of things. To that view, actualities seem to float in a wider sea of possibilities from out of which they are chosen; and, somewhere, indeterminism says, such possibilities exist, and form a part of the truth.

The only difference between James and many-worlds is that in many-worlds actualities do not float in a WIDER sea of possibilities; the seas are of the same size for they are identified with each other. One could, for instance, take a classical phase space and declare that all initial conditions are equally real and refuse to discriminate between any of the them. Each initial condition is a world, and that is that. Who is to say that one is more real than the other? [[The standard many-worlder usually objects to me at this point by saying something like, "But there is no notion of superposition there!" So I ask, "What does that mean?" and "What role does it play?": they always fall flat. The reason is they don't know what role it plays; they likely would have been predisposed to many-worlds even if they had known no quantum mechanics. This is one reason I believe there is a lot of truth in the assessment you gave in the last note.]] If you want to see to what ridiculous ends one can take this point of view – even to the point of completely forgetting about physics when one speaks of the many worlds – have a look at Max Tegmark's article, "Is 'the theory of everything' merely the ultimate ensemble theory?" *Ann. Phys.* **270**, 1–51. You can also find the article on the net at http://www.sns.ias.edu/~max/toe.html. I don't think there is any better technical argument against many worlds than to read this paper (though certainly Mr. Tegmark wouldn't see it that way!!). Who needs quantum mechanics to have many worlds?

But that's a little bit of an aside; let me get back to Charlie. The other day he said to me – and this is a direct quote – "It is the fabric of possibilities that is real." That's his way of describing the many-worlds point of view. He sees it that the task of science is to delimit what is possible and not go beyond that. Anything beyond that is a kind of religion, or chauvinism as he calls it. There is a way in which I am mildly in agreement with this, but I don't see any way of grounding the word "possibility" in a way that does not take into account what is "known" ... and for that one needs a knower (always). The real extent of my mild agreement is that I don't believe the *actual* outcomes of quantum mechanical experiments, or the mechanism by which they arise, are contained within the structure of the theory: it is a theory of what is possible, nay, probable, but based on what one knows. Wave functions don't collapse because of any mechanism in the world that changes them abruptly, they collapse when and only when our knowledge changes. [[Can a dog collapse a wave function? Dogs don't use wave functions. Heck I didn't collapse a wave function until I was at least 20 years old. We should never confuse the world with what little bit we know of it.]]

It seems to me that there must be a sense in which the world just is – i.e., the sort of thing that Charlie and Howard are striving for – but that way of describing the world is not open to the methods of science. How can one or even a community of observers bootstrap themselves to such a point of view? And indeed even if it could be done, what would it mean for the world that "is" to be nothing more than a list of possibilities without actualities (or, to be more accurate to Charlie's point of view, a list of possibilities that are all equally actual)? All that is just too high falutin' for me ... and as I tried to explain above, simply too empty to be a useful or interesting guide to the world.

The beauty I see in quantum mechanics is precisely the beauty of Copenhagen. As I tried to say it crisply to Bill Wootters the other day: the great lesson of quantum mechanics is that the world can be moved. There is a reason we are stuck with a physics that is "the ability to win a bet" instead of a physics that is a static portrait of "what is." If the world can be moved, we simply can't ask for more; it would be self-contradictory. There has to be some room left within the physical theory; it can't be a closed book if the book itself isn't closed.

But I ramble. Let me get back to some of the questions you asked. What should we do with all these ideas? I'm certainly not averse to our writing something together. If we can make the points clearly, it could be a good service to the community. Any more concrete ideas as to how we could go about it? Do you have an outline in mind?

OK, that's enough for now. Ping me your thoughts on all that I said here and I'll pong you back. (Don't forget to look at the further stuff below.)

Below follows the never completed note including the Stapp quote:

Thanks again for the long note putting some thoughts in order about the nice conversation we had with Charlie April 4 (Easter).

I think I do agree with you that:

Herbal Treatment 3: *[I]t seems that there is a great crossroads in the next century, whether we continue as a society to ignore the creation of reality in & by science or we choose to use that creativity for good purposes.*

Do you have anything concrete worked out along these lines? Have you written anything expanding upon this sentence? Is there a particular part of Muddling Through that I should take a look at?

Anyway, I thought in this note I'd try to add my two cents to your valiant summary of our discussion chez Dyer. The most important thing I could see, from my perspective, was the unnerving similarity between the following two thoughts (one Charlie's and one mine).

1) Chris: The quantum mechanical formalism does not address a process or mechanism by which the singular, particular measurement outcomes in experiments come about. That aspect of our world is just not contained within the formalism and cannot be, but that is no blemish to the theory.

2) Charlie: The quantum mechanical formalism does not address a process or mechanism by which the singular, particular measurement outcomes in experiments come about. That aspect of our world is just not contained within the formalism and cannot be, but that is no blemish to the theory.

See the difference? If not, then read one more time.

The only great distinction as far as I could tell was in the intonation with which they were presented! Actually, that's an exaggeration, but these sentences did have two vastly different thoughts behind them.

The reason I would say such a sentence (in my intonation) is because it seems clear to me that quantum theory is about and only about our knowledge. It is the best we can say – the most we can predict – about the outcomes of our prodding of the world. It is the best estimate we can give of the world's reaction to our interventions. I don't condemn the weatherman when he knows no physics or chaos theory, when he knows not the details of how rain is actually made; his predictions are useful nonetheless and it is worth my while to pay him for his services. So too, quantum mechanics should not be blamed for not providing the omniscience we have come to expect from classical physics: it cannot predict the outcomes of my prods with complete certainty because the world itself does not know how it will react (nor can it know in advance if or how I will prod). That it gives no mechanism is because it is about our knowledge; that it cannot give a mechanism is because there is none there to be found. "But what about the weatherman? We can imagine one that is infinitely good, can't we? One that is on the mark each and every time? This says your assessment of quantum theory need not be correct." Well, we can imagine unicorns too. Just because a theory is about knowledge, it does not automatically mean that it can be improved. (And that statement need not be a renunciation of the goals of science.)

A controversy is raging today about the power of our minds. Intuitively we know that our conscious thoughts can guide our actions. Yet the chief philosophies of our time proclaim, in the name of science, that we are mechanical systems governed, fundamentally, entirely by impersonal laws that operate at the level of our microscopic constituents.

The question of the nature of the relationship between conscious thoughts and physical actions is called the mind-body problem. Old as philosophy itself it was brought to its present form by the rise, during the seventeenth century, of what is called "modern science". [...] The central idea is that the physical universe is composed of "material" parts that are localizable in tiny regions, and that all motion of matter is completely determined by matter alone, via local universal laws. This *local* character of the laws is crucial. It means that each tiny localized part responds only to the states of its immediate neighbors: each local part "feels" or "knows about" nothing outside its immediate microscopic neighborhood. Thus the evolution of the physical universe, and of every system within the physical universe, is governed by a vast collection of local processes, each of which is 'myopic' in the sense that it 'sees' only its immediate neighbors.

The problem is that if this causal structure indeed holds then there is no need for our human feelings and knowings. These experiential qualities clearly correspond to large-scale properties of our brains. But if the entire causal process is already completely determined by the 'myopic' process postulated by classical physical theory, then there is nothing for any unified graspings of large-scale properties to do. Indeed, there is nothing that they *can* do that is not already done by the myopic processes. Our conscious thoughts thus become prisoners of impersonal microscopic processes: we are, according to this "scientific" view, mechanical robots, with a mysterious dangling appendage, a stream of conscious thoughts that can grasp large-scale properties as wholes, but exert, as a consequence of these graspings, nothing not done already by the microscopic constituents.

17 August 1999, "A Bernstein Off the Earth?"

Where've you been? Did you fall off the world? Most importantly, did you get the long note I wrote you a couple of weeks ago concerning THE conversation chez Dyer? I think it's my best attempt yet to pin down what I hate about many-worlds. (It also pins down what I hate about modal interpretations of quantum mechanics.) I would love to hear your comments. Also, I'd be interested in hearing further ideas about our pursuing this issue in the public eye, i.e., by writing some of it down in a paper. (Remember the only rule I have in writing papers is that the author lists be alphabetically ordered.)

I wish you had been at the foundations conference in Maryland last week. In my talk (which I think went quite well this time), I took a strong stance on the interpretation of QM. Namely I stated very explicitly that the most one can hope for in interpreting the state vector is that it corresponds to a state of knowledge. If you use state vectors, you can collapse them; if you don't, then you can't. They're just not physical entities; they don't exist in and of themselves. In fact I started off the whole talk with a slide of that wonderful quote by Bruno de Finetti (suitably modified to the quantum context):

My thesis, paradoxically, and a little provocatively, but nonetheless genuinely, is simply this:

QUANTUM STATES DO NOT EXIST.

The abandonment of superstitious beliefs about the existence of Phlogiston, the Cosmic Ether, Absolute Space and Time, ..., or Fairies and Witches, was an essential step along the road to scientific thinking. The quantum state, too, if regarded as something endowed with some kind of objective existence, is no less a misleading conception, an illusory attempt to exteriorize or materialize the information we possess.

(the ghost of Bruno de Finetti)

One thing that was really funny was how in conversations, these ideas were referred to variously as "the knowledge interpretation of quantum mechanics" or "Fuchs's interpretation." Finally, at the last lunch, just as everyone was parting, I turned to Lucien Hardy and Robert Garisto (an editor at PRL) and said, "You know, I never said this because I knew that I would be able to convey the ideas more effectively if I didn't ... but the point of view I've been advocating is nothing other than the Copenhagen interpretation."

I had some really superbly productive conversations with Lucien Hardy. If you ever get the chance, get to know him well; he is a scholar in the best sense. I think he's one of the few already poised to help us stretch and tone our point of view. (He has even admitted that he finds it somewhat attractive.)

Write me some time (soon)!

14 September 1999, "A Fleck of Fleck"

This morning while reading Markus Fierz's book *Girolamo Cardano: 1501–1576, Physician, Natural Philosopher, Mathematician, Astrologer, and Interpreter of Dreams*, I came across a passage of Cardano's that reminded me of something I had sent you about Fleck. In particular, recall that I was struck by the following two passages in Fleck:

Name as a property – p. 136: A name here has a completely different significance from what it has today. It is not an arbitrary, conventional designation or one that arose by historical accident. The meaning is inherent in the name, and its investigation constitutes an integral part of acquiring knowledge about what it names. The name ranks as a property of its object of reference.

and

Meaning as a property of the object – p. 137: We are thus confronted with ideograms, or graphic representations of certain ideas and certain meanings. It involves a kind of comprehending where the meaning is represented as a property of the object illustrated.

Fleck was spurred to these passages by a reading of Fontanus's epitome of Vesalius's *Anatomy.* I think it was published in 1642.

Anyway it appears that Cardano was already fighting this thought-style when he wrote *De Libris propriis*, ca. 1562. In a passage offering some basic suggestions about how books should be written, Cardano writes:

Pay attention to the things Galen was criticized for, and don't think that they will bring you praise. Always remember: Words are there to describe things, not things to illustrate words.

I wonder how widespread this practice – the one of Fontanus, i.e., that of thinking of the symbol as a property – was by the time Cardano wrote this? I can tie these thoughts to one other thread by telling you something I read in a paper by Charles Enz [C. P. Enz, "The Wavefunction of Correlated Quantum Systems as Objects of Reality," in *Vastakohtien todellisuus: Juhlakirja professori K. V. Laurikainen 80-vuotispäivänä*, edited by U. Ketvel, *et al.* (Helsinki U. Press, 1996), pp. 61–76.] It has to do with a concept that Pauli called the "idea of the reality of the symbol." Enz writes:

For a Chinese or a Japanese the reality of symbols is exemplified by the Chinese character representing his name. Indeed for a Chinese *a calligraphic character thus is, after all, like the cipher of his identity.* In the same way as the Chinese characters have to be appreciated not as static pictures but by recreating in one's mind the dynamics of the calligrapher's mind, body and brush drawing the successive strokes, the understanding of a specific quantum phenomenon – which, as I tried to show, is most accurately described in terms of the wavefunction ψ – is a **dynamical process**. This suggests that the quest for quantum reality may be best characterized as **active realism**. Indeed, in order to understand the **symbolism of** ψ one has to recreate in one's mind the **properties expressed by** ψ much in the way Chinese characters have to be read.

I think this is perhaps closely connected to something much more technical I wrote Howard the other day ... but you can discuss that with Howard.

01 September 2000, "The Allure of Texas"

To help answer a question you asked a long time ago. From J. L. Casti, *Paradigms Lost: Images of Man in the Mirror of Science*, (William Morrow and Co., New York, 1989):

Texas may call itself the Lone Star State but Texans have always done things in a big way, so when the agenda item is reality generation no one will be surprised to find that the "lone star" is magically transformed into an entire universe of glowing objects, the centerpiece being nothing less than the meaning of meaning itself. The chief architect of this Texas-sized version of reality is John A. Wheeler, director of the Center for Theoretical Physics at the University of Texas at Austin.

The heart of the Austin Interpretation championed by Wheeler is the idea of a reality created by the observer through exercise of the measurement option. The Austin school believes that we are wrong to think of the past as having a definite existence "out there." The past exists only insofar as it is present in the records we have today. And the very nature of those records is dictated by the measurement choices we exercised in generating them. Thus, if we chose to measure an electron's position yesterday in the lab and recorded the resulting observation, then that electron's position from yesterday exists but its velocity doesn't. Why not? Simply because we chose to measure the position and not the velocity.

Because this very act of *choosing* is always involved in what we measure, Wheeler feels that the act of observation is "an elementary act of creation." ...

We should hasten to note that the Austin Interpretation champions an *observer*-created reality, not a consciousness-created one. The Austin view, while differing from Copenhagen in significant ways, still accepts some of the crucial aspects of Bohr's position. Most important, the two schools agree that scientists can communicate unambiguously only about the final results of measurement. For Wheeler, the essence of existence (reality) is meaning, and the essence of meaning is communication defined as the joint product of all the evidence available to those who communicate. In this view meaning rests on action, which means decisions, which in turn force the choice between complementary questions and the distinguishing of answers. Putting all these links together, out pops the Austin Interpretation of reality generation by exercise of the quantum measurement option.

09 September 2000, "More Fleckulation"

The "Fleck materials" book is quite nice. There are several articles by Fleck himself, one of them previously unpublished:

1) Some Specific Features of the Medical Way of Thinking [1927]
2) On the Crisis of 'Reality' [1929]
3) Scientific Observation and Perception in General [1935]
4) The Problem of Epistemology [1936]
5) Problems of the Science of Science [1946]
6) To Look, To See, To Know [1947]
7) Crisis in Science [unpublished, 1960]

It also includes 14 articles about Fleck's ideas and their context in Polish philosophy. Finally, it contains a full bibliography of Fleck's writings.

R. S. Cohen and T. Schnelle, *Cognition and Fact: Materials on Ludwik Fleck*, (D. Reidel, Dordrecht, 1986).

You should have a look at it if you get a chance.

26 October 2000, "Activating or Catalyzing?"

Thanks a million for the notes! I'll incorporate them soon.

I've had a tremendously difficult time trying to come up with a name for this document. Here below are the ones I've tried. Any votes? And please explain why you think what you think.

The Undetached Observer:
The Activating Observer:
The Catalyzing Observer:
The Malleable Reality:
The Malleable Substrate:

Resource Material for a Paulian–Wheelerish Conception of Nature

29 November 2000, "Anecdote"

Too bad you're not coming to Vienna. I'm letting the soul (but not the libido) of Schrödinger slowly seep into me.

Let me tell you a quick anecdote while I have your ear. It's one you'll appreciate given your great respect for Ludwik Fleck. I was talking to Experimentalist X the other night, asking him to evaluate his various students so I could tumble over in my head whether any should be approached about employment at Bell Labs. Sadly, one of his students didn't get the highest of recommendations. The main point was that the student didn't seem to appreciate the difference between making an experiment happen and keeping himself/herself busy with work. "If a part doesn't work, don't waste time trying to fix it – kill it, and get a new one," he said. "It can be tough in the lab. It's almost as if you have to look at your equipment and say, '*I will you to work*'. You have to command nature. And Student Y just doesn't seem to have that."

Have you ever heard anything more marvelous come directly from the mouth of an experimentalist?

7

Letters to Doug Bilodeau

22 August 1999, "Your Article"

I am writing to let you know that I enjoyed your article quant-ph/9812050, "Why Quantum Mechanics is Hard to Understand," very much. I was in particular very pleased with your Section IV "Dynamics vs. History." It may interest you to know that Markus Fierz expressed a similar point of view in his article: "Does a physical theory comprehend an 'objective, real, single process'?" in *Observation and Interpretation in the Philosophy of Physics*, edited by S. Körner (Dover, NY, 1957), pp. 93–96. I will append the full text below for your enjoyment. [See note to Rüdiger Schack, dated 29 August 1999.]

Most importantly, I would like to ask you for the final coordinates of your paper. Was it published somewhere? I would like to cite it in an upcoming paper.

28 August 1999, "Not Much"

I don't have a good idea about where to send your paper. I think chances are that *AJP* will be tougher than *FP*. In any case, it had an impact on me. My old advisor used to say of our joint papers, "If three people read this paper, we'll be doing OK!" If you know of two others with yours, you're probably doing OK.

I like this focus on the concept of "object." I look forward to anything you might have to say about it. I have some slides that I use as a rallying cry to the idea that quantum information theory may have some impact on quantum foundations. The first is a list of axioms for quantum mechanics. It starts off with Axiom 0: Systems exist. Then I point out the stark contrast between this long list of rather abstract looking axioms and the ones of special relativity: for overemphasis, I have a slide that only lists,

$$c \text{ is constant}$$
$$\text{physics is constant}$$

Then I joke that until we can reduce quantum mechanics to such a simple crisp statement, we will do things like have opinion polls about people's various interpretations. Then I put up a slide titled "The Jim Hartle 1968 (section IV) Interpretation of Quantum Mechanics (suitably modified)" and say that I usually can't even vote because I'm never represented. Then I read a very Copenhagenish sounding passage from Jim's old paper, changing a few words here and there. The upshot is that a quantum state should be understood as a state of knowledge not a state of nature. And so I finally return to the slide with the axioms, and I put an overlay over it. Beside each axiom EXCEPT Axiom 0, it says, "Give an information theoretic justification." This, I say, is the deepest duty of quantum information theory.

So further elucidation of the concept of object would be great. I do think that Kant still has a lot of useful stuff hidden in his system. If you can wait about a month I'll send you a large compilation of things that I've had to say about this; you might enjoy it.

The paper I alluded to is one by Carl Caves, Rüdiger Schack, and myself titled "Bayesian Probability in Quantum Mechanics." It will have a companion, more technical paper, titled "On Unknown Quantum States." We'll be citing you in the first.

What is your situation? Are you a student? Postdoc? Professor? (None of the above?)

Anyway, again, thank you for the long note.

Doug's Reply

Object is a practical, functional concept, not ontological or constitutive in a mechanical sense. On the other hand, paradoxically, our whole concept of "objective" reality is based on "objects" (naturally) and all we can know or understand of the "real" structure of the physical world must come by way of them. Hence the need for Kantian subtlety. I know that's not very clear, but I'm working on it. For now, I just want a general quantum mechanical method applicable to all cases which spells out the relationship between objects and systems. So I'm in the process of going through a number of simple but real-world applications of QM, trying to find an optimal general system for describing what's going on. What people usually do is fall back on the concept of particle, and treat the particles as classical objects whenever possible, sliding back and forth between quantum and classical properties as needed. But that is neither clear nor consistent nor applicable to all cases.

I think you're right that information theory is tied to quantum foundations. But not in the sense that Wheeler meant with his *"It from Bit"* slogan, if I understand him. Information (like objects themselves) is a feature of the way we experience and interact with the world. I like your simple summary of relativity. We do have to do the same with QM – the opposite of complex formalizations like quantum logic. But even Axiom 0 is not without difficulties. "Exist" is a loaded word. There are phenomena correctly described by quantum systems, but people tend to confuse a system with an object or a component of a mechanical world.

06 September 1999, "Labor Day Lounging"

Bilodeau-ism 1: *One of the positive responses was from Tom Siegfried, a science writer for the Dallas Morning News. He wrote a column about the paper which appeared Jan. 4 of this year.*

Tom is a nice guy; I've met him a couple of times at conferences. He does seem to take a real interest in the foundations of quantum mechanics. That has been both good and bad. One of the bad times was when he wrote a couple of articles saying that Chris Adami and Nicolas Cerf had "solved the measurement problem in quantum mechanics." Take a look at `quant-ph/9806047` for instance and some of the connected papers and tell me if you don't think that was a misjudgment.

I would enjoy having a look at this article of his. If you have a copy of the words in your machine, please forward them to me.

Bilodeau-ism 2: *But even Axiom 0 is not without difficulties. "Exist" is a loaded word.*

Agreed. The point mostly was to make a point: to show that there must be some solid background for the remainder of the theory (just as Rosenfeld expresses in the passage below).

Bilodeau-ism 3: *BTW, the Rosenfeld paper says some very good things. As I read over it again, I realized how much it influenced me.*

I know of that paper too; it influenced me in the same way. In fact I thought it expressed itself so clearly that I scanned a passage in and sent it to David Mermin. The result was quite positive; it turned his head a little too. I'll attach that below. [See note to David Mermin, dated 17 April 1998, titled "How Do I Sleep?"]

06 September 1999, "Another Request"

I just had a look at your webpage. If you have your paper "Physics, Machines, and the Hard Problem" written in TEX or LATEX or MS Word, I would appreciate getting an electronic copy to have a read through.

Also I had forgotten to tell you … another *large* source of lucid writings by Rosenfeld can be found in:

L. Rosenfeld, *Selected Papers of Léon Rosenfeld*, edited by R. S. Cohen and J. J. Stachel, Boston Studies in the Philosophy of Science, Vol. 21, (D. Reidel, Dordrecht, 1979).

05 November 1999, "Setting the Initial Condition"

First let me apologize for not replying to your nice letters before now. This move to Los Alamos has just been so very hectic that I've hardly had time for anything enjoyable lately. But fate imposed a little break upon me the other day: I became infected with salmonella Monday, and that put me out of active commission for three days … enough so that I had to cancel giving an invited plenary talk (!) at the New England Section APS Fall Meeting this week. (I can still see my career crumbling before me because of my misadventures with Albuquerque restaurants!)

But every cloud has a silver lining – so they say – and mine was the opportunity to read or reread everything that you've sent me. I remain quite impressed by your collection of thoughts about quantum mechanics. I think there is quite some correlation between our points of view (to the extent that mine is yet firmly defined). Do you come across the same impression, or do you see great places where we diverge? I would very much like to hear your thoughts on my collection of ramblings, dreams, and desires. As you are (to use your words) "intensely dedicated to this program of clarification," I am intensely dedicated to have some completely new physics come out of finally taking completely seriously the proper point of view about QM. So your thoughts will be immensely welcome.

One question right away though. In Footnote 3 of your *JCS* paper, you write:

When I write of the physicist's 'freedom' of action in setting up experiments and controlling parameters, I am not taking a position on the philosophical question of 'freedom of the will'. I mean here only a pragmatic freedom which is independent of the physical entities being observed.

Can you explain in more detail what you meant by this? Can you think of any other references that I might read that influenced you on this point? In that regard, it may be helpful if you could comment on my notes to Howard Barnum (starting with the one titled "It's All About Schmoz"), to John Preskill (starting with the one titled "Two Rabbis in a Bar," but only right after Preskillism 3). (Somewhat tangentially, but also connected to this query, you might also enjoy the note to Herb Bernstein titled "Epiphenomena Chez Dyer.")

Have I answered all the questions that you've asked me? Most likely not, but let me now try ... or at least add a little meat to some of my previous answers.

Bilodeau-ism 4: *I am curious what is the subject of the paper for which you wanted the citation, if you don't mind describing it. I am encouraged by what appears to be a recent convergence of experimental and theoretical focus on phenomena which will help bring about a genuine and substantial clarification of fundamental concepts.*

Actually, as it stands I now have three projects going on that have some connection to the view we share about quantum mechanics. The first is an "opinion piece" for *Physics Today* that I'm writing with Asher Peres. Its title will be "Quantum Mechanics Needs No Interpretation." We're presenting that as something of a rebuttal to two articles that *PT* published within the last year: one endorsed Bohmianism, and one endorsed Consistent Histories. This may be the "project from hell," I don't know – Asher and I certainly diverge on some of our opinions about quantum mechanics. It will be nice to see what part of our opinions can be put into a coherent whole. Perhaps the greatest divergence between us is attitude: I believe I am much more inclined than Asher to see the clarification of quantum mechanics as recognizing that we are on the tip of a great iceberg. I think he's more inclined to see it as the closing of a book. This article should be finished by the end of November.

The second project (with Caves and Schack) is one that must be finished before I go to the Naples meeting on "Chance in Physics" just after Thanksgiving. I've been titling my talks about it "On Unknown Quantum States," but I don't yet know what we'll be calling the paper. A recent talk abstract for the idea goes as follows:

There is hardly a paper in the field of quantum information theory that does not make use of the idea of an "unknown quantum state." Unknown quantum states can be protected with quantum error correcting codes. They can be teleported. They can be used to check whether an eavesdropper is listening in on a communication channel. But what does the term "unknown state" mean?

In this talk, I will make sense of the term in a way that breaks with the vernacular: an unknown quantum state can always be viewed as a known state – albeit a mixed state – on a larger "multi-trial" Hilbert space. The technical result is a quantum mechanical version of the de Finetti representation theorem for exchangeable sequences in probability theory: a density operator on an infinite tensor product of complex Hilbert spaces has complete exchange symmetry if and only if it can be expressed as a convex combination of identical product states. Interestingly, this theorem fails for real Hilbert spaces. The implications of this theorem for the point of view that quantum states represent nothing over and above one's knowledge of a quantum system will be discussed.

In particular, one result is that quantum states *only* have a good interpretation as "states of knowledge, not states of nature" if the Hilbert spaces of QM are over a complex number field instead of the reals. This seems to say that *complex* quantum mechanics is crucial for the understanding you and I have. This is one example of the thing I call for in the samizdat I sent you: pick an axiom of quantum mechanics and "give an information theoretic justification for it."

The final project (again with Caves and Schack) is a much larger one and it probably won't be finished until next spring. The tentative title is "Bayesian Probability in Quantum Mechanics (The Unexpurgated Version)." Here's a tentative abstract:

We show that – despite their being specified by fundamental physical law – quantum probabilities are best understood within the Bayesian approach to probability theory. In that approach, probability always quantifies a state of knowledge, obtaining an operational definition only through a subject's consistent betting behavior. The distinction between classical and quantum probabilities lies not in their definition but in the nature of the information they encode. In the classical world, *maximal* information about a physical system is complete in the sense of providing definite predictions for all possible questions that can be asked of the system. In the quantum world, maximal information is *not* complete and cannot be completed. This distinction provides a novel way to define the meaning of quantum indeterminism and randomness. Through this we find a stronger connection between probability and frequency than can be justified classically. Finally we reconsider the notion of an "unknown quantum state" – an oxymoron within the Bayesian approach. The solution to this conundrum is found in a quantum version of de Finetti's representation theorem for exchangeable sequences.

There, I hope that gives you something of a feeling about what I'm up to.

Bilodeau-ism 5: *Also, I've started to look at your papers and a few others in quantum computing/information theory. Any suggestions for recent general reading in that area?*

I'm very glad to hear that. As I see it, quantum mechanics has always been about information ... but we're only now looking for the proper tools with which to express it correctly. (But be aware, that is a minority opinion within my subfield; most people in the field seem to be followers of 'many worlds'.) Anyway, a good place to start learning quantum information theory is John Preskill's lecture notes for his course at Caltech. You can obtain them at his website http://www.theory.caltech.edu/people/preskill/index.html. Also there you can find a paper of his "The Future of Quantum Information" where you can see how much his opinion about all this diverges from mine. And you can find some nice public lectures and other links.

OK, it's getting late and I should get back to bed: complete recovery is not upon me yet. I look forward to hearing from you when you get a chance.

Doug's Reply

I do think that our views are very close; at least we are motivated by many of the same insights. The main difference is that I am still leery about "information" as a foundational concept. But I have an open mind about it. I think most of the conceptual confusion in QM today is related to the inadequacy of the idea of "particle", which carries with it too much of the old Cartesian notion of the geometrical basis of physical existence. Feynman has greatly advanced our understanding of quantum physics, but his attachment to the particle idea has persisted and made it more difficult to advance further conceptually. Perhaps the concept of information can help to clarify the dynamical rather than geometrical/ontological nature of the quantum. [. . .]

To answer your first question: When I wrote the comment in the *JCS* paper about freedom of will, I was saying that to establish causal connections or verify patterns in phenomena, it is necessary that the actions of the experimenter not be correlated with the contingent details of the phenomena under investigation. To know what "free will" in an absolute sense means is a problem more subtle and difficult even than quantum mechanics. It is possible that a lack of correlation or the idea of measurement in general will turn out to imply some kind of property of ontological independence in the observer, but for the purpose of doing physics, I think it is sufficient to assume that the behavior of the observer/experimenter is not determined by some external agency which also controls the phenomena being observed – e.g., that the brightness of a star does not change just because I decide to look at it or vice versa. In any case, the observer can be that way and still be free or deterministic, I think. The whole question of free will is perhaps not well posed. I have read some of the writings you mention, but will say more in a couple of days when I have had a chance to digest it better.

Re: your prospective article for *Physics Today*. I have long thought it would be possible to blow Bohmian mechanics and many-worlds out of the water with a simple analysis of what physics does and why those viewpoints were introduced and how they fail to accomplish what they set out to do. I think they fail pretty drastically and cover up their failures with obviously feeble rationalizations. I'll try to say more about that in a couple of days, too. Consistent histories I'm not sure that I really understand at all. Accounts I've seen in several papers seem really opaque and poorly motivated to me, and I don't see that it means anything except in simple cases where it reduces to ordinary calculation of Feynman amplitudes. So I don't know how to critique it.

21 March 2000, "You Might Enjoy"

I discovered that Tom Siegfried has a book now. I even discovered that he makes a brief mention of me in it. He writes, "Chris Fuchs (rhymes with books) applied Landauer's principle ..."! I wrote everyone in my family, "Now that's a reporter! He even cared that the reader get the pronunciation of our name!" You might enjoy it. The opening sentence is, "John Wheeler likes to flip coins." Unfortunately, he didn't seem to report your work: it looks like he may have written most of it before he discovered your paper.

Here's the reference,

T. Siegfried, *The Bit and the Pendulum: From Quantum Computing to M Theory – The New Physics of Information*, (Wiley, New York, 2000).

22 March 2000, "Wheeler"

Yes I do know John. I first met him in his undergraduate course "Great Men, Great Minds of Science." He paid attention to me a little because I made a 105 on his final exam. The five bonus points came from completing the sentence "No elementary quantum phenomenon is a phenomenon until ..." Strangely no one else in the class (of 50 or so) got the bonus! Anyway, then I did a semester-long research course with him, mostly under the guidance of one of his students. The project wasn't so interesting (it was in computational general relativity): much better was getting to hang around the rest of his research group who were doing things with quantum mechanics.

Also John is my academic great-grandfather. The lineage goes back like this: Fuchs–Caves–Thorne–Wheeler–Herzfeld. Look at the letter I wrote to Asher Peres, 05 April 1998 titled "Other Things."

I like to accumulate things like this. For instance, I have an "Einstein number" of three. This is because I have written papers with Peres who has written papers with Rosen who has written papers with Einstein. I have a Bohr number of five or better: Fuchs–Caves–Thorne–Wheeler–Bohr. And I have a Pauli number of four, I believe. That also gives me a Heisenberg number of five or better.

15 April 2000, "Pierce Cringes"

Thank you for the wonderful, wonderful note. You have indeed given me a lot of food for thought with this one. I will comment in depth on it once it has all sunk in better.

But in the meantime, let me make a quick comment on one of your points that brought a smile to my face.

Bilodeau-ism 6: *5. Redevelop the foundations of information theory and its connections to physics in light of the above.*

There is a slide I've made to open some of my talks on quantum mechanical channels. It's a quote from a paper by J. R. Pierce commemorating the first 25 years of information theory (*IEEE Trans. Inf. Theory*, vol IT-19, 3–8 (1973)). After reading it to the audience, I say, "The amazing thing is that even after over 25 more years, we still don't know how to answer Pierce's question." But with you, I should emphasize another point! Here's the quote:

I think that I have never met a physicist who understood information theory. I wish that physicists would stop talking about reformulating information theory and would give us a general expression for the capacity of a channel with quantum effects taken into account rather than a number of special cases.

08 June 2000, "Resend?"

Could you do me a favor and send me another copy of

D. J. Bilodeau, "Physics, Machines, and the Hard Problem," *J. Consc. Stud.* **3**, 386–401 (1996).

My last version was all marked up with notes about what parts I had wanted to enter into my computer. Now, I'll just have to read it again!

Bilodeau-ism 7: *I don't know that I would agree that the past is as malleable as the future. But I do believe strongly that the meaning of the past is in part dependent on what we do in the future.*

I didn't say I believed it; I said I wanted to believe it. But, in any case, I am intrigued by your last remark. Could you expand on it?

Like so many times when I've written you, I'm flying across the waters again.

Despite the fire, I went on with my "Quantum Foundations in the Light of Quantum Information" meeting. It went quite well, I think. People really got into the spirit, and I think something tangible will come of it. I'll forward on to you the pre-meeting problem set I had sent to all the attendees. One of the questions was answered fairly easily already, but there's still work to be done on the others. Also, it turns out that everyone contributed problems just as I had hoped. So the total list is substantially longer now. Perhaps the best thing that came out of the meeting for me is that Schumacher, Schack, and I developed a novel argument for linearity based purely on inference issues (related to the quantum de Finetti representation).

Doug's Reply

[Referring to: "I didn't say I believed it; I said I wanted to believe it. But, in any case, I am intrigued by your last remark. Could you expand on it?"]

It's a little like when I'm trying to draw something, and my hand slips and makes an unintended mark on the paper, but then afterwards the shape of the mark looks suggestive and may end up as part of a newly-imagined picture. Events are far more complex than how we conceptualize them. Correlations are more extensive and subtle than we generally expect (e.g., "karma"?). Looking back, I think we can find ways in which accidents become opportunities, and in which our own apparently random, mindless actions take on an effectively intentional nature if we have the insight and imagination to incorporate them into a larger "life-pattern". Maybe what we thought was accidental wasn't really. So part of the trick in life is to keep our minds open to the unexpected, and in particular be ready to turn the seemingly random into positive and constructive additions to our world-building. Seems to me to work that way, anyway. There is a kind of new-ageish formula in meditation or prayer I've heard which says, "This or something better." I.e., we should be careful not to let our wishing for a certain outcome exclude an even better possibility. (There's another good formula – "for the highest good of all concerned" – there may exist win-win solutions we would never see if our imaginations were locked into a zero-sum state space. We reshape the world with imagination, intention, and expectation. Much of the process is unconscious. We have already done much work of which we might not even be aware. But the meaning of a seed depends on what we do with it.

8

Letters to Gilles Brassard

18 November 1997, "Bolt from the Blue"

I saw Adrian Kent's four papers on the server early this morning, and I was reminded of Léon Rosenfeld's description of Bohr's reaction to the EPR paper:

This onslaught came down upon us as a bolt from the blue. Its effect on Bohr was remarkable. We were then in the midst of groping attempts at exploring the implications of the fluctuations of charge and current distributions.... A new worry could not come at a less propitious time. Yet, as soon as Bohr had heard my report of Einstein's argument, everything else was abandoned ...

I suspect the air in Montréal will be just as this when you all awake today. How I wish I were there to see the excitement!

I hope for the sake of our field, and a generally more exciting world, that Kent is right! As you all know, this sceptical physicist (i.e., me) was always a little wary of the Mayers–Lo–Chau strong claim of having considered all possible quantum protocols. I would like to hear the Montréal verdict once it is in.

A good day to all!

31 August 1999, "Subliminal Messages"

Quantum Foundations in the Light of Quantum Cryptography
Quantum Foundations in the Light of Quantum Cryptography
Quantum Foundations in the Light of Quantum Cryptography
Quantum Foundations in the Light of Quantum Cryptography
Quantum Foundations in the Light of Quantum Cryptography
Quantum Foundations in the Light of Quantum Cryptography
Quantum Foundations in the Light of Quantum Cryptography
Quantum Foundations in the Light of Quantum Cryptography
Quantum Foundations in the Light of Quantum Cryptography
Quantum Foundations in the Light of Quantum Cryptography
Quantum Foundations in the Light of Quantum Cryptography

Quantum Foundations in the Light of Quantum Cryptography
Quantum Foundations in the Light of Quantum Cryptography
Quantum Foundations in the Light of Quantum Cryptography

23 January 2000, "I See Why Bit Commitment!"

Now on a more positive note. (This note was started just after I wrote you the negative one about .... Now, however, I am in the comfort of my office at home ... having coffee, thinking Sunday thoughts.) Let me just say that I had a bit of an epiphany in the shuttle bus at Dulles Airport: for the first time I have understood why you want to take *both* the EXISTENCE of secure key distribution and the NONEXISTENCE of bit commitment as pillars in your sought-after derivation of QM. You have been thinking more deeply than me since the beginning!

Let me place at the end of this note a little piece from my samizdat. [See letter to Greg Comer, 22 April 1999, titled "Fuchsian Genesis."] It sort of presents what I've been trying to get at in a dramatic way: it may be my best presentation of the idea of why quantum key distribution has something to do with the foundations of quantum mechanics. But more to the present point, let me tell you about a second way I use to get the point across. I've used this slide in a few talks. (I'll place a PostScript file of it in the next mail.) It consists of five frames with the following little story.

In the first frame God starts to speak to Adam at a time just before Genesis, "Adam, I am going to build you a world. Do you have any suggestions?"

Adam: Mostly I don't want to be alone. I want to have friends ... and enemies to spice things up ... and generally just plenty of people to talk to.

God: Done. I'll give you a world populated with loads of other people. But you ask for a bit of an engineering feat when you ask to be able to talk to them. If you want to communicate, the world can't be too rigid; it has to be a sort of malleable thing. It has to have enough looseness so that you can write the messages of your choice into its properties. It will make the world a little more unpredictable than it might have been for me – I may not be able to warn you about impending dangers like droughts and hurricanes anymore – but I can do that if you want.

Adam: Also God, I would like there to be at least one special someone – someone I can share all my innermost thoughts with, the ones I'd like to keep secret from the rest of the world.

God: Now you ask for a tall order! You want to be able to communicate with one person, and make sure that no one else is listening? How could I possibly do that without having you two bifurcate into a world of your own, one with no contact whatsoever with the original? How about we cut a compromise? Since I'm already making the world malleable so that you can write your messages into it, I'll also make it sensitive to unwanted eavesdropping. I'll give you a means for checking whether someone is listening in on your conversations: whenever information is gathered from your communication carriers, there'll be a reciprocal loss in what you could have said about them otherwise. There'll be a disturbance. Good enough? You should be able to do something clever enough with that to get by.

Adam: Good enough!

God: Then now I'll put you in a deep sleep, and when you awake you'll have your world.

Adam: Wait, wait! I overlooked something! I don't want an unmanageable world, one that I'll never be able to get a scientific theory of. If whenever I gather information about some piece of the world, my colleagues lose some of their information about it, how will we ever come to agreement about what we see? Maybe we'll never be able to see eye to eye on anything. What is science if it's not seeing eye to eye after a sufficient amount of effort? Have I doomed myself to a world that is little more than chaos as far as my description of it goes?

God: No, actually you haven't. I can do this for you: I'll turn the information–disturbance tradeoff knob just to the point where you'll still be able to do science. What could be better? You have both privacy and science.

So Adam fell into a deep sleep, and God set about making a world consistent with his desires. And, poof(!), there was QUANTUM MECHANICS.

That's the tale. But now I see the crucial spot of outlawing bit commitment within it. God could have supplied Adam with a set of impenetrable boxes (and keys to open them) where he could place his information whenever he wanted some secrecy. A bit commitment protocol could certainly be used in that secondary fashion. But God chose to make all information open for all the world to see: he just left the possibility of an imprint whenever someone has a look.

Now for the old piece from the samizdat. Have fun.

22 March 2000, "It from Bit"

I was just lying in bed thinking about our conference and planning the sorts of things that I will say in my "Setting the Flavor of the Meeting" talk. One thing that occurred to me is that I will most surely start off with a slide of a quote from one of John Wheeler's letters (written to Carroll Alley) [See note to Max Tegmark dated 18 January 2001, item 62.]. I'll place it below; you ought to read it. I don't think there's anything more appropriate to start the meeting with.

In that connection, I was thinking that it is precisely because of him and his influence that at least four participants would even dream that quantum information would have something to say about the foundations. (Bernstein, Fuchs, Schumacher, Wootters, that is.) Also in that connection I remembered one of John's talks in 1994. I wrote up a little story about it once; let me insert that here:

John Wheeler (a long-time professor at Princeton and later at the University of Texas) was a great advocate that information theory had something deep to say about quantum mechanics. He always exuded this air of urgency about him: "It is imperative for us to understand the meaning of quantum mechanics in the grand scheme of things! We must make as many mistakes as we can, as fast as we can, so that we can hope to obtain an understanding within our lifetime!" The last time I saw him was in 1994 at a little conference in Santa Fe – the one, in fact, where Peter Shor's factoring algorithm was announced. (Actually, come to think of it, I saw him one time later that same year ... at his 83rd birthday festschrift.) Anyway, at Santa Fe, Wheeler gave a talk (probably titled "How Come the Quantum?") that he closed with a slide depicting Planck's head (maybe etched on a coin or something). I remember, he said (roughly), "In 1900 Planck discovered the quantum. The end of the century is drawing near. We only have six years left to understand why it is that it's here. Wouldn't that be a tribute?!"

Well, this year is the 100th anniversary of Planck discovering the quantum of action. Here's the punch line. What would you think about inviting John to the meeting as a token participant? Just a thought, really.

15 May 2000, to the attendees of the Montréal Meeting, "Problem Set Coming"

In the next email, you will find a longer letter from that I started a few days ago and brought to something of a closing point this morning. It is a problem set I wrote up to indicate the sort of playfulness I hope you will all join in on in drawing up your set of "concrete problems" for our meeting.

I apologize that I was not able to write things in as much detail as I would have liked. However, I still hope you can catch on to the drift of most questions. As some of you know, my house and everything in it burnt up in the Los Alamos fires last week. Life has become more hectic than I ever imagined it could be.

Still I look forward to a fun and truly productive meeting starting Wednesday. I think we're all finally in a position to really take quantum mechanics by the tail. There are untold treasures out there if we'll all just take the trouble to look for them.

15 May 2000, "Problem Set Based on Information–Disturbance Foundation Quest"

I hope you've had a chance to think about the request Gilles and I made in our invitation letter: namely, to compile a list of concrete problems whose solutions might shed some light on the foundations of quantum theory. What we were thinking in particular is that no point of view about quantum foundations is worth its salt if, at this stage, it doesn't raise as many questions as it answers. Why should we buy into a point of view if it doesn't lead to more fun or, at the very least, something more concrete than a stale philosophical satisfaction?

With that in mind, I've decided to grease the gears a bit by giving you a preview of some of the problems motivated by my particular ish-ism. If you haven't yet created a set of your own problems (based on your ish-ism of course), I hope this will give you a flavor of what we were thinking when we made our request. Certainly the more varied the sets of problems everyone brings, the greater the chance we have for making some real progress!

The point of view I'm likely to represent at our meeting is, I think, best captured (though perhaps a little flamboyantly) by a manifesto I wrote a couple of years ago. Let me reproduce that here as an introduction and motivation to the problems that follow.

Genesis and the Quantum

In the beginning God created the heaven and the earth. And the earth was without form, and void; and darkness was upon the face of the deep. And the Spirit of God moved upon the face of the waters. And God said, Let there be light: and there was light. And God saw the light, that it was good; and God divided the light from the darkness. And God called the light Day and the darkness he called Night. And the evening and the morning were the first day. ... [And so on through the next five days until finally ...] And God saw everything that he had made, and, behold, it was very good. And there was evening and there was morning, a sixth day. Thus the heavens and the earth were finished, and all the host of them.

But in all the host of them, there was no science. The scientific world could not help but *still* be without form, and void. For science is a creation of man, a project not yet finished (and perhaps never finishable) – it is the expression of man's attempt to be less surprised by this God-given world with each succeeding day.

So, upon creation, the society of man set out to discover and form physical laws. Eventually an undeniable fact came to light: information gathering about the world is not without a cost. Our experimentation on the world is not without consequence. When *I* learn something about an object, *you* are forced to revise (toward the direction of more ignorance) what you could have said of it. It is a world so "sensitive to the touch" that – with that knowledge – one might have been tempted to turn the tables, to suspect a priori that there could be no science at all. Yet undeniably, distilled from the process of our comparing our notes with those of the larger community – each expressing a give and take of someone's information gain and someone else's consequent loss – we have been able to construct a scientific theory of much that we see. The world is volatile to our information gathering, but not so volatile that we have not been able to construct a successful theory of it. How else could we, "Be fruitful, and multiply, and replenish the earth, and subdue it?" The most basic, low-level piece of that understanding is quantum theory.

The *speculation* is that quantum theory is the unique expression of this happy circumstance: it is the best we can say in a world where *my* information gathering and *your* information loss go hand in hand.[1] It is an expression of the "laws of thought" best molded to our lot in life. What we cannot do anymore is suppose a physical theory that is a direct reflection of the mechanism underneath it all: that mechanism is hidden to the point of our not even being able to speculate about it (in a scientific way). We must instead find comfort in a physical theory that gives us the means for describing what we can *know* and how that *knowledge* can change (quantum states and unitary evolution). The task of physics has changed from aspiring to be a static portrait of "what is" to being "the ability to win a bet."

This speculation defines the larger part of my present research program.

A. Some Concrete Problems

Problem #1: Pre-Gleason, or Why Orthogonality?

Andrew Gleason's 1957 theorem is an extremely powerful result for the foundations of quantum theory. This is because it indicates the extent to which the Born probability rule

---

[1] Why is that a happy circumstance? Because it implies in part that the book of Nature may not yet be a written product. "The world can be moved."

and even the state-space structure of density operators are *dependent* upon the theory's other postulates. Quantum mechanics is a tighter package than one might have first thought.

The formal statement of the theorem runs as follows. Let $\mathcal{H}_d$ be a (complex or real) Hilbert space of dimension $d \geq 3$, and let $S(\mathcal{H}_d)$ denote the set of one-dimensional projectors onto $\mathcal{H}_d$. We shall suppose that whatever a "quantum measurement" is, it always corresponds to some complete orthogonal subset of $S(\mathcal{H}_d)$. Particularly, within each such orthogonal set, the individual projectors are the theoretical expressions for the possible outcomes of the measurement associated with it.

Assume now that it is the task of the theory to assign probabilities to the outcomes of all conceivable measurements. Suppose all that we know of the way it does this is the following: There exists a function

$$p : S(\mathcal{H}_d) \longrightarrow [0, 1] \tag{8.1}$$

such that

$$\sum_{i=1}^{d} p(\Pi_i) = 1 \tag{8.2}$$

whenever the projectors Π_i form a complete orthonormal set. It might seem a priori that there should be loads of functions p satisfying such a minimal set of properties. But there isn't. Gleason's result is that for any such p, there exists a density operator ρ such that

$$p(\Pi) = \mathrm{tr}(\rho\Pi). \tag{8.3}$$

In words, Gleason's theorem derives the standard Born probability rule *and*, in the process, identifies the quantum state–space structure to be the density operators over $\mathcal{H}_d$. Moreover, he gets this from assumptions that are ostensibly much weaker than either of the end results. This theorem is quite remarkable in that it requires no further conditions on the class of allowed functions p beyond those already stated. In particular, there is not even an assumption of continuity on the functions p.

A question on my mind is to what extent, if any, does the structure of this theorem support an information–disturbance foundation for quantum mechanics? I think this might be fruitfully explored by thinking in the following way. The assumptions behind Gleason's theorem naturally split into two pieces. (A) The questions that can be asked of a quantum system *only* correspond to orthogonal projectors onto $\mathcal{H}_d$. A consequence of this is that there is no good notion of measuring two distinct questions simultaneously – that is, there is no good notion of an AND operation for two measurements. And (B), it is the task of physical theory to give probabilities for the outcomes of these questions, and we can say at least this much about the probabilities: They are *noncontextual* in the sense that, for a given outcome, it does not matter which physical question (i.e., which orthogonal set) we've associated it with. This is the content of the assumption that the probability rule is of the form of a "frame function" (a function satisfying Eqs. (8.1) and (8.2)).

It seems to me that the first assumption to some extent captures the idea that information gathering is invasive. If you gather some information and I gather some other information, there is no guarantee that we can put the two pieces of information into a consistent picture: my information gathering has disturbed the relevance of the information you've already gathered. The second assumption, however, appears to be more of the flavor that nevertheless such information gathering is not *too* invasive. For otherwise one might imagine the probabilities for a measurement's outcomes to depend upon the full specification of the orthogonal set used in its definition. The Born probability rule clearly has a much weaker dependence on the measurement than it might have had.

A question whose answer could bolster (or discourage) this point of view is the following. Why is the invasiveness of quantum measurement specifically captured by identifying measurement outcomes with orthogonal sets of projectors? Hilbert space has a lot of structure; why single out precisely the orthogonal projectors for defining the notion of measurement? To get a handle on this, we could try to see how it might have been otherwise.

As a wild example, consider an imaginary world where quantum measurements are not only associated with orthogonal projectors, but with the projectors onto *any* complete linearly independent set of vectors. This would be a notion of measurement that made use solely of the linear structure of $\mathcal{H}_d$, eschewing any concern for its inner product. What kinds of probability rule can arise for such a notion of measurement? In particular, can one have an interesting "noncontextual" probability rule in the spirit of Gleason's theorem? More precisely, what kind of functions p can satisfy Eqs. (8.1) and (8.2) but with the summation in the latter equation satisfied for any linearly independent set?

Well, it's not hard to see that the only noncontextual probability rule that works for all "measurements" of this kind would have to be the trivial probability assignment of $1/d$ for each outcome, no matter what the measurement. To give an example of how to see this, visualize three linearly independent unit vectors v_1, v_2, and v_3 in R^3 and imagine assigning them probabilities p_1, p_2, and p_3. Hold v_1 and v_2 fixed and rotate the third vector whichever way you wish. As long as it doesn't fall on the two lines spanned by v_1 and v_2, then the projector associated with it must always be assigned the same probability, namely p_3. Now do the same thing with vector v_2, holding v_1 and v_3 fixed. This will make almost all vectors on the unit sphere associated with projectors of probability p_2, proving that $p_2 = p_3$. Finally, one does the same trick with v_1, proving that $p_1 = p_2 = p_3 = 1/3$.

The lesson is simple: if every linearly independent complete set of vectors in $\mathcal{H}_d$ constituted a measurement, one could not hope to retain a noncontextual probability assignment for measurement outcomes without making the world an awfully dull place!

But maybe this version of the game is just too dumb. So, let's try to spice it up a bit by explicitly using the inner product structure of $\mathcal{H}_d$, but in a nonstandard way. Again consider R^3, the smallest Hilbert space on which Gleason's standard theorem can be proved. Suppose now that a "measurement" corresponds to any three vectors with a fixed angle relation between themselves. What I'm thinking of here is to start with three vectors v_1, v_2, and v_3 whose angles (moving around them cyclically) are α, β, and γ. Now rigidly rotate that structure in all possible ways to generate all possible measurements. Are there any interesting *noncontextual* probability rules – again in the spirit of Gleason – that one can associate with this notion of "measurement?"

Here I don't know the answer. But I do know of some special cases where one again gets only the trivial assignment $p_1 = p_2 = p_3 = 1/3$. For instance, take the cases where $\alpha = \beta = \gamma$ and α is such that if we rotate around v_1, v_2 will fall back upon itself after an odd number of "clicks" – what I mean by a click here is rotating v_2 into v_3 and so on ... click, click, click. What happens if we run through such a "clicking" process? Well, v_1 must be constantly associated with the same probability value p_1 by the assumption of noncontextuality. But then by that same assumption, as v_2 rotates into the old v_3, it must pick up the probability p_3. And so on it will fluctuate up and down: p_2, p_3, p_2, p_3 ... until it finally falls upon its original position. If this happens in an odd number of clicks, then it will have to be the case that $p_2 = p_3$ or the assumption of noncontextuality would be broken. Similarly we can see that the whole circle generated by rotating v_2 and v_3 around v_1 must be "colored" with the same probability value. Finally, run through the same process but by rotating about the vector v_2. This will generate a second circle that intersects with the first. From that and the assumption of noncontextuality, it follows that $p_1 = p_2 = p_3 = 1/3$ and this will be true of any triad by the very same argumentation.

Another case where one can see the same effect is in the single qubit Hilbert space C^2. There the game would be that any two vectors with a fixed angle α between them would constitute a measurement. Thinking about the Bloch-sphere representation of C^2, one can use the argument similar to the one above to see that whenever $\alpha \neq 90°$ the only possible noncontextual probability assignment is $p_1 = p_2 = 1/2$ for all possible measurements.

Conjecture 1: *In the case of R^3, whenever one of the angles α, β, or γ is not identically $90°$, then the only possible noncontextual probability assignment for measurement outcomes will be the trivial one $p_1 = p_2 = p_3 = 1/3$.*

How to tackle such a problem? I think it may not be too hard actually, especially if one assumes that the noncontextual probability assignments, whatever they are, must be continuous functions. The starting point would be to try to trace through Asher Peres's derivation of the standard Gleason theorem in his textbook. There, something will surely fail when one looks at the expansion of the proposed "frame functions" in terms of spherical harmonics. (I hope someone will bring Asher's book with them to the meeting: I would bring mine, but I don't have it anymore.)

What's to be learned from this problem? I'm not quite sure, but I think mostly it will help reinforce the idea that our standard notion of quantum measurement is not simply an arbitrary structure. It's there for a reason, a reason we still need to ferret out.

Problem #2: Wootters Revamped with POVMs

Bill Wootters in his Ph.D. thesis explored an alternative derivation of the quantum probability rule. His work was based on the hope that it could be obtained via an extremization principle much in the spirit of the principle of least action in classical mechanics. (I believe he may talk about this very problem at the meeting.) The quantity extremized was the Shannon information a measurement reveals about a system's preparation, under the assumption that one has many copies of the system all with identical measurements.

Specifically the scenario was this. Consider a *real* Hilbert space of dimension d and a fixed orthogonal basis within that space. One imagines that one has possession of N copies of a quantum system with that Hilbert space, all with precisely the same quantum state $|\psi\rangle$. Which quantum state? One drawn randomly with respect to the unique unitarily invariant measure on the rays of $\mathcal{H}_d$, only one doesn't know which. The fixed orthogonal basis represents a measurement that one will perform on the separate copies in an attempt to ascertain the unknown preparation.

[**NOTE:** *Here, through the remainder of this problem set, the words were thrown together hurriedly after the fire.*]

Anyway, Bill's attempt of a derivation didn't really work so nicely for complex Hilbert spaces. The question here is, can we make it work after all, if we start thinking of POVMs as a primitive notion of measurement in its own right.

Specifically, the first thing we must ask is does there always exist an informationally complete set of rank-one POVM elements all of equal weighting on $\mathcal{H}_d$? And it might be even nicer if that set could be taken to have precisely d^2 elements. That is, for each d, does there exist a set of d^2 projectors $|b\rangle\langle b|$ and a positive number g such that

$$g \sum_{b=1}^{d^2} |b\rangle\langle b| = I? \tag{8.4}$$

We know that there does exist such a set if $d = 2$ or $d = 3$. When $d = 2$, just take any four states corresponding to the vertices of a regular tetrahedron on the Bloch sphere. For the case $d = 3$, Bill has explicitly worked out an example that perhaps he can remind us of. Also I remember some vague murmurings by Armin Uhlmann that "of course" they exist in all dimensions. But still, we should treat the existence in all d as an open question – I'm not sure how much of Armin's talk was statement of known fact, how much was conjecture, and how much was faith.

If such a set exists always, then we can ask of it precisely the same question that Bill did in his thesis. Assume we don't yet know the quantum probability law: we only know that there is some function f for which the probability p_b is given by

$$p_b = f\left(|\langle \psi | b \rangle|^2\right), \tag{8.5}$$

when the system's "unknown" preparation is $|\psi\rangle$. What function f extremizes the information we gain about $|\psi\rangle$ when we have only one copy of the system available? What function f extremizes the information when we have a very large number of copies available?

What does this have to do with my manifesto? Perhaps nothing. But I have always felt that Bill's attempt at derivation was missing something in that nowhere in it did it make use of the idea that quantum measurements are invasive beasts: it talked about information, but it didn't talk about disturbance. If Bill's derivation does turn out to work nicely by the addition of POVMs, then maybe that will be some motivation for me to rethink my ish-ism.

Problem #3: Post-Gleason, or Should I Think von Neumann Is Special?

For a long time, I have disliked the tyranny of thinking of von Neumann measurements as more fundamental than other POVMs. Here's a question that might break some of that orthodoxy.

Suppose such a set of informationally complete POVMs as described in the last problem exists. Let us think of the class of all "primitive" measurements on $\mathcal{H}_d$ as those that can be gotten from acting the unitary group on that set. For instance, for a single qubit, the primitive measurements would correspond to all possible regular tetrahedra drawn on the Bloch sphere.

Let us now imagine a notion of "frame function" as in Problem #1 for these kinds of measurement.

Question 1: Is there a Gleason-like theorem for these structures? And, in particular, does the extra freedom of having d^2 outcomes to play with simplify the proof of Gleason's result?

Question 2: By use of the Church of the Larger Hilbert space, can we construct an arbitrary POVM with this notion of primitive measurement? That is, is there a kind of Neumark extension theorem for this notion of measurement?

Problem #4: Where Did Bayes Go?

From my point of view, quantum states are best interpreted as states of knowledge, not states of nature. Quantum mechanics is **mostly** a "law of thought" in that it provides a firm method of reasoning and making probabilistic estimates in light of the fundamental **physical** situation that the world is "sensitive to our touch."

With that in mind, I have to ask myself why doesn't wavefunction collapse look more like Bayes' rule for updating probabilities under the acquisition of new information. Recall Bayes' rule for when we acquire some data D about a hypothesis H:

$$P(H|D) = \frac{P(H)P(D|H)}{P(D)}. \tag{8.6}$$

On the other hand, when we perform an efficient POVM $\{E_b\}$ and find outcome b, we should update our quantum state ρ according to

$$\rho \longrightarrow \tilde{\rho}_b = \frac{1}{p_b} U_b E_b^{1/2} \rho E_b^{1/2} U_b^{\dagger}, \tag{8.7}$$

where U_b is some unitary operator and $p_b = \mathrm{tr}\rho E_b$.

Forgetting about the unitary U_b for the present discussion, notice the difference in expression of these two "collapse" rules. Bayes' rule involves states of knowledge alone: it is constructed solely of probabilities. Quantum collapse, on the other hand, appears to involve two distinct kinds of entity: density operators and POVMs. Can we put it into a form more reminiscent of Bayes' rule and perhaps learn something in the process.

Here's one way that I think might be fruitful. For each density operator ρ and each POVM $\{E_b\}$, we can construct a canonical decomposition or refinement of ρ: just multiply the equation $I = \sum E_b$ from the left and the right by $\rho^{1/2}$. We get,

$$\rho = \sum_b p_b \rho_b, \tag{8.8}$$

where

$$\rho_b = \frac{1}{p_b} \rho^{1/2} E_b \rho^{1/2}. \tag{8.9}$$

Note that with this, and just a little bit of algebra, we can rewrite the collapse rule (again forgetting about the U_b) to be

$$\tilde{\rho}_b = \rho^{-1/2} \left(\rho^{1/2} \sqrt{\rho^{-1/2} \rho_b \rho^{-1/2}} \rho^{1/2} \right)^2 \rho^{-1/2}. \tag{8.10}$$

This expression is, I think, quite intriguing. This is because it turns out that the quantity in the large parentheses above,

$$G(\rho_b, \rho) \equiv \rho^{1/2} \sqrt{\rho^{-1/2} \rho_b \rho^{-1/2}} \rho^{1/2} \tag{8.11}$$

has been characterized independently in the mathematical literature before. It appears to be the most natural generalization of the notion of "geometric mean" from positive numbers to positive operators. Here are some references:

1. W. Pusz and S. L. Woronowicz, "Functional Calculus for Sesquilinear Forms and the Purification Map," *Rep. Math. Phys.* **8**, 159–170 (1975).
2. T. Ando, "Concavity of Certain Maps on Positive Definite Matrices and Applications to Hadamard Products," *Lin. Alg. App.* **26**, 203–241 (1979).
3. F. Kubo and T. Ando, "Means of Positive Linear Operators," *Mathematische Annalen* **246**, 205–224 (1980).
4. M. Fiedler and V. Pták, "A New Positive Definite Geometric Mean of Two Positive Definite Matrices," *Lin. Alg. App.* **251**, 1–20 (1997).

Can someone bring these references? As you know, my copies don't exist anymore. (If you only have time to copy one, perhaps the last one is best … as it gives a large summary of all things known. Jozsa may be most interested in the first reference by Pusz and Woronowicz. Kubo and Ando would be the third most useful.)

Something fun to do is to rewrite the classical Bayes' rule in a similar form as this revamped quantum rule. Indeed the standard geometric mean crops up in precisely this way. Now try to further the analogy if you can.

One possible fact I seem to recall is that the operator geometric mean can be character-ized in the following way. Start with ρ_b and ρ and consider the following matrix, where X is also a positive semidefinite matrix:

$$\begin{pmatrix} \rho & X \\ X & \rho_b \end{pmatrix}. \tag{8.12}$$

Then $X = G(\rho_b, \rho)$ is the matrix that maximizes the above matrix in the sense of making it as large as it can be in the standard matrix ordering sense.

Can we learn something about why the collapse rule takes the form it does from this exercise. (Perhaps Jeff Bub can give us an introduction to his other characterization of the Lüders collapse rule.)

B. Some Not-So Concrete Problems

Proto-Problem #1: Computing Power vs. Error Correctability

We know (suspect) that quantum computing gives us a speed up over classical computing. But we also know that we have to strain slightly harder to get error correction and fault tolerance for it. Imagine now the set of all computational models (whatever that might mean). Within that set will be both classical computing and quantum computing, but also a lot of other things. Could it be the case that quantum computing hits some kind of happy medium, the one where the ratio of speed-up to error correction resources is best? (You can see that this is directly motivated by my parable: could the speed-up of quantum computing be due to this world's wonderful sensitivity to our touch?)

Proto-Problem #2: Entanglement Monogamy and Schrödinger's Insight

Can we think of a way of viewing quantum entanglement as a secondary effect? The primary effect being information–disturbance tradeoff. Let me just cut and paste an old email here. [See the note to Todd Brun, titled "Information Theoretic Entanglement," dated 8 June 1999 (other letters), and the note to Howard Barnum, titled "It's All About Schmoz," dated 30 August 1999.]

Proto-Problem #3: Down with Beables, Up with Dingables

Quantum logicians (and presumably Jeff Bub), like to think that quantum mechanics is about "what is," i.e., beables not observables (a phrase coined by Bell). The only way they can do that is by introducing a kind of logic about the "facts of the world" that is different from our usual logic of AND, OR, and NOT. I, on the other hand, like to think that that kind of change of logic is a strong indicator that we just can't get a notion of a "free-standing reality" within physics.

So the question on my mind is what kind of algebraic structure (if any) does my parable indicate/motivate for a purely algebraic approach. I've already said, I don't think AND makes any sense at all in such a world. Do the other notions from standard logic still make sense though?

I realize, I've left too much vague in this question. Talk to me at the meeting.

Proto-Problem #4: Intrinsic Characterization of Complete Positivity

I really don't like the Church of the Larger Hilbert space. Let's see how far we can get toward CPMs and POVMs without ever having to assume it. Is there a physically motivated criterion for CPMs that makes no reference to the Church.

Proto-Problem #5: Doing Gleason with Algebraic Numbers

Problem motivated by Bennett remark. Recently Pitowsky and Meyer have raised some interesting questions about Gleason's theorem when the number field of quantum mechanics is restricted to the rationals. It might be useful for our understanding of QM to ferret out what is essential and what is not in its formulation. Cabello and Peres completely discount Pitowsky and Meyer because their world has no superposition principle. But who cares? Well, maybe if Gleason still works in the minimal world with a superposition principle (i.e., algebraic number fields), we should think harder about the meaning of the "superposition principle."

Proto-Problem #6: Challenge to Everettistas

- Would your interpretation still work if the number field of $\mathcal{H}_d$ were the reals instead of the complexes? If it were the rationals instead?
- Would your interpretation still work if the time evolution of the universe as a whole were nonlinear instead of unitary?
- Would your interpretation still work if the collapse rule of QM were anything different from the standard one?
- Would your interpretation still work if _____? [You fill in the blank, challenge yourself.]

I understand that I'm being belligerent, but I *suspect* your answers to each of these will be "yes." Cf. Any of David Lewis's philosophical works on the modal logics and the plurality of worlds, or Max Tegmark's paper `gr-qc/9704009`, "Is the 'theory of everything' merely the ultimate ensemble theory?" The tentative conclusion I draw from this is that the Everettista has a contentless interpretation.

17 June 2000, and to Montréal Meeting participants, "Memories of Montréal"

I hope you all had a nice and thought-filled time at our little meeting in Montréal last month. I know that I did: in a way it was a dream come true for me. I was especially pleased with all the interaction I got a chance to see – Bill and Ben working out a set of foil theories with which to compare quantum mechanics ... David and Rüdiger finding some common ground between Peierls' remarks and the Dutch book argument ... Rüdiger, Ben, and I concocting a Bayesian argument for the linearity of time evolutions ... Charlie cautiously admitting a tear in his Everettista fatigues because of Ben's talk ... and so on. (I'll place the full list of talks below.)

Perhaps I'm just being mystical, but I feel strongly that we'll find the greatest things (and technologies) to come out of quantum mechanics when we finally grasp the parts of it that make us feel the most uncomfortable. The theory is begging us to ask something new and profound of Nature.

As most of you know – I think everyone but Charlie and Herb, that is – there was a general consensus at the end of the meeting that all the participants should write up their questions and distribute them to the rest of the crowd. This was considered to be a first step toward the writing of a set of articles for a special issue of *SHPMP*. (*SHPMP* stands for the journal *Studies in History and Philosophy of Modern Physics* which Jeremy Butterfield has kindly offered as a home to our articles. Also it turns out that Jeff and David are on its editorial board.) I hope everyone who hasn't started that project will do so soon, before the magic of the meeting starts to fade from your soul. If we could churn some articles out by the end of the summer that would be great.

So mainly I send this note out as a reminder to all.

Workshop on Quantum Foundations in the Light of Quantum Information

Centre de Recherches Mathématiques Université de Montréal

16–19 May 2000

Gilles Brassard and Christopher A. Fuchs, organizers

17 May

Gilles Brassard, Université de Montréal
Quantum Foundations in the Light of Quantum Cryptography

Christopher A. Fuchs, Los Alamos National Laboratory
Quantum Foundations in the Light of Quantum Information

William K. Wootters, Williams College
Quantum Mechanics from Distinguishability?

Benjamin W. Schumacher, Kenyon College
Doubting Everett

18 May

N. David Mermin, Cornell University
Pre-Gleason, Post-Peierls & Compumentarity

Herbert J. Bernstein, Hampshire College
Why Quantum Mechanics?

Richard Jozsa, University of Bristol
Foundations of an Interpretation of Quantum Mechanics in the Light of Quantum Computing

Charles H. Bennett, IBM Research at Yorktown Heights
Entanglement-Assisted Remote State Preparation

<u>19 May</u>

Rüdiger Schack, University of London – Royal Holloway
    Quantum Gambling and Bayesian Probability in Quantum Mechanics

Patrick M. Hayden, Oxford University
    Two Lessons from the Heisenberg Representation, or
    How I Learned to Stop Worrying & Love Non-Commutativity

Jeffrey Bub, University of Maryland
    Some Reflections on Quantum Logic

26 July 2000, "Interaction without Interactoids?"

Remember my excitement in Mykonos about how the question "What is real about a quantum system?" should be answered with the reply "The ZING! that gives rise to quantum key distribution. That is, what is real is nothing more than the full set of information–disturbance curves." This is probably wacky, but now I'm having dreams of connecting up that idea with gravity. That the quantum "sensitivity to touch," the ZING!, might be a kind of energy in its own right, and therefore couple to gravity. Like I said, kind of wacky, but it has me extremely excited.

9

Letters to Jeffrey Bub

03 April 2000, "Foundations in the Light of Information"

Below is a slightly modified version of the invitation letter I sent everyone. (The only change is an update on the attendance.) Looking forward to seeing you again. (BTW, I hope you'll take the problems list seriously.)

Dear friends,

Gilles Brassard and I are organizing a little meeting in Montréal from Wednesday May 17 through Friday May 19 on the subject of "Quantum Foundations in the Light of Quantum Information and Cryptography." We hope you can come. Mainly the purpose of the meeting is to gather a group of our friends who think that some aspects of quantum foundations are a little more mysterious than they ought to be and, importantly, are intrigued by the idea of applying quantum information to the task of cleaning things up. The atmosphere of the meeting should be quite relaxed with plenty of time for discussion and/or private brooding: we only ask that you be prepared to give a 20 to 30 minute talk (though longer talks are welcome too) expressing your point of view and any avenues for progress you foresee. Attendants are also encouraged to compile and share a list of concrete problems whose solution would tell us something novel about the quantum foundations.

A lucky few attendees will stay in the Rockledge Apartments near the University, where one living room will be made available for some of the more informal discussions. (This venue has been quite productive in the past: it helped give rise to quantum teleportation, for instance.) The remainder of the attendees will stay at Hotel Chateau Versailles in the downtown area (a short bus ride from the university). Attendants may also consider a Saturday night stay-over May 20 for saving on airfare and perhaps talking some more. Living expenses will be provided. Travel assistance may be available for those in need.

The list of potential attendees is presently this:

Invitee	Status	Known Orientation
1) Charles Bennett	(confirmed)	Many-Worlds-ish
2) Herb Bernstein	(confirmed)	Copenhagen-ish
3) Gilles Brassard	(confirmed)	-ish
4) Jeffrey Bub	(confirmed)	Quantum-Logic-ish
5) Chris Fuchs	(confirmed)	Quantum-Mechanics-is-a-Law-of-Thought-in-a-World-Sensitive-to-the-Touch-ish
6) Lucien Hardy	(confirmed)	Bohm-ish
7) Patrick Hayden	(confirmed)	Local-Action-ish
8) Richard Jozsa	(confirmed)	Nonlocal-Action-ish
9) David Mermin	(confirmed)	Correlation-without-Correlata-ish
10) Ben Schumacher	(confirmed)	Terribly-Unhappy-ish
11) Umesh Vazirani	??	??
12) Bill Wootters	(confirmed)	Wavefunction-Straddles-the-Fence-between-Objective-and-Subjective-ish
13) Arthur Zajonc	??	Light-of-Nature/Light-of-Mind-ish

Please let us know at your earliest convenience whether you can attend.

Best regards,

Chris and Gilles

30 September 2000, "Commit the Bit!"

Bubism 1: *I downloaded your mammoth 'Notes on a Paulian Idea' and have been dipping into it. Really fascinating stuff. I resonate to your idea that we live in a world where there are some funny constraints on gathering and communicating information – and that this ultimately involves an implicit ontological claim. The world is like THIS and not like that (say, a classical world). This raises the question of what kind of science we can have in such a world. We presumably have the science – quantum mechanics – but the issue is what attitude we should have to this science, how we should interpret it, given the constraints on information. This puts the interpretation problem in a new light. I might even change my mind about some things.*

Have you had a chance to think about the emails I sent you at Gilles' urging?

Bubism 2: [From a letter originally to Gilles Brassard.] *I'm sorry if I've been instrumental in publicizing this conjecture [. . .], but I do think that this is too intriguing a*

conjecture to keep under wraps. I think there's a good chance the conjecture is correct (just a gut feeling again), and if not then it will likely turn out that what you want is `key distribution + no bit commitment + X`*, where this X is something well-defined and interesting. To me, this is what is so fascinating about the work on quantum information and quantum cryptography: the foundational significance of the results. In fact, your conjecture – whether or not it's correct, or just close to correct – is the reason people interested in foundations of physics should be interested in bit commitment, and it provides the rationale for submitting the paper to Foundations of Physics. So I think if I cut this out of the paper, it's a really big cut. I can easily replace the paper with a new version, but I can't force myself to stop thinking about key distribution and bit commitment in the light of this conjecture. Of course, you are not asking me to do that, but it feels rather odd for me to censor the paper in this way and not mention something that I think is crucial. I think the idea should be out there for people to think about – and realistically, there's no way to keep the idea within a small closed circle. So I do hope you'll change your mind.*
[…]

I should also mention that I have been invited to give a number of talks this semester: in Spain the beginning of October (a talk in a 'magisterial session,' whatever that means, that is a tribute to John Bell), at Johns Hopkins and at Yale in November. I have also organized a symposium on quantum information and quantum computation at the biennial Philosophy of Science Association meeting in Montréal the first week of November. In all of these talks I will be talking about quantum entanglement, nonlocality, and the significance of the work on quantum cryptography, and I had intended bringing in your conjecture. So I would have to censor these talks as well, and it will feel something like telling a joke without the punch line. This is what makes the subject interesting philosophically.

I've finally had a chance to read all the notes you sent me. I was especially pleased to learn how seriously you take the idea Gilles and I wish to see pursued. I really don't have a problem with you mentioning our speculation in your *Foundations of Physics* paper or in your upcoming talks. The idea needs a significant amount of refinement before anything can come of it, and probably the only way that will ever happen is to have more of our community thinking about it and participating in an open semi-random walk to the truth. So, with Gilles' countenance, I would encourage you to do just what you had planned.

Now, how to cite it properly? I've been spouting the half about "the existence of quantum key distribution being part of the foundations of quantum mechanics" since I was working for Gilles as a postdoc in 1996. I've been able to find at least two relevant passages in the Paulian-idea samizdat: 10–16 December 1997 in a letter to David Mermin and 6 July 1998 in a letter to Rolf Landauer. You can cite that document and list my website as the source: I'll have a Bell Labs website by the end of the week if you can wait that long. [By the way, prompted by the fire, I've been toying with the idea of posting an extended version of the samizdat on `quant-ph` itself. Especially, if I can get someone respectable like David

Mermin to legitimate it with a Foreword. Do you have any opinion on this silly idea?] Also though, I've got three papers that allude to the idea. Here they are:

C. A. Fuchs, "Information Gain vs. State Disturbance in Quantum Theory," *Fortschritte der Physik* **46**(4,5), 535–565 (1998). [Reprinted in *Quantum Computation: Where Do We Want to Go Tomorrow?*, edited by S. L. Braunstein (Wiley–VCH Verlag, Weinheim, 1999), pages 229–259.]

See the last section in particular. (That was actually placed on quant-ph in 1996; it took a long time to appear properly.) Also,

C. A. Fuchs, "Just *Two* Nonorthogonal Quantum States," in *Quantum Communication, Computing, and Measurement 2*, edited by P. Kumar, G. M. D'Ariano, and O. Hirota (Kluwer, Dordrecht, 2000), pages 11–16.

And, finally, for the most thorough discussion:

C. A. Fuchs and K. Jacobs, "An Information Tradeoff Relation for Finite-Strength Quantum Measurements," submitted to *Physical Review A*. See quant-ph/0009101.

To the best of my recollection, Gilles first mentioned the half about bit commitment to me at the Newton Institute workshop in Cambridge last summer as we were discussing these earlier ideas. Some time after that I had a fairly stirring epiphany when I realized that I *really* did want to incorporate Gilles' idea into my program. I'll dig up the note explaining that and forward it to you in a minute.

It is a pretty speculation, isn't it? I don't think Gilles really has anything to worry about: even if you do manage to get other people thinking about it, I think it'll be a slow process to settle the issue. We'll have plenty of chance to make a technical contribution if we so desire.

Jeff's Reply

Just a quick note to say thanks for your input on this, and for the various references. I've just returned from Spain and have a backlog of things to take care of for tomorrow.

Re your query about posting the samizdat on quant-ph, I think that's an excellent idea. I must say your question: 'What if God made the world so that information gathering and exchange are characterized by a specific sort of limited privacy (= the quantum world)?' is really resonating with me.

I now see my book as showing that if you want to interpret quantum mechanics as giving you an ontology with a "detached observer," to use Pauli's phrase, then that ontology will be like Bohm's theory, in the sense that it will (i) violate Lorentz invariance, and (ii) involve a mechanism that will prevent you from seeing the violation. But this would violate Einstein's dictum: "God is subtle, but not malicious." There would be physically different situations associated with alternative space-like separated measurements that would be in principle empirically inaccessible. (In Bohm's theory this would happen because you could never localize a Bohmian particle more precisely than the localization given by a quantum state. So you could never control the particle position precisely enough to instantiate the two different situations.)

So it seems to me that you have two ways to go. You could say that the way to understand quantum mechanics is "like Bohm's theory" (in a sense that would include modal interpretations and a bunch of other things, but not "collapse" theories), or you could say that the ontology is "veiled" (d'Espagnat's term) because of something about the way in which information works.

Now one might say that God is just as malicious this way, in "veiling" the ontology. But my sense is that the "information" view is really different. To continue with the theological metaphor, there would seem to be no point to the Bohmian sort of "veiling," which then becomes malicious. Or putting it differently, the question "why should it be like that?" has no interesting answer. The difference between the "Bohmian" view and the "information" view is something like the difference between the Lorentz-FitzGerald theory (in which you have real spatial contraction because of motion through the ether, but you can never measure this) and relativity.

So the interesting question (philosophically) now becomes: What sort of science can we have if the world is like this, i.e., if the world has the feature that information gathering and exchange are constrained in a certain way? This makes the Paulian view you push in the samizdat seem more sophisticated than the Bohmian or Einsteinian alternative.

My gut feeling was always with Einstein and against Pauli as far as the "detached observer" issue is concerned. That is, it always seemed to me that you should look for a physics in which the observer is irrelevant. So if you end up with a physics in which there appears to be an essential reference to the observer, then you are on the wrong track and have to think again. But what if the way the world is put together means that the kind of science possible is restricted in some way because of constraints on information? That strikes me as a perfectly good ontological (= observer-free) possibility – it has a sort of Kantian flavour, but with a new twist.

30 October 2000, "Curiosity"

I was in a bit of a conversation with Richard Jozsa and I forwarded to him your email from October 9 where you said that the Paulian idea is "resonating" with you. This was his reply:

Those remarks by Bub are very interesting. I think I would actually go along with much of what he says in spirit, but I cannot accept your view (if I am representing you not incorrectly) that physics is about the state of knowledge of an observer, rather than about an "objective" reality. I would interpret Bub's use of the term "information" as corresponding to something objective too, in contrast to just the state of knowledge of some observer.

Now I'm just curious to know how you react to this, especially his interpretation of your use of the term "information." My own (snippy) reaction to Richard was this:

Somehow you always miss the point. I thought Jeff expressed what I'm seeking quite adequately: the point is that deep within quantum theory we can hope to find an ontological statement. But that direct ontological statement will refer to our interface with the world: the world is wired in such and such a way that the surface terms in the theory (the density operators) can only refer to our subjective states of knowledge. That does not negate that there is a real world out there and that we strive to say useful things about it.

Jeff's Reply

Here's what I would say. The position I'm moving to (following reading your Paulian manifesto and re-thinking some things) is this:

It may be that our universe (i.e., objective reality) is structured in such a way that there are certain constraints on gathering and communicating information. These constraints seem to be expressed as a limitation on privacy (the possibility of secure key distribution and the impossibility of secure bit commitment, if you and Gilles are right). I take this as a perfectly objective statement about the ontology of our universe: ontologically, the situation is such that information acquisition and communication are constrained in a certain way.

Now, if this is the case, then we have to ask the question about what sort of science we can have in such a world. That is, there are constraints on epistemology, given that the ontology is of such and such a sort. (There's a very Kantian flavour to this, by the way.) I would now say that the science we can have in such a world is a science about the sort of information we can have, and how this information can be moved around.

Classical physics can be interpreted 'descriptively,' as saying what there is in the world, what sorts of physical system there are, what their properties are, and how these change as the systems interact. This is descriptive in an observer-free sense – the whole story can be told without mentioning an observer.

Now, I always thought that it was important to keep to the same observer-free format in formulating quantum mechanics. In my book I showed that if you want to do this with quantum mechanics, then you are going to end up with something like Bohmian mechanics. That means that the theory will not be Lorentz invariant, but it will incorporate a mechanism for hiding this: empirically, you won't be able to see the violation of relativity. I think this is unsatisfactory because it makes God malicious and not subtle. (You recall Einstein's quote: "God is subtle, but not malicious.") Bohmian quantum mechanics is to standard quantum mechanics like the Lorentz–FitzGerald theory is to special relativity.

But it may be that the way quantum mechanics is formulated, in terms of a notion of state that refers to preparation and measurement – our manipulations of the world and the information we obtain from such manipulations – rather than in terms of a notion of state that specifies a list of properties at a particular time as in classical mechanics, is just the right way to formulate a physical theory because of the way the world is wired, as you put it. It may be that an observer-free physics is not an option for our world because of the way the world is wired. So we have to re-think what a fundamental, objective theory of mechanics should look like if the world is wired in this way, how it should be formulated and what it should be about. I would say that it should be about the sort of information we can have, and how this information can be moved around.

I think all this is perfectly "objective." It's just that the nature of objective science has to be re-thought if the ontology has this sort of structure (involving constraints on information).

10 December 2000, "The Oyster and the Quantum"

I have been terribly rude to you and I hope you will accept my apology! When someone else takes one's ideas seriously (as you have done with mine), the last thing one should do is return only silence!! More seriously, there was a time when I welcomed and even craved letters like the last one you sent me. But since arriving at Bell Labs, the duties and the travels have been falling upon me so quickly that I am afraid I am getting buried in an avalanche. I haven't been able to reply to *any* of the more thoughtful emails in my mailbox.

But, all that said, thank you for your email! Let me try to make up a little for my silence now. I am in the last few days of a visit to Vienna, and I think I may have finally wrangled my delinquent email account back to a manageable level. So, if you reply to the present note, I will do my best to return a prompt response. (It is most helpful that the Christmas holidays are coming up soon.)

Where do I begin? First off, I don't know that I can say what I'm trying to get at any more cleanly than in the little blurb "Genesis and the Quantum" I placed in my problem set for the Montréal conference. The only thing that competes with that might be the Introduction I wrote for `quant-ph/0009101` "Information Tradeoff Relations for Finite-Strength Quantum Measurements" and the important footnote 27 therein. If you haven't read that Introduction, please try to have a look at it.

Maybe one further stab at it comes from the silly abstract I wrote for the Caltech meeting last month.

Title: The Oyster and the Quantum

Abstract: I say no interpretation of quantum mechanics is worth its salt unless it raises as many technical questions as it answers philosophical ones. In this talk, I hope to convey the essence of a salty, if not downright briny, point of view about quantum theory: The deepest truth of quantum information and computing is that our world is a world wildly sensitive to the touch. When we irritate it in the right way, the result is a pearl. The speculation is that this sensitivity alone gives rise to the whole show, with the quantum calculus portraying the best agreement we can come to in such a world. True to form, I will ask more questions than I know how to answer. However, along the way, I will give a variant of Gleason's theorem that works even for rational Hilbert spaces – even two-dimensional ones – and prove some constraints on how quantum states change under measurement.

But let me try to clarify that further by explicitly addressing what you wrote me.

Bubism 3: *I have a question about your conjecture that quantum mechanics is equivalent to (can be derived from?) the possibility of secure key distribution and the impossibility of secure bit commitment.*

Secure key distribution makes it possible for two people to communicate privately, in the sense that they can discover any attempt to eavesdrop on their private communication. As I understand you, you see the impossibility of secure bit commitment as a limitation on this privacy – a limitation that makes science possible.

What I don't get is how this makes science possible. What is the connection?

Most importantly, I would warn that you not take the statement "QKD + NQBC = QM" too literally. The ingredients I see as crucial to the endeavor are instead: (1) That none of us can ever completely hide the effects of our interactions with the world. There are no lock boxes for information. Or, in the conception of `quant-ph/0009101`, every agent in the world has it within his power to partially align his state of knowledge about a system with any other agent's (even if only ever so slightly). And (2) that there are ways to write information onto a system so that any surreptitious gathering of the information will necessarily decrease the writer's predictability about how the system will react to his further interactions with it.

Item (1) sounds a lot like the nonexistence of bit commitment. Item (2) sounds a lot like the existence of QKD. But I don't know that I would really carry it any more literally than that (though I have certainly contemplated it). The main point is that in order for science to exist we have to be able to come into alignment – at least to some extent – concerning our predictions on how systems will react to our further interactions with them. On the other hand, the existence of the information–disturbance tradeoff (especially in quantum key distribution protocols) expresses that there is a limit to the extent with which we can come into alignment. The latter effect has to be generally weaker than the former effect or we would be sunk in the game of science.

That is very roughly the idea, and I just don't know how to make it more precise than that yet. For whatever reason, however, I do have a tremendous faith that this is at least the right direction with which to chip away at the problem of understanding the quantum world.

What I want to know, for instance, is WHY the information–disturbance curve in my paper with Asher [*PRA* **53**, 2038 (1996)] doesn't rise more steeply than it does in quantum mechanics. If it did, would we not be able to have science? Or, look at the property I proved in Appendix A of quant-ph/0009101. This says that for each quantum measurement (i.e., positive operator valued measure), there is a way of performing the measurement so that on average I will be left with more predictability about how the system will react to further interventions upon it. It need not have been so. One could imagine other worlds. What goes wrong in those worlds?

Bubism 4: *It has just occurred to me that I may have misunderstood what you have in mind by a "limited privacy."*

I don't know that I ever used the term "limited privacy." I think you were the one who has always emphasized that phrase (and probably also invented it).

Bubism 5: *The way in which private communication is secured in our quantum world is not this way. Rather, Alice and Bob can communicate privately by exploiting the fact that an eavesdropper can always be detected, to any level of security. Now this feature of the world depends on such things as "no cloning" and "no information gain without disturbance." But these features essentially (?) involve entanglement. If you have entanglement, then you can't have bit commitment.*

I don't know how much these features really require entanglement *per se*. Somehow I imagine that one need only deal with strictly weaker features of the tensor product structure in order to see them arise. See for instance our paper "Nonlocality without Entanglement" [*PRA* **59**, 1070 (1999)]. Here is a set of states that have a novel information–disturbance tradeoff in them with respect to local observers even though they are orthogonal and have no entanglement.

Moreover, going back to the scenario that Asher and I explored. Imagine that we equip quantum mechanics with a kind of discrete time evolution that never entangles systems; it always leaves them in product states (but is otherwise inner-product preserving at those discrete time steps). With it one would have a no-cloning theorem and an information–disturbance principle to boot. But never any entanglement.

Overall, I'm inclined to think that the key ingredient is the noncommutivity of states. It seems to me that entanglement is only a derivative concept. (I can actually make this notion precise – especially within this context – but you're going to have to wait a few weeks for the new paper I'm writing up.)

Bubism 6: *You want to take quantum mechanics as a theory of the way information is represented and the limitations on the communication of information, and not a description of the behavior of particles, as in classical mechanics. Granted, the way the world is hard-wired might impose limitations on the gathering of information and the exchange of information – limitations expressed precisely by the "limited sort of privacy" we have (i.e., secure kd, but no secure bc), hence by science necessarily taking the form of quantum mechanics. But how does it follow from this that we must interpret quantum mechanics as a theory of information, and not as a descriptive theory in the sense of classical mechanics?*

It doesn't. You're completely on track there. Do you remember my slide where I listed the axioms of quantum mechanics in Montréal? In my presentation I said how I'm always struck by the stark contrast between that list of axioms and the ones we take for our other cornerstone theory of the world (referring to special relativity): (1) the speed of light is constant, and (2) physics is the same in all frames. The debate over the foundations of quantum mechanics will not end until we can reduce the theory to such a set of crisp physical statements – I believe that with all my heart. However, just as special relativity will always be interpretable in Lorentz's way, quantum mechanics will likely always be interpretable in Bohm's way. There's nothing we can do about that.

What I'm really searching for is just a polite way to say, "Ahh, blow it out your butt. You can believe that Lorentzian way of looking at things if you want to, but why when you have this absolutely simple alternative conceptual structure?"

What I'm looking for is just a couple of crisp physical statements – contradictory appearing even, just as Einstein's – that can characterize what quantum mechanics is all about. Something like (but more precise than):

1. The effects of our interventions into the world are nondiminishable. And
2. But still we have science; all the world is not simply a dream of our own concoction.

Once we get that cleared up, we'll finally be ready to move to the next stage of physics, much like Einstein was ready to move on to general relativity once he had reduced the Lorentz contractions to the two statements above.

Jeff's Reply

Thanks for your long letter. I've printed out several of your papers that I want to study. I'm particularly intrigued by your 'Nonlocality Without Entanglement' paper. I'll get back to you shortly with some observations – I've been snowed under with end of term duties.

I understand what you say about QKD and information trade-offs, and I think there is definitely something deep and important there. But I'm still a bit puzzled by what you say about bit commitment.

You'll probably be surprised to hear me saying that I now agree entirely with your "blow it out your butt" comment on Bohm and modal interpretations, and your reference to Lorentz here. I don't think these approaches shed any light at all on what's going on in quantum cryptography and quantum computation – in fact, they obscure things, as far as I can see.

11 January 2001, "Thanks"

Thanks a million for [...]

So, let me reward your kindness with the pleasure of a question. I wonder whether you've thought about this before. Suppose I have two noncolorable sets of rays (in the sense of Kochen and Specker). One set lives in one vector space and the other lives in another. The dimensions of these two spaces can be anything you like, say d and f. Suppose also there are m rays in one set and n rays in the other. Now imagine forming all possible tensor product rays from the two sets. What I mean by this, really, is to form the set of all tensor products of projection operators and then to turn those into rays. Anyway, there will be mn such rays, and they will live in a df-dimensional vector space.

The question is this. Can we say anything about the colorability of such a set? That is, more prosaically, is the tensor product of two noncolorable sets also noncolorable? Or, are there counterexamples? Conversely, one can ask is the tensor product of two colorable sets always colorable? I haven't thought very hard about these questions, so they may be trivial to answer, or they may be hard. The main point is, what can one say about the "interaction" between KS theorems and the tensor-product structure of quantum mechanics?

14 January 2001, "Barnum and KS"

Bubism 7: *I think the answer to your question is: yes. I say this on the basis of the following reasoning (so if there is a flaw in the reasoning, then I'm not sure):*

I think your reasoning is sound. Thanks. I'm just playing with several ideas in my head about how to better pose a question about what connection (if any) there is between KS theorems and the existence of an information–disturbance tradeoff. And the question I asked you came up somewhere in the middle of that. What I'm grappling for is something like the following. (Before I go further, though, please, please forgive my vagueness.) Presumably I can construct various *noncontextual* hidden variable theories with an information–disturbance tradeoff property. BUT, we know that these noncontextual theories CANNOT match quantum mechanics in every detail. So the question is this, what general statements can be made of their information–disturbance properties? The guess of course in my mind is that they will not be able to match quantum mechanics in ALL I–D properties. And that is because I hope sincerely that quantum mechanics is nothing other than the conjunction of all its I–D properties.

18 January 2001, "The Short Answer"

OK, so I won't wait for the weekend. Let me just try to give you a short-ish answer and see if I can get away with it.

Bubism 8: *What I'm really saying is this: You have embeddability in the 2-D special case. But you still have the feature of an information–disturbance tradeoff. So perhaps this is not all there is to quantum mechanics in the general case. Or perhaps in the general case the information–disturbance tradeoff is qualitatively different from the situation in the 2-D case.*

Indeed, I have quite understood that the 2-D case might be a special one. I'm sorry I didn't mention that in my note to you. And, in fact, if what I were trying to get at didn't start until $d = 3$, I don't think I would be so bothered. Remember the fascination I have for Gleason's theorem? So, it might appear that there is something essential about quantum mechanics that does not kick in until $d = 3$. In other words, I might also bank on that to fulfill my vague information–disturbance thoughts.

But that's the cheap answer. A deeper answer relies (1) on some partially unpublished work, and (2) on some terrain in quantum information that you're probably not all that familiar with (yet). And the combination of these two things is going to make it difficult to talk about this in email. Let me give you the very roughest of sketches, enough maybe to whet your interest, and then I'm going to have to leave it at that for now.

The main point of departure is that whenever the words "information" and "disturbance" are used, they should not be used in the fashion you did in your previous note to me. Instead one should harken back to the quantum cryptographic usage of the terms. I tried to lay this out very carefully in my recent paper with Jacobs. (It's all in the Introduction; so as far as this point is concerned, all you need to read is the Introduction.) The next step in the process is to realize, as Asher and I did in our *PRA* '96, that the optimal information–disturbance tradeoff comes about only by a POVM-equipped Eve. For a given amount of information, projective measurements are generally too harsh.

So what? Well, you also have to remember that I was never imbued with the idea that projective measurements are somehow more fundamental than general POVMs. That's because it never seemed convincing to me that measurements reveal "properties" and, thus, that measurements should obey the algebraic properties one would deem necessary for that. So, I want to think of POVMs as every bit as fundamental as anything else. You say, "To measure a three-outcome POVM on a qubit, you MUST imagine interacting the qubit with a larger ancilla." I say, "Humbug! If you admit any mystery about standard von Neumann measurements, why lay such stock in them for the POVM case? How does adding the word 'ancilla' take any of the mystery out of POVMs? Just take them as equally fundamental and be done with it."

So, let me ask, can I always simulate arbitrary POVM measurements with a noncontextual hidden variable theory? NO, of course not (because I can't even simulate the von Neumann measurements). But what's interesting is that I cannot even do it when $d = 2$. (This last statement is the partially unpublished part of what I'm telling you. I say "partially" because it is joint work with Caves and Renes that has not been published. When I showed it off to David Mermin, however, he faintly remembered having seen something like that ... and then ultimately remembered Paul Busch. So, we came across it independently, but we three will now never get any credit for it. Too bad.)

Putting all this together, it is not obvious to me – and I'm willing to bet it's wrong – that one can even simulate the curves Asher and I found. Or, at the very least, the curves for some similarly defined, but more complicated problem. So that's the sort of direction I'm thinking in. And, I think it gives hope to the program. But as I also said earlier, it wouldn't shock me either if a fundamental change in the I–D question comes about when $d = 3$.

Did that whet your interest? Now the problem is, you're going to ask me even more questions! And I'm up to my knees in the day to day fire-fighting that has come with my new job. (Oops, I should call it a "more senior position.") So, if I end up being silent for a while on things other than organizational, please don't take offense.

10

Letters to Carlton Caves

08 April 1996, "Hartle/Coleman Comments"

I like the grass fire analogy: I agree we should stomp it out if we can. However, as with most debates, I imagine we'll just have to wait until it falls by the wayside ... or wait until someone gives it a name like the Modern Interpretation of Quantum Mechanics with Sugar on Top, so that everyone else will just discount it and move on.

Do you know whether Coleman has published any of this? Wojciech first told me about it at least a year ago (and told me I should contact Coleman concerning it to "help your career along"). If you have a copy of the paper, please send it my way. I've been at a little disadvantage in thinking about this because I cannot find my copy of Hartle's paper and the library has been shut down since Wednesday evening; I'll be able to get a new copy Tuesday morning. I know I liked the paper very much after turning semi-Bayesian because of what it said in words (even though I never agreed with the mathematics); however, it certainly did have a frequentist streak running through it ... and that's what's bringing us back now.

I think what I'll do in this note is just write down various thoughts as they come to me; I've got a few pieces of junk spread out on my desk to prod me along. I can see I'm going to repeat much that you've already written me, but I want to have it written down for myself. Later we can see what comes of any of it. (By the way, I am sure that I will split many infinitives in this letter; I hope you have the patience.)

Point 1) I think there is much to be said for funneling the problems of this point of view into their loose use and confusions of "definitely" and "with probability 1." If I understand you correctly, both Hartle and Coleman would like to take it as an axiom of quantum theory that: if a quantum system is prepared in a state $|s\rangle$ which is a normalized eigenvector of the Hermitian operator $\hat{S}$, i.e.,

$$\hat{S}|s\rangle = s|s\rangle, \tag{10.1}$$

then the eigenvalue s will *definitely* be the outcome of a measurement corresponding to $\hat{S}$. If, on the other hand, $|s\rangle$ is not an eigenstate of $\hat{S}$, then what we can say about the outcome (i.e., its probability or whatever) is yet to be determined. In particular, they would like to determine just exactly *what can be said* from the simple axiom above (the "Eigenstate Axiom").

They do this by talking (ostensibly) about what happens upon the repeated measurement of $\hat{S}$ on identical preparations $|\psi\rangle$ that are not eigenvectors of $\hat{S}$. (Let's keep things simple and suppose that $\hat{S}$ is 2×2 and its eigenvalues and eigenvectors are labelled by 0, 1 and $|0\rangle$, $|1\rangle$, respectively.) Thus they consider an infinitely large quantum system prepared in the pure state

$$|\psi^{\infty}\rangle = |\psi\rangle|\psi\rangle|\psi\rangle \cdots . \tag{10.2}$$

Presumably what they *would like* to do – by defining probability as frequency – is show that a measurement of the operator

$$\hat{S}^{\infty} = \hat{S} \otimes \hat{S} \otimes \hat{S} \otimes \cdots \tag{10.3}$$

on $|\psi^{\infty}\rangle$ can *only* yield a "completely random" string of outcomes $\vec{s} = s_1 s_2 s_3 \cdots$ with the "correct" frequency of 0s and 1s, i.e.,

$$f_0 = \lim_{n\to\infty} \frac{1}{n} \sum_{i=1}^{n} \delta_{0,s_i} = |\langle 0|\psi\rangle|^2, \tag{10.4}$$

$$f_1 = \lim_{n\to\infty} \frac{1}{n} \sum_{i=1}^{n} \delta_{1,s_i} = |\langle 1|\psi\rangle|^2. \tag{10.5}$$

However, instead of doing this, they move on to something else first. Hartle defines the "frequency" operator

$$\hat{F}_0^{(n)} = \sum_{s_1 \cdots s_n} \left(\frac{1}{n} \sum_{i=1}^{n} \delta_{0,s_i} \right) |s_1\rangle \cdots |s_n\rangle \langle s_n| \cdots \langle s_1|, \tag{10.6}$$

and similarly for $\hat{F}_1^{(n)}$. (The first summation in this is taken over all possible strings of outcomes $\vec{s}_n = s_1 s_2 \cdots s_n$ of length n.) He then lets

$$\hat{F}_0^{\infty} = \lim_{n\to\infty} \hat{F}_0^{(n)}, \tag{10.7}$$

etc. I presume – but certainly cannot know, since I have not seen the paper – Coleman defines a projector corresponding to the set of random outcome strings in the following way. Let $\mathcal{S}_m^{(n)}$ be the set of all strings $\vec{s}_n = s_1 s_2 \cdots s_n$ of length n that pass "all" randomness tests (*with respect to the probability distribution induced by a measurement of $\hat{S}$, i.e., the numbers $|\langle 0|\psi\rangle|^2$ and $|\langle 1|\psi\rangle|^2$*) at confidence level m. (There are various ways of making this notion formal; I'll come back to this a little later.) Then take

$$\hat{\Pi}_m^{(n)} = \sum_{\vec{s}_n \in \mathcal{S}_m^{(n)}} |\vec{s}_n\rangle \langle \vec{s}_n|, \tag{10.8}$$

and finally let

$$\hat{\Pi}_{\text{rand}} = \lim_{m \to \infty} \lim_{n \to \infty} \hat{\Pi}_m^{(n)}. \tag{10.9}$$

This projector should correspond to the subspace spanned by all truly "random" outcome sequences.

Let us assume, for the moment, that there are no real mathematical difficulties in defining the operators above in their infinite limits ... or, at the very least, that if there are difficulties in the definitions I gave, they can be out-maneuvered by somebody smarter than me. Hartle then goes on to show that

$$\hat{F}_0^\infty |\psi^\infty\rangle = |\langle 0|\psi\rangle|^2 \, |\psi^\infty\rangle \quad \text{and} \quad \hat{F}_1^\infty |\psi^\infty\rangle = |\langle 1|\psi\rangle|^2 \, |\psi^\infty\rangle. \tag{10.10}$$

Coleman presumably shows

$$\hat{\Pi}_{\text{rand}} |\psi^\infty\rangle = |\psi^\infty\rangle. \tag{10.11}$$

The conclusions Hartle and Coleman would like us to draw from these nifty equations is that, indeed(!), a measurement of $\hat{S}^\infty$ can *only* yield a "completely random" string of outcomes $\vec{s} = s_1 s_2 s_3 \cdots$ with the "correct" frequency of 0s and 1s. They base this on the Eigenstate Axiom even though, as far as I can see, there is no completely clean cut connection between measurements of the operator $\hat{S}^\infty$ and the constructed operators $\hat{F}_0^\infty$, $\hat{F}_1^\infty$, and $\hat{\Pi}_{\text{rand}}$.

Putting that objection aside, however, this is still a *silly* conclusion (as you've already pointed out). For just let us construct two other projectors:

$$\hat{\Pi}' = \hat{\Pi}_{\text{rand}} + |000\cdots\rangle\langle 000\cdots|, \tag{10.12}$$

$$\hat{\Pi}'' = \hat{\Pi}_{\text{rand}} - |\text{rand}\rangle\langle\text{rand}|, \tag{10.13}$$

where the first projector contains one clearly nonrandom outcome string, and the second has one component of the original $\hat{\Pi}_{\text{rand}}$ taken away. What are we to make of these? Since, surely,

$$\hat{\Pi}' |\psi^\infty\rangle = \hat{\Pi}'' |\psi^\infty\rangle = |\psi^\infty\rangle, \tag{10.14}$$

should we take it to mean that $000\cdots$ is – contrary to our earlier conclusion – a possible outcome? Or, instead, should we take it that the excluded random string in $\hat{\Pi}''$ is *not* possible after all?

The problem lies in asserting that something *definitely does happen* contingent upon its having *probability 1* in someone's mind's eye. On this, I do agree. What we have seen above is a dangerous mix of "sloppy" frequentism and confidence that quantum mechanics is about unpredictable thingies. A strict frequentist (such as von Mises or Reichenbach) would *not* base his belief in a random string occurring in an infinite measurement repetition on a probability-1 statement like the Law of Large Numbers. Rather, he would only apply probability to phenomena that are already "known" to be repeatable and which have definite limiting relative frequencies (but are lawless in pattern otherwise). That is to say, the frequentist only applies probability statements to individual terms in a random infinite sequence or *kollectiv.*

> *"Erst das Kollectiv, dann die Wahrscheinlichkeit"*
>
> (*R. von Mises*)

I used to be attracted to the idea that individual quantum measurement outcomes (as opposed to classical coin tosses) really were elements of kollectivs. Apparently so was Paul Benioff [1,2]. The main thing I learned from that silly business was that if one really wants this to be the case, then one cannot hope to find evidence for it in the present structure of quantum theory. It has to be taken as a *new* axiom – it is no more derivable from quantum mechanics than the kollectiv is derivable from a theory in which "probability" makes its appearance first. (Benioff wrote several papers – I cited two above, but there are many more – in which he tried to do just that, i.e., replace some of the standard axioms with ones about kollectivs, but I don't think anything came of it.)

Point 2)   A few things about "mathematical foppery." In two of Benioff's papers (which are about standard quantum theory instead of the stuff mentioned above) [3,4], he worries about "maverick worlds" in the Everett interpretation. To get at the questions he's concerned with, he does something that has a lot of the flavor of what Coleman seems to be talking about. Namely, he plays with the mathematics of constructing a state like $|\psi^\infty\rangle$, but which only has random outcome sequences in its expansion. He makes it pretty clear that there are a lot of difficulties in defining such a thing rigorously. If you're really worried about whether Mr. Coleman's mathematics are on the up and up, these papers are probably worth a second look.

Also in these papers, Benioff points out that the notion of a "random" string is not a God-given concept: there are several distinct notions of random string definable within the limits of the phrase "strings that pass *all* statistical tests." There is the one due to Martin-Löf, but there are also others (some of which are also due to Martin-Löf but less well-known) that essentially depend on what kinds of oracle you allow your Turing machines to have. Benioff found this intriguing because he hoped that quantum theory would give a means of defining the "true" notion of randomness. Does Coleman give any real justification for the notion of randomness he uses?

One last thing that may be of real interest for the interpretational question. You say,

Cavesism 1: *He stated his results in terms of a spin-half particle and measurements of z-spin, but he claims to have the general result, and it's not hard to believe that he does. ... where randomness is defined algorithmically (or, equivalently, in terms of all the statistical tests of randomness).*

Let us restrict ourselves to the standard Martin-Löf definition of randomness in terms of statistical tests. Then if we are talking about randomness with respect to a fair coin toss there is indeed a simple relation between Martin-Löf's definition and one in terms of algorithmic complexity, namely an infinite string $\vec{s} = s_1 s_2 \cdots s_n \cdots$ is random if and only if there exists a constant c such that for all n

$$K(s_1 s_2 \cdots s_n) \geq n - c. \tag{10.15}$$

(Here $K(\cdot)$ is the self-delimiting version of complexity of Chaitin and Levin.) However, if you try to define randomness for a biased coin in the same way, things become much more sticky! For instance, you might have thought that for a coin biased with probability p for heads, a string is random if and only if there exists a constant c such that for all n

$$K(s_1 s_2 \cdots s_n) \geq n H(p) - c, \tag{10.16}$$

where $H(p)$ is the Shannon information of the distribution $\{p, 1 - p\}$. But that ain't the case. Instead it can be shown [5] that $\vec{s}$ is random with respect to $\{p, 1 - p\}$ in the Martin-Löf sense if and only if there exists a constant c such that for all n

$$K(s_1 s_2 \cdots s_n) \geq -\log \mu[s_1 s_2 \cdots s_n] - c, \tag{10.17}$$

where $\mu[s_1 s_2 \cdots s_n]$ is the *probability* for the string $s_1 s_2 \cdots s_n$.

The point here is that even if one uses algorithmic complexity to define what one means by randomness, one still has to start off with the notion of probability and build on top of that. If Coleman really has a general result *and* defines randomness in the relatively standard way of Martin-Löf, then he must make use of the quantum probability expressions $|\langle 0|\psi \rangle|^2$, etc., at the outset ... and in a way that is not equivalent to inserting a simple parameter as in Eq. (10.16). That, it seems to me, can hardly be viewed as deriving the quantum mechanical probability law!

So, after all and independently of the general problem listed in Point 1, I am sceptical of what Coleman has to say. Nevertheless, this brings me back to Hartle's original work.

Point 3) What can be salvaged of the Eigenstate Axiom point of view? There is something very interesting in Hartle's paper if it can be made to work out in the proper way. (This is an idea inspired by a paper of Ballentine [6], which itself is very frequentist in its orientation.)

Normally the Law of Large Numbers can be made use of to sharpen up imprecise statements about single events (i.e., that the event will occur with such and such probability) into definite statements of probability 1 (like the probability of such and such frequency is one). If we take the Eigenstate Axiom to be as you would like, i.e.,

Axiom: If $|\psi\rangle = |s\rangle$, then a measurement of $\hat{S}$ will reveal outcome s with unit probability,

then Hartle's demonstration does hint at something that would be great. Namely, that the structure of quantum mechanics is such that it allows us to derive statements with less than unit probability from something that itself is specified with probability one. In other words, the structure of quantum mechanics may be such as to allow us to pursue the Law of Large Numbers in reverse order. In this way it would look like the MaxEnt principle in setting up priors ... but from the Eigenstate Axiom alone and not the standard full structure of the theory. After all, it is intriguing that the quantity $|\langle s|\psi\rangle|^2$ seems to pop out of his mathematics from nowhere.

The problem, though, in carrying this out is in 1) doing it *without* identifying probability and frequency, and 2) meeting my old objection that $\hat{S}^\infty$ and $\hat{F}_s^\infty$ are not related in a completely obvious way.

I'll leave you with that for now ... my wrist is really starting to hurt. I'm sure I'll have more to say after I get a chance to read Hartle's paper (and maybe Coleman's). Have any new developments come about since your last note to me? Keep me informed.

1. P. Benioff, "Possible strengthening of the interpretative rules of quantum mechanics," *Physical Review D*, **7**(12), pp. 3603–3609, 1973.
2. P. Benioff, "Some consequences of the strengthened interpretative rules of quantum mechanics," *Journal of Mathematical Physics*, **15**(5), pp. 552–559, 1974.
3. P. A. Benioff, "Finite and infinite measurement sequences in quantum mechanics and randomness: The Everett interpretation," *Journal of Mathematical Physics*, **18**(12), pp. 2289–2295, 1977.
4. P. Benioff, "A note on the Everett interpretation of quantum mechanics," *Foundations of Physics*, **8**(9/10), pp. 709–720, 1978.
5. M. van Lambalgen, "von Mises definition of random sequences reconsidered," *Journal of Symbolic Logic*, **52**(3), pp. 725–755, 1987.
6. L. E. Ballentine, "Can the statistical postulate of quantum theory be derived? – A critique of the many-universes interpretation," *Foundations of Physics*, **3**(2), pp. 229–240, 1973.

12 April 1996, "Fopping Math"

Last night, I dug into the black hole of quantum information that is my file cabinet and found a couple more tidbits that might be of interest to you. The first is a paper by Farhi, Goldstone, and Gutmann [1] that claims to do the argument of Hartle "right." The second is a paper by Mittelstaedt [2] that I, unfortunately, cannot trace the source of. (Perhaps it was one of the conference proceedings he edited and so it can be found in the library there. Actually it's a pretty good article all around about the "measurement problem.") I still haven't been able to get a copy of Hartle's old article – the *AJP*s here don't go back that far – I may have to go to McGill to get one.

To tell you the things I want to tell you, let me build a little notation again. First we start off with the state corresponding to n preparations

$$|\psi^n\rangle = |\psi\rangle|\psi\rangle\cdots|\psi\rangle, \tag{10.18}$$

where

$$|\psi\rangle = c_0|0\rangle + c_1|1\rangle, \tag{10.19}$$

and $|0\rangle,|1\rangle$ are eigenvectors of the observable $\hat{S}$ that we are talking about. Also let

$$|\psi^\infty\rangle = \lim_{n\to\infty} |\psi^n\rangle. \tag{10.20}$$

Define $\hat{S}^{(n)}$ and $\hat{S}^\infty$ similarly. Recall the definitions of the frequency operators $\hat{F}_0^{(n)}$, $\hat{F}_1^{(n)}$, $\hat{F}_0^\infty$, and $\hat{F}_1^\infty$ from my last note.

Now Farhi and company point out, and I think rightly so (though it's hard to tell without getting steeped in mathematics), that the actual statement proven by Hartle, i.e.,

$$\lim_{n\to\infty} \left\| \hat{F}_s^{(n)} |\psi^n\rangle - |\langle s|\psi\rangle|^2 |\psi^n\rangle \right\| = 0, \tag{10.21}$$

$s = 0, 1$, cannot be construed as an eigenvalue equation for the operator $\hat{F}_s^\infty$. They claim that this is really a statement about finite n, saying: for any $\epsilon > 0$, there exists a finite n such that

$$\left\| \hat{F}_s^{(n)} |\psi^n\rangle - |\langle s|\psi\rangle|^2 |\psi^n\rangle \right\| < \epsilon. \tag{10.22}$$

Therefore no conclusions based on the "Eigenstate Axiom" can be drawn from it.

To "remedy" this situation, they then work to find frequency operators $\mathcal{F}_s^\infty$ on the (nonseparable) Hilbert space $\mathcal{H} = H \otimes H \otimes H \otimes \cdots$ such that

$$\mathcal{F}_s^\infty |\psi^\infty\rangle = |\langle s|\psi\rangle|^2 |\psi^\infty\rangle. \tag{10.23}$$

The idea being, of course, that the Eigenstate Axiom *really* can be applied to this beast. But, boy, you want to talk about "mathematical foppery," you should see the lengths they go to here to get at this. (I have not attempted to disentangle it.)

In any case, that might have represented some clarification for me if I had not seen quite a jumble of other inane statements throughout the rest of the paper. For instance, consider the following from their "Conclusions" section. (Emphases were added by me.)

To what extent is the system of an infinite number of copies physically realizable and can the measurement of the frequency have physical meaning? This is like asking whether an infinite sequence of coin flips can be physically realized and what does the Strong Law of Large Numbers say in this case? Clearly a coin cannot actually be flipped an infinite number of times any more than someone can count to infinity; yet we can make sensible statements that refer to these operations. The Strong Law of Large Numbers says that if you did flip a coin an infinite number of times you would, *with absolute certainty*, find the proportion of heads to be one-half. Our quantum statement is that if you construct an infinite number of copies of [the system] and measure the frequency of the outcomes you would, *with absolute certainty*, find the frequency to be [the quantum mechanical prediction].

This, of course, is just the sort of problem with the interpretation you've been pointing out all along. However, I find this paragraph even a little more telling: since these fellas don't have a proper understanding of the implications of the classical SLLN, their quantum conclusions seem doomed from the outset.

If they want to say that an outcome string with the "correct" frequencies will occur *with absolute certainty* just because it is a member of a set of probability -1 ... and that an outcome string with "incorrect" frequencies will NOT occur *with absolute certainty* just because it is a member of a set of probability -0 ..., then they must give a criterion for why some probability -0 sets are "more probability -0" than others. (Recall Mr. Orwell.) For, every individual outcome string in this event space is, literally, of probability zero. This is the sort of thing that I called "sloppy frequentism" the other day. Only some extra criteria like von Mises idea of the kollectiv can make this silliness go. Things can certainly be no better in the quantum world.

How do you combat this sort of thing? I haven't a clue: no more than I would know how to turn a strict nonbeliever into a Bayesian. And the more I think about it, the less I'm inclined to bother.

However, as I said last time, I am more than willing to reëxamine what might be salvaged of Hartle's derivation. In the words of Mittelstaedt, "There is, however, no obvious interpretation of this result." (Which, by the way, has been generalized by Ochs [3] to situations where the preparations are not absolutely identical and may even be entangled.)

Among other things, though, I'm feeling pretty stupid about this thing of the connection between $\hat{F}_s^{(n)}$ and $\hat{S}^{(n)}$. What am I missing? Let's lay this to rest. If I think of a measurement of $\hat{S}^{(n)}$ as *only* revealing its eigenvalues, then indeed there is a clear-cut connection between the two operators: reading off the eigenvalues of $\hat{S}^{(n)}$ amounts to reading off the eigenvalues of $\hat{F}_s^{(n)}$. However, if I think of the repeated measurement of $\hat{S}$ in more physical terms of actually carrying out the measurements one by one ... then there are no real degeneracies in its spectrum: each outcome corresponds to a *distinct* string of length n. In this sense, the projectors onto the spaces spanned by different frequencies are a coarse-graining of the strings generated by a measurement of $\hat{S}^{(n)}$. To some extent, I see two different beasts; how should I adjust my glasses?

I suppose even if you cannot satisfy me that the questions are the same, it doesn't really matter. I continue to be intrigued that the standard probability expression $|\langle s|\psi\rangle|^2$ seems to pop out of nowhere in Hartle's derivation.

Suppose we take the Eigenstate Axiom to be along the lines of:

Axiom: For the measurement of a given Hermitian operator $\hat{S}$, the closer (in the Hilbert space sense) the preparation $|\psi\rangle$ is to being an eigenstate of $\hat{S}$, the more *likely* the measurement will reveal the appropriate eigenvalue of $\hat{S}$.

Then, without anything else being said about the probability interpretation, Hartle's result would reveal the quantity $|\langle s|\psi\rangle|^2$ popping up in a very elementary way. I sort of like that. However, it still seems to me that you need something more to get down to a probability assignment for the single measurement case. If one has a strong inclination or belief for one set of frequencies over another in a large number of repetitions and no other prior information, does that give a warrant to set a prior that is independent, identically distributed, and matching the expected frequency in the single measurement case? It seems to me that that is likely, and might actually be justified on Bayesian grounds. Or maybe even MaxEnt grounds – something like, independent distributions for the individual measurements will lead to the highest overall Shannon info.

In this sort of program though, I'm also a little bothered by another thing. I am accustomed to thinking of the eigenvalues of measurement operators like $\hat{S}$ as being more or less arbitrary labels for the outcomes. Why should anything be different in this case? Could there be some Cox-like freedom for rescaling probabilities (via arbitrary functions of the "plausibilities") that we are missing here?

Perhaps what is more important is that, for any basis $\{|s\rangle\}$, the strings "random" with respect to the measure $|\langle s|\psi\rangle|^2$ are close to $|\psi^\infty\rangle$, while all other strings are not. And this might be why a Coleman-type derivation might be better than one of a Hartle type. However, making the derivation rigorous is something else. (Just look at Benioff's *JMP* paper for the difficulties in expanding $|\psi^\infty\rangle$ in the right way so that the nonrandom components can be excluded.)

But that leads me to ask to what extent Coleman has really carried this sort of thing out ... or, alternatively, to what extent Benioff *already* has carried it out (given this new perspective). It would really be great if you'd take a look at the two old Benioff papers on the Everett interpretation.

Oh, finally as a side note, intrigued by these ideas, you can imagine that I was elated when I ran across – completely by accident – Ref. [4]. The game that this guy tries to play is to drop the standard axioms of probability and reconstruct them (much as Cox did in another way). In particular, he holds on to the Weak Law of Large Numbers and a couple other simpler qualitative things and tries to get all the rest (including Bayes' rule) out of it. I haven't studied the paper in detail yet, but its similarity (in the classical domain) with Hartle/Coleman might be worth fleshing out.

That's enough for now.

1. E. Farhi, J. Goldstone, and S. Gutmann, "How probability arises in quantum mechanics," *Annals of Physics*, **192**, pp. 368–382, 1989.
2. P. Mittelstaedt, "The objectification in the measuring process and the many worlds interpretation."
3. W. Ochs, "On the strong law of large numbers in quantum probability theory," *Journal of Philosophical Logic*, **6**, pp. 473–480, 1977.
4. R. E. Neapolitan, "A limiting frequency approach to probability based on the weak law of large numbers," *Philosophy of Science*, **59**, pp. 389–407, 1992.

15 April 1996, "No More Math"

Ahh, like Baron Frankenstein, you create a monster and then want nothing to do with it! OK, I can see you're tiring of the "Hartle/Coleman" issue, and that's all right. I just want to put down some last thoughts; then I'll wait until you come back to discuss it more. Mostly, I'm afraid that you got the wrong impression from my last note: I certainly do NOT want to see this problem mathematized further. I think that would be the wrong way to tackle it.

The constructive idea that you should take from my last note (until we meet again) is the following. I think there may be something in the mathematical piece of Hartle's original derivation that is worth using for our own purposes. Namely, that the quantitative quantum mechanical probability expression may have its roots in a much more qualitative statement ... just as Cox showed that the standard probability axioms can be derived from more qualitative statements. The qualitative statement in this case is not Hartle's "Eigenstate Axiom" but rather something much more Bayesian: regardless of the size of the Hilbert space, the closer the system's state vector is to an eigenstate of the Hermitian operator being measured, the more likely that eigenstate will correspond to the measurement outcome. Period. Then Hartle's derivation demonstrates that we can make ever more likely predictions about the outcome of a measurement of the frequency operator. Presumably, by Bayesian methods, we can then use that knowledge as prior information for setting up a probability assignment for the outcomes in an individual trial. That assignment will be the one the old axioms of quantum theory told us it would be anyway.

That's it; that's the idea. Have a good CEPI and I look forward to hearing back from you when you have some time.

21 December 1997, "Falling Off the Earth"

I'll send you the final version of the proposal I sent off to [a gazillion colleges]. The last section has been significantly restructured. You'll probably like it less, but – at least for the present – I like it much more. I've presented it all much more confidently, added some details about MaxEnt, and got even more outlandish at the end. (I guess now, not only have I come out of the closet, but I've shut the door behind me.) Believe it or not, this was done on Hideo's advice this time. I've also gotten rid of any hint of the long-term/short-term thing: now the last section looks to be purely motivational.

Carl's Replies

Cavesism 2: *I am very impressed by your proposal. My main warning is that if I'm swept away, more sensible people will probably not be.*

Cavesism 3: *I just finished reading your revised proposal a bit more carefully. It's just terrific, and as I said, deserves to be out there where people can read it and appreciate it.*

Cavesism 4: *I understand fully your reasons for not wanting the proposal displayed. If you change your mind at some point and display it on your own home page, let me know so that we can link to it. It's such a terrific way of saying this stuff about using quantum mechanics to do things instead of just putting up with its limitations. Of course, all of us theoretical physicists have long been thankful that the world isn't classical, but that was just because quantum mechanics gave us something to do. Now there is a better reason for preferring a quantum world.*

[NOTE: With these flattering compliments, Carl suggested that I post this oddity of a research proposal to a webpage – either to my own or to his Information Physics page at UNM. For various reasons, I declined at the time. It won't be much more out in the open in the middle of this book, but it seems worthwhile to record it here.]

The Structure of Quantum Information

Dedicated to a good teacher.

I. Orientation

The world we live in is well-described by quantum mechanics. What should we make of that? In a way, the answer to this question was once less positive than it is today. For although quantum theory is a tool of unprecedented accuracy in predicting and controlling the phenomena about us – and by way of that is the basis of our technological society – the intellectual lesson we have come to derive from it has been one largely of limitations. The best place to see this attitude at work is in a standard presentation of the Heisenberg uncertainty relations. It is almost as if the world were holding something back that we really had every right to possess: the task of physics, or so it was believed, is simply to sober up to this fact and make the best of it.

In contrast to this textbook lesson, the last five years have seen the start of a significantly more positive, almost intoxicating, attitude about the basic role of quantum mechanics. This is evidenced no more clearly than within the small, but growing [3], community of workers in *Quantum Information Theory* [4] and *Quantum Computing* [5].[1] The point of departure in both these disciplines is not to ask what limits quantum mechanics places upon us, but instead what novel, productive things we can do in the quantum world that we could not have done otherwise. In what ways can we say that the quantum world is fantastically better than the classical world?

The two most striking examples of this so far have been quantum cryptographic key distribution [6,7] and Shor's quantum factoring algorithm [8,9]. In the case of the first example, one sees that quantum mechanics allows two communicators to transmit to each other a random cryptographic key in such a way that eavesdropping on the transmission can be excluded out of hand. This is impossible in the classical world because there is no direct connection between the information that can be gathered about a physical system's state and the disturbances induced upon that state in the process [10,11]. Without the indeterminism of quantum mechanics, two-party data transmissions would remain forever vulnerable to clever or powerful eavesdropping techniques.

In the case of the second example, one sees that algorithms designed for computers built of unabashedly quantum components – that is, components that can remain coherent with each other throughout the computation – can factor large integers exponentially faster than anything written for standard classical computers. To give a quick example of what this means in real terms, consider a 600-digit number that is known to be the product of two (secret) primes. The number of computational steps required of a classical computer to crack it into its two components is of the order of 10^{34}. In contrast, the corresponding

---

[1] For quick reference, two recent articles on the subject can be found in *Physics Today*: Oct. 1997, pp. 19–21, and Oct. 1995, pp. 24–30. Some WWW links can be found in John Preskill's "Physics 229" homepage at http://www.theory.caltech.edu/people/preskill/ph229. Also see Oxford University's Centre for Quantum Computation homepage at http://www.qubit.org/.

number of steps for a quantum computer is only 10^{11}. Quantum computing can give 23 orders of magnitude greater efficiency in this problem!

These two examples are the most outstanding of the class, and there is well-founded hope that they are the tip of a technological iceberg. However, I believe there is a similarly founded hope that they are also the small tip of a *physical* iceberg. Looking at quantum mechanics through the eyes of these two fields cannot help but lead to greater and deeper – and perhaps the deepest – insights into its structure and ultimate use. These are the insights that could poise physics for the great breakthroughs that will surely come about next century, even in disciplines as far-flung as quantum gravity and quantum cosmology.

But this is my grand vision. Little will come of it if it is not preceded by years of more realistic, more concrete exploration of the *structure of quantum information*. This, as part of the accumulating results of the communities just described, is the subject of this research proposal.

To give some indication of the wider context that flows into the specialization of Quantum Information and Quantum Computing, one need only note that, by its very makeup, it must call upon the expertise of standard communication theory, cryptography, computation theory, number theory, signal processing, and various branches of statistical mechanics. Dreams of possible experimental implementations have called upon the quantum optics community [12,13], the ion trap community [14,15], the NMR spectroscopy community [16,17], and to a smaller extent solid-state physics [18,19].

The particular aspect of Quantum Information Theory that has been my focus the last four years is closely allied to the well-established tradition in mathematical physics pioneered by the likes of Holevo [20], Lieb [21], Lindblad [22], and Uhlmann [23]. It is my intent to strengthen and build upon the connections between that tradition and this upstart field, which already in many ways is its continuation. I hope this becomes apparent in the details that follow.

II. Research Proposal

My research interests might be described as *tria juncta in uno*.[2] The conjunction of these three topics, for the most part, exhausts what is presently meant by "Quantum Information Theory."

• **Sending Classical Information on Quantum Mechanical Channels.** People encode "classical" information – like the stories in today's newspaper – into the states of quantum systems for a simple reason: to get it from one place to another. Since the world is quantum mechanical, this, in the last analysis, is exactly what one always does in transmitting information. Strangely enough however, literally almost all of modern information theory (as exhibited in the 44 existent volumes of *IEEE Transactions on Information Theory*) has ignored this fact in any but the most trivial ways.

---

[2] Not to worry, I won't pretend that I knew this phrase before looking in my thesaurus! ... But when you learn something like this, you've just got to use it! Apparently this phrase is the motto of "The Most Honourable Order of the Bath," a particular British order of chivalry.

Once one takes it seriously that physical information carriers are quantum mechanical, one can ask a whole host of questions that could not have been asked before. For instance, can it help the receiver to collect many separate transmissions before performing the quantum measurement required to decode them [24,25]? That is to say, can collective quantum measurements on separate signals be more powerful than individual measurements [26,27]? Can it ever help to entangle separate transmissions – as with Einstein–Podolsky–Rosen pairs – before sending them through the channel [28]? Can one help evade a channel's noise by encoding the signals in nonorthogonal quantum states [29,30], in spite of the fact that the classical analog of this corresponds to sending noisy signals?

Of course, the answer to all these questions is "yes." And this is enough to demonstrate that these lines of thought are not trivial. However, the work remaining before a theory as coherent as classical information theory can emerge is legion. In particular, the counterpart of the most basic question of all classical information theory – What is the capacity of a discrete memoryless channel? – has yet to be solved.

The most exciting prospect of this set of questions for physics is the potential it holds for giving new and unique and very strongly motivated measures of "correlation" between two subsystems of a larger whole. Shannon's solution of the channel capacity question brought with it a measure of correlation (the "mutual information") of a generality greatly exceeding the scope of its motivation [31]. Its physical applications have ranged from information theoretic versions of the Heisenberg relations [32] to a final solution of the old Maxwellian demon problem [33,34]. One can expect no less for a quantum measure of classical correlation. In particular, the importance of uniquely quantum measures of correlation for quantum statistical mechanics has been emphasized recently by Lindblad [35] and Schack and Caves [36].

• **Information Gain vs. Quantum State Disturbance in Quantum Theory.** The engine that powers quantum cryptography is the principle that it is impossible to gather information about a quantum system's unknown state without disturbing that system in the process. (This is so even when the state is assumed to be one of only two nonorthogonal possibilities.) This situation is often mistakenly described as a consequence of the "Heisenberg uncertainty principle" but, in fact, is something quite distinct [10,11] and only now starting to be studied in the physical literature. A more accurate account of the principle is that it is a feature of quantum mechanics that rests ultimately on the unitarity of the theory, and may be seen as a quantitative extension of the so-called "no-cloning theorem" [37,38,39]. In contradistinction, the Heisenberg principle concerns our inability to "get hold" of two classical observables – such as a position and momentum – simultaneously. It thus concerns our inability to ascribe *classical* states of motion to *quantum* systems – that has very little to do with the issue of encoding information in and retrieving information from the quantum states themselves.

Because this way of looking at "information gain" and "disturbance" for quantum systems is itself purely quantum mechanical and does not rely on antiquated classical notions, it holds the possibility of giving the best understanding yet of those things the founding fathers (like Heisenberg, Pauli, and Bohr) labored so hard to formulate. But what is the unifying theme? What are the directions to take? One can elaborate upon the direction defined by practical quantum cryptography [40,41,42] or one can take a more direct route inspired by the original no-cloning theorem [43,44,45]. Each method is begging for a more systematic account than has been afforded by these simple preparatory explorations.

A novel approach, and one which I have turned my attention to recently, is to seek out the connection between quantum entanglement measures and the information–disturbance principle [45,46,47]. The main point about this line of thought is that in the scenario of quantum cryptography, any would-be eavesdroppers must become entangled with the information carriers traveling between the legitimate users. Can one read the tradeoff between information and disturbance directly from something to do with entanglement itself? Perhaps by a sort of "entanglement conservation" principle? These are the sorts of question that first require progress in the next research topic.

- **Quantifying Quantum Entanglement: Separating It from Classical Correlation.**
The preoccupation of classical information theory is to make the correlation between sender and receiver as high as possible. This is what communication is about. But it is only part of the story in Quantum Information Theory. The quantum world brings with it a new resource that senders and receivers can share: quantum entanglement, the stuff Einstein–Podolsky–Rosen pairs and Bell-inequality violations are made of. This new resource, of all the things mentioned so far, is the most truly "quantum" of quantum information. It has no classical analog, nor might it have been imagined in a classical world.

What is quantum entanglement? It is *not* probabilistic correlation between two parts of a whole. Rather it is the *potential* for such a correlation. In a quick portrayal:

classical correlation –

Alice and Bob entered a lottery for which they were the only players. They have not opened their "winnings" envelopes yet, but the messages in them say that one is the winner and one is the loser. Which is which, they do not know – they only know the correlation – but the answer is there, objectively existent, without their looking.

quantum entanglement –

Alice and Bob will eventually perform measurements on the EPR pair their envelopes contain and the outcomes *will* be correlated. However, before the measurements are performed, there are no objectively existent variables already there. Different measurements can and will lead to different correlations.

In a certain sense, entanglement is a kind of *all-purpose correlation* just waiting to be baked into something real – a quantum "Martha White's Flour" [48]. The uses for this all-purpose correlation are manifold within Quantum Information Theory. Beside the applications above [6,28], there is also quantum-state teleportation [49], quantum superdense coding [50,51], error-correction for quantum computers [52], entanglement-assisted multi-party communication games [53], better control of atomic frequency standards [54,55], and the list goes on.

The deepest set of questions here, and the largest focus of my present research [56,57], concern quantifying this newly recognized physical essence in an application-independent way [58,59,60]. As an example, take an EPR pair, half of which has been transmitted through a noisy (decohering) quantum channel. Because of the noise, the final state of this bipartite system is no longer a pure state: it is described by a mixed state density operator. Some of the correlation there is still potential or all-purpose, but some – because the decoherence has helped promote it to a more tangible status – is simply classical correlation. How do we quantify the amount of each? What, if anything, is the exchange rate between the two? With some time, creativity, and hard work, we will one day have these issues under control.

- **Synthesis.** In some ways the project of Quantum Information Theory can be likened to the beginning of thermodynamics. It is not our place to develop the question "What is heat, work, energy?" but instead "What is correlation, indeterminism, entanglement?" No informed judgment of the historic question could have been made before the development of a quantitative theory of thermodynamics, and it will be likewise with our field. Whereas the fruits of the old question were the mechanical theory of heat and its corollary of atomism, we do not yet have a firm grasp of where our field is leading. It is clear, however, that it is going somewhere and somewhere fast; at the very least, its applied, technological innovations can neither be denied nor safely ignored.

What are correlation, indeterminism, entanglement? This is what the three research areas above are trying to make quantitative. Each contains within itself a little seed of the others; each sheds light on the structure of quantum information.

III. Wider Seas

Eight years after the inception of classical information theory, Claude Shannon, its founder, warned [61]:

Although this wave of popularity is ... pleasant and exciting for those of us working in the field, it carries at the same time an element of danger. While we feel that information theory is a valuable tool in providing fundamental insights into the nature of communication ... it is certainly no panacea .... Seldom do more than a few of nature's secrets give way at one time.

History has borne Shannon out: information theory is not a panacea. But, cure-all or not, the field has had a great impact on applications that can hardly be said to resemble the original one, that of describing communication over noisy channels [62]. One need only look at information theory's influence on fields as far ranging as biology, economics, and psychology [63], to see this point.

What is it that we can expect of *Quantum* Information Theory once it is complete and coherent? What more might it say about a *few* of natures secrets? With the reader's indulgence, I will attempt to express some of my present views on the question. These have to do with the "grand vision" and "physical iceberg" of the Introduction – the real sources of my day-to-day motivation.

The year 1957 is significant in physical thought because it marks the penetration of information theory into physics in a systematic way – into statistical mechanics in particular [64]. This refers to the *Maximum Entropy* or "MaxEnt" program for statistical mechanics set into motion by E. T. Jaynes [65]. With the tools of information theory, one was able for the first time to make a clean separation between the purely *statistical* and the purely *physical* aspects of the subject matter.

Perhaps it would be good to present a mild example of this. Because of MaxEnt, a standard statistical mechanical ensemble – like the canonical ensemble – can finally be seen for what it really is: an expression of the physicist's *state of knowledge* (specified, of course, by the experimental parameters under his control). Though this reveals a subjective element in statistical mechanics, the ensemble is not arbitrary. Two physicists working on a single experiment and possessing identical data – if true to their states of knowledge – will derive the same distributions for the system's variables. The *structure* of the canonical distribution, with its exponential form, is due purely to the kind of knowledge the experimenter possesses – in this particular case, the expectation value of some observable and nothing else. That is to say, the canonical distribution's form is a theorem of the laws of inference, *not physics*. The physics of the system rests solely in its Hamiltonian and boundary conditions. This conceptual separation between the physical and the statistical can be fruitful. With it, one can, for instance, derive the second law of thermodynamics in an almost trivial way [66].

In contrast to this, quantum theory is at its heart a statistical theory of *irreducibly* statistical phenomena – this is the great lesson of the Kochen–Specker theorems and the Bell inequalities [68]. What can this possibly mean for the issue just explored? With due attention to the success of the MaxEnt program in the last 40 years [67], one can hardly feel it unreasonable to ask: What part of the formal structure of quantum mechanics is forced upon us by physics alone – i.e., that the theory be about irreducibly statistical phenomena – and what part is forced upon us as a consequence of the form *any* theory must take in light of that subject matter [69,70]? George Boole called probability theory a "law of thought" because it specifies the rules with which we should think when we come upon situations where our information is incomplete [71,72]. What part of quantum mechanics is simply "law of thought," and what part is *irreducibly* physics?

A mature Quantum Information Theory is likely to be uniquely stationed to contribute to this question, or at least to test whether anything might come of it. The quantitative statements it *will* possess for the information–disturbance tradeoff and the correlation–entanglement dichotomy should be of just the right flavor for such a thing. Both threads explore the difference between probabilities that can be improved upon because they correspond to lack of knowledge and probabilities that are more the nature of "potentialities" for which no improvement can be had.

Once this issue is settled, one may finally hope for a simple, crisp statement of what our quantum theory is all about [73]. And once that is in hand, who knows what the limits might be? To place the issue within an historical context, one can speculate how long it would have taken to stumble across general relativity if it had not been for the compelling vision that Einstein found lying behind the Lorentz transformations. The equations were there with Lorentz, but the essence of it all – and the simple picture with which progress could be made – came with Einstein's special relativity.

Comparable to this opportunity for fundamental physics, one might imagine a similar blossoming of opportunity for other endeavors. A classification of quantum theory's content in the way suggested above could distill mathematical structures that other, extra-quantum-mechanical, efforts [74,75,76] might use to their advantage. After all, this is exactly the sort of thing that happened with the MaxEnt program: its applications range from observational astronomy to pharmaceutical studies to artificial intelligence[3] [65]. Are there fields beside quantum physics that encounter situations where the maximal knowledge of something can never be made deterministically complete [77,78,79]? If so, then they will plausibly find novel use for the *mathematics*[4] of quantum physics and Quantum Information Theory.

The main point that I would like to impress with this final speculation – even allowing that the details above be taken with a grain of salt – is that the structure of quantum mechanics is an amazingly beautiful intellectual edifice. It would be a shame if its only application were for quantum mechanics itself.

IV. Summary

There is a grand adventure in front of the physics community called Quantum Information and Quantum Computing. Its hallmark is to view quantum theory in a way little before explored, in a way that accentuates the positive. Delimiting the structure of quantum information may well be a key to great progress in fundamental physics – but that may be some time in the coming. In the meantime there is much solid work to be done exploring entanglement, information vs. disturbance, and the information-carrying capacities of quantum mechanical systems. This is my preoccupation; this is the field of research I call my home. I thank you for your consideration.

V. References

1. *Note 1*: This bibliography makes no attempt to be complete or to consistently cite original references. Its main purpose is to give the reader a more detailed introduction to the issues discussed in this text.
2. *Note 2*: All "LANL e-print" listings refer to the Los Alamos National Laboratory E-print Archive at http://xxx.lanl.gov.
3. G. Taubes, "All Together for Quantum Computing," *Science* **273**, 1164 (1996).
4. C. H. Bennett, "Quantum Information and Quantum Computation," *Phys. Tod.* **48**, No. 10, 24 (1995).
5. A. M. Steane, "Quantum Computing," *Rep. Prog. Phys.* **61**, 117 (1998).
6. C. H. Bennett, G. Brassard, and A. K. Ekert, "Quantum Cryptography," *Sci. Am.* **267**, No. 10, 50 (1992).

---

[3] A particularly cogent example of the use of these methods in artificial intelligence can be found at Microsoft Research Division's *Decision Theory & Adaptive Systems Group* homepage: http://www.research.microsoft.com/dtas/.

[4] Please note that I *did* say "the mathematics of" here.

7. A. Muller, H. Zbinden, and N. Gisin, "Underwater Quantum Coding," *Nature* **378**, 449 (1995).

8. D. P. DiVincenzo, "Quantum Computation," *Science* **270**, 255 (1995).

9. A. Ekert and R. Jozsa, "Quantum Computation and Shor's Factoring Algorithm," *Rev. Mod. Phys.* **68**, 733 (1996).

10. C. A. Fuchs and A. Peres, "Quantum State Disturbance vs. Information Gain: Uncertainty Relations for Quantum Information," *Phys. Rev. A* **53**, 2038 (1996).

11. C. A. Fuchs, "Information Gain vs. State Disturbance in Quantum Theory," *Fortschr. Phys.* **46**, 535 (1998).

12. Q. A. Turchette, C. J. Hood, W. Lange, H. Mabuchi, and H. J. Kimble, "Measurement of Conditional Phase Shifts for Quantum Logic," *Phys. Rev. Lett.* **75**, 4710 (1995).

13. A. Furusawa, J. L. Sørensen, S. L. Braunstein, C. A. Fuchs, H. J. Kimble, and E. S. Polzik, "Unconditional Quantum Teleportation," *Science* **282**, 706 (1998).

14. C. Monroe, D. M. Meekhof, B. E. King, W. M. Itano, and D. J. Wineland, "Demonstration of a Universal Quantum Logic Gate," *Phys. Rev. Lett.* **75**, 4714 (1995).

15. R. J. Hughes, *et al.*, "The Los Alamos Trapped Ion Quantum Computer Experiment," *Fortschr. Phys.* **46**, 329 (1998).

16. D. G. Cory, A. F. Fahmy, and T. F. Havel, "Ensemble Quantum Computing by NMR Spectroscopy," *Proc. Natl. Acad. Sci. USA* **94**, 1634 (1997).

17. N. A. Gershenfeld and I. L. Chuang, "Bulk Spin-Resonance Quantum Computation," *Science* **275**, 350 (1997).

18. D. Loss and D. P. DiVincenzo, "Quantum Computation with Quantum Dots," *Phys. Rev. A* **57**, 120, (1998).

19. B. E. Kane, "A Silicon-Based Nuclear Spin Quantum Computer," *Nature* **393**, 133 (1998).

20. A. S. Holevo, "Statistical Decision Theory for Quantum Systems," *J. Multivar. Anal.* **3**, 337 (1973).

21. E. H. Lieb, "Convex Trace Functions and the Wigner–Yanase–Dyson Conjecture," *Adv. Math.* **11**, 267 (1973).

22. G. Lindblad, "On the Generators of Quantum Dynamical Semigroups," *Comm. Math. Phys.* **48**, 119 (1976).

23. A. Uhlmann, "The 'Transition Probability' in the State Space of a *-algebra," *Rep. Math. Phys.* **9**, 273 (1976).

24. A. S. Holevo, "The Capacity of Quantum Communication Channel with General Signal States," *IEEE Trans. Inf. Theor.* **44**, 269 (1998).

25. B. Schumacher and M. D. Westmoreland, "Sending Classical Information via Noisy Quantum Channels," *Phys. Rev. A* **56**, 131 (1997).

26. A. S. Holevo, "On the Capacity of Quantum Communication Channel," *Prob. Inf. Transm.* **15**, 247 (1979).

27. A. Peres and W. K. Wootters, "Optimal Detection of Quantum Information," *Phys. Rev. Lett.* **66**, 1119 (1991).

28. C. H. Bennett, C. A. Fuchs, and J. A. Smolin, "Entanglement-Enhanced Classical Communication on a Noisy Quantum Channel," in *Quantum Communication, Computing and Measurement*, edited by O. Hirota, A. S. Holevo, and C. M. Caves (Plenum, New York, 1997).

29. C. A. Fuchs, "Nonorthogonal Quantum States Maximize Classical Information Capacity," *Phys. Rev. Lett.* **79**, 1163 (1997).

30. C. A. Fuchs, P. W. Shor, J. A. Smolin, and B. M. Terhal, "Quantum-Enhanced Classical Communication," to be submitted to *Phys. Rev.* A; preliminary draft available upon request.

31. T. M. Cover and J. A. Thomas, *Elements of Information Theory*, (John Wiley & Sons, New York, 1991).

32. M. J. W. Hall, "Information Exclusion Principle for Complementary Observables," *Phys. Rev. Lett.* **74**, 3307 (1995).

33. C. H. Bennett, "The Thermodynamics of Computation – a Review," *Int. J. Theo. Phys.* **21**, 905 (1982).

34. W. H. Zurek, "Algorithmic Randomness and Physical Entropy," *Phys. Rev.* A **40**, 4731 (1989).

35. G. Lindblad, "Quantum Entropy and Quantum Measurements," in *Quantum Aspects of Optical Communications*, edited by C. Bendjaballah, O. Hirota, and S. Reynaud (Springer-Verlag, Berlin, 1991).

36. R. Schack and C. M. Caves, "Information-Theoretic Characterization of Quantum Chaos," *Phys. Rev.* E **53**, 3257 (1996).

37. W. K. Wootters and W. H. Zurek, "A Single Quantum Cannot Be Cloned," *Nature* **299**, 802 (1982).

38. C. H. Bennett, G. Brassard, and N. D. Mermin, "Quantum Cryptography Without Bell's Theorem," *Phys. Rev. Lett.* **68**, 557 (1992).

39. H. Barnum, C. M. Caves, C. A. Fuchs, R. Jozsa, and B. Schumacher, "Noncommuting Mixed States Cannot Be Broadcast," *Phys. Rev. Lett.* **76**, 2818 (1996).

40. C. A. Fuchs, N. Gisin, R. B. Griffiths, C. S. Niu, and A. Peres, "Optimal Eavesdropping in Quantum Cryptography. I. Information Bound and Optimal Strategy," *Phys. Rev.* A **56**, 1163 (1997).

41. B. A. Slutsky, R. Rao, P.-C. Sun, and Y. Fainman, "Security of Quantum Cryptography Against Individual Attacks," *Phys. Rev.* A **57**, 2383 (1998).

42. J. I. Cirac and N. Gisin, "Coherent Eavesdropping Strategies for the 4-State Quantum Cryptography Protocol," *Phys. Lett.* A **229**, 1 (1997).

43. V. Bužek and M. Hillery, "Quantum Copying: Beyond the No-Cloning Theorem," *Phys. Rev.* A **54**, 1844 (1996).

44. N. Gisin and S. Massar, "Optimal Quantum Cloning Machines," *Phys. Rev. Lett.* **79**, 2153 (1997).

45. D. Bruß, D. P. DiVincenzo, A. Ekert, C. A. Fuchs, C. Macchiavello, and J. A. Smolin, "Optimal Universal and State-dependent Quantum Cloning," *Phys. Rev.* A **57**, 2368 (1998).

46. D. Deutsch, A. Ekert, R. Jozsa, C. Macchiavello, S. Popescu, and A. Sanpera, "Quantum Privacy Amplification and the Security of Quantum Cryptography Over Noisy Channels," *Phys. Rev. Lett.* **77**, 2818 (1996).

47. B. Schumacher and M. D. Westmoreland, "Quantum Privacy and Quantum Coherence," *Phys. Rev. Lett.* **80**, 5695 (1998).

48. L. Flatt and E. Scruggs, "The Martha White Theme," on *Flatt & Scruggs: 1948–1959* (Bear Family Records, 1994).

49. C. H. Bennett, G. Brassard, C. Crépeau, R. Jozsa, A. Peres, and W. K. Wootters, "Teleporting an Unknown Quantum State via Dual Classical and Einstein–Podolsky–Rosen Channels," *Phys. Rev. Lett.* **70**, 1895 (1993).

50. C. H. Bennett and S. J. Wiesner, "Communication via One- and Two-Particle Operators on Einstein–Podolsky–Rosen States," *Phys. Rev. Lett.* **69**, 2881 (1992).

51. K. Mattle, H. Weinfurter, P. G. Kwiat, and A. Zeilinger, "Dense Coding in Experimental Quantum Communication," *Phys. Rev. Lett.* **76**, 4656 (1996).

52. D. Gottesman, *Stabilizer Codes and Quantum Error Correction*, Ph.D. Thesis, California Institute of Technology, 1997; LANL e-print quant-ph/9705052.

53. R. Cleve and H. Buhrman, "Substituting Quantum Entanglement for Communication," *Phys. Rev.* A **56**, 1201 (1997).

54. J. J. Bollinger, W. M. Itano, D. J. Wineland, and D. J. Heinzen, "Optimal Frequency Measurements with Maximally Correlated States," *Phys. Rev.* A **54**, R4649 (1996).

55. S. F. Huelga, C. Macchiavello, T. Pellizzari, A. K. Ekert, M. B. Plenio, and J. I. Cirac, "On the Improvement of Frequency Standards with Quantum Entanglement," *Phys. Rev. Lett.* **79**, 3865 (1997).

56. C. H. Bennett, D. P. DiVincenzo, C. A. Fuchs, T. Mor, E. Rains, P. W. Shor, J. A. Smolin, and W. K. Wootters, "Quantum Nonlocality without Entanglement" *Phys. Rev.* A **59**, (1999); LANL e-print quant-ph/9804053.

57. D. P. DiVincenzo, C. A. Fuchs, H. Mabuchi, J. A. Smolin, A. Thapliyal, and A. Uhlmann, "Entanglement of Assistance," to appear in *The 1st NASA International Conference on Quantum Computing & Quantum Communications (NASA QCQC'98)*, edited by C. Williams (Springer-Verlag, Berlin, 1998); LANL quant-ph/9803033.

58. C. H. Bennett, G. Brassard, S. Popescu, B. Schumacher, J. A. Smolin, and W. K. Wootters, "Purification of Noisy Entanglement and Faithful Teleportation via Noisy Channels," *Phys. Rev. Lett.* **76**, 722 (1996).

59. C. H. Bennett, D. P. DiVincenzo, J. A. Smolin, and W. K. Wootters, "Mixed State Entanglement and Quantum Error Correction," *Phys. Rev.* A **54**, 3825 (1996).

60. W. K. Wootters, "Entanglement of Formation of an Arbitrary State of Two Qubits," *Phys. Rev. Lett.* **80**, 2245 (1998).

61. C. E. Shannon, "The Bandwagon," *IEEE Trans. Inf. Theor.* **IT-2**, No. 3, 3 (1956).

62. C. E. Shannon, "A Mathematical Theory of Communication," *Bell Sys. Tech. J.* **27**, 379, 623 (1948).

63. J. Campbell, *Grammatical Man: Information, Entropy, Language, and Life*, (Simon & Schuster, New York, 1982).

64. E. T. Jaynes, "Information Theory and Statistical Mechanics," *Phys. Rev.* **106**, 620 (1957).

65. A convenient source of information about this program can be found in the dozen or so conference proceedings all published under the same title, *Maximum Entropy and Bayesian Methods* (Kluwer, Dordrecht).

66. E. T. Jaynes, "Gibbs vs. Boltzmann Entropies," *Am. J. Phys.* **33**, 391 (1965).

67. E. T. Jaynes, "Where Do We Stand on Maximum Entropy?," in *The Maximum Entropy Formalism*, edited by R. D. Levine and M. Tribus (MIT Press, Cambridge, MA, 1979).

68. N. D. Mermin, "Hidden Variables and the Two Theorems of John Bell," *Rev. Mod. Phys.* **65**, 803 (1993).

69. C. M. Caves and C. A. Fuchs, "Quantum Information: How Much Information in a State Vector?," in *The Dilemma of Einstein, Podolsky and Rosen – 60 Years Later*, edited by A. Mann and M. Revzen, *Ann. Israel Phys. Soc.* **12**, 226 (1996).

70. C. M. Caves, C. A. Fuchs and R. Schack, "Bayesian Probability in Quantum Mechanics," to be submitted to *Am. J. Phys.* Preliminary draft available upon request.

71. G. Boole, *An Investigation of the Laws of Thought*, (Dover, New York, 1958).

72. E. T. Jaynes, *Probability Theory: The Logic of Science*. This massive book was unfortunately not completed before Prof. Jaynes' death. Preprints are available at http://bayes.wustl.edu/.
73. C. Rovelli, "Relational Quantum Mechanics," *Int. J. Theor. Phys.* **35**, 1637 (1996).
74. I. Pitowsky, "From George Boole to John Bell – The Origins of Bell's Inequality," in *Bell's Theorem, Quantum Theory and Conceptions of the Universe*, edited by M. Kafatos (Kluwer, Dordrecht, 1989).
75. I. Pitowsky, "George Boole's 'Conditions of Possible Experience' and the Quantum Puzzle," *Brit. J. Phil. Sci.* **45**, 95 (1994).
76. W. Segal and I. E. Segal, "The Black-Schole Pricing Formula in the Quantum Context," *Proc. Natl. Acad. Sci. USA* **95**, 4072 (1998).
77. S. Watanabe, "A Model of Mind-Body Relation in Terms of Modular Logic," *Synthese* **13**, 261 (1961).
78. A. Peres and W. H. Zurek, "Is Quantum Theory Universally Valid?," *Am. J. Phys.* **50**, 807 (1982).
79. J. L. Heilbron, *The Dilemmas of an Upright Man: Max Planck as Spokesman for German Science*, (U. California Press, Berkeley, 1986), pp. 127–128.

22 December 1997, "Bohr was a Bayesian"

It dawned on me recently that Bohr was actually a Bayesian. Anyway, if you view him that way, some of the crap he said miraculously starts to make sense.

It's also dawned on me recently that I need to get to bed. SuperShuttle arrives at 5:30 AM and it's already past 1:15!! Kiki and I fly to Texas tomorrow, i.e., today. However, I'd like to talk to you Tuesday about the "Bohr from Bayes" program (i.e., "A Bayesian Derivation of the Quantum Probability Law"). E-mail me the times you plan to be in the office Tuesday and I'll give you a call when it is most convenient on me. (Remember I'm on vacation!)

In the meantime, I'll forward on (for your holiday enjoyment) some of the ramblings I sent to David Mermin. The beginnings of the bald assertion [sic] above start up somewhere about Merminition 12.

20 September 1998, "Dropping In"

I read the Beller article in *Physics Today*. In fact, I've read several of her articles before: she writes very well. She of course has a point about Bohr's intractable language; I've spent many hours myself trying to make some sense of it all. To the people with less patience than I, I'm sure it's not obvious that they should struggle to find some meaning there. That's exactly why someone has to get in and say something reasonable about (a modern-day version of) the "Copenhagen interpretation" before things get out of hand. It's starting, for instance, to look like the editor of *Physics Today* has some Bohmian leanings: why else would he/she allow the two Goldstein articles and the Beller article within such proximity of each other?

Carl's Reply

I agree with you that it is very disturbing that *Physics Today* has run all those articles endorsing Bohmian mechanics. It's caused me to get some books and articles about Bohmianism and to think about it a little.

The main thing I got out of the Beller article was that wonderful "cock-a-doodle-doo" quote from Einstein, which shows that Einstein was not a naïve realist. He believed that the ONLY reality we have is the one we construct from our perceptions and, more sophisticatedly, from our scientific theories and models. Our everyday experience leads ineluctably to the notion that there is a real world out there, in which we can often make reliable predictions of the future based on past observation and which proceeds on its way without much noticing our existence. Einstein was quite right in maintaining that any fundamental scientific theory ought to allow construction of that real world. But, of course, our perception of an everyday "effective reality" does not mean that our theory of the microsopic world must have an objective reality, since microscopic phenomena do not contribute directly to our perception of an everyday effective reality.

I am confident that the Bayesian interpretation of quantum mechanics is up to the challenge. All probabilities are ignorance probabilities. There are certain situations where we are free to ask whatever question we want to a microscopic system, and then our questions and the answers we get help to create the Universe in which we live. For macroscopic phenomena, however, the laws of physics, with their position-dependent potentials, combine with decoherence to restrict the set of questions we can ask, thus giving rise to an emergent reality. The statistical predictions of quantum mechanics within this emergent reality are just what we need to construct an effective classical world, in which we make inferences among alternatives based on the probabilities we assign; moreover, we can often make reliable predictions based on past observations, as in a deterministic, classical world.

This emergent reality is just what is needed for our everyday life and also for the other sciences, such as biology, geology, astronomy, etc. Bohmians, postmodernists, and Bohr and other physicists who tried to apply complementarity to social and political questions – all make the basic mistake of reductionism – i.e., thinking that our quantum-mechanical picture of the microscopic level must naïvely correspond to our picture of the macroscopic level. As far as the higher, macroscopic level of everyday life is concerned, the only function of the lower, microscopic level of quantum mechanics is to give rise to the emergent realistic structure of our classical world. It's simply wrong to think that lack of realism at the microscopic level must give rise to lack of realism at the macroscopic level. This is the reason that postmodernists can't use the lack of realism of quantum mechanics to discredit objectivity and truth in our everyday lives and in the study of history, politics, and culture.

Bohmians, it seems to me, make just the opposite mistake. Convinced that macroscopic reality can only come from microscopic reality, they make up a microscopic reality for quantum mechanics. This microscopic reality is truly bizarre – an acausal web in which everything in the Universe influences everything else. Who needs this Bohmian microscopic reality? It is utterly unlike any kind of reality that is demanded by our perceptions. Moreover, it is completely irrelevant to and unnecessary for the emergent reality provided by quantum mechanics. Why bother with it?

As you can see, I have been bothered not only by Bohmians, but also by postmodernists. If we're going to say that microscopically there is no objective reality, then how do we avoid falling into the trap of saying that everything is subjective? The answer, it seems to me, is that the absence of objective reality in quantum mechanics does not invalidate the emergent reality of our everyday lives.

From the Bayesian view the two great questions – What basis? and Which alternative within that basis? – are answered. Decoherence gives us the basis (I know there are problems here, but bear with me), and once you have a basis, the notion that there is one actual alternative within that basis is simply what the whole idea of an emergent reality means. In contrast, at the microscopic level, where one can ask different questions, the idea that there is one actual alternative for each question is inconsistent with the statistical predictions of quantum mechanics.

24 September 1998, "Emerging Reality"

I'm just about on my way to a long weekend in Texas.

I've just printed out your motivational note so that I can chew upon it over the weekend. In the mean time, would you try to say a little more precisely what you mean by the phrase "emergent reality?" Please, please, please. (I honestly don't understand the phrase.) Also, what papers/books have you been reading about Bohmianism? I keep coming back to the idea that maybe there is a way to punch a hole in it, like you suggested: that would be wonderful. The only weak point that I know of is in getting the initial distributions right, i.e., via some damping mechanism or such (the first attempt was Bohm, "Proof that Probability Density Approaches $|\psi|^2$ in Causal Interpretation of Quantum Theory," *PRA* **89**, pp. 458–466, 1953). I wonder how good those arguments are?

27 January 1999, "Bohmian Computers"

It dawned on me today after talking to Carl that I had forgotten an old conversation I once had with none other than the famous Howard Barnum. The place was a path near the Villa Gualino in Torino; the time was about four years ago. The subject was precisely Bohmian computers. It, like so many subjects in the field still to be explored by the mediocre, had already been visited by the mind of Howard Barnum. I remember him asking whether the existence of Bohmian mechanics already proved that there existed a classically efficient algorithm for factoring.

19 February 1999, "A Bright Young Man"

I was taken aback, shocked even! Can you believe what I heard one of Jeff's graduate students mutter here: "Physics is the ability to win a bet." The fellow's name is Joseph Robert Buck, Jr. Watch out for him; a proto-Bayesian is a good thing.

02 March 1999, "Caves History"

By the way, I noticed in reading your old paper "In Defense of the Standard Quantum Limit," that you had at sometime in the deep past read Schrödinger's 1935 paper on entanglement (the one translated into English by Trimmer). If I can make a recommendation, I would suggest you have a look at it again while rooting through those Galavotti/Jeffrey papers. The guy is really fixated on the point that the thing that sets the quantum world apart from the classical one is not that maximal information is incomplete, but that maximal information of the whole does not correspond to maximal information of the parts. (You may recall I made this point to you before in a note titled, "Oh Hibernia.") [See note to Rüdiger Schack, dated 25 January 1999.]

03 May 1999, "Complex H Space"

One of your references should be Chapter 1 of that book *Quaternionic Quantum Mechanics* (I think it's by Steven Adler, but you'll have to check the library or something for the exact references.)

The other references I can think of are Wootters' two papers below. And there's at least one by John Wheeler with some discussion of the match between uniform distributions in probability space and Hilbert space (it works for complex but not for reals). Wheeler's paper was in the special issue of the *IJTP* in 1982. I think it was titled "The Computer and the Universe."

1. W. K. Wootters, "Quantum mechanics without probability amplitudes," *Foundations of Physics*, vol. 16(4), pp. 391–405, 1986.
2. W. K. Wootters, "Local accessibility of quantum states," in *Complexity, Entropy and the Physics of Information* (W. H. Zurek, ed.), (Redwood City, CA), pp. 39–46, Addison-Wesley, 1990. Santa Fe Institute Studies in the Sciences of Complexity, vol. VIII.

10 June 1999, "Gleasonization"

Cavesism 5: *If we're radical probabilists instead of rational Bayesians, how can [we] expect someone to adopt the quantum rule?*

I guess I agree with this. That's why I wrote the sentence:

Quantum mechanics, despite all its objectivist trappings, is a theory of how we should behave in light of the facts the world presents to us – that is, in light of the answers the world gives to the questions we ask.

in my last attempt at an introduction to our paper. (Have a look at it again when you get a chance.)

But I'm starting to get a little more dubious about the statement:

Cavesism 6: *That's why Gleason's theorem is the greatest triumph of Bayesian reasoning: probs that are consistent with the Hilbert space structure must be derived from a density operator.*

My worry comes from a couple of fronts. The first is the way Gleason's theorem is intimately tied to the field over which the quantum mechanical Hilbert spaces are defined. It's a consequence of David Meyer's recent result (and Adrian's extension) that Gleason's theorem does not hold for Hilbert spaces over a rational field (real or complex). (This is a question I had posed to Itamar Pitowsky a couple of years ago, and then to David ... but he already knew the answer.) This new fact troubles me a little bit.

Aside: Though, really, it is probably just telling us that that kind of discrete structure doesn't capture the essence of quantum mechanics. It makes me curious what discrete structures would still give rise to Gleason. One idea that pops up – and one directly tied to the importance of distinguishability – is whether the theorem can be proved if one "rationalizes" not the field for the vector space, but instead the *angles* between the vectors. "The valid questions that quantum systems can be asked correspond to a given standard orthonormal basis and all bases that can be gotten from it by rational rotations. Or perhaps by rotations that are rational fractions of pi; this assumption might be more natural. The theory must then give probabilities for the outcomes." – that's the sort of thing I'm thinking.

The other front – and this one is more important – comes from the suspicion that the assumptions behind Gleason's theorem are not purely Bayesian in motivation. There's something extra there. I probably can't make this more clear than by tacking on a note I wrote to Howard the other day. [See note to Howard Barnum dated 2 June 1999.] Probabilities to do with the Hilbert space structure, yes. But why probabilities that are context independent (in the sense below). That part strikes me as having nothing to do with Bayesianism.

10 September 1999, "The Dangers of Probabilismo"

Cavesism 7: *This seems somehow relevant to Chris's idea that Hamiltonians are subjective, whereas I prefer to think that they are the objective part of our description of the world.*

I believe I understand your worry (fear). I too feel that there must be some solid bedrock in quantum mechanics somewhere ... but I haven't been convinced that the Hamiltonian is it. The point is, as best I can tell, each and every argument we have used for the subjectivity of the wave function can also be used for the unknown CPM.

1. An unknown Hamiltonian can't be measured on a single shot. If you want to know the Hamiltonian you can't ask the system it lives on, you have to ask the maker.
2. The decomposition of the CPM is not unique. (Somebody once told me that a similar argument was the most damning for the quantum state.)
3. Let half of an EPR pair be the control bit entering in a controlled unitary gate. The action on the target bit can be made to go one way or the other, according to our distant measurement.
4. Probabilities are subjective, and quantum states are just compendia of probabilities. The Hamiltonian's main (sole?) purpose within the theory is just to evolve those compendia. That makes it look more like a law of thought than a law of nature (see note 5 below).

And – now this one is a lot more speculative – 5) Time evolutions are linear. Just, as you pointed out to me, this is something of an analogy to Bayes' rule. I wonder whether this suggests that Hamiltonian evolution might be enforced solely by probabilistic coherence à la Dutch book.

As Herb likes to say, "We need a little reality." But I think it's pretty important that we hit the right one when we do.

12 September 1999, "CPM Extreme Points"

Here's the reference (finally):

Seung-Hyeok Kye, "On the convex set of completely positive linear maps in matrix algebras," *Math. Proc. Camb. Phil. Soc.* **122**, 45–54 (1997).

23 September 1999, "The Ontological Tide"

Cavesism 8: *When you deny the dynamics an ontological status, you will leave everyone far behind, because you will be left with no ontology at all.*

No, I wouldn't want to go that far. And I don't think we have to. I can tell now, you've never quite gotten this information–disturbance bit I've been preaching. (It might not be right, but it does seem to be a way out.) So, let me try again: read the little manifesto below that I wrote up earlier this year. [See note to Greg Comer dated 22 April 1999.] It drips of ontology … but an ontology that gets shunted away every time we try to get our grubby hands on it. THAT statement is an ontological one in and of itself.

23 September 1999, "The Ontological Rip Tide"

Continuing,

Cavesism 9: *Here's a thought. I think there is no more power[ful] principle for dynamics than the statement that we never lose information unless we voluntarily throw it away. Either classically or quantum mechanically this means that we can lose information about a system if it interacts with another system about which we don't have maximal information. The new ingredient quantum mechanically is entanglement – Schrödinger's "maximal information about the whole is not maximal information about the parts." This means that we can lose information about a quantum system if it interacts with another system even though we have maximal information about the whole.*

If you think about it, you might notice a similarity between this and the manifesto I just sent you. Here's how I put it to Todd Brun a few months ago: … [See note to Todd Brun dated 8 June 1999 (Other Letters).]

Be careful of the rip tide you're swimming.

27 September 1999, "Cleaning the Plate" and "Last Scrap"

Cavesism 10: *On the ontology question, I wasn't convinced by your information–disturbance document. Look as hard as I could, I didn't find the slightest hint of what the underlying ontology is supposed to be.*

> Whereof one can not speak, thereof one should not speak.
> *(L. Wittgenstein)*

Cavesism 11: *Scientists have a way of not liking being told that something is outside their speculation.*

Thus my reason for testing the idea that information gathering and information disturbance go hand in hand as the "mechanism" (a real property of the world) to ensure that maximal information is not complete. That is the proposed mechanism for blocking the speculation. I had precisely this conversation with Lucien Hardy a couple of months ago. The I vs. D "mechanism" seemed to placate him a bit: I drew strength from that.

Cavesism 12: *Once we get to this point, we proceed to the points in your second note, where I am glad to see that we converge instead of diverging and where I'm happy to concede that you have long appreciated the point, and I am only now starting to catch up.*

I hope you didn't get the impression that I was worried about some piddling issue of priority. I only wanted to take the opportunity to let your own words push you in the right direction. The dates on the passages I sent you were there because I cut and pasted them from my master file "Notes on a Paulian Idea" (soon to be made available to some of our sympathetic colleagues).

It is my plan to attain nirvana by becoming egoless.

Cavesism 13: *Information–disturbance is the fundamental principle – beyond just chanting the slogan that "maximal information is not complete" – that says that you can go ahead and gather all the information you want – i.e., draw from the well – but you must produce a disturbance to be consistent with the slogan – i.e., the well is more than just a source of information, it's also a destroyer of information.*

I'm not sure I quite like the way you present things here. Remember, in my world it takes at least two scientists to tango ... and a lot more than that to come to some consensus about a theory. It is information gathering and INFORMATION disturbance that accompany each other. But we might be in agreement about the broad sweep of things, especially the part about consistency.

24 March 2000, "Title and Abstract?"

How about the one below? This was the talk I gave at MIT. I'll try to remember roughly what I said then.

Title: Quantum Information, Quantum Channels

Abstract: Most physics students, with their first lesson on the Heisenberg uncertainty principle, are given a subliminal message: quantum mechanics is a limitation. The attitude is, "Quantum mechanics is something we deal with because we have to, but wouldn't the world have been so much better if we could just measure a particle's position and momentum simultaneously?" This talk is about the counterpoint to that attitude. Recent advances in the fields of quantum computation, quantum cryptography, and quantum information theory show that the physical resources supplied by the quantum world are anything but a limitation. With these new resources we can do things almost undreamt of before, from the secure distribution of one-time pads for use in cryptography to the factoring of large numbers with a polynomial number of steps. The magic ingredient in all this is something called quantum information. I will illustrate the subtle strangeness of this new kind of information and the nice effects it buys with several concrete examples drawn from my own work in quantum cryptography and quantum channel-capacity theory.

30 June 2000, "Greece"

In Greece right now. Hitting naked breast overload, but otherwise I'm OK. Strangely, I've spent a lot of time with Roland Omnès the last couple of days, and he's starting to strike me as really OK. In the right light, his view of QM (even including consistent histories) may not be so different from ours. It's been a most useful time. Too bad though that he still seems to be stuck in thinking that taking measurement as a primitive in the theory is the same thing as taking measurement as a primitive in the world. (Though a few things seem to hint that he doesn't believe that either.)

Thanks for the Jaynes quotes too. I'm hoping to get a chance to read them and think about them on my four hour boat ride to Athens tomorrow.

Carl's Preply

Motivated by Chris's reading of Jaynes's article, "Predictive Statistical Mechanics," I read the three articles that seem to have his most direct statements about quantum mechanics. The quotes I mined follow.

From E. T. Jaynes, "Probability in Quantum Theory," in *Complexity, Entropy and the Physics of Information*, edited by W. H. Zurek (Addison-Wesley, Redwood City, CA, 1990), pp. 381–403:

Abstract: "For some sixty years it has appeared to many physicists that probability plays a fundamentally different role in quantum theory than it does in statistical mechanics and analysis of measurement errors. It is a commonly heard statement that probabilities calculated within a pure state have a different character than the probabilities with which different pure states appear in a mixture, or density matrix. As Pauli put it, the former represents '... eine prinzipielle *Unbestimmtheit*, nicht nur *Unbekanntheit*'. But this viewpoint leads to so many paradoxes and mysteries that we explore the consequences of the unified view, that all probability signifies only human information. We examine in detail only one of the issues this raises: the reality of zero-point energy."

p. 382: "Today we are beginning to realize how much of all physical science is really only *information*, organized in a particular way. But we are far from unravelling the knotty question: 'To what extent does this information reside in us, and to what extent is it a property of Nature?'"

p. 385: "Let me stress our motivation: if quantum theory were not successful pragmatically, we would have no interest in its interpretation. It is precisely *because* of the enormous success of the QM mathematical formalism that it becomes crucially important to learn what that mathematics means. To find a rational physical interpretation of the QM formalism ought to be considered the top priority research problem of theoretical physics; until this is accomplished, all other theoretical results can only be provisional and temporary.

"This conviction has affected the whole course of my career. I had intended originally to specialize in Quantum Electrodynamics, but this proved to be impossible. Whenever I look at any quantum-mechanical calculation, the basic craziness of what we are doing rises in my gorge and I have to try to find some different way of looking at the problem, that makes physical sense. Gradually, I came to see that the foundations of probability theory and the role of human information have to be brought in, and so I have spent many years trying to understand them in the greatest generality."

pp. 386–387: "Einstein's thinking is always on the ontological level traditional in physics, trying to describe the realities of Nature. Bohr's thinking is always on the epistemological level, describing not reality but only our information about reality. The peculiar flavor of his language arises from the absence of all words with any ontological import; the notion of a 'real physical situation' was just not present and he gave evasive answers to questions of form: 'What is really happening?' ...

"Although Bohr's whole way of thinking was very different from Einstein's, it does not follow that either was wrong. In the writer's view, all of Einstein's thinking – in particular the EPR argument – remains valid today, when we take into account its ontological character. But today, when we are beginning to consider the role of information for science in general, it may be useful to note that we are finally taking a step in the epistemological direction that Bohr was trying to point out sixty years ago."

p. 387: "...Our present QM formalism is a peculiar mixture describing in part laws of Nature, in part incomplete human information about Nature – all scrambled up together by Bohr into an omelette that nobody has seen how to unscramble. Yet we think the unscrambling is a prerequisite for any further advance in basic physical theory, and we want to speculate on the proper tools to do this.

"We suggest that the proper tool for incorporating human information into science is simply probability theory – not the currently taught 'random variable' kind, but the original 'logical inference' kind of James Bernoulli and Laplace. For historical reasons explained elsewhere, this is often called 'Bayesian probability theory.'"

p. 390: "We would like to see quantum theory in a similar way; since a pure state ψ does not contain enough information to predict all experimental results, we would like to see QM as the process of making the best predictions possible from the partial information that we have when we know ψ. If we could either succeed in this, or prove that it is impossible, we would know far more about the basis of our present theory and about future possibilities for acquiring more information than we do today."

From E. T. Jaynes, "Predictive Statistical Mechanics," in *Frontiers of Nonequilibrium Statistical Physics*, edited by G. T. Moore and M. O. Scully (Plenum Press, NY, 1986), pp. 33–55.

"We think it unlikely that the role of probability in quantum theory will be understood until it is generally understood in classical theory and in applications outside of physics. Indeed, our fifty-year-old bemusement over the notion of state reduction in the quantum-mechanical theory of measurement need not surprise us when we note that today, in all applications of probability theory, basically the same controversy rages over whether our probabilities represent real situations, or only incomplete human knowledge.

"If the wave function of an electron is an 'objective' thing, representing a real physical situation, then it would be mystical – indeed, it would require a belief in psychokinesis – to suppose that the wave function can change, in violation of the equations of motion, merely because information has been perceived by a human mind.

"If the wave function is only 'subjective,' representing a state of knowledge about the electron, then this difficulty disappears; of course, by definition, it will change with every change in our state of knowledge, whether derived from the equations of motion or from any other source of information. But then a new difficulty appears; the relative phases of the wave function at different points have not been determined by our information; yet they determine how the electron moves.

"There is no way quantum theory could have escaped this dilemma short of avoiding the use of probability altogether. Not only in Physics, but also in Statistics, Engineering, Chemistry, Biology, Psychology, and Economics, the nature of the calculations you make, the information you allow yourself to use, and the results you get, depend on what stand you choose to take on this surprisingly divisive issue: are probabilities 'real'?

"But in quantum theory the dilemma is more acute because it does not seem to be merely a choice between two alternatives. The 'subjective' and 'objective' aspects are scrambled together in the wave function of an electron, in such a way that we are faced with a paradox like the classical paradoxes of logic; whatever stand you take about the meaning of the wave function, it will lead to unacceptable consequences.

"To achieve a rational interpretation we need to disentangle these aspects of quantum theory so the 'subjective' things can change with our state of knowledge while the 'objective' ones remain determined by the equations of motion. But to date nobody has seen how to do this; it is more subtle than merely separating into amplitudes and phases.

"As many have pointed out, starting with Einstein and Schrödinger fifty years ago and continuing into several talks at this Workshop, the Copenhagen interpretation of quantum theory not only denies the existence of causal mechanisms for physical phenomena; it denies the existence of an 'objectively real' world.

"But surely, the existence of that world is the primary experimental fact of all, without which there would be no point to physics or any other science; and for which we all receive new evidence every waking minute of our lives. This direct evidence of our senses is vastly more cogent than are any of the deviously indirect experiments that are cited as evidence for the Copenhagen interpretation.

"Now let's look at the mind-boggling problem from a different side. A single mathematical quantity ψ cannot, in our view, represent incomplete human knowledge and be at the same time a complete description of reality. But it might be possible to accomplish Bohr's objective in a different way. What he really wanted to do, we conjecture, is only to develop a theory which takes into account the fact that the necessary disturbance of a system by the act of measurement limits the information we can acquire, and therefore the predictions we can make. This was the point always stressed in his semipopular expositions. Also, in his reply to EPR he noted that, while there is no physical influence on S, there is still an influence on the kinds of predictions we can make about S.

"On deep thought, it will be seen that whenever we allow probabilities to become 'physically real' things, logical consistency will force us, eventually, to regard the objects as 'unreal.' If we are to reach Bohr's goal while at the same time keeping our objects real we must recognize, with Laplace, Maxwell, and Jeffreys, that whenever we use probability it must be as a description of incomplete human knowledge, as it was in classical statistical mechanics."

From E. T. Jaynes, "A Backward Look to the Future," in *Physics and Probability: Essays in Honor of Edwin T. Jaynes*, edited by W. T. Grandy, Jr. and P. W. Milonni (Cambridge University Press, Cambridge, England, 1993), pp. 261–275.

p. 269: "... throughout the history of quantum theory, whenever we advanced to a new application it was necessary to repeat this trial-and-error experimentation to find out which method of calculation gives the right answers. Then, of course, our textbooks present only the successful procedure as if it followed from general principles; and do not mention the actual process by which it was found. In relativity theory one deduces the computational algorithm from the general principles. In quantum theory, the logic is just the opposite; one chooses the principle to fit the empirically successful algorithm."

pp. 270–271: "What has held up progress in this field for so long? Always our students are indoctrinated about the great pragmatic success of the quantum formalism – with the conclusion that the Copenhagen interpretation of that formalism must be correct. This is the logic of the Quantum Syllogism:

The present *mathematical formalism* can be made to reproduce many experimental facts very accurately.

Therefore

The *physical interpretation* which Niels Bohr tried to associate with it must be true; and it is naïve to try to circumvent it.

Compare this with the Pre-Copernican Syllogism:

The mathematical system of epicycles can be made to reproduce the motions of the planets very accurately.

Therefore

The theological arguments for the necessity of epicycles as the only perfect motions must be true; and it is heresy to try to circumvent them.

In what way are they different? The difference is only that today everybody knows what is wrong with the Pre-Copernican Syllogism; but (from the frequency with which it is still repeated) only a relatively few have yet perceived the error in the Quantum Syllogism."

p. 272: "One of the principles of scientific inference – which has always been well understood by the greatest scientists – is that it is idle to raise questions prematurely, when they cannot be answered with the resources available. For Isaac Newton it would have been foolish to raise questions that were not foolish for Erwin Schrödinger 250 years later; for Gregor Mendel it would have been foolish to raise questions that were not foolish for Francis Crick 100 years later. By 'foolish' we mean 'without hope of success'. Of course, we all enjoy indulging in a little free speculation about the future of science; but for scientists to expend their serious professional time and effort on idle speculation can only delay any real progress."

07 September 2000, "Critical Letters and Reply"

Cavesism 14: *Liked the "effective reality," too.*

[NOTE: This sentence of Carl's refers to my article with Asher Peres: C. A. Fuchs and A. Peres, "Quantum Theory – Interpretation, Formulation, Inspiration: Fuchs and Peres Reply," *Phys. Tod.* **53**(9), 14, 90 (2000).]

But I had even managed to get that into the original:

We do not deny the possible existence of an objective reality independent of what observers perceive. In particular, there is an "effective" reality in the limiting case of macroscopic phenomena like detector clicks or planetary motion: Any observer who happens to be present would acknowledge the objective occurrence of these events. However, such a macroscopic description ignores most degrees of freedom of the system and is necessarily incomplete.

Keep in mind, though, that when I speak of an effective reality I think I mean something quite different from you. [See "Carl's Reply" to my note "Dropping In," dated 20 September 1998.] In particular, what I have in mind seems to have nothing to do with the particular properties of the Hamiltonian or dynamics. That's where we part mostly. My "effective reality" is, instead, a function of our ignorance and the crudeness of our experimental intervention capabilities. It is about the idea that the information–disturbance tradeoff disappears under such conditions. The *Physics Today* reply makes that a little more explicit:

They could do that because this aspect of carbon is part of the "effective reality" quantum theory produces in some regimes of our experience. Indeed, this "effective reality" forms the ground for all our other quantum predictions simply because it is the part of nature that is effectively detached from the effect of our experimental interventions. But, if one tries to push this special circumstance further and identify an overarching "reality" completely independent of our interventions, then this is where the trouble begins and one finds the *raison d'être* of the various "interpretations."

13 November 2000, "Bush vs. Bohr"

From the *New York Times* this morning:

Throughout the primaries and general election, the Bush campaign had been unusually confident, partly because Mr. Bush and his aides genuinely seemed to think they would win. But advisers also said they were trying to convey a sense of inevitability to Mr. Bush's candidacy. Since the election, the Bush team has done much the same thing, striking the posture of victory, which helps explain why Mr. Bush spoke last week about his planning for the transition and his aides leaked the names of potential top cabinet members.

I do think the parallels between Bush's behavior and consistent-history quantum mechanics are very real.

01 February 2001, "An Ed/Ellen Jaynesian"

You are an Ed Jaynesian! I just came to realize this in rereading the quotes of his that you jotted down.

"To achieve a rational interpretation we need to disentangle these aspects of quantum theory so the 'subjective' things can change with our state of knowledge while the 'objective' ones remain determined by the equations of motion. But to date nobody has seen how to do this; it is more subtle than merely separating into amplitudes and phases."

(Saying Ed Jaynesian, I keep thinking of the Ellen Jaynesians in *The World According to Garp* and wonder if you too will eventually cut out your tongue when you realize that Hamiltonians are just the pure states of time evolution.)

01 February 2001, "My Jaynes"

On the other hand, I should classify myself a Jaynesian too. Perhaps a mild Jaynesian, for my passion only extends so far as:

"We would like to see quantum theory in a similar way; since a pure state ψ does not contain enough information to predict all experimental results, we would like to see QM as the process of making the best predictions possible from the partial information that we have when we know ψ. If we could either succeed in this, or prove that it is impossible, we would know far more about the basis of our present theory and about future possibilities for acquiring more information than we do today."

26 February 2001, "Death of a Salesman"

I know the difference between science and philosophy. You said that you "would never, ever believe that Hamiltonians are not the real properties of systems." To the extent that I know you too, I know you didn't mean that. You only meant that *without* firm scientific evidence placed upon your plate you would not believe it. That's my burden, I understand, not only with respect to you but, more importantly, for judgment day. Maybe I'll believe I'm wrong ultimately, but if so, I think we still both have to ask ourselves the source of the "miracle" that something unquestionably a state of knowledge can masquerade so easily as a state of nature at times. The evidence below is nothing new – the essential was in an old email to you – but maybe it looks more respectable now. [See quant-ph/0012067, "Storage of quantum dynamics on quantum states: a quasi-perfect programmable quantum gate," by G. Vidal and J. I. Cirac.]

11

Letters to Greg Comer

The ghost of my father sat in front of me
    sprinkling salt into its beer.
The floor, wooden and sole soaked slick
I could just push my feet and rub my fingers.
"Give me more life," I whispered.

I wanted to touch the flakes of rust
    on the cooler, mingled with sweat.
Bottle caps everywhere; it didn't understand
    the boredom.
My only sound was that of a screen door.
"My son will be a professor of physics in three years."

14 February 1995, "Turtle Wax"

Let me wax just a little. A few quotes by Wolfgang Pauli on quantum mechanical randomness and objective reality. (Taken from the English translation of a paper titled "Matter.")

"Like an ultimate fact without any cause, the *individual* outcome of a measurement is, however, in general not comprehended by laws. This must necessarily be the case ..."

"In the new pattern of thought we do not assume any longer the *detached observer*, occurring in the idealizations of this classical type of theory, but an observer who by his indeterminable effects creates a new situation, theoretically described as a new state of the observed system. In this way every observation is a singling out of a particular factual result, here and now, from the theoretical possibilities, thereby making obvious the discontinuous aspect of the physical phenomena."

"Nevertheless, there remains still in the new kind of theory an *objective reality*, inasmuch as these theories deny any possibility for the observer to influence the results of a measurement, once the experimental arrangement is chosen."

Objectivity grounded upon randomness!! Hmm, I like that.

04 March 1995, "Saturday Slush"

Let me give you another Pauli quote ... as is my habit of late. This one is from an article titled "The Theory of Relativity and Science" and is really good.

"In spite of this, Einstein held firmly to the narrower concept of reality of classical physics; from this point of view a description of nature which permits single events not determined by laws was bound to appear to him "incomplete." He combined with this a regressive longing, not indeed for the old mechanistic idea of the point-mass, but for his geometrical conception of the field in the general theory of relativity. Motivating his attitude, he frankly explained that to depart from the narrower reality concept of physics before quantum mechanics seemed to him to be getting perilously close to a point of view in which it is impossible to discriminate sufficiently clearly between dream or hallucination and "reality." As against this, the objective character of the description of nature given by quantum mechanics has appeared to the rest of us to be adequately guaranteed by the circumstance that its statistical laws describe reproducible processes, and that the results of observation, which can be checked by anyone, cannot be influenced by the observer, once he has chosen his experimental arrangement."

Man o' man I like that. In all my years of reading silly things about quantum mechanics, I have never found anybody better than Pauli at putting forth the crucial point in such a clear fashion. Even Wheeler in his Game of Twenty Questions, didn't make it quite this clear. The world in some very real sense is a construct and creation of thinking beings simply because its properties are so severely tied to the particular questions we ask of it. *But* on the other hand, the world is not completely unreal as a result of this; we generally cannot control the outcomes of our measurements. It is precisely because the outcome of the individual event cannot be determined by law that the world still has a trace of reality. If we actually could control the outcomes of measurements, *then* the world would as well be a "dream or hallucination." ... But we can't!

Do you remember the following words from a poem I wrote last year:

"Einstein must have foreseen the evil in our dear quantum. I don't envy the pain; the blood so very frightening. No evil in *A*-bomb, but rather the evil of no image; the evil of no substrate; the burning people in Waco; my father tied to a bed. The poor roaming wisdom of E."

Well Mr. Pauli confirms my suspicion of what Einstein feared most in quantum mechanics.

07 March 1995, "Please, Please No More!"

One LAST time, for the sake of a little more clarity! I promise I won't send any more quotes along these lines unless I find something significantly new. But, in the mean time, Mr. Pauli again; this time from "Albert Einstein and the Development of Physics."

"We often discussed these questions together, and I invariably profited very greatly even when I could not agree with Einstein's views. 'Physics is after all the description of reality,' he said to me, continuing, with a sarcastic glance in my direction, 'or should I perhaps say physics is the description of what one merely imagines?' This question clearly shows Einstein's concern that the objective character of physics might be lost through a theory of the type of quantum mechanics, in that as a consequence of its wider conception of objectivity of an explanation of nature the difference between physical reality and dream or hallucination becomes blurred.

The objectivity of physics is however fully ensured in quantum mechanics in the following sense. Although in principle, according to the theory, it is in general only the statistics of series of experiments that is determined by laws, the observer is unable, even in the unpredictable single case, to influence the result of his observation – as for example the response of a counter at a particular instant of time. Further, personal qualities of the observer do not come into the theory in any way – the observation can be made by objective registering apparatus, the results of which are objectively available for anyone's inspection. Just as in the theory of relativity a group of mathematical transformations connects all possible coordinate systems, so in quantum mechanics a group of mathematical transformations connects the possible experimental arrangements.

Einstein, however, advocated a narrower form of the reality concept, which assumes a complete separation of an objectively existing physical state from any mode of its observation. Agreement was unfortunately never reached."

27 January 1996, "Philo"

The philosophy of it all my friend. I just heard a song on the radio that I had not heard since the Chapel Hill days – it drew my thought to you. I remember writing you almost exactly three years ago about my arrival in the land of the quantum. I'm writing you from the very same machine that I did then. So much is the same and so much is different.

Philosophy in every turn I used to say. Can you believe I've secured a full time job playing with quantum mechanics. I get paid to explore it and pick it apart in any way I wish. I get paid to keep dreaming about this lovely little structure.

The present project is finding bounds on the minimal resources required to transpose a quantum state from here to there. It's just another way to explore how quantum a set of little quantum thingies are. Paul Simon is the moral guide.

When I get back to Montréal two projects are on the burner. Another calculation to gauge the tradeoff between information gain and disturbance in quantum measurement. And another attempt to derive much of quantum theory from information theoretic principles – an extension of Bill Wootters work as John Wheeler's graduate student.

So many ways to profiteer from the madness of trying to build a quantum computer. That's OK though – there really is hope that all the efforts won't be wasted along the way.

I'm having a cup of coffee; it's a little after 8:00 in the evening. Kiki is in Montréal just going to bed about now. The world is turning beneath me and I feel it; it all seems very mystical tonight.

Where am I going with this? I don't know. I just wanted to say that the quantum is alive and well. And that coming out here panned out. And that in some ways it's just the way it was three years ago.

I turn within myself.

20 February 1996, "The Mont-Royal Sweats"

She said: How much does information weigh?
I said: How heavy is a kiss?

Thanks for the tidbits and epiphanies.

I do like it up here very much (except when I have to walk the dogs every morning and every evening!). We have a really big apartment; it's about twice as big as our last house. And for some reason it has more the feel of an "academic" home – a place where thoughts have a chance of being made. It's a shame I'll have to move before I know it; I was really looking forward to a two year stay in Montréal.

I'm glad you're getting to teach a "Modern Physics" course. (I presume that means ideas about the Bohr atom and a little special relativity?) Is this your first time in such an endeavor? I would like to think that when I have the same opportunity I'd break the bounds of protocol ... and not mention even once any crap about wave–particle dualism or any other such vague ideas. Rather I'd focus on quantum mechanical randomness at the outset and not even attempt to get past two and three-level systems. But of course such blasphemy will never be; I know that I'll be bound by what the other professors in the department need. Ahh, the dreams in my life.

Gilles just wrote a semi-popular account of what new things are happening in quantum computing; if it's any good (I haven't read it yet), I'll send it your way.

This quantum information business, do you know what I like about it? I very much like it because it puts the focus just where it should be: that quantum theory is a theory of information and predictability ... and probably nothing more. It gives us new handles for studying and manipulating this idea. That's what I like, and that's why I've decided to stick with it. When I'm old and grey, I'll come back to the deeper question of what's really going on with the quantum phenomena themselves. (⟵ That's probably a lie; since when have I ever been able to stop thinking about the deeper structure of the quantum?)

The silly theorem I'm trying my damnedest to prove right now is another simple example of this sort of stuff. I have a set of pure quantum states ψ_i on some n-dimensional Hilbert space that occur with probabilities p_i. Then I consider "coding" them in any way I like onto the states of a d-dimensional Hilbert space with $d < n$; i.e., I imagine making any one-to-one correspondence between the ψ_i and a similar set of vectors on the smaller Hilbert space. These "code" vectors are later to be "decoded" by an automatic device that obeys the laws of physics, i.e., has a unitary interaction with the d-dimensional space. The final output is a set of new states ρ_i (which are generally mixed because we may ignore the final state of the device itself). The question is: what is the largest average inner product that can be made between the ψ_i and the ρ_i by such a procedure? I would like to say that that number is bounded above by the sum of the largest d eigenvalues of the density matrix formed from the ensemble of ψ_is ... but I just can't show it!!! And it's driving me crazy.

My pet idea of late is that the foundation of unitary dynamics for quantum theory can be replaced with the idea that nonorthogonal quantum states cannot be cloned or broadcast.

There is a certain theorem originally due to Wigner that makes this idea quite plausible. His theorem is that any evolution on Hilbert space that preserves inner products must either be unitary or anti-unitary. Cloning nonorthogonal quantum states requires that certain inner products must decrease. So if you outlaw cloning at the outset, then you outlaw inner-product-decreasing evolutions. Well it turns out that if you outlaw such evolutions you also outlaw inner-product-increasing evolutions (this I've proven, it's not hard). So all you're left with is inner-product-preserving evolutions. Given that and Wigner's theorem, quantum mechanics is almost uniquely pinned down. It's a very pleasant idea I think … because it is the no-clonability of nonorthogonal states that gives quantum information theory all its oomph.[1]

27 February 1996, "Kant Cola"

Just thought I'd take a little coffee break before I get started up for the morning. I've got to start going through (with a fine-toothed comb) a first draft of a paper that Howard Barnum, Ben Schumacher, Richard Jozsa, and I are coauthoring. (Ben wrote the first draft of this one.) The subject is on the full fledged converse to the quantum coding theorem; namely, if less than a von Neumann's entropy worth of qubits are sent per transmission, then a quantum signal can be reconstructed with vanishingly small error. ⟵ I know that doesn't mean much to you, but that's the subject.

The weather up here is very nice today. It's a little below freezing, but the sun is out and there's not a bit of wind.

Kiki and I went into the Outrèmont area Friday evening in search of an interesting restaurant … and what a find we made! Let me tell you a strange little story. During the summer of 1985, I was reading a book by C. F. von Weizsäcker titled *The Unity of Nature*. Most of the book was about quantum mechanics and Kantian philosophy. Apparently it spurred me to have the following dream. I was in a little hole-in-the-wall joint somewhere in Austin; my old friend David was there, also John Simpson and Marshall Burns. The place really stood out in my mind because of the Bohemian feel to it: dark, smoky, mystical almost. The night wasn't filled with much of interest: David only wanted to talk about getting drunk, John only wanted to talk about finding a girl, and Marshall only wanted to talk about philosophy. In those days John didn't drink alcohol, so, at some point, when he asked for a drink, I thought we'd be out of luck. But upon looking around, I saw a refrigerator in the middle of the bar near the pool table. We walked over to it and took a look. It was filled with all different sorts of vegetable drinks. John grabbed one, and I looked through it for something more interesting. At the very back, I found one lone can of "Kant Cola." That was written on the label, along with a small portrait of Immanuel Kant. I opened the can, took a drink, … and, for a miraculous moment, I understood all

---

[1] NOTE: I later found a flaw in this argument. The problem is, cloning transformations do not describe enough kinds of "inner-product-decreasing" maps. See C. A. Fuchs, "Information Gain vs. State Disturbance in Quantum Theory," *Fort. der Phys.* **46**(4,5), 535–565 (1998) for a correct solution to the problem.

the intricacies of the world – I understood the necessity of quantum mechanics. When I came out of my trance, the can was empty and I knew that I would never see the light again. Then I awoke. I was so taken with this dream that the next day I sketched out the layout of the joint and made a record of the dream. That was over ten years ago.

So back to the restaurant of Friday. The place was called "City Pub"; it was such a strange little place: dark, smoky, mystical almost. The food was excellent – far better than it should have been for the price. Each option in the place was only $4.99. I had steak, fries, and a vegetable. Kiki had potato soup, quiche, fries, and veggie. They had a special on beer, three for the price of one (so we had six). The music was some sort of strange mesh of things that I suspect you'd only hear in some little bar in Germany where everyone wears black. Anyway, we had quite a time there. However, just a little while before leaving I started to note how similar this place was to the place in my dream 10 years ago. I told Kiki the whole story. Then I looked around and – strangely enough – there was a refrigerator in the middle of the room near the pool table! I was so taken by this that upon my way to the restroom, I took a look into it. What a disappointment: it only contained beer. However, the restroom did have a surprise for me. In the middle of all the graffiti (about Québec's hoped-for independence) was something written in bold black letters:

De nobis ipsis silemus
(E. Kant)

That made my evening. I wrote down the words so that I wouldn't forget and went home to look up my old notes on the dream. Sure enough, there were similarities in the layout of the two places, and moreover, I saw that the name of the original place in my dream was "Hole in the Wall Pub." Very strange. I asked Rüdiger Schack to translate the words for me, and he came up with "About ourselves we remain silent."

That's the story. I suppose I should be off to work.

06 March 1996, "Rub Me"

How would I respond? "If upon learning of quantum mechanics, you are not left with your head spinning ... then you haven't understood a thing." (Paraphrase of Bohr)

A question quite similar to this is what got me interested in quantum theory in the first place. In my sophomore year of high school, in my chemistry class, I saw some of these wonderful pictures of electron clouds around a hydrogen nucleus. And there were lumps of the cloud that were disconnected. So I asked the teacher what is the meaning of this cloud. She said, "It's where an electron can be." I said, "Then, for any given atom, a more accurate picture would only have one lump. Because there's no way then that the electron could get from one lump to the other." She said, "No, you're wrong." I said, "No, I'm right. Convince me otherwise." She couldn't of course, and I thought, "Ahh, what an idiot." But then I went off to college and thought long and hard about these silly quantum thingies ... and came to the conclusion that I was after all the one who was wrong! I've never quite gotten out of the befuddlement.

Maybe the best thing I can do is quote Charlie Rasco: "You just have to realize that *electrons are not little fucking billiard balls!*"

A better answer, perhaps, is what I actually believe. In the problem of the square well, like in all other quantum mechanical problems, the state vector we ascribe to the system encapsulates just exactly what we can predict about the outcomes of our probings on the system and it quantifies how much we are surprised about the outcome we actually find. If we are honestly able to build an honest-to-god square well potential and have a particle in it in its first excited state (and we are able to do so repeatedly), then we will never find the particle at the center. Period. If we ever find it there, the quantum state we ascribe the system should be updated to take into account that we have not been preparing the system as we had said we would. Any attempt to build a more "classical" picture beyond this – one that explains why some particles are found to the left of the center and why some are found to the right – is dangerous business. Chances are well beyond the 99.999% mark that there will be an inconsistency in any such fix-up you try. Just try it and see: you will find that (in one way or other) you have sneaked in superluminal signalling into physics, or broken the Second Law of Thermodynamics, or even worse, given up on a world in which there is anything beyond what Allah wills.

26 March 1996, "Savagery"

Silly, silly science; why did we ever get involved!

When I first started finding books on quantum mechanics in my freshman year in high school, I thought, "What dull stuff!" I preferred to devote my time to things about relativity, wormholes, and black holes. It was only after getting to college that my interest in quantum mechanics took off. It happened the first week I was there, after reading Heinz Pagels' *The Cosmic Code*. It was then that I started to take seriously that my old questions in my chemistry class (that I told you of) didn't have such satisfactory answers (and it wasn't just the ignorance of my old chemistry teacher).

So the upshot is: be patient, I guess. Maybe they'll decide it's interesting after they leave your nest.

27 March 1996, "Strange Comment"

Here's an interesting comment I thought you might like to hear. I was talking to one of the computer scientists interested in quantum cryptography here the other day [Jeroen van de Graaf]. I made a comment that the number of baryons in the universe is estimated to be roughly 10^{80}. He said (quite seriously), "Is that all?" "That's just an 80 digit number ... it can be factored on a classical computer."

01 April 1996, "Puppy Scoops"

I read a wonderful little piece on Einstein's ontology yesterday while sitting in a coffee shop – Chez Dick, I love the name! (Actually I read a much more detailed account in

Fine's book a year or so ago, but had apparently forgotten the force of it. The source this time was Aage Petersen's book.) The point is this. Mr. E did *not* take the existence of an objective (real) world independent of the knowers of it as an a priori given. So, though E was not a dyed-in-the-wool Baptist, he was not a dyed-in-the-wool *realist* either. His argument was better than a simple straight-up *belief* in a real world. He saw clearly that: it is *not* the case that there are no ontological systems for coordinating and organizing our experiences other than a "real world." Rather, he only contended that the idea of a real world *is* a simple explanation. Moreover, at least until the advent of quantum theory, it was an idea that served its purpose very well – allowing all science hitherto to be based on it. His only point was that he would be very hard pressed to give up the idea of a "real world" and – as of yet – had not accumulated enough evidence to force himself to do so.

04 May 1996, "You May Enjoy"

For some reason or other I received the note attached below. You may enjoy. I found particularly interesting the comment:

Diffeomorphism invariance is, in my view, precisely the expression of the statement that the theory may make no implicit reference to reference frames outside of the system.

Hmm? But what if quantum phenomena really require "measurement" to come into being? I.e., something outside the system to register their reality?

Ahh, it's like talking to a brick wall with some of these guys.

Anyway, my present attitude is summarized on p. 27 of Asher's book. "Other authors introduce a wave function for the whole Universe. In this book, I shall refrain from using concepts that I do not understand."

19 July 1996, "Old Moods"

I'm in one of those old "want to derive quantum mechanics" moods. Gloomy rainy days always do that sort of thing to me – they're always so conducive to deeper thoughts. I have it stuck in my head that "quantum mechanics is a law of thought" and the silly idea won't go away.

Today, on a more mundane level, I've been working on whether entanglement can be used to enhance the classical information capacity of quantum mechanical channels. The quantum is so fun. I've done some decent preliminary work; now all I need to do is get to Yorktown Heights (IBM Research) and get these guys to do a little simulation and see what happens.

Always I come back to the silly questions about what is at the heart of the matter. Is it "entanglement", i.e., that a maximal state of knowledge about a composite system is not a maximal state of knowledge about its parts? Is it that nonorthogonal quantum states cannot be identified with certainty? Is it that information cannot be gained about nonorthogonal alternative states of a quantum system without disturbing it? And, what do all these questions have to do with the idea that the structure of quantum mechanics is a "law of thought" and nothing more? (The physics is in assigning Hamiltonians and boundary conditions.)

Rainy days and love songs. I wrote Sam a longish note on all this "law of thought" business today. It's like I really feel it today, for one reason or the other. [See note to Sam Braunstein, dated 19 July 1996 (Other Letters).]

I noticed this morning that yet another algorithm has come out for something that can be done in an interesting way on a quantum computer. I like to see that sort of thing! Yes indeedy.

What would the world be without our friend the quantum. I wouldn't want to imagine really; I suspect it would be a mess ... or perhaps even nonexistent. But who really knows. My friend Richard Jozsa thinks that all of quantum mechanics is hogwash. He likes working on quantum computing because it gives a good vehicle for fleshing out all the interesting consequences of the theory. However, if he had to take a bet, it would be that when we actually attempt to build an interesting quantum computing device, we'll find that it simply won't work. QM will play out before then ... or so he's willing to bet.

Me, I take the opposite religion ... that quantum mechanics is here to stay, just as probability theory is here to stay. Who would you call conservative, me or Richard?

16 August 1996, "Hot Pavement and Slushy QM"

Well, you know, I really knew that you probably weren't going to become too enthralled by the variational problem. Like so many of my old girlfriends, you were just swept away by the lust of the moment ... forgetting how hard it is to make a relationship go. Tsk, tsk. At the very least, I forgive you ... just as with my old girl friends.

More seriously, do what you can, when you can, and just make sure you're having fun. I've been tied up with so many things myself. Bennett, Smolin, and I have made really quite a discovery. And we've been trying to consolidate that into something even more powerful, maybe even a bombshell.

The question is this: suppose you have a noisy fiber optic cable and you need to transmit one bit of information down it. Your resources are two and only two photons. Can it help in this simple transmission problem to entangle the photons into an EPR pair before sending them along their way? That is, can the bit (0 or 1) be identified better at the receiver if the photons are entangled first at the transmitter. The answer is, *wonderfully*, yes it does help. Entanglement can really help protect information as it's being transported. We now have explicit examples. The next question is whether we can up the overall information carrying capacity of the channel via entanglement. If this is true, then we'll have a bombshell.

By the way, someone here has lent me a copy of a new book by Chris Isham, *Lectures on Quantum Theory: Mathematical and Structural Foundations*. It looks to be really very good and very concise, saying just what needs to be said and not much more. It's only 220 pages and I hear pretty cheap. I think I would recommend it to someone, as yourself, interested in getting a tight introduction back into the game.

With that, let me leave you for the weekend, by quoting the first sentence of the last paragraph of Isham's book. "The central issue in all this is really the phenomenon of quantum entanglement, and its striking contrast with the reductionist concepts of Western philosophy."

18 August 1996, "Morning Thoughts"

A gloomy Sunday morning in Montréal, coffee in hand, Tony Bennett on CD. I'm reading a little book called *Night Thoughts of a Classical Physicist* and I came across something that made me think of you.

Voigt had decided on a career in physics rather than in music, since a musician had to be absolutely first rate and a physicist could get along on less.

29 August 1996, "Gerry Niewood on Saxophone"

I'm courageously entering this letter without any topic of conversation planned whatsoever. I'm feeling very strange tonight, watching my family fall apart because of a very poorly written will. I don't think I'll talk about that though.

As you've probably guessed, I do have Simon and Garfunkel on in the background: *Concert in Central Park*. And of course, quantum theory is on my mind too. It's always on my mind when I'm troubled. It's so mystical, so mysterious. I think I take religious comfort in it. I worked hard at it today, exploring how to shuffle things about in Hilbert space – you know the problem. "Just weary to my bones." I wish I could get a better handle on the problem; it eludes me and I just don't know what to expect. I know what I want to be true, but I just don't know what to expect.

Almost a religion. That concept has really been taking me lately: I am a priest, a student of the holy scriptures.

01 September 1996, and to David Baker, "The Iconostasis"

Presently, I am reading a biography of C. G. Jung; yesterday, I came across a little tidbit that I found quite intriguing. Let me repeat it:

The core of the Pueblo religion was that the deity sun needed the assistance of his sons, the Pueblos, who lived on the roof of the world. Their religious practices helped the Sunfather through his daily course. They believed that if their religion were to die out the sun would die, too, and the world be left in eternal night. . . . And he [Jung] found some solace in the thought that the central idea of the Pueblo religion – that God requires the cooperation of his creatures to carry out and perfect His existence – had long been familiar to him through the writings of medieval Christian mystics like Meister Eckhart and through his own meditations on the mystery of the God–Father–Son relationship.

As you both know, I've long been interested in the idea of a "participatory universe" (Wheeler's big U with an eye at the top looking at the other side) because of quantum mechanics. However, I had not realized that a similar idea can be found in more established religions ... much less in some older versions of Christianity!

Do either of you have any thoughts or pointers on this? I presume Jung wrote something up on this; I'll look that up. How about Joseph Campbell? Does he have much to say on the Pueblo Indians? Who the heck was Meister Eckhart?

Greg's Reply

Joseph Campbell got interested in world religions in the first place by studying the American Indians. So yes, he has very much to say about American Indian religions, and I assume he talks about the Pueblos – but it has been so long since I read Campbell I can't remember many particulars. [...]

Campbell has this wonderful picture in one of his books: a photograph of a relief on a wall somewhere in India, I think. The relief has 4 brothers at the bottom, asleep, and Vishnu, the cosmic-dreamer whose dream is the universe, at the top and also asleep. Campbell's interpretation is this: by construction, the brothers exist because of Vishnu's dream. The amazing thing is Vishnu exists because of the dream of the 4 brothers. The first time I saw this I immediately thought of Wheeler and his capital U. If you like, I can give you a reference for Campbell's book. I'm willing to bet that this example from the Indian culture predates anything in Christianity. It would be interesting to know what the Dead Sea Scrolls say, though, since they contain the earliest written records of Christianity.

I seem to recall other cultures where the Sun God needs help to get up and around the Earth each day, for instance, maybe the Egyptians or Aztecs? But, I don't know if they considered it a central idea that the God would die if the religion died.

03 September 1996, "A Procrastinating Plunk"

It's amazing how I can always find time to procrastinate when I have to. I told you I was reading Jung's *Synchronicity* yesterday, right? Well, I did read it finally, and I must say, I was quite disappointed. In all, I think I like Pauli's meager writings on the concept much more than Jung's. In fact, I now think what I had been running around calling "Jung's" synchronicity concept is really Pauli's: whatever it is that Jung himself is talking about now appears to be much closer to ghosts and goblins than I had imagined. It seems only to have been a fancy way of talking about astrology, ESP, and psychokinesis – things he was evidently taken by.

There were a few tidbits in the book that I really liked the sound of, but they were pretty scarce on the whole.

31 May 1997, "Long Day"

This week was a pretty good week. As predicted, I did get to meet Mermin and had nice ample time to get to talk to him about quantum mechanics. He's struggling hard to formulate his own interpretation of quantum mechanics; he figures that since he's over 60 now, he has the right to do so. For a while he was calling the set of thoughts the "Ithaca Interpretation of Quantum Mechanics," but that ran into a little trouble from the locals; now he's jokingly calling it the "75 Hickory Road Interpretation." I'm not sure what to make of it; it's still a little too vague for me to formulate any strong opinions. The main point seems to be the idea that "if we can sweep all the problems of interpretation under a single rug called *objective probability*, then that would be progress." Then the issue would be to just figure out what the hell "objective probability" could possibly be. I'm very slightly sympathetic to this, but I now pretty strongly believe that there can never be "objective probability," and would rather he focused on searching for a notion of "objective indeterminism." For me, probability encapsulates a state of knowledge; it's a good notion of indeterminism that seems so hard to define. Especially one that seems so crucially dependent upon the information processing capabilities of those beings that work to describe the world.

Something rather strange about Mermin cropped up this week. The guy actually thinks that philosophers of the French literary-criticism tradition (such as Derrida) are worth wading through. Also that they likely have something to say of use in understanding the quantum! Of all places, I think I was least likely to look there!!!

09 June 1997, "Dictionaries and Their Problems"

How are you my friend? It's been so long since I've written you anything of substance, I almost wonder if I can still remember how! Lately, I've once again taken to reading about Bohr's (and the other founding father's) thoughts on the epistemological and ontological lessons of quantum mechanics. I suppose part of my reason for getting back to these things is just a general tiredness of looking at equations; maybe it's a form of procrastination – papers need writing, papers need revising, papers need refereeing, talks need preparing ... and I'm getting a little tired of it all.

In any case, the exercise is having its own payoff. Maybe I'll share a little with you. Remember I told you that Mermin suggested that Derrida's mumblings shouldn't be written off? I guess I'm starting to think he was right (though I have to admit that I haven't yet read any of Derrida's own writings, only commentaries). It seems that the focal point of Derrida's thought centers around none other than your "problem of the dictionary"! Let me try to give you something of a flavor of how these things might be connected to the quantum. My starting point has been an excellent essay by John Honner titled, "Description and Deconstruction: Niels Bohr and Modern Philosophy" (found in *Niels Bohr and Contemporary Philosophy*, edited by Jan Faye and Henry J. Folse (Kluwer, Dordrecht, 1994), pp. 141–151). I hope you enjoy the quotes:

"Derrida undermines the notion that words and signs can capture present experience: our tracing of experience always discloses a supplement, a 'difference'. This attack is equivalent to a subversion of the notion of strong objectivity and correspondence theories of truth. For the deconstructionist, the foundations for knowledge are never securely laid: words do not correspond exactly to the world. 'Presence' can never present itself to a present consciousness, and hence experience is always and already constituted as a text. [I use 'text' loosely here, of course, meaning any collection of signs – discourse, mathematical equations, pictures, poems, prose, drama, hand-waving – used to trace and express insight and experience.] A text is a collection of signs and any sign presumes a presence which it represents, but the sign is not the same as that which it represents. In signifying our awareness of a presence a move is made from the presence to sign. By the word 'presence' Derrida is indicating something like substance, essence, or object, but he rejects such 'totalising' categories as these, for such terms assume more about the presence than perhaps we are entitled to assume. The term may 'trace' the presence, but a remainder is always left over."

Speaking and writing are, according to Derrida, "linear" activities which lock us into space and time. "The great rationalisms of the seventeenth century," as Derrida describes them, fall into the trap of objectivity and neglect the timelessness of self-presence. The linearity of the words limits the conditions for the use of language: "If words and concepts receive meaning only in sequences of differences, one can justify one's language, and one's choice of terms, only within a topic [an orientation in space] and an historical strategy." Here we have a curious serendipity. Our usage of words is tied, arguably, to the re-identifiability of particular objects, which itself implies those bastions of classical physics, the conservation of position and momentum and an absolute space-time framework. And it was precisely these bastions that Bohr attacked. As I have argued elsewhere, Bohr's fundamental arguments entail a provocative hint at a link between the given character of ordinary language and a deterministic–mechanistic view of the workings of nature. For Bohr, classical physics is the inexorable result of the use of language based on the identification of experienced material particulars; or, vice versa, the use of language based on identification of experienced particulars will ultimately lead to a sense of the persisting presence and movement of material objects in space and time, and hence to principles of conservation, causal change, and continuous space-time frameworks.

23 September 1997, "Airy Nothing"

I'm nearing the end of the flight and feeling a little philosophical. I hope you'll let me entertain you for a while. Lately I've been thinking about the airy nothings of quantum mechanics again. It's been a long time since I've done that to any extent – it's sort of refreshing.

Indeterminism and entanglement. The first is an old friend, that you know. The second, though, every day takes my heart a little more. In a certain way, indeterminism couldn't live without entanglement: the EPR argument would have triumphed over indeterminism if entanglement hadn't *also* led to a necessary violation of Bell inequalities. I believe in the ultimate indeterminism of quantum mechanical measurement outcomes just because of the experimental confirmation of Bell inequality violations and the experimental confirmation of Special Relativity. I've said this to you before (probably three years ago), but now it's starting to weigh on my mind more heavily. If I want to understand quantum indeterminism, then I must also understand entanglement: the argument goes in just that order.

Luckily for me, I think, the field of Quantum Information is especially suited to that purpose. Viewing entanglement as a new resource is the main thing on everybody's mind. In fact, I'm starting to feel that the situation we're in can be likened to the beginning of thermodynamics. What is heat, energy, work? No one knew at the outset; some thought them fluids, some thought them vital forces much like the soul, and so on. However, one thing did become clear eventually: no informed judgment on that fundamental question stood a chance until there existed a quantitative theory of thermodynamics. Without that, one could have never come across the mechanical theory of heat and the corollary of atomism that it led to.

So what is this thing called entanglement? What is its use? That we're just starting to understand. If I had to put it in a phrase right now, I would say it is "all-purpose correlation." Alice and Bob come to me and say, "Give us a little correlation, something that we can both have and no one else can possess. We think we're going to need it pretty badly tomorrow." I say, "Sure, no problem, just tell me which variables you'll be needing correlated and I'll do the trick for you." They say, "Sorry, we don't know which ones we'll need correlated yet. A lot of that will depend upon what we actually encounter tomorrow."

In the classical world, Alice and Bob would have been out of luck. But because the world is quantum, I actually can do something for them. I can give them a little "all-purpose correlation." And it turns out that that stuff can be really useful for several tasks. (In fact, we're finding ever more uses all the time.)

Thus, in a certain way, I'm starting to be impressed that "entanglement" shares a strong analogy to "energy." Both fulfill similar roles in our engineering endeavors: they are "all-purpose" essences that can be used for various beneficial tasks. Once we understand that in real depth, I think we'll finally put a dent in this question of "How come the quantum."

12 November 1997, "Mad Girl's Lovesong"

I found these lines from a little poem on the web a minute ago. I thought you might enjoy. They're from

"Mad Girl's Lovesong" by Sylvia Plath:

> I shut my eyes and all the world drops dead
> I lift my eyes and all is born again.
> (I think I made you up inside my head)
>
> The stars go waltzing out in blue and red
> And arbitrary darkness gallops in
> I shut my eyes and all the world drops dead

19 November 1997, "Poesy of the Quantum"

> "Many are poets but without the name,
> For what is poesy but to create"
> (*Lord Byron*)

If so, then we all be poets in this quantum world!

25 November 1997, "Flatt and Scruggs"

By the way, thanks for the words to the Martha White Theme! I don't think I knew of the song's existence before that ... or at least I don't remember it. I was referring to an old Tennessee Ernie Ford television commercial for the flour.

Because of this little gem that you gave me, I now make the following citation in my research proposal:

L. Flatt and E. Scruggs, "The Martha White Theme," on *Flatt & Scruggs: 1948–1959* (Bear Family Records, 1994).

What do you think those Princeton boys will make of this when they see it?

17 December 1997, "It's a Wonderful Life"

Good holidays to you. This morning, as I was driving to work, it dawned on me that roughly this day 10 years ago, I was conferred my degrees at the University of Texas. Time does fly.

It made me think of a little anecdote about John Wheeler that I heard from John Preskill a few days ago. In 1972 he had Wheeler for his freshman classical mechanics course at Princeton. One day Wheeler had each student write all the equations of physics s/he knew on a single sheet of paper. He gathered the papers up and placed them all side-by-side on the stage at the front of the classroom. Finally, he looked out at the students and said, "These pages likely contain all the fundamental equations we know of physics. They encapsulate all that's known of the world." Then he looked at the papers and said, "Now fly!" Nothing happened. He looked out at the audience, then at the papers, raised his hands high, and commanded, "Fly!" Everyone was silent, thinking this guy had gone off his rocker. Wheeler said, "You see, these equations can't fly. But our universe flies. We're still missing the single, simple ingredient that makes it all fly."

Merry Christmas.

04 January 1998, "Lazy Weekend"

There's so much that needs doing, but still I feel compelled to a lazy weekend. I'm sitting here listening to the rain drizzle down outside, listening to the Sunday blues radio show, and thinking of you. My two dogs are at my feet, contemplating the meaning of life in their own small way. Kiki is preparing for her first day back at school tomorrow: 28 kindergarteners freshly back from a half-month of play ... ouch.

Lately I've been thinking about the program of Law Without Law again. Perhaps it's just a fancy form of procrastination. I keep dreaming of the day when all this will become immaculately clear, and we will have the start of a new physics. I guess I've been saying this for eight years, but it seems that it really must be just around the corner: I think we're almost at a point where the possibilities in our world will open up like a blooming flower.

Ever more I am compelled to believe that the ontology of Wheeler's "game of twenty questions, surprise version" is not only a central lesson of quantum theory, but actually the singular principle upon which the detailed structure of the theory is built. The "fact" that *my* information-gathering yields a disturbance to *your* predictions is the only "physical" (or ontological) statement that the theory makes; all the rest of the structure is "law of thought" subject to that consideration. To put it another way, quantum theory is a theory of "what we have the right to say" in a world where the observer cannot be detached from what he observes. It is that and nothing more.

The central issue then becomes: what are the further implications of this "lack of detachedness" for observers? Now that we know that it is actually the essence of quantum phenomena, what can we do with it? Quantum cryptography is a nice applied example of that line of thought. But there's got to be so much more. It seems to me that we're almost poised in the same way that Einstein was when he finally formulated the physical/ontological observation that "maybe it's not coincidental or accidental that gravitational and inertial mass are numerically the same." It was then just a question of counting the time until something wonderful came out of its asking.

Anyway, thanks for lending me your ear this afternoon. Did I send you the notes I wrote on David Mermin's Ithaca Interpretation of QM? I don't think I did ... or at least I can't find a record of it. In case I didn't – and, in case you're interested(!) – I'll forward them on to you following this note. I say the same things there that I did above, but perhaps in slightly more detail. (See especially the stuff following Merminition 12.)

Greg's Reply

I am amazed to receive again remarks about Wheeler's game of twenty questions, for just last night I was exposed to something of particular importance. I was watching an interview of Paul McCartney by David Frost. Of course it was fascinating. But ... the most fascinating was this: Frost was asking Paul about the source of some of the Beatles most famous work. Paul mentioned that his most played song, *Yesterday*, came to him in a dream; almost no effort on his part. The killer was this: he likened the creation of the *Sgt. Pepper* album to a Mike Leigh – do you know this guy? – play, where one does not start with a script! You start off by asking the actors (I suppose) "... are you a dentist, or a Bob? And then you work it up from there." (The quote is from Paul last night.) Sounds like twenty questions to me. The actors randomly decide who they are, and through self-consistency an undeniable plot emerges. I think I was most excited by this because it was a Beatle, and that now a Beatle was professing what may be a fundamental fact about physical law, something the physics profession itself is unwilling to admit. Out-laws forever! Out-side-the-laws, through their own designs, leading to laws.

19 January 1998, "Happy Quotes"

I forgot, I was going to send on some quotations that I found on the net the other day (on some homepage concerned with the existence of God or something). Anyway they're pretty nice; I append them below. I especially like the last one by Heisenberg. It gets at exactly what I've been trying to get at with my motto, "Bohr was a Bayesian."

Also, let me add to that one I heard around here the other day:

"The measurement problem refers to a set of people."

(Hideo Mabuchi)

Oh yeah, – one more thing. I found a really nice interview with John Wheeler on the web the other day. It's at http://www.bigear.org/vol1no4/wheeler.htm. It's worth taking a look at.

God is subtle, but he is not malicious.

(Albert Einstein)

God does not play dice.

(Albert Einstein)

Anyone who is not shocked by quantum theory has not understood it.

(Niels Bohr)

No development of modern science has had a more profound impact on human thinking than the advent of quantum theory. Wrenched out of centuries-old thought patterns, physicists of a generation ago found themselves compelled to embrace a new metaphysics. The distress which this reorientation caused continues to the present day. Basically physicists have suffered a severe loss; their hold on reality.

(Bryce DeWitt, Neill Graham)

The observer appears, as a necessary part of the whole structure, and in his full capacity as a conscious being. The separation of the world into an "objective outside reality" and "us," the self-conscious onlookers, can no longer be maintained.

(Walter Heitler)

One is led to a new notion of unbroken wholeness which denies the classical analyzability of the world into separately and independently existing parts. The inseparable quantum interconnectedness of the whole universe is the fundamental reality.

(David Bohm)

Some physicists would prefer to come back to the idea of an objective real world whose smallest parts exist objectively in the same sense as stones or trees exist independently of whether we observe them. This, however, is impossible.

(Werner Heisenberg)

I remember discussions with Bohr which went through many hours till very late at night and ended almost in despair, and when at the end of the discussion I went alone for a walk in the neighboring park I repeated to myself again and again the question: "Can nature possibly be as absurd as it seemed to us in these atomic experiments?"

(Werner Heisenberg)

If we ask, for instance, whether the position of the electron remains the same, we must say "no;" if we ask whether the electron's position changes with time, we must say "no;" if we ask whether the electron is at rest, we must say "no;" if we ask whether it is in motion, we must say "no."

(J. Robert Oppenheimer)

A pragmatist is concerned with results, not reality.

(*J. Robert Oppenheimer*)

Pragmatism is an intellectually safe but ultimately sterile philosophy.

(*J. Robert Oppenheimer*)

The "paradox" is only a conflict between reality and your feeling of what reality "ought to be."

(*Richard Feynman*)

No elementary phenomenon is a phenomenon until it is an observed phenomenon.

(*John Archibald Wheeler*)

What we learn about is not nature itself, but nature exposed to our methods of questioning.

(*Werner Heisenberg*)

I want to know how God created this world. I am not interested in this or that phenomenon, in the spectrum of this or that element. I want to know His thoughts; the rest are details.

(*Albert Einstein*)

I don't like it, and I'm sorry I ever had anything to do with it.

(*Erwin Schrödinger*)

26 January 1998, "Lunch Break"

By the way, it dawned on me about mid-morning that my remark about time was prompted by something I heard in the X-Files last night. A time traveler said something like, "Can you imagine the horror of a world with no past and no future?"

General relativity and its four-dimensional manifold? I've never been able to shake the feeling that part of the story is still missing. Like Wheeler said, "The equations of physics that we have can't fly ... but still our world flies!" David Mermin dug up something interesting recently in Rudolf Carnap's autobiography:

Once Einstein said that the problem of the Now worried him seriously. He explained that the experience of the Now means something special for man, something essentially different from the past and the future, but that this important difference does not and cannot occur within physics. That this experience cannot be grasped by science seemed to him a matter of painful but inevitable resignation.

I'll be willing to bet that Einstein was wrong on this count, just as he was with quantum mechanics. (And that perhaps the two wrongs weren't/aren't unconnected.) It is not that the difference *cannot* occur within physics, but rather that our *present* physical description is not capable of doing it.

07 February 1998, "Getting Nerves"

Thanks for your "test-taking" message of the other day. It may not be of much comfort to you now, but your pain is a source of thought about *what is* and *what is not*, and, to that end, serves us both. You've caused me to think a little more than I would have otherwise this week.

In case it's of use to you, I'll attach below a note I wrote my friend Herb (Bernstein) [See note dated 10 December 1997 to Herb Bernstein.] alerting him to an article by Sylvan Schweber. Take a look at it if you get a chance (and can find it). You're not the only one trying to reconcile the stuff on the inside with the stuff on the outside.

I wish you could have been at my colloquium though; I think it was one of my best performances yet. (And I think this is my first physics colloquium ... unless the talk I gave at your school was a colloquium.) They say I had an attendance of over 70 people; I just know that the lecture hall was pretty full. I tried very hard to convey the feeling that physics is changing ... not because the equations are changing, but because we are finally starting to feel at home in the quantum world. We are starting to understand that quantum mechanics is a gift from God, not a devil in the machine.

03 March 1998, "Japanese Kitchen"

I'm in Tokyo right now. I just thought I'd drop in on you for a moment to say hello. Last night at dinner I had the nice opportunity of spreading the word of our old friend John Wheeler to our Japanese colleagues. I told them the story of the Game of 20 Qs (surprise version). This came up because Prof. Hirota told me that when asked why quantum communication devices can perform better than classical ones, he answers that, "It is because the quantum world is 'noncausal'." When I pressed him further to find out what exactly he meant by the word "noncausal," it turned out that he was essentially talking about 20 Questions (surprise version). So that was very pleasing.

Hirota, by the way, is a great fan of Giacometti. He has his work plastered all around the laboratory.

31 March 1998, "What a Lunchtime Bohr"

Remember I once wrote you the following lines:

There you have it in any case. Concerning the "free choice" in quantum measurement: Dirac voted "nature," Heisenberg voted "man," Bohr voted "neither," and Wheeler voted "both."

with the point being that I don't believe that history bears too much hope that Bohr really would have fallen into agreement with Wheeler's game-of-twenty-questions ontology.

Anyway, I just ran across another quote that has me thinking along the same lines again. Let me record it:

The circumstance that, in general, one and the same experimental arrangement may yield different recordings is sometimes picturesquely described as a "choice of nature" between such possibilities. Needless to say, such a phrase implies no allusion to a personification of nature, but simply points to the impossibility of ascertaining, on accustomed lines, directives for the course of a closed indivisible phenomenon. Here, logical approach cannot go beyond the deduction of relative probabilities for the appearance of the individual phenomena under given conditions.

It is so difficult to figure out what the fellow is really saying.

16 April 1998, "Just the Way It Had To Be"

Well I'm very pleased that you started off your last note with an allusion to our wonderful quantum theory, but honestly there was no potentiality here. In this case, it has been clear for years that the world was going to turn out just the way it did. Your getting tenure was an example of the classical world at its very best: if the initial conditions are fixed and pure, the trajectory has no choice but to be exactly what it is. Your initial conditions made themselves known that very first year we were together in North Carolina. It's quite hard for a seeker of the truth to hide in anonymity. Congratulations to you and your family! Life now will certainly be a little more secure for all of you.

But now since you started it, let me do pick up on your allusion. What does the life of the academic hold? I'm not so sure in all detail – especially as I am not one myself – but a certain general outline is clear. It comes from something that the scientific world could not properly take seriously until the quantum. Léon Rosenfeld, I think, said it best, "We are not merely contemplating the world, but acting upon it and thereby modifying its course."

05 May 1998, "Expected"

Thank you for thinking of me when reading about Ed Jaynes' demise. Sadly though, I had already expected it and come to grips with the sadness. I found out last week from Larry Bretthorst at Washington U. that Ed had just been removed from dialysis and wasn't expected to live for more than a few days. He was a truly great man. The most unfortunate thing for mankind is that he never got the chance to finish his already massive book on "probability theory as extended logic." We, the children of this beautiful idea, will have to forge ahead without our beacon.

27 May 1998, "My Head, My Head"

Yikes my head is hurting right now. But that's not why I write you.

I enjoyed your musings about strong gravitational fields and matter. I wish I had something interesting to say back to you on that account, but so far I'm empty. I have had some musings lately on matter, but I guess there's not much of a connection. I keep toying with the notion of matter – more specifically, energy – as a "resource." That is to say, if I think of the resources of classical (non-general-relativistic) physics required to get some task done, then they all seem to boil down, in one form or other, to one single, simple "substance." Namely energy. Be it elbow grease, petroleum jelly, or electromagnetic jelly, they all boil down to some manifestation of energy-expenditure. Now, here's where things get exciting. With our new understanding of quantum information theory, we have come to realize that "entanglement" is a physical resource in its own right. One can use it for teleportation, to reduce the communication complexity of some games, for quantum cryptography, to protect signals in noise, etc., etc., etc. This I think is a resource *sui generis*. I.e., it is not reducible to energy. And in this way, I think we have something here truly distinct from classical physics. But on the other hand, it does appear to be intimately tied to energy through Landauer's erasure cost. Namely, if I want to turn some non-concentrated entanglement into a concentrated form (like pure EPR pairs), then I generally have to perform some quantum measurements, and thus gather some random noise from the quantum nothingness. If I try to think of this entanglement concentrator as a finite, but closed system, I am stuck. This is because its memory will eventually get full with all the bits generated from the measurements. Thus it ultimately has to erase, and that links our erasure process back to thermodynamics. Thus entanglement is distinct from energy (unless you tell me it couples to the gravitational field!!!), but still it appears that one cannot get away from the use of energy to make it useful in the first place.

But, as I say, my head hurts and this is only making it worse.

18 July 1998, "Short Note"

It's Saturday morning and I'm sitting in the little village of Benasque. My second week here has just passed away and I'm starting my third. I've spent a lot of time collaborating with Carl Caves trying to push our point of view about quantum mechanics to its logical conclusion. We are trying to derive that Hilbert space has to be a *complex* vector space (rather than a real one) simply from the idea that quantum states are states of knowledge and nothing more. So far, there is a piece that's working and a piece that's not. Keep your fingers crossed.

Time translation symmetry. It's always struck me that beyond all else, there is one fundamental fact of all our experience and that is that time flows. No law of physics captures this. (I say this fully aware of all the tracts you can find on the "arrow of time" in any library.) Physics is missing something, and something very fundamental. So I'm glad you're thinking about these things. Keep me filled in on all your thoughts. In loose connection to this, I'll forward you a note I wrote Rolf Landauer a couple of weeks ago. I doubt that "information is conserved" in any useful sense: every time one reaches into the quantum well, a little information is created.

28 July 1998, "Macaroni"

The last time I was in Europe (last month), I met up with Charlie Bennett as usual. I don't know how it happened, but we got into a conversation about coin tossing (real coin tossing, not the philosophical issue). Charlie said, "You know, there's a really easy method for predicting someone else's tosses; I'll show you." I said, "Oh bull." Then he pulled a quarter from his pocket and gave it to me. I flipped; he called heads. It was heads.

So I did it again. He called heads. It was heads. I said, "Oh you just got lucky again." "Well, do it again," he says. So I flip. He calls heads. It was heads!! In astonishment, I say, "OK what's the trick you have up your sleeve? Did you give me a fixed coin?!?" He bursts out laughing, "No, not at all. One time in eight you get lucky, and when you do, it's a really good joke!"

But still, synchronicity happens. The evidence is our two stories. I'll never forget mine; maybe you'll never forget yours.

07 September 1998, "The Quotable Pauli"

Well I'm a little angry because I can't get connected to the web this morning and I have no intention of going in to the office on Labor Day (I'm fooling myself into believing that I deserve a day of rest). So who knows when you'll finally get this note. But here it is ... exactly what you've been waiting for: The Quotable Pauli.

I include the first two quotes for the sake of definition, i.e., for his meaning of "detached observer." But the last quote, now the last quote, that's the real reason I'm sending this note. Three cheers for Pauli! (By the way have I ever told you about my secret desire to get a parrot? I'd name it Pauli and teach it to say that it wants a cracker.)

... [I]t seems to me quite appropriate to call the conceptual description of nature in classical physics, which Einstein so emphatically wishes to retain, "the ideal of the detached observer." To put it drastically the observer has according to this ideal to disappear entirely in a discreet manner as hidden spectator, never as actor, nature being left alone in a predetermined course of events, independent of the way in which the phenomena are observed.

(Wolfgang Pauli, in a letter to Bohr (15 February 1955))

In the new pattern of thought [quantum mechanics] we do not assume any longer the *detached observer* ... but an observer who by his indeterminable effects creates a new situation, theoretically described as a new state of the observed system. In this way every observation is a singling out of a particular factual result, here and now, from the theoretical possibilities, thereby making obvious the discontinuous aspect of the physical phenomena.

(Wolfgang Pauli, in "Matter" (1954))

Indeed I myself even conjecture that the observer in present-day physics is still too completely detached, and that physics will depart still further from the classical example.

(Wolfgang Pauli, in "Phenomenon and Physical Reality" (1957))

09 October 1998, "Billiards and the Eight"

Speaking of things common to us, I've started sitting in on Kip Thorne's GR class. I've only attended two lectures so I haven't started to learn much of substance yet. The thing that really strikes me is the way these lectures have taught me something about myself. Before these lectures, I hadn't appreciated how far my world-view has come to diverge from the one of classical physics. Everything Kip says has such a foreign feel to me. He speaks of particles, fields, and world lines; he talks about what "things" are "doing." Strange, really strange, how classical physics has come to be so far from my heart. Physics has become such a "Wheelerian game of twenty questions" for me that I guess I've forgotten the old point of view.

That old point of view ... that the world was made of little billiard balls. At best, it's made of Magic Eight Balls.

17 November 1998, "A Break in Pace"

And to comment on another thing in that same letter: boy, how I wish I could have been at that symposium on Jaynes! I am now Bayesian through and through. But, concerning quantum mechanics, I am a Bayesian of a quite different flavor than usual. Most thorough Bayesians just don't believe quantum mechanics – Jaynes didn't believe quantum mechanics. Witness this quote by A. J. M. Garrett (that I'm stealing from our paper):

The nondeterministic character of quantum measurement can, and should, be taken to imply a deeper "hidden variable" description of a system, which reproduces quantum theory when the unknown values of the variables are marginalised over. Differences in measurements on identically prepared systems then represent differences in the hidden variables of the systems. Not to seek the hidden variables, as the Copenhagen interpretation of quantum mechanics arbitrarily instructs, is to give up all hope of improvement in advance, and is contrary to the purposes of science.

This wave of thought is what Carl, Rüdiger and I are trying to set right in our paper "Bayesian Probability and Quantum Mechanics."

17 November 1998, "Heresy!"

Kimble is really pounding on me to get that *Physics Today* proposal out, but I just can't let this confusion go! I will not have my name shamed!

Comerism 1: *I am a bit confused about the passage on quantum mechanics.*

Probably because I don't write nearly as clearly as I would like to think that I do!

Comerism 2: *Do I understand correctly that you now believe in some type of hidden variables?*

Yeaks!!! Never, never give that thought another ounce of attention. I couldn't, I wouldn't, I would never think that. All the evidence is in; it's only been hard for some to cope with the verdict!

Comerism 3: *Or, do you believe that we still need to dig deeper into quantum mechanics, contrary to what the Copenhagen teachings would have us do?*

Oh we certainly need to dig deeper, always. But my understanding of Copenhagen (in any of the 30 versions) has never been that they commanded us to stop thinking. Bohr very much attempted to express a line of thought that demonstrated – in a certain sense – that quantum mechanics is the end of the line. He just wasn't so good at it. He didn't think that it was an arbitrary edict that the end of the line was here, as Garrett would have us think.

Comerism 4: *Which wave of thought is it that should be expanded on?*

I said "set right" meaning that it should be corrected, not "expanded upon"! This is where your confusion is coming from.

By saying that I am Bayesian through and through, I mean that I understand that probability quantifies a state of knowledge: it is *never* a physical property outside of someone's head. This statement even applies to the probabilities generated by quantum mechanics. Now, most Bayesians – myself *not* included – also take it that whenever one's state of knowledge is not maximal (i.e., one assigns non-0-1 probabilities to something), then that knowledge can be improved upon. Moreover, one ought always look to improve it. That is what Garrett is trying to state clearly. But, from my point of view, this last statement is not compelled at all by the Bayesian doctrine. It is simply a throwback from classical thought, where one always has it that nonmaximal information can be made complete (with sufficient effort). In the quantum world this is just not the case: maximal information is not complete and cannot be completed!

05 April 1999, "Pasadena Visit"

Charlie, Herb Bernstein, and I had the most wonderful conversation last night about the foundations of quantum mechanics. Herb and I are much alike, thinking that reality (in part) is created by quantum questions and quantum answers. Charlie, though, is a strict many-worlder. There were some good sparks a flyin'. Most interesting though was the deep similarity in some of the things we were all saying. For the first time, it sort of made me believe that the only question is one of language. In the near future I hope to construct a little essay about this.

A little sad news came out of the discussion. I mentioned that I had wanted to write Rolf Landauer again about his slogan "Information is Physical" to tell him about the parts I like and the parts I don't, and to find out what his response is. Charlie said it wouldn't do any good: Rolf is on his deathbed. They've found cancer in his brain, removed part of it, and given him only a month or so to live. I missed my chance with Jaynes; now I've missed my chance with Rolf. The finiteness of life.

11 April 1999, "One I Like"

Is there no way out of the mind?
(Sylvia Plath (1932–63),
U.S. poet, Apprehensions.)

22 April 1999, "Fuchsian Genesis"

In the beginning God created the heaven and the earth. And the earth was without form, and void; and darkness was upon the face of the deep. And the Spirit of God moved upon the face of the waters. And God said, Let there be light: and there was light. And God saw the light, that it was good; and God divided the light from the darkness. And God called the light Day and the darkness he called Night. And the evening and the morning were the first day. ... [Day 2], [Day 3], [Day 4], [Day 5] ... And God saw everything that he had made, and behold, it was very good. And there was evening and there was morning, a sixth day. Thus the heavens and the earth were finished, and all the host of them.

But in all the host of them, there was no science. The scientific world could not help but STILL be without form, and void. For science is a creation of man, a project not yet finished (and perhaps never finishable) – it is the expression of man's attempt to be less surprised by this God-given world with each succeeding day.

So, upon creation, the society of man set out to discover and form physical laws. Eventually an undeniable fact came to light: information gathering about the world is not without a cost. Our experimentation on the world is not without consequence. When *I* learn something about an object, *you* are forced to revise (toward the direction of more ignorance) what you could have said of it. It is a world so "sensitive to the touch" that – with that knowledge – one might have been tempted to turn the tables, to suspect a priori that there could be no science at all. Yet undeniably, distilled from the process of our comparing our notes with those of the larger community – each expressing a give and take of someone's information gain and someone else's consequent loss – we have been able to construct a scientific theory of much that we see. The world is volatile to our information gathering, but not so volatile that we have not been able to construct a successful theory of it. How else could we, "Be fruitful, and multiply, and replenish the earth, and subdue it?" The most basic, low-level piece of that understanding is quantum theory.

The *speculation* is that quantum theory is the unique expression of this happy circumstance: it is the best we can say in a world where *my* information gathering and *your* information loss go hand in hand. It is an expression of the "laws of thought" best molded to our lot in life. What we cannot do anymore is suppose a physical theory that is a direct reflection of the mechanism underneath it all: that mechanism is hidden to the point of our not even being able to speculate about it (in a scientific way). We must instead find comfort in a physical theory that gives us the means for describing what we can *know* and how that *knowledge* can change (quantum states and unitary evolution). The task of physics has changed from aspiring to be a static portrait of "what is" to being "the ability to win a bet."[2]

This speculation defines the large part of my present research program.

Why do I say all this? Because I wanted to say it to myself and use you as the other side of my brain. Today I'm going to lunch with Roger Penrose and there is no doubt that the

---

[2] The nice phrase "physics is the ability to win a bet" is due to J. R. Buck (a grad student at Caltech) circa 19 February 1999.

topic of the interpretation of quantum mechanics will come up. How best to express what needs expression? That's the question on my mind this morning: this is iteration #1.

27 June 1999, "Raining Down in Cambridge"

"Anyone who considers arithmetical methods of producing random digits is, of course, in a state of sin."

(John von Neumann)

As you can see, I'm in Cambridge now. In fact, I'm looking out my window to all the rain that's coming down – what else would you expect? Man, I love it here! I haven't felt so productive in ages. Carl Caves and I are on the tail of a really good theorem (useful to interpreting quantum mechanics in a Bayesian way) and the old English pubs just can't be beat. The Isaac Newton Institute too is just wonderful: I have never been to a place more conducive to thinking. Any place that actually puts chalk boards in the john so that you can continue your discussions in there if need be, well … is just alright in my books!!! Ben Schumacher and I have been working on some quantum coding theorem stuff. And Vladimir Bužek and I have been able to some extent to put into equations some of the ideas in the "genesis" note I sent you. (He calls that note my "manifesto.") All these things are very exciting for me.

You asked a question in your last note. Well, I really like that quote of von Neumann. It's true.

Comerism 5: *How is it that the mathematical formalism of quantum mechanics – which I take here to be simply the solving of a partial differential equation – can model for me random results?*

The continuous time-evolution part of the theory concerns how our state of knowledge changes when we're not shoving a system about to see how it will react (i.e., we're not asking it a question, we're not making a measurement on it).

Comerism 6: *Does randomness enter when we take the output of the partial differential equation and interpret that output as a probability?*

Yep, that's the whole story. The outcomes of our measurements are governed by no laws (much less differentiable ones). The probability is there because there is ignorance or impredictability … sometimes more so, sometimes less so. But there's always some. The randomness in quantum mechanics is that no matter how much information we have about how a system will react to our prods, it is never enough. Maximal information is not complete … that's the slogan of the Albuquerque interpretation.

You, by the way, once asked about whether the information–disturbance point of view about quantum mechanics would nullify the many-worlds point of view. I never answered you. Sorry about that. Unfortunately I don't think it will. But part of that is because the many-worlds point of view is much less well-defined than its proponents would have us believe. The real question is which point of view about quantum mechanics is the most ripe for leading to new things. Which one will carry us through to a greater appreciation of the possibilities in our world? The many-worlds point of view is a dead end in that regard. Understanding that the world is, at its core, malleable to our hands? I think that that will ultimately be the greatest gift this century has to give to the next.

28 August 1999, "No Podunks Here"

Comerism 7: *It's amazing at times the synchronization that occurs. Just last week, Bill Thacker was looking around for an explanation for why one needs complex vector spaces. He found one book that said it was because one wants wavelike behavior, but linearity in the Hamiltonian, which therefore requires an i in the Schrödinger equation (so that if one "squares" the Schrödinger equation, then the appropriate sign will come out in front of the time derivatives). It will be very interesting to see what you and Caves have to say.*

Interesting synchronicity with Mr. Thacker. Below I'll attach some references I sent to Carl for the use of his funding agent; Bill will also be able to make good use of them. The discussion in Steven Adler's book is useful and good, but I'm not too impressed by the arguments there. The one he likes the most is this. If you suppose that time evolutions must be unitary, then the generators of the unitary transformations (the "Hamiltonians") cannot be observables in real-Hilbert-space QM. But I say, "So what? Why need they be observable?" I don't think Wheeler's favorite argument (actually it's one of Wootters', but never published by he himself) is so good either; though I fluctuate on that. The argument that Caves, Schack and I have is much more akin to the Wootters argument listed below. At least the two arguments have the same mathematical source. The basic effect comes about from this: for a Hilbert space of dimension d, the dimensionality of the vector space of Hermitian operators over the base space is d^2 for complex Hilbert spaces, but only $d(d+1)/2$ for real ones. This means that the two theories have quite different behaviors as concerns the conglomeration or concatenation of separate systems. In the first case, the operator space of the tensor product is isomorphic to the tensor product of the original operator spaces. In the second case, the operator space of the tensor product is "larger" than the tensor product of the operator spaces. Now we are able to give an information theoretic reason why one should want the former over the latter: it's actually required if one takes it seriously that quantum states are states of knowledge and NOT states of nature.

The way I see it is this. It seems to me that the only reasonable understanding of the wave function is that it is information and nothing else. It is a symbol that stands for what one knows, the most one can know ... and makes no commitment to ontology outside of the information-disturbance relations that one can derive from the mathematical structure that is built around it. But one cannot stop there, as our Copenhagen forefathers did. (That's why there's so damned much confusion in the field.) Instead, we in the quantum information field are obliged to give an information theoretic reason for each and every axiom of quantum mechanics. When we find an axiom or theorem for which we can't do that – my personal favorite for a candidate is information-disturbance – then we will have identified the crucial ontology in the theory. That's my take. And I think this new theorem (we call it the quantum de Finetti theorem) is a small step in that direction.

25 November 1999, "Thanks and Giving"

I hope you and your family have a wonderful Thanksgiving today.

My latest problem is that I've got it in my head that if I'm going to make any real contribution to physics, it will be vicariously, through guidance to some good graduate students. I have a big research program planned out now, one that I know I can't pull off by myself ... and one that I know has enough open ends that lots of outside inspiration will be needed (the inspiration of some hard thinking students). This firmed up in me during my teaching at the summer school in Italy this summer. I started to collect this set of groupies there, and I have to tell you it did wonders for the ego. I started to think (for the first time), "You know, I could do this. I have a role in this community."

But let me go back to this dream before taking off. You know Asher and I are writing this thingy for *Physics Today*. Well, we were having a bit of a fight about the meaning of the wavefunction. He likes to say that wavefunction is just a symbol that stands for a system's preparation. I on the other hand like to say that it stands for the information we have about how a system will react to measurement. So I got a little smart with him. In one passage he had written that if we know a quantum state then we can prepare as many copies of that state as we wish. I retaliated by saying the following:

Just for one last emphasis: suppose a pure state were a pretty good approximation to what we could say about all the observables associated with the radius of the universe. How would we produce arbitrarily large numbers of that state?

Ultimately I won, but one thing he said in retaliation to my retaliation caught my eye. I love it! I leave you with a quote of it for wishes of a happy Thanksgiving.

ANSWER:
It's just a matter of money. As Archimedes answered to the king: give me firm support, and I shall move the Earth.

The lesson of the quantum is that the world can be moved.

17 December 1999, "Trinity"

Comerism 8: *Forgive my ignorance, but isn't there some famous nuclear explosion site with "Trinity" in the name?*

Yes, the Trinity test was the first test of an atomic device, July 1945. July 15, I think. The bomb was a plutonium bomb. They tested a plutonium bomb, because by that time there was overwhelming confidence that the uranium bomb would work. Trinity is out in the White Sands missile base – I've been there once (they only open it to the public two days per year). At my house, because of the old fellow that I lived with one summer Jack Aeby, I have a shard of glass that was formed from the sand below that explosion. It is a mineral that was unknown to the world before the explosion; it has an official name: trinitite. When you touch it, you really do find yourself thinking, "I am become death, the destroyer of worlds."

23 January 2000, "Darling Quotes"

Let me send you some quotes from David Darling's book. You'll see the connection to what you wrote me in your holiday letter.

From D. Darling, *Equations of Eternity: Speculations on Consciousness, Meaning, and the Mathematical Rules that Orchestrate the Cosmos*, (Hyperion, New York, 1993):

The interface between mathematics and everyday reality appears sharp and immediate at this point: one sheep, one finger, one token; another sheep, another finger, another token, and you can take away tokens or add them, as you can with your fingers. The tokens – the numbers – are just abstracted fingers; the operations for dealing with the tokens are just the abstracted raising or lowering of the fingers. You make a one-to-one correspondence between the tokens and whatever it is you want to reckon, and then forget about the fingers.

At first, it seems clear from this that mathematics must be somehow already "out there," waiting to be discovered, like the grain of the stone. One sheep add one sheep makes two sheep. Two sheep add two sheep makes four sheep. That is certainly the practical end of the matter as far as the shepherd and the merchant are concerned. But already, even in this most simple mathematical maneuver, something strange has happened. In saying "one sheep add one sheep" we seem to be implying that any two sheep will always be identical. But that is never the case. Physically, the first sheep is never exactly equal to the second: it may be a different size, have different markings. It takes only one molecule to be out of place between the two, and they are not identical. Indeed, because they are in different places they are inevitably not the same on that basis alone. We have extracted a perceived quality to do with the sheep – namely, their "oneness," their apartness – and then merged this quality by means of another abstraction – the process of addition. What does it mean, physically, to "add" things? To put them together? But then what is "putting together" two sheep? Placing side by side, in the same field – what?

All this may seem like nit-picking. But on the contrary, it brings us back to the central mystery – the relationship between the inner and the outer, the world of the rational mind and the world "out there." In the physical world, no two sheep are alike. But, more fundamentally, *there are no "sheep."* There are only some signals reaching the senses, which the left brain combines and then projects as the illusion of a solid, relatively permanent thing we call a sheep.

Like all objects, sheep are fictions: chimeras of the mind. It is our left hemispheres, having through natural selection evolved this skill for extracting survival-related pieces of the pattern, that trick us into seeing sheep, trees, human beings, and all the rest of our neatly compartmentalized world. We seek out stability with our reasoning consciousness, and ignore flux. We shut our eyes to the continuous succession of events if those events seem not to substantially affect the integrity of what we see. So, through this classifying and simplifying approach we make sections through the stream of change, and we call these sections "things." And yet a sheep is not a sheep. It is a temporary aggregation of subatomic particles in constant motion – particles which were once scattered across an interstellar cloud, and each of which remains within the process that is the sheep for only a brief period of time. That is the actual, irrefutable case

and

When you think, the cosmos thinks – not in some nebulous, poetical sense, but literally. Your brain is a product and a process of the cosmos. When you look out and form pictures in your mind and try to make sense of those pictures, you are the universe trying to make sense of itself. Forget that we sometimes imagine ourselves to be tiny bipeds on a ball of rock somewhere in the awful depths of space–time doing cosmically trivial things such as changing a diaper, or mowing a patch of grass, or brewing the next cup of coffee. The incredible and undeniable fact is that we are the thinking, reasoning components of the universe – a realization that makes it slightly less astonishing that our brains might have some role to play at the cosmic level.

25 January 2000, "Fear"

Have we made progress in our 11 years of discussion? I've certainly made huge gobs of progress from the process. I started out 11 years ago thinking (or believing in the back of my mind at least) that we were just inches away from a final understanding of the universe. Today I think we're not even close, haven't even scratched the surface of "reality." That's why I wrote John Preskill (the 10 September 99 note), "...I don't see physics as an expression of our great knowledge of the world ... but more accurately as an expression of our great ignorance." Seeing what needs to be done is progress!

27 January 2000, "Chili Cheese Fries"

Comerism 9: *I think what impressed me about Pi [the movie] was how much I disagreed with the main premise – that "mathematics is the language of nature."*

You know, maybe I should see it again, now that these issues are so much clearer in my mind. (I think I saw it in Pasadena about three years ago.) Reading that article by Banville and almost simultaneously having a conversation with Bill Wootters about the limitations of a dog's understanding of the world made a big turning point in my life. [See note to Herb Bernstein, dated 2 January 1999.]

07 February 2000, "The Fine Line"

Comerism 10: *He has suggested that I put my work on the lanl archive. I am reluctant to do this because I failed to get the work published in good refereed journals; I think it sets the wrong precedent.*

My buddy Jeff Nicholson and I have had long fights about this issue. He stands more where you do now; I tend toward the opposite extreme. Your career has been punished simply because you had some bad luck of the draw in having truly awful referees. Why should that propagate?

We had a similar situation in our field. Stephen Wiesner wrote a wonderful little paper circa 1971 that essentially contained the first quantum cryptography protocol. Moreover it contained a whole new (and important!) classical cryptographic protocol called "one out of two oblivious transfer." He submitted it to the *IEEE Transactions on Information Theory*. (His father was an important information theorist and actually president of MIT.) The paper was rejected in a scathing way. Wiesner stuck it in his desk, tuned in, turned on, and dropped out. The paper had to wait until 1984 to be revived by Charlie Bennett and Gilles Brassard. These days, Wiesner lives in a hut in the desert somewhere in Israel. Not completely cause and effect here, I'm sure, but things just didn't need to go that way. Now this year, there's some hope that Wiesner will win the Quantum Communication award that Shor and Kimble won last year.

If I were you, I would stick the paper on the archive. Or better yet, have X submit it for you, with a note explaining that it was submitted and rejected by Journal Y, along with the date of that event.

20 February 2000, "Rabbits on the Moon"

I've just come into my office out the crisp night air of Haifa and an early sighting of a beautiful full moon. And like it has been the case for the last two years, but never before that, I couldn't help but marvel at the rabbit on the moon: his head and ears in the top left corner, and his bushy tail in the bottom right. A Japanese friend [Masahide Sasaki] mentioned its existence two years ago; the legend is part of their culture. Now it's always there whenever I look. Where was it for the previous 32 years of my life? When I stop to think about this, I just can't shake the idea that so much of the order we see in the world around us is simply placed there by ourselves. For so much of my life the man in the moon was a perfectly adequate metaphor – now I can't see it at all.

The world is an organism, the ancient civilizations said. No, the world is a mechanism, Laplace said: I can see no organism. Now that my eyes are opened, what could be clearer? The world is a _____, quantum information said. Now that I see it, I can never turn back.

25 June 2000, "Waxing Surfboards"

I'm sitting on my balcony waiting for the Grecian sunrise and contemplating my finite existence. I thought about titling this note "Waxing Philosophical," but somehow the surfboards intruded into my mind more forcefully. The scene is quite beautiful: to my left are some typically Greek-styled beach houses, straight in front of me is a little island peeking over the waters, and above that is the shining sliver of the morning's moon.

Yesterday I met quite an interesting guy, Arkady Plotnitsky. He's in the English Dept. faculty at Purdue University, but his sideline is on interpreting Bohr. And from what I could tell yesterday, he's quite accomplished in that respect: there are a load of similarities between our points of view. I had tried to read his book *Complementarity* about three years ago, but gave up in frustration. In fact, I had even contemplated making fun of the book during one of my journal club talks at Caltech as one of the prime examples of the new science war gibberish. When I told David Mermin that story, he quickly defended Plotnitsky and told me that I really should take him seriously, that he is impressive (much more so in person than in his book). And David's really right on this count.

I don't believe I'm going to see the real sunrise; it looks like it'll be coming up over the hill to my left.

Waxing surfboards philosophical.

28 July 2000, "Great Quote"

If triangles made a god, they would give him three sides.

– Charles de Montesquieu (1689–1755), French philosopher, lawyer,
  Lettres persanes, letter 59 (1721).

10 August 2000, "Heavy Information"

Comerism 11: *In the meantime, Bekenstein once wrote a paper with Schiffer where they pose the question if information had "real" physical qualities. For instance, does it weigh? I can get the reference if you are interested.*

Yeah, I would like those references if you can dig them up. I think there were a few papers. I remember the consensus of Caves and the Great W (that's what Carl calls Bill Unruh) was that there was something wrong in those papers – they had technical reasons.

I myself don't have a strong feeling that "information" should have any physical (perhaps a better word would be ontologic) properties. For, whenever information is a meaningful concept, we have explicitly put ourselves into the picture. The information, whenever there is any, resides in our heads, not in the objects of our attention. However, I do think that information-theoretic questions can go a long way toward elucidating properties of the critters we use as information carriers. For instance, that ultimate cap on how much I can predict versus how much you can predict if we imagine the system in a measurement context: is that telling us that there is a property we should be mindful of when we contemplate using that object for getting something done? (Well, yes, both in quantum cryptography and quantum computing: that "sensitivity to the touch" is a *resource* that doesn't exist in the classical world, and it can do things for us if we use it properly.) Is this extra something that quantum systems have, something we've been missing (however subtle) in our accounting of how objects affect the spacetime around us?

Yeah, I do want to see the Bekenstein–Schiffer papers again.

11 August 2000, "The Mathematics of Experiment"

Comerism 12: *Is it reasonable to say that the experimental process is also a form of mathematics? (See, now that you opened the window of possibilities, I'm trying to ease my conscience.) Actually, I now think of mathematics as just a search for, or creation of, logical structures.*

I like your last sentence. The only way I would change it is to substitute for "logical structures" something like "structures congenial for our reasoning in light of the information we dredge up from the world." (It's longer winded but it better captures my present attitude.)

18 September 2000, "A Non-Randy Non-Bugger"

Comerism 13: *So, this thought is keeping me from my grant: Our laws of physics are not about the real world, never will be? They're only rules that describe our interface with the real world? And what would it matter if even the real world of Einstein existed? Because it's not about the "real world" but rather how we are allowed to interface with it, and perhaps it with us?*

I was especially intrigued by this point of yours. My friend Herb Bernstein also likes to make it (over and over). By complete chance I happened to read a very nice expression of the same thought the other day in an article on a "feminist approach to teaching quantum mechanics." The article was quite good actually. But I've been thinking about writing the author to say that what she has written has nothing to do with a "feminist reading of Bohr." It's just a *careful* reading of Bohr ... and I'm qualified to say that as I have *both* the same reading of Bohr *and* I'm likely one of the most chauvinistic men she's ever met. How do you think she'd react?

Anyway, here's the passage below. The part that's especially like your comment is in the last paragraph.

K. Barad, "A Feminist Approach to Teaching Quantum Physics," in *Teaching the Majority: Breaking the Gender Barrier in Science, Mathematics, and Engineering*, edited by S. V. Rosser (Teachers College Press, New York, 1995), pp. 43–75.

The Newtonian worldview is compatible with an objectivist epistemology, in which the well-prepared mind is able to produce a privileged mental mirroring of the world as it exists independently of us human beings. That is, what is "discovered" is presumed to be unmarked by its "discoverer." The claim is that the scientist can read the universal equations of nature that are inscribed in [God's] blackboard: Nature has spoken. Paradoxically, the objects being studied are given all the agency, even and most especially when they are seen as passive, inert objects moving aimlessly in the void. That is, these cultureless agents, existing outside of human space–time, are thought to reveal their secrets to patient observers watching and listening through benignly obtrusive instruments. Notice that agency is not attributed to human beings; once all subjective contaminants have been removed by the scientific method, scientists simply collect the pure distillate of truth.

The Newtonian worldview is still so much a part of contemporary physics culture that it infects the teaching of post-Newtonian physics as well. That is, the stakes are so high in maintaining the mirroring view of scientific knowledge that quantum physics is presented as mysticism.

and

Notice that particular experimental arrangements can be used to give more or less definite meaning to each of the complementary variables, but due to the lack of object–instrument distinction ... it is not possible to assign the value obtained to the object itself. The "property" being measured in a particular experimental context is therefore not "objective" (that is, a property of the object as it exists independently of all human interventions), but neither is it created by the act of measurement (which would belie any sensible meaning of the word *measurement*). Bohr speaks of this "interaction" between "object" and "instrument" as a "phenomenon." The properties then are properties of phenomena. That is, within a given context, classical descriptive concepts can be used to describe phenomena, our intra-actions within nature. (I use the term *intra-action* to emphasize the lack of a natural object–instrument distinction, in contrast to *interaction*, which implies that there are two separate entities; that is, the latter reinscribes the contested dichotomy. ... That is, the ambiguity between object and instrument is only temporarily contextually decided; therefore, our characterizations do not signify properties of objects but rather describe the intra-action as it is marked by a particular constructed cut chosen by the experimenter (see Barad, 1995 for more details).

The notion of "observation" then takes on a whole new meaning according to Bohr: "[B]y an experiment we simply understand an event about which we are able in an unambiguous way to state the conditions necessary for the reproduction of the phenomena" (quoted in Folse, 1985, p. 124). According to the analysis of the previous section, this is possible because, in performing each measurement, the experimenter intervenes by introducing a constructed distinction between the "object" and the "measuring device" (e.g., deciding whether the photon is part of the object or the instrument). The claim is that unambiguous, reproducible measurements are possible through the introduction of constructed cuts. Notice that "[n]o explicit reference is made to any individual observer": Different observers will get the same data set in observing any given phenomenon. Therefore, reproducibility, not some Newtonian notion of objectivity denoting observer independence, is the cornerstone of this new framework for understanding science.

For Bohr, the uncertainty principle is a matter of the inadequacy of classical description. Unlike the "mirroring" representationalism inherent in the Newtonian–Cartesian–Enlightenment framework of science, scientific concepts are not to be understood as describing some independent reality. A post-Newtonian framework sees these constructs as useful (i.e., potentially reproducible) descriptions of the entire intra-action process (the phenomenon, which is context dependent by definition), not of an isolated object. The implications of this finding are profound. In Bohr's own words:

> The extension of physical experience in our own days has ... necessitated a radical revision of the foundation for the unambiguous use of elementary concepts, and has changed our attitude to the aim of physical science. Indeed, from our present standpoint, physics is to be regarded not so much as the study of something a priori given, but rather as the development of methods for ordering and surveying human experience.
>
> (*Bohr, 1963, p. 10*)

In other words:

> These facts not only set a limit to the extent of the information obtainable by mea-
> surements, but they also set a limit on the meaning which we may attribute to such
> information. We meet here in a new light the old truth that in our description of
> nature the purpose is not to disclose the real essence of [physical objects] but only
> to track down, so far as it is possible, relations between the manifold aspects of our
> experience.
>
> *(Bohr, 1963, p. 18)*

and

Bohr's philosophy of physics involves a kind of realism in the sense that scientific knowledge
is clearly constrained, although not determined, by "what is out there," since it is not separate
from us; and given a particular set of constructed cuts, certain descriptive concepts of science
are well-defined and can be used to achieve reproducible results. However, these results cannot
be decontextualized. Scientific theories do not tell us about objects as they exist independently
of us human beings; they are partial and located knowledges. Scientific concepts are not simple
namings of discoveries of objective attributes of an independent Nature with inherent demarcations.
Scientific concepts are not innocent and unique. They are constructs that can be used to describe
"the between" rather than some independent reality. (Why would we be interested in such a thing
as an independent reality anyway? We don't live in such a world.) Consideration of mutually
exclusive sets of concepts produces crucial tensions and ironies, underlining a critical point about
scientific knowledge: It is the fact that scientific knowledge is socially constructed that leads to
reliable knowledge about "the between" – which is just what we are interested in. This shifting
of boundaries deconstructs the whole notion of identity: Science can no longer be seen as the
end result of a thorough distillation of culture. There is an author who marks off the boundaries
and who is similarly marked by the cultural specificities of race, history, gender, language, class,
politics, and other important social variables. Reproducibility is not a filter for shared biases. In
stark contrast to the objectivist representationalism that is usually transmitted to students, the new
framework inspired by Bohr's philosophy of physics is robust and intricate. In particular, there is an
explicit sense of agency and therefore accountability. And so I refer to this Bohr-inspired framework,
which shares much in common with central concerns in contemporary feminist theories, as "agential
realism."

06 October 2000, "The Evolution of Thought"

Comerism 14: *Remember my epiphany from a month ago? Well, it has now prompted the
following response: If wave functions don't collapse, because they don't exist, then why is
there the Schrödinger equation, i.e., why do these things that don't exist evolve?*

My state of mind exists, and it evolves (even when I'm gathering no new information).
Schrödinger evolution, I think, ultimately just reflects the differing rates at which internal
clocks run (the system's and mine). The deeper questions are why linearity? Why unitarity?
I'm not very satisfied with all the answers I've heard to date.

14 October 2000, "On the Mark"

Just a very quick note.

Comerism 15: *Or is linearity perhaps a necessary consequence of a description of nature that is not about nature, but rather about how observers interface with nature?*

That's exactly the direction I've been heading, and the sort of thing I had been wanting to write to you. So you've anticipated me and saved me a little trouble. Let me try to give you a smidgen of the flavor. Ben Schumacher, Rüdiger Schack, and I have a little argument for the linearity of time evolutions that goes like this.

Suppose you and I walk into a room and agree that we have a gazillion (i.e., an infinity of) copies of some physical system. The only thing is we disagree about the overall quantum state of the lot. I say it is ρ; you say it is σ. (This means nothing more than that we differ in our probability assignments for how the systems will react to our potential measurements.) The question is, under what conditions can we be sure to converge in opinion if you start making measurements system by system and I look over your shoulder to see the results too. Caves, Schack and I answered that question last year with the "quantum de Finetti theorem" (still need to publish the damned thing): remember it only worked for complex, not real, vector spaces. It states that the necessary and sufficient condition that we will ultimately converge in opinion is that our initial density operator assignments have the property of exchange symmetry. (Also we need the good Bayesian assumption that nothing ever be assigned probability absolutely zero; there should always be some ϵ no matter how tiny. But that's a technical point that doesn't concern the present discussion.) What this means is that if we interchange any two of the gazillion systems, my state should evolve from ρ to ρ and yours should evolve from σ to σ. This gives some operational meaning to the statement that all the systems are the same. But beside that, you should also see that in order to come to a tighter agreement we have to have at least some initial agreement. Otherwise no amount of empirical evidence will sway us otherwise. Anyway, these are the necessary and sufficient conditions.

Now let us ask what might be a reasonable form for the notion of "identical physical evolutions" in this context? I say it is this: if we have conditions so that we can converge in opinion before the evolution takes place, then we should also be able to converge in opinion after the time evolution. Technically, time evolution should preserve exchange symmetry. Also if we time evolve our converged opinion it should be identical to what we would converge to after the time evolution. Well, using the quantum de Finetti theorem, what Ben, Rüdiger, and I proved is that these conditions imply linearity. Linearity can be viewed as coming from the idea that science is about our converging in opinion.

We still don't have a good argument for unitarity (or more carefully complete positivity at the level of density operators), but that's the sort of direction I'm heading. There'll eventually be a way to crack it: I'm confident.

OK, now I need to go whack some weeds and write some recommendation letters for an Oxford boy (a Rhodes scholar even, but I try not to hold that against him).

14 October 2000, "More Evolution"

Comerism 16: *Pretty simple questions, right?!? I guess you get to ask me what happens to my "worldview" if the equivalence principle or causality or spacetime are discovered not to exist, right?*

I'd be especially curious to know what would happen if we had no equivalence principle. How strange would our world be then? What would break?

OK, now I really must go whack weeds.

16 October 2000, "More Linearity"

Comerism 17: *Thanks for your replies! I was pretty much shocked that you've actually got some proofs going on the question of linearity.*

Oh, I've got a load of things like that that are still unpublished. That's why my collaborators are starting to hate me. X, for instance, gets especially miffed that I have an extra outlet to bring fame and fortune to myself that he doesn't: I go to loads of meetings and talk about all these things (privately and when I'm on the stage) and they sort of become folklore knowledge. I really need to work hard to remedy that. But getting at the bottom of things is so much more interesting than writing things up!!

So let me spend a little bit of time this morning trying to corrupt you. Lately (or maybe for years) I've got it in my head that the time evolution part of quantum mechanics is the least interesting part of the theory. In fact – I conjecture – it is so uninteresting as to be identical with classical Hamiltonian time evolution … *when* coupled in terms of the Liouville equation. The italicized *when* is important. What I'm toying with is that it would be interesting to give a direct derivation of the Liouville equation from fundamental principles (and forget the usual track of starting with Hamilton's or Lagrange's equations and *then* moving to Liouville's). That is to say, what general principles can we give that would uniquely pin down the form of the evolution of probability distributions over a phase space? Then we would see that once Newton agreed that physics should be scientific (in the sense that we would all ultimately come into agreement) and that the "observables" in the theory should be x and p, then he had absolutely no choice but to write down the precise equation that he did, i.e., $F = ma$.

One thing that is already clear to me is that the argument I presented in the last email also works for classical physics: one just relies on the classical de Finetti theorem where we relied on the quantum one. So the classical Liouville evolution must be linear if it is going to preserve convergence of opinion in exchangeable situations. But that's still a far cry away from getting the full Liouville equation.

So, I find myself wondering what extra principles are needed even there. Here's one idea that crossed my mind last year or so. I'll just cut and paste from my samizdat: [See notes to Asher Peres, dated 23 September 1999, 26 September 1999, and 27 September 1999.]

My description of the Wigner theorem in that is not quite right. Here's what I should have said:

If we have a bijection from unit vectors to unit vectors on a Hilbert space that preserves all [absolute values of] inner products, then [one can redefine the overall phase assignments of] that mapping [so that it becomes] a unitary linear map or an antiunitary antilinear map. If we further suppose that mappings are continuously connected to the identity, then we are left with the unitaries.

Oh, here's another entry, this time written to Rüdiger Schack: [See note to Rüdiger Schack, dated 22 September 1999.]

Anyway, would you be interested in thinking about this problem with me? You certainly have a greater knowledge of classical mechanics than I do. I'm not quite sure that "overlap preservation" will do the trick. (Is that completely equivalent "volume preservation?" Doesn't one need more than that?) Maybe one needs something more than one needed in the quantum case. (For instance, I suspect overlap preservation for all functions, not just normalized probability distributions will do the trick – because then one has a full vector space not just a cone – but that wouldn't be interesting from the physical side in this case.) So the question is, if extra assumptions are needed, can one give them a knowledge-theoretic justification?

Finally a vaguer question, but one that I know will pique you is: if this does work out, how does the result mesh with general relativity. Can one generally give a Liouville form for the Einstein equations? (Probably in the foliable cases for sure. But what about the rest?) Can we learn something about the equivalence principle from considerations such as this? (That's the sort of thing I was trying to push you toward yesterday.)

16 October 2000, "Three References"

Glad to hear your positive response. Have a look at the introduction and conclusions sections of any of these papers and tell me what you think. (Uhlhorn's paper might be the most notable in that he relaxes Wigner's assumptions significantly but still recovers the result – that's the sort of thing I'm banking on in the classical case. But you might feel the most comfortable with Bargmann's for a first shot.)

1. V. Bargmann, "Note on Wigner's Theorem on Symmetry Operations," *J. Math. Phys.* **5**, 862–868 (1964).
2. C. S. Sharma and D. F. Almeida, "A Direct Proof of Wigner's Theorem on Maps Which Preserve Transition Probabilities between Pure States of Quantum Systems," *Ann. Phys.* **197**, 300–309 (1990).
3. U. Uhlhorn, "Representation of Symmetry Transformations in Quantum Mechanics," *Arkiv För Fysik* **23**, 307–340 (1963).

29 October 2000, "Poincaré Singularities"

Have a think about this little passage from an essay Poincaré wrote sometime before 1912 (when he died). What do you think he would have made of the big bang or black holes?

From H. Poincaré, "The Evolution of Laws," in his book *Mathematics and Science: Last Essays (Dernières Pensées)*, translated by J. W. Bolduc, (Dover, New York, 1963), pp. 1–14:

Mr. Boutroux, in his writings on the contingency of the laws of Nature, queried, whether natural laws are not susceptible to change and if the world evolves continuously, whether the laws themselves which govern this evolution are alone exempt from all variation. ... I should like to consider a few of the aspects which the problem can assume.

and

In summary, we can know nothing of the past unless we admit that the laws have not changed; if we do admit this, the question of the evolution of the laws is meaningless; if we do not admit this condition, the question is impossible of solution, just as with all questions which relate to the past. ...

But, it may be asked, is it not possible that the application of the process just described may lead to a contradiction, or, if we wish, that our differential equations admit of no solution? Since the hypothesis of the immutability of the laws, posited at the beginning of our argument would lead to an absurd consequence, we would have demonstrated *per absurdum* that laws have changed, while at the same time we would be forever unable to know in what sense.

Since this process is reversible, what we have just said applies to the future as well, and there would seem to be cases in which we would be able to state that before a particular date the world would have to come to an end or change its laws; if, for example, our calculations indicate that on that date one of the quantities which we have to consider is due to become infinite or to assume a value which is physically impossible. To perish or to change its laws is just about the same thing; a world which would no longer have the same laws as ours would no longer be our world but another one.

27 November 2000, "A Lesson in Physics?"

From today's *New York Times*:

This view of what judges do when they interpret statutes may prove a hard sell even for Justice Scalia, who wrote in a concurring opinion in a 1991 case, James Beam Distilling Company v. Georgia: "I am not so naïve (nor do I think our forebears were) as to be unaware that judges in a real sense 'make' law. But they make it as judges make it, which is to say as though they were 'finding' it – discerning what the law is, rather than decreeing what it is today changed to, or what it will tomorrow be."

A lesson in physics?

15 January 2001, "Good Turns of Phrase"

I just ran across this and I wanted to get in my hard-disk archive. So, I pick on you.

From: Richard Rorty, "Phony Science Wars," [Review of *The Social Construction of What?* by Ian Hacking], *Atlantic Monthly*, November 1999.

The stalemate that Hacking brilliantly describes but does not try to break is between many scientists' intuition of the inevitability of quarks and many philosophers' suspicion that the claim of inevitability makes sense only if the idea of the intrinsic structure of reality makes sense. This teeter-totter between conflicting intuitions is, Hacking rightly says, a genuine intellectual problem. Which answer one gives to his third question – about the source of the stability of the most reliable bits of science – is likely to be a matter of which side of the seesaw has most recently descended.

These alternating intuitions have been in play ever since Protagoras said "Man is the measure of all things" and Plato rejoined that the measure must instead be something nonhuman, unchanging, and capitalized – something like The Good, or The Will of God, or The Intrinsic Nature of Physical Reality. Scientists who, like Steven Weinberg, have no doubt that reality has an eternal, unchanging, intrinsic structure which natural science will eventually discover are the heirs of Plato. Philosophers like Kuhn, Latour, and Hacking think that Protagoras had a point, and that the argument is not yet over.

The most vocal and inflamed participants in the so-called science wars are treating the latest version of this fine old philosophical controversy as a big deal. In the very long run, perhaps, it will prove to be one. Maybe someday the idea of human beings answering to an independent authority called How Things Are in Themselves will be obsolete. In a thoroughly de-Platonized, fully Protagorean culture the only answerability human beings would recognize would be to one another. It would never occur to them that "the objective" could mean more than "the agreed-upon upshot of argument." In such a culture we would have as little use for the idea of the intrinsic structure of physical reality as for that of the will of God. We would view both as unfortunate and obsolete social constructions.

12

Letters to Charles Enz

12 June 1999, "Interest in Pauli"

I have come across a reference to a paper of yours that I have not been able to obtain. May I ask of you to send me a reprint of it? The paper is:

C. P. Enz, "Wolfgang Pauli and the Role of the Observer in Modern Physics," in *Philosophy of the Natural Sciences: Proceedings of the 13th International Wittgenstein Symposium*, edited by P. Weingartner and G. Schurz (Verlag Holder–Pichler–Temsky, Vienna, 1989), pp. 110–119.

I have read several of your other papers on Pauli's thought and have been uniformly pleased with each. I am especially interested in his ideas concerning the demise of the "detached-observer" notion within quantum theory. (I would also be interested in any of your other works on the subject that may have been written since your 1992 article, "Wolfgang Pauli between Quantum Reality and the Royal Path of Dreams." This is the latest one that I am aware of.)

About myself, I am a theoretical physicist at Caltech who specializes in quantum information theory and quantum computing. I am well acquainted with your colleague Nicolas Gisin at the University of Geneva.

21 August 1999, "One More Article?"

Upon my return from Cambridge a couple of weeks ago, I found in my mailbox your package full of wonderful papers! Thank you so much. I have gobbled them all down except the one written in German (unfortunately I don't know a word of German).

I wonder if I can impose upon you one more time. I noticed a citation to still another of your papers of which I was not aware: "The wave function of correlated quantum systems as objects of reality" in a book with a Finnish title (Helsinki University Press, 1996), pp. 61–76. May I obtain a copy of that also? (If you happened to write it in TEX or LATEX, you could just email the file and I could view it that way; I am much more interested in the thoughts than in having an official reprint.)

My deep interest in Pauli comes from his interest and thought about the "engaged observer." In particular, part of my research program is to see how far the idea that the engaged observer is not only consistent with the structure of quantum theory but actually compels that structure (i.e., the standard axioms of the theory). For the technical side of this research, I take as my starting point many of the concepts and techniques we have learned from quantum cryptography and quantum information theory. But on the philosophical side I have Pauli, Fierz, yourself and perhaps some other disciples that I am not yet aware of.

09 June 2000, "Another Request"

I wonder if I might trouble you again with another request for your papers on Wolfgang Pauli. (Do you remember my earlier requests that you so nicely fulfilled?) My house was one of the unfortunate ones that burned in Los Alamos during the forest fire last month. Consequently my wife and I lost almost everything we owned, including all our books and all our papers.

If you could replenish my supply of all your papers, I would be ever so grateful.

I know that I had at least the following papers of yours on Pauli, but I believe there were several more that I had not yet recorded into my computer.

1. C. P. Enz, "W. Pauli's Scientific Work," in *The Physicist's Conception of Nature*, edited by J. Mehra (D. Reidel, Dordrecht, 1973), pp. 766–799.
2. C. P. Enz, "The Space, Time and Field Concepts in Wolfgang Pauli's Work," in *Symposium on the Foundations of Physics, Joensuu, 1987*, edited by P. J. Lahti and P. Mittelstaedt (World Scientific, Singapore, 1987), pp. 127–145.
3. C. P. Enz, "Wolfgang Pauli and the Role of the Observer in Modern Physics," in *Philosophy of the Natural Sciences: Proceedings of the 13th International Wittgenstein Symposium*, edited by P. Weingartner and G. Schurz (Verlag Holder–Pichler–Temsky, Vienna, 1989), pp. 110–119.
4. C. P. Enz, "Book Review: *Beyond the Atom: The Philosophical Thought of Wolfgang Pauli. By K. V. Laurikainen*," *Found. Phys.* **20**, 1025–1028 (1990).
5. C. P. Enz, "Wolfgang Pauli between Quantum Reality and the Royal Path of Dreams," in *Symposia on the Foundations of Modern Physics 1992: The Copenhagen Interpretation and Wolfgang Pauli*, edited by K. V. Laurikainen and C. Montonen (World Scientific, Singapore, 1992), pp. 195–205.
6. C. P. Enz, "The Wavefunction of Correlated Quantum Systems as Objects of Reality," in *Vastakohtien todellisuus: Juhlakirja professori K. V. Laurikainen 80-vuotispäivänä*, edited by U. Ketvel, *et al.* (Helsinki U. Press, 1996), pp. 61–76.

13

Letters to Henry Folse

17 January 2001, "First Contact"

I am writing this letter to find out if you are interested in attending a conference in Växjö, Sweden this summer titled "Quantum Theory: Reconsideration of Foundations." I do this because I have read a few of your papers and parts of your book on Bohr, and I have been extremely impressed. By trade, I am a practitioner of the new field of quantum information theory and computing – and indeed the main session I am organizing at the Växjö meeting will be about quantum foundations reconsidered in that light – but your description of Bohr's thought meshes so well with what I'm seeking in physics that I'd like to get you there too. PLEASE NOTE that this is not exactly a real invitation yet: I am still lobbying for a more philosophical contingent at the meeting. But, I would very much like to hear your reaction just as soon as possible. If things work out, and you would like to come, I will recontact you very soon with further details.

The main reason I want you at our meeting is that I have this "madly optimistic" (Mermin called it) feeling that Bohrian–Paulian ideas will lead us to the next stage of physics. That is, that thinking about quantum foundations from their point of view will be the *beginning* of a new path, not the end of an old one. I think you find this kind of thought attractive too, so I think the conference will be quite a natural vacation for you.

In this regard, allow me to place several pieces of (used!) information below. (Please don't be offended by this: It just seemed like the simplest way to get it to you quickly.) What you will find is the following. [. . .]

07 February 2001, "Heartfelt Thanks"

Yesterday I received the package you sent me. I can't express how grateful I am. I never imagined that you would be sending such a wealth of material! Indeed I didn't even know that you had more than three or four articles (outside your book) on the subject!

As I promised, I should have them digested before I meet you. I am so happy that you're coming to the meeting.

16 February 2001, "Is It Complete?"

On top of the last question I sent you, I wonder if I can ask another favor of you? Could you scan down the list below and note whether it is complete with respect to your quantum writings? If any glaring mistakes jump out at you, please let me know. If you could also fill me in with the details of the last two entries, that would be great.

1. H. J. Folse, "The Copenhagen Interpretation of Quantum Theory and Whitehead's Philosophy of Organism," *Tulane Stud. Phil.* **23**, 32–47 (1974).
2. H. J. Folse, "The Formal Objectivity of Quantum Mechanical Systems," *Dialectica* **29**, 127–?? (1975).
3. H. J. Folse, "A Reinterpretation of Democritean Atomism," *Man and World* **9**, 393–417 (1976).
4. H. J. Folse, "Complementarity and the Description of Experience," *Int. Phil. Quart.* **17**, 377–399 (1977).
5. H. J. Folse, "Quantum Theory and Atomism: A Possible Ontological Resolution of the Quantum Paradox," *Southern J. Phil.* **16**, 629–640 (1978).
6. H. J. Folse, "Kantian Aspects of Complementarity," *Kant-Studien* **69**, 58–66 (1978).
7. H. J. Folse, "Platonic 'Atomism' and Contemporary Physics," *Tulane Stud. Phil.* **27**, 69–88 (1978).
8. H. J. Folse, Jr., "Complementarity, Bell's Theorem, and the Framework of Process Metaphysics," *Process Studies* **11**, 259–273 (1981).
9. H. J. Folse, *The Philosophy of Niels Bohr: The Framework of Complementarity*, (North-Holland, Amsterdam, 1985).
10. H. J. Folse, "Complementarity and Scientific Realism," in *Foundations of Physics: A Selection of Papers Contributed to the Physics Section of the 7th International Congress of Logic, Methodology and Philosophy of Science*, edited by P. Weingartner and G. Dorn (Verlag Hölder–Pichler–Tempsky, Vienna, 1986), pp. 93–101.
11. H. J. Folse, "Niels Bohr, Complementarity, and Realism," in *PSA 1986: Proceedings of the Biennial Meeting of the Philosophy of Science Association, Vol. I*, edited by A. Fine and P. Machamer (Philosophy of Science Association, East Lansing, MI, 1986), pp. 96–104.
12. H. J. Folse, "Complementarity and Truth," transcript of talk given at Valamo Monastery in Finland (1987).
13. H. J. Folse, "Niels Bohr's Concept of Reality," in *Symposium on the Foundations of Modern Physics 1987: The Copenhagen Interpretation 60 Years after the Como Lecture*, edited by P. Lahti and P. Mittelstaedt (World Scientific, Singapore, 1987), pp. 161–179.
14. H. J. Folse, "Realism and the Quantum Revolution," in *Abstracts of the 8th International Congress of Logic, Methodology, and Philosophy of Science, Vol. 4, Part I*, (Inst. of Philosophy of the Academy of Sciences of the USSR, Moscow, 1987), pp. 199–200.
15. H. J. Folse, "Complementarity and Space-Time Description," in *Bell's Theorem, Quantum Theory and Conceptions of the Universe*, edited by M. Kafatos (Kluwer, Dordrecht, 1989), pp. 251–259.
16. H. J. Folse, "What Does Quantum Theory Tell Us About the World?," *Soundings* **72**, 179–205 (1989).

17. H. J. Folse, "Bohr on Bell," in *Philosophical Consequences of Quantum Theory: Reflections on Bell's Theorem*, edited by J. T. Cushing and E. McMullin (U. Notre Dame Press, Notre Dame, IN, 1989), pp. 254–271.
18. H. J. Folse, "Bohr's Framework of Complementarity and Pauli's Philosophy," in *Kohti uutta todellisuuskäsitystä. Juhlakirja professori Laurikaisen 75-vuotispäivänä* (Towards a New Conception of Reality. Anniversary Publication to Professor Laurikainen's 75th Birthday), edited by ?? (Yliopistopaino, Helsinki, 1990), pp. 91–99.
19. H. J. Folse, Jr., "Complementarity and the Description of Nature in Biological Science," *Bio. Phil.* **5**, 211–224 (1990).
20. H. J. Folse, "Laudan's Model of Axiological Change and the Bohr-Einstein Debate," in *PSA 1990: Proceedings of the Biennial Meeting of the Philosophy of Science Association, Vol. I*, edited by A. Fine, M. Forbes, and L. Wessels (Philosophy of Science Association, East Lansing, MI, 1990), pp. 77–88.
21. H. J. Folse, "Metaphysical Awakening in Philosophy of Quantum Physics: A Review Article," *Int. Stud. Phil.* **23**, 89–98 (1991).
22. H. J. Folse, "Complementarity and Our Knowledge of Nature," in *Nature, Cognition and System II: Current Systems-Scientific Research on Natural and Cognitive Systems*, Volume 2: On Complementarity and Beyond, edited by M. E. Carvallo (Kluwer, Dordrecht, 1992), pp. 51–66.
23. H. J. Folse, "The Environment and the Epistemological Lesson of Complementarity," *Environmental Ethics* **15**(4), 345–353 (1993).
24. H. J. Folse, "Bohr's Framework of Complementarity and the Realism Debate," in *Niels Bohr and Contemporary Philosophy*, edited by J. Faye and H. J. Folse (Kluwer, Dordrecht, 1994), pp. 119–139.
25. A. Dotson and H. Folse, "Bearers of Properties in the Quantum Mechanical Description of Nature," *Int. Stud. Phil. Sci.* **8**, 179–194 (1994).
26. H. J. Folse, "Essay Review: Niels Bohr and the Construction of a New Philosophy," *Stud. Hist. Phil. Mod. Phys.* **26**, 107–116 (1995).
27. H. J. Folse, "The Bohr–Einstein Debate and the Philosopher's Debate over Realism versus Anti-Realism," in *Realism and Anti-Realism in the Philosophy of Science*, Boston Studies in the Philosophy of Science, Vol. 169, edited by R. S. Cohen, R. Hilpinen, and Q. Renzong (Kluwer, Dordrecht, 1996), pp. 289–298.
28. H. J. Folse, "Ontological Constraints and Understanding Quantum Phenomena," *Dialectica* **50**, 121–136 (1996).
29. H. J. Folse, "Realism, Idealism, and Representation in the Description of Nature," in *Avartuva ajatus: Julkaisutoimikunta*, edited by U. Ketvel, A. Airola, T. Kallio-Tamminen, S. Laurema, K. Rainio, and J. Rastas (Luonnonfilosofian seuran, Espoo, 1999), pp. 73–77.

09 March 2001, "The Right Choice"

Sorry for my long delay in getting back to you. I'm off to Japan in about five hours for two weeks, and *trying* to get things tied up here before leaving has made the week pretty hectic.

Anyway, the most important message I wanted to tell you is that I'm gettin' damned happy I invited you to Växjö. Reading your papers has been a really pleasant experience.

I think I hit my first dozen last night at about this time. (But maybe the bigger question is, why am I up at this time?!?) Below, I'll place my Folse compendium as it stands – every word typed in lovingly with my own little fingers! Of course, I have a few quibbles with some things I've read, but I think I'll save my comments until I've read the complete body of work. 27 years is a long time, and you could well have changed your mind about some things: I'll give you the benefit of the doubt for now.

Folse-ism 1: *I should certainly warn you that very few in this community agree much with my reading of Bohr.*

I couldn't care less about that: I like it (or most of it), and that's all that counts for me. Besides, I've read Bohr myself – fairly carefully I've always thought – and your view significantly coincides with my memory of that.

Folse-ism 2: *I'm more concerned about my lack of knowledge of anything about information theory.*

You need not be too concerned, but of course it wouldn't hurt you to do a little reading on the side if you've got some time. Somewhere below, you wrote:

In describing the phenomena of observational interactions, quantum theory describes them as being caused by the interactions between the observing systems and microsystems. The fact that we can form no representation, no mechanical picture, of the atoms on which the mechanistic description of the phenomenal world is based, hardly reveals that we are ignorant of what these entities *are*. Rather it testifies to what we have learned about them – that they cannot be so represented – through explorations of the atomic phenomenon in which their strange behavior is revealed to human experience.

It is my strongest opinion that the great fruits of quantum information and computing get at precisely this point ... and in spades! The point is that this "nonrepresentability" in actual fact boils down to a positive statement rather than a negative one. So it would do you well to learn a little about our field. (And, luckily, most of what we do is not abstruse stuff: it's just basic quantum mechanics, viewed mostly from a new point of view with a new set of goals in mind.) Where to start? Maybe a good place would be three "recent" *Physics Today* articles:

1. Gottesman and Lo, "From Quantum Cheating to Quantum Security," *PT* November 2000 (don't have the page numbers).
2. Preskill, "Battling Decoherence: The Fault-Tolerant Quantum Computer," *PT* June 1999, p. 24.
3. Bennett "Quantum Information and Computation," *PT* October 1995, p. 24 (interesting coincidence).

If you get that far, let me know, and I'll suggest a couple of really *mild* technical articles that'll be worth their weight in gold in insight.

But why do I think it would do you well? Because I think you have an honest heart. And, because while I believe Bohr and his gang certainly started to point us in the right direction,

I think we have a long, long technical way to go before we can claim a particularly deep understanding of the quantum structure. Here's how I put it in exasperation to David Mermin once:

What's your take on this passage? Can you make much sense of it? What does he mean by "providing room for new physical laws?" What "basic principles of science is he talking about?" What five pages of derivation are lying behind all this business?

It nags me that Bohr often speaks as if it is clear that the structure of quantum theory is derivable from something deeper, when in fact all the while he is taking that structure as given. When did he ever approach an explanation of "Why complex Hilbert spaces?" Where did he ever lecture on why we are forced to tensor products for composite systems? It's a damned shame really: I very much like a lot of elements of what he said, but as far as I can tell all the hard work is still waiting to be done.

The issue in my mind is *not* to *start* with complex Hilbert space, unitary evolution, the tensor product rule for combining systems, the identification of measurements with Hermitian operators, etc., etc., and *showing* that Bohr's point of view is *consistent* with that. Instead it is to start with Bohr's point of view (or some variant thereof) and see that the *precise* mathematical structure of quantum theory *must* follow from it. Why complex instead of real? Why unitary, rather than simply linear? Indeed, why linear? Why tensor product instead of tensor sum? And, so on. When we can answer *these* questions, then we will really understand complementarity.

I'm banking my career on the idea that the tools and issues of quantum information theory are the proper way to get at this program.

OK, I've got to get some sleep. I have this dream that I'm going to get work done all the way to Japan. But if I don't get some sleep, I'll certainly be kidding myself. (By the way, I will most certainly be in email contact while I am away, so write if you wish. But don't send any viruses!)

14 March 2001, "The Archive Again"

Folse-ism 3: *If you're planning to visit the archive at the Bohr Institute, you should get in touch with the head archivist, Dr. Finn Aaserud. When I last visited, they weren't really set up for the "drop in" visitor.*

Thanks again for the Archive information. Ben Schumacher and I, as we were hiking down the old Hakone trail, were getting more and more excited about the possibility of spending two days in the archive after Sweden. Can you tell me how to get in touch with the person you speak of?

Ben told me a wonderful story yesterday. Somehow the new *Star Wars* movie came up in the conversation (Charlie Bennett and John Smolin were also in the conversation). I said I had never seen the movie. John said it was awful. Ben said it was wonderful. I suspect I'd have to side with John, just seeing what I've seen of the previews and reviews, but the surprising thing was that Ben said, "When the movie was over, I was really depressed."

"For about 10 days after the movie, I was really depressed because I wanted to be a Jedi. To learn the ways of the Force. I'd never have a chance to live that wondrous life, and that depressed me." But then he went on to say, "But about two months later I was in London and I saw the play *Copenhagen*. And that got me going all over again. And it almost depressed me, *but* then I realized, 'Wait! I am a Jedi!' Learning the ways of the quantum world *makes* me a Jedi!' "

14

Letters to Bob Griffiths

10 February 1997, "Sunday Afternoons and Consistent Histories"

I've finally completed to my satisfaction your challenge of taking a plunge into "consistent histories." And I've had great fun! In particular, I read all or most of each of the following:

1. "Consistent Histories and the Interpretation of Quantum Mechanics,"
2. "Quantum Interpretation Using Consistent Histories,"
3. "Correlations in separated quantum systems: A consistent history analysis of the EPR problem,"
4. "The Consistency of Consistent Histories: A Reply to d'Espagnat,"
5. "A Consistent History Approach to the Logic of Quantum Mechanics,"
6. "Consistent Histories and Quantum Reasoning."

I don't think I've read everything, but I'll trust that I haven't missed anything too significant – at least for a novice education in the subject.

I say I had great fun because, before this, my only introduction to consistent histories had been through various talks by Gell-Mann, Hartle, and Zurek. And, I can't say I ever found any of those satisfactory – they all made it a very confusing muddle to me. I never could figure out the point, i.e., what actually was being proposed as a "solution" to the quantum interpretation problem. Consequently, I often just fell asleep or thought about other things or both. In contrast, I found all your papers logically crisp.

I now understand what you mean when you say that consistent histories is a very natural extension of the standard quantum mechanics. Apart from the "ontological aspects" of the interpretation, as far as I can tell, there is nothing questionable about it whatsoever. It appears to me to be a valuable tool, and I regret not having practiced with it before.

What I mean by this particularly is that I have known for quite some time how to play a certain game when I restrict my attention to a single orthogonal set of projectors on a Hilbert space. For instance, I know that I can build a device to "clone" the quantum states associated with that set; more than that, I know that I can build a device to "broadcast" any density operator diagonal in that representation. Basically, I can play the games that I might have imagined playing in the classical world. What I have learned from your papers is that there are a whole host of new games of "classical inference" within (consistent) "frameworks" that I should contemplate. And that may well lead to more economical ways of viewing problems than I could have anticipated with textbook quantum mechanics. Better yet, it may lead to new problems! (I think you and Chi-Sheng are carrying that aspect of it forward quite nicely.)

All that said, my main purpose in writing this note is to learn even more – especially about those aspects of the "interpretation" that I am not so disposed to agree with. I hope you'll bear with me, especially in those cases when I will – of necessity – be fairly vague. (It can be very difficult to debate a point when you feel that your opponent is wrong, but you also know that you yourself do not know what is right!)

As I say, I believe your consistency conditions and the examples you give for what can be done with them are technically unassailable. Concerning the "foundational value" of consistent histories, however, I am not completely enthusiastic. First let me state outright that I am fairly agnostic about most things: interpretations of quantum mechanics are no exception. I am perfectly willing to work both sides of the railroad track, so to speak. I've seen how believing in "many worlds" helped David Deutsch search for interesting applications of quantum parallelism. I've seen how believing that quantum theory is likely to be secondary and derived from some deeper theory of communication helped lead Bill Wootters to the quantum teleportation idea. I've now seen how consistent histories led you and Chi-Sheng to some nice things. The point is I don't have too much reason to be too hard-headed about any one point of view, nor too much passion, but I would like to express my troubles.

Several times over in your papers you say things similar to one line I'll quote from your last paper:

The principal thesis of the present paper is that the major conceptual difficulties of nonrelativistic quantum theory ... can be eliminated, or at least tamed, by taking a point of view in which quantum theory is a fundamentally *stochastic* theory, in terms of its description of the time development of a physical system.

Presumably what you mean by this is that, within a *single* consistent family of histories – or framework as you call it – one and only one history is actually the case. Quantum theory ascribes a "weight" or probability to each history to signify that even complete knowledge of the initial and final conditions, D and F, does not carry enough information for us to predict precisely which history within the family is actually the case. A consequence of this is that quantum theory concerns an indeterministic state of affairs. What I mean by the phrase "is actually the case" is the common everyday usage of the notion, i.e., something that is an objective reality and observer-independent. "Regardless of the two opposing rulings, it is a fact that O. J. Simpson either did or did not commit the murders; we simply may never know which is actually the case."

In equal measure you then state a fundamental requirement of the consistent histories interpretation:

A meaningful description of a (closed) quantum mechanical system, including its time development, must employ a single framework.

And finally you qualify the consistent histories formalism and interpretation with a discussion of the following sort.

... once we have agreed that quantum mechanics is a stochastic theory in which the concept of "true" corresponds to "probability one", then precisely because probabilities (classical or quantum) only makes sense within some algebra of events, the truth of a quantum proposition is necessarily labelled, at least implicitly, by that algebra, which in the quantum case we call a framework. The existence of incompatible quantum frameworks is no more or less surprising than the existence of non-commuting operators representing dynamical variables ...

Let me use these quotes and my extra – but presumably mild – interpretive work above as a point of departure. I believe you are dead on the mark with the very last statement, i.e., that the existence of incompatible frameworks is no more or less surprising than the existence of incompatible (noncommuting) observables in standard quantum mechanics. Moreover, I completely endorse the systematic scheme consistent histories gives us for making Wittgenstein's famous statement precise in the quantum context, "Whereof one cannot speak, thereof one must be silent." Whenever a statement of the form $A \wedge B$ cannot be formed, it is simply meaningless.

What I cannot see is how consistent histories makes the "objectivity" of "quantum properties" more palatable or believable than the single event formalism that we learn in standard quantum mechanics. That is to say, if I take the quantum interpretation problem to be that of answering the questions

1. Where do the unpredictable events in quantum phenomena come from?
2. Are they always "there" independent of the framework I employ in my description of the system?

then I do not see how consistent histories has moved us any closer to a solution. I don't even see how it has "tamed" the problem.

I do not say these things because I am *not* willing to revise the set of "logics" I might entertain for reasoning about our world (as was apparently the case with d'Espagnat). The problem I have is that I do not see these changed rules of logic as a *call* for revising my intuition for what can be said of the "real" in the world (or the "True" to use your terminology), as you suggest your reader do. Rather, I view that as exactly the nub of the problem for the "stochastic" interpretation of quantum theory you endorse. (Here I am using stochastic in the sense I defined above, which I think is an adequate version of what you have in mind.) If proposition A is True in one framework, and proposition B is True in an incompatible framework, then I do not see the meaninglessness of the proposition $A \wedge B$ as a call for revising my notion of "being the case." Rather, it does push me very hard to understand the (incompatible) True propositions as something other than completely objective states of affairs.

Essentially all I'm saying here is that I don't see how consistent histories has tamed any of the strangeness of the logic you refer to; your papers provide no convincing arguments in this regard (or I am a poor reader!). I believe that it is precisely this strangeness about quantum events (i.e., decompositions of the identity operator) that has brought so many to interpret them as referring to something *other* than completely independent and objective states of affairs. This is the sort of thing that led to what I understand as the "Copenhagen interpretation": the meanings of the propositions A and B are defined solely in terms of the experimental procedures required for eliciting their truth values; the proposition $A \wedge B$ is meaningless precisely when there exists no experimental procedure for eliciting the truth values of A and B simultaneously.

I believe I understand your concern that so much of quantum mechanics has, in the past, been tied conceptually to the notion of measurement. It would be a very strange world indeed that floated around in superposition until Henri Becquerel and his laboratory appeared in one branch of the wave function! More seriously, the notion of a real world, independent of what we think and see, is beyond a doubt *almost* indispensable for our coordination in this world. As I once wrote my friend Greg Comer:

From where and of what utility comes this notion of an "objective world" independent of man, woman, animal, and plant? My opinion is that (in the end) "objective reality" is posited for nothing more than to have a device for coordinating the various experiences common to all those who communicate. Mark Twain once wrote, "If you tell the truth, you don't have to remember anything." It seems to me that that pretty much sums it up. The objective truth saves us from having to make sure that the stories we tell are consistent; it saves us from having to remember all aspects of the past. It gives us a means for determining whether someone's behavior is insane. It gives us a means by which to determine the guilt or innocence of an accused murderer. Is there any other real motivation for positing an objective reality? I can think of none … but why should I, the reasons listed above should be powerful enough for anyone involved in the sciences.

The problem of course with a measurement-based interpretation of quantum mechanics is that much of this "simplest view of the world" seems to fade away. The fighting of that, I believe, may have been your motivation for seeking out consistent histories.

If this is the case, then I share your troubles. However, it seems to me that I haven't yet seen a convincing way of getting past these things by revising our notion of *conjunction*, i.e., $\wedge$, for "objective" but incompatible propositions. I, in my own way, have tried to grapple with these difficulties by trying to make sharper what quantum probabilities are about. For instance, Carl Caves and I in our paper "Quantum Information: How Much Information in a State Vector" (I gave you a copy) tried to promote the following idea.

"Quantum probabilities" are ignorance probabilities; they express ignorance about the outcomes of potential measurements. *What is different in quantum physics is not the status of probabilities, but rather the nature of the alternatives.* In classical physics, probabilities are concerned with *actualities*: "One of the alternatives actually occurs, but since I don't know which, I assign probabilities based on what I do know." The probabilities that describe intrinsic quantum unpredictability – the "quantum probabilities" – express ignorance about *potentialities* that are actualized by measurement: "I know one of these alternatives *would* occur if I enquired about that set of alternatives, but since I don't know which, I assign probabilities based on what I do know."

(For a more extended discussion of this, see the bottom of page 22 through the top of page 26 of that article.) I would change the language of some of this now if I could – Carl likes the word "potentiality" more than I do – but I think the intention is clear. One would like to divorce the language of quantum mechanics from measurement *per se*. Our attempt in this passage was through emphasizing that quantum probabilities are about "relational contingencies" more than actualities: they are about what would occur if such and such situation were at hand.

This, at least formally, is not so different from what I saw many times over in your papers on consistent histories: the physicist chooses *any* framework consistent with his initial and final data, D and F; thereafter he can derive probabilities for the elements within that framework. The real difference appears only to come in the "ontological status" we would give those elements. I think some of your very nice examples to do with incompatible frameworks – like the one to do with Eqs. (3.4)–(3.8) of Ref. (2) above – will allow me to sharpen up my phrase "such and such situation at hand" substantially. However, I don't see how my new found knowledge about consistent histories will change my point of view drastically. (I welcome you to try to convince me otherwise!)

The problem with the focus on contingencies – which is the point of view I am happiest with right now – is that you will immediately ask, "What of the world before physicists were here making their measurements? What element in quantum theory reflects the clockwork of the world that ticks away independently of man?" The only answer I am capable of giving right now is, "I am not sure *anything* in quantum mechanics reflects that." Given my point of view, I would think that the hypothetical questions I just gave you are much like, "What in Bayesian probability theory reflects the operation of the weather before weathermen appeared on the scene?" Perhaps it's much like Heisenberg once said:

If the system is closed, ... [it] is then represented by a vector in Hilbert space. The representation ... so to speak, contains no physics at all.

When I'm feeling sceptical I sometimes think that perhaps there's not a heck of a lot more to it than your "quantum bicycle riding."

On a side note, let me talk about one more thing of an even-more philosophical flavor than the things we've discussed so far; after that, I'll return to some further technical comments about consistent histories. In your last *PRA* paper, you ask somewhat rhetorically at the beginning, "Do quantum measurements reveal pre-existing properties of a measured system, or do they in some sense create the properties they reveal?" Near the end of it you ask, "... does quantum theory itself specify a unique framework? ... if the answer is 'no', ... [does this somehow imply] that physical reality is influenced by the choices made by the physicist?" The truth of the matter is, to a nonnegligible extent, I like to think that there is some truth to the wilder of these alternatives. Quantum indeterminism is certainly one of the most wonderful things found in this century. If this indeterminism had not been of such a variety as to allow us more free reign over our destiny or destruction, I think I would be quite disappointed in it. When my wife and I have a philosophical discussion on just this point, I like to say that the invention of the electric coffee maker is just as real a creation as anything else in the world; if she can accept that, then why can she not accept that a photon's polarization (or what we have a right to say about it) is also created by our actions? I am not joking when I think these may be somewhat on the same footing. The question then turns to whether quantum theory always requires an intervention on man's part for the properties of the world to come into being. In that regard, I am much, much more sceptical!! (I refer you back to my talk of clockworks above.) However, just to emphasize again, I think there is more than a grain of truth in John Wheeler's "Game of 20 Questions, Surprise Version" and it is the quantum muddle that first led us to take it seriously. (If this doesn't completely frighten you off, I can send you a couple of short little pieces of my correspondence that makes these ideas clearer.)

Whew, this note is growing! Let me try to wrap it up soon. Back to more technical points. Several times in your papers you have made mention of "quantum logic" and contrasted your system to that. Always, however, you point to systems like Birkhoff and von Neumann's to do with orthomodular lattices. It doesn't appear that you are aware that there is still another approach to "quantum logic" that is very much of the flavor of the system you construct in the consistent history interpretation. These systems are sometimes known as "partial Boolean algebras"; a simple introduction to what is known about them can be found in R.I.G. Hughes book *The Structure and Interpretation of Quantum Mechanics* (pp. 192–194). In a nutshell,

A partial Boolean algebra is thus a set of Boolean algebras pasted together in a consistent way, so that, where two or more Boolean algebras overlap, their operations agree with each other.

These objects are much like logical versions of a manifold in general relativity: instead of being locally Lorentzian, they are locally Boolean. For instance, when two propositions are not contained within compatible Boolean algebras their conjunction is simply deemed meaningless, just as in your system. Apparently this logical structure predates the more famous Birkhoff and von Neumann one by several years, in the work of Strauss. Hughes also wrote a review article which is pretty good, *J. Symbolic Logic* **50**:558–566 (1985). For the same reason that I find your system interesting, I think these systems and the questions logicians are asking of them are quite nice. For instance, one can ask whether the set of partial Boolean algebras is isomorphic to the set of finite dimensional Hilbert spaces. That is, once one has decided that physics has to do with partial Boolean structures, is one led ineluctably to a Hilbert space description of physical phenomena?

On a different subject, I think it would be very nice to see if the consistent history idea can be made to be more general than it already is. By this, I do *not* mean something like Isham suggests, i.e., to use non-tensor-product histories; that sort of generalization strikes me as taking away the prettiness of the history idea. Instead, as you can probably guess, I am thinking of using more general decompositions of the identity for the elementary events in your histories. In particular, it would be very nice if something like the consistency conditions could be worked out for general POVMs. My inclination is that POVMs can be viewed as measurements that are every bit as elementary as von Neumann measurements. If that is the case, then I don't see why they can't also be fit somehow into the consistent history framework for closed systems. Do you have any thoughts on this?

Finally in Section 6.2 of your original article, you question the uniqueness of your "weight" function. I think that is an intriguing question. As you probably know, Jim Hartle in his 1968 *AJP* article, has a very nice pseudo-derivation of the Born probability rule. (In my opinion, he somewhat confuses "probability" with "frequency"; however, his train of thought does appear to lead nicely from qualitative statements to the quantitative rule – as you say of the Everett efforts to the same effect, the starting point is not completely devoid of probability concepts. Carl, Howard Barnum, and I plan to write an article putting in our two cents worth on this subject soon.) In any case, have you attempted to pin down your weight function by similar techniques?

I guess that's more than enough for now. My left hand is hurting pretty badly! I hope you find some of this commentary useful. Your reactions, good or bad, would be appreciated. In particular, if you can give me some sounder reasoning for the "objectivity" of your histories, I hope you will attempt it. I hope you don't write me off as a real quack!

15

Letters to Adrian Kent

30 August 1998, "Desiderata"

I just read over your four desiderata for an acceptable "über-quantum theory" for the third time. (When I finish this note, I'll read them over for the fourth time ... at the very least for simile's sake.) They're not so uninspired! (... to use your words, that is.) But I do hope you will keep your promise of writing a longer exposition.

In the mean time, let me ask you for one point of clarification about the connection between your Desideratum #1 and my latest happy thoughts.

Kentism 1: *1. A theory in which some elementary, and precisely defined, components constitute the basic statements from which propositions about an external and objective reality can be built. Maybe trajectories à la Bohm, maybe collapse events à la GRW, maybe events of more general form à la consistent historians, maybe the evolution of the classical gravitational field in some (unknown) theory that links classical metric and quantum matter. Maybe something quite different from any of the above. But, at any rate, something which allows us to talk about what happened in the past, in the absence of observers, and which allows us to resolve the measurement problem by replacing the vague notion of measurement by something precise.*

As I tried to convey to you in Spain, I believe that I am basically a realist at heart. The evidence seems conclusive (to me, at least) that there were things in this world long before there were physicists grappling with the quantum measurement problem. There were things in this world long before there was intelligent life at all. However – and this may be where we disagree the most – I don't see why "reality" should be construed as something absolutely static. It is the essence of a nondeterministic world that there is real change within it. But I think it goes much further than that. Because I see the coffee maker as having no less of an ontological status than a rock (or an electron for that matter), I am struck with the happy thought that man's actions do contribute – in part! – to the actual construction of reality.

Part of my quest for the last three years has been to find some precise counterpart to all that within quantum theory itself. And I surely think it is there. The most clear-cut example of it is in the existence of quantum key distribution schemes. As I see it, these examples are trying to tell us something very deep about the story of quantum theory; I think they capture the ontological content of the theory. It is this idea that I would like to sharpen up to the point of being useful.

Here is my question. Suppose in a few years I could come up with a clear, precise statement of what it is that I'm trying to get at. Something in essence that takes away the vagaries of the statement: "The world in which we live happens to have a funny property. It is that *my* information gathering about something you know, causes *you* to loose some of that knowledge ... and this happens even in the case that you know all my actions precisely. Physical theory, and quantum mechanics in particular, is about what we can say to each other and what we can predict of each other in spite of that funny property." Would that constitute something that fulfills your Desideratum #1? It would be my claim that Hilbert space vectors, hermitian operators, unitary operators, and even consistent histories for that matter, all lie on the side of the subjective in the theory: they are all pieces of the theory that concern what we can ask, what we can say, and what we can predict. (They have no ontological content ... and it is a sad thing that people have been searching for so much ontology in them for so long.) On the other hand, the "funny property" does strike me as having some ontological content. And it does have a (rather ethereal) counterpart in the theory. Would pinpointing that precisely meet your Desideratum #1? Especially, if in excising the vagary in quotes above, we will be forced to acknowledge measurement as a primitive for the theory? (Please note the distinction between a primitive for the theory and a primitive for the world in itself.)

30 August 1998, "Well Look At That"

Just a short note this time. Right after writing you about coffee makers and electrons earlier today, I ran across the following little passage in Heisenberg's book *Physics and Beyond*.

A few hundred yards away, a large liner was gliding past, and its bright lights looked quite fabulous and unreal in the bright blue dusk. For a few moments, I speculated about the human destinies being played out behind the lit-up cabin windows, and suddenly Wolfgang's questions got mixed up with it all. What precisely was this steamer? Was it a mass of iron with a central power station and electric lights? Was it the expression of human intentions, a form resulting from interhuman relations? Or was it a consequence of biological laws, exerting their formative powers not merely on protein molecules but also on steel and electric currents?

The coincidence was too much for me to pass up, and so I found myself copying the words into the computer. (You know how I am.)

02 September 1998, "Vague and Vagary"

Thanks for the appraisal. Let me chew upon it for a while. Certainly clarity is what I strive for ... and you're helping me in that regard. You're right, I never do refer to "consciousness." That word frightens me. I don't even really have an interest in what it means. So I would hope that I'm *not* trying to build a picture of the world based on it – I don't think I am. Besides that, though, I do worry a little bit that you and I sometimes use the same words when we mean different things: I'll try to sort that out.

I apologize for the continued use of "slogans" in summarizing what it is that I'm trying to get at: unfortunately, they are all I've got right now!! But let me leave you with one more before I go to my chewing. It's part of the conversation I'm having with you in my head right now, and I'm having a hard time staying quiet; maybe, just maybe, it's the pithiest version yet for expressing what I mean by saying that I'm a "realist." As I see it: Quantum theory is nothing beyond WHAT WE HAVE THE RIGHT TO SAY in a world where information gathering necessarily causes disturbance. (Disturbance to what? To each other's descriptions, nothing more.) Where's the reality in this? Well it has to be there, or it would be awfully hard to imagine how it is that we come to intersubjective agreement in an only-partially predictable world. But that reality, we have every right to believe, must be even more numinous than Kant's noumenon!

Kentism 2: *I wouldn't query this if I didn't know your fondness for linguistic discussions. But since I do: do you really mean "vagaries" here? Rather than, say, "vagueness".*

Vagary: a whimsical, wild, or unusual idea or notion. (Webster's Encyclopedic Unabridged) Try again and tell me if it still doesn't make sense. The last thing I want to do is be vague (though I'm often forced to be!).

17 January 1999, "Cigar and the Greeks"

On another subject, over the weekend I read Schrödinger's *Nature and the Greeks*, *Science and Humanism*, and *Mind and Matter*. That helped me recall that you once placed this little quote at the end of one of your letters to me:

In fragment D 125 … [Democritus] introduces the intellect in a contest with the senses. The former says "Ostensibly there is colour, ostensibly sweetness, ostensibly bitterness, actually only atoms and the void"; to which the senses retort: "Poor intellect, do you hope to defeat us while from us you borrow your evidence? Your victory is your defeat." You simply can not put it more briefly and clearly.

(Erwin Schrödinger, Nature and the Greeks)

I didn't understand then why you put it there, and I guess I don't even understand it now. What struck you about this? I've gotten the impression over and over again that your inclination is that of a "naïve realist" (no disrespect intended). Have I missed something?

20 January 1999, "Clarity"

Kentism 3: Nature and the Greeks, *yes, I'm glad it is propagating. Wonderful book, even though he does go off on extended rambles before pulling himself and his argument together at the end, just when you'd almost abandoned hope.*

*Am I a naïve realist? I really should get my philosophical self-definitions straight –
except that I know that e.g., where consciousness is concerned I fit none of the standard
philosophical positions, most of which seem to me elaborately silly.*

*But well, to try to answer: it seems to me that consciousness is a genuine phenomenon,
that qualia are at least as real as tables, and that anyone genuinely interested in understanding the world ought to be at least as interested in the inner world as the outer. Moreover, that while consciousness obviously seems to inhere in particular material objects
(brains), its properties are not, according to our present understanding, reducible to known
facts about the material world. It's pretty much a complete mystery. So I'm on the side of
the senses, and of Schrödinger, on this one – unless they press further and start to deny the
existence of the atoms and the void (which I hadn't understood them as doing – what was
your impression?).*

Thanks for clarifying where you lean for me. My initial impression of the Democritus
fragment was that it was pressing further (at least for rhetorical purposes). But now that
I've read your point of view, I see that it need not have been. So I guess it's hard to say
where it was really going without looking at the context surrounding it. Have you read an
extended version of the fragment in another source?

Actually, the more I sit here thinking about it, the more I think your reading is likely to
be on the right track. Democritus ultimately invented a very fine set of atoms with which
to compose the consciousness. Atoms understanding atoms: silly or not, it was a physical
mechanism for explaining cognition.

20 January 1999, "Atomic Souls"

Kentism 4: *Does your account of Democritus's very fine cognitive atoms (which is new
to me but seems – making obvious allowances for era – to be in characteristic good taste)
come from original, translation, or commentary – and if t or c, can you recommend?*

You can find a discussion of them in Schrödinger's book itself. It's in Chapter 6, "The
Atomists," more particularly "main feature" number five (pp. 78–80 of the Canto edition).

03 February 1999, "Noumenal Mental Atoms"

Kentism 5: *By the way, I still think, contra Schrödinger, that Democritus' picture of the
soul as built of finer atoms is a rather good one – for its time, of course, of course. D.
saw the mind–body problem as a real problem, saw (apparently) that it has no very good
resolution in his atomic model of the material world, but saw also that it could have a
physical solution if you accept that mind and matter are both reducible, but to qualitatively
different things. A characteristically sharp insight, maybe, not (as S. suggests) a lapse of
taste brought about by the metaphor of the soul as breath.*

*And – though it's desperately unfashionable to say so, and I trust won't be taken as a
wildly naïve comment: I do appreciate how strong the counterarguments are – we still are
not absolutely sure Democritus is wrong on this. (Are we? Are you, I mean?)*

Did Democritus have the right intuition? Actually I sort of hope he did too. I wouldn't really want my world made of two fundamentally different things. But my realism is much closer to Kant's than yours (appears to me to be): if mind and matter are made of the same stuff, then I suspect they'll both be just as noumenal.

05 February 2001, "The Eerie Parallel"

[NOTE: The following Kentism refers to a letter consisting solely of the exclamation, "Thanks for the pointers. But I'm not quite sure what functionalism is!!"]

Kentism 6: *If I understand right, functionalists believe that statements about mind states (conscious thoughts and perceptions, if you prefer) are equivalent to statements about physical actions. For example, on this view, to be angry is precisely the same thing as to display (perhaps subtle and well hidden) signs of anger.*

It's a view which has little to recommend it except a pleasing sense of answering a deep question with no work, and so naturally has become very widely held and respected among philosophers of mind. In this, it eerily parallels the Everett interpretation. It is only fitting that the two should be combined into a grander exercise in question-begging.

At times in the past I've found myself wanting to be able to write like Mark Twain, and like William James. Both had this way of putting things – very different ways – that made me shiver from seeing the truth in their thoughts. Today I found myself wanting to be able to write like you!

16

Letters to Rolf Landauer

15 February 1998, "Law Without Law ... Or Something Like That"

This morning I read one of the papers you gave me last week ("Information is Inevitably Physical") and was reminded of your interest in the idea that the laws of physics may not be written in stone once and for all ... but instead may be contingent upon the particular details of the universe itself ... and indeed may be evolutionary à la Peirce, Wheeler, Lee Smolin, etc. At least one version or other of this kind of thing has intrigued me for some time (essentially since meeting John Wheeler in 1985). It's been sort of a hidden hobby, somewhat connected with separating the wheat from the chaff in quantum mechanics. My personal writings on the subject are now pretty voluminous, though not yet tight enough in substance that I show them to too many: most are in the form of emails to colleagues such as you.

Anyway, if you're interested, perhaps I'll compile a small compendium and send it your way when I get back from Tokyo March 15 – I'm travelling essentially every moment until then. Maybe this will help us open a dialog to pinpoint where some real progress might be made.

In the meantime let me tip you off to another article on the subject that I ran across recently: S. S. Schweber: "The Metaphysics of Science at the End of a Heroic Age," in *Experimental Metaphysics*, edited by R. S. Cohen, M. Horne, and J. Stachel (Kluwer, Dordrecht, 1997), pp. 171–198. Among other things, it is nice in that it contains a good bibliography of several previous efforts in the direction (that at least I wasn't aware of).

Rolf's Reply

By all means send me your thoughts; especially if I do not have to guarantee, ahead of time, how much energy I devote to them.

I am not sure, either, how much energy I can devote to tracking down all the related literature. I started saying my stuff in 1967, in an article which was published in a journal which is far from obscure, and has been reprinted (admittedly in an obscure journal). I have the delusion that it is the job of later arrivals to make contact with my work, rather than the other way around. But I am not inflexible about that.

06 July 1998, "Evolution and Physics"

I remembered a few days ago that I had promised to start something of a dialog with you about evolutionary notions of physical law.

The thing that brought this back to mind was a nice endorsement I got from Carl Caves on some of the stuff I've written. (I'll place a piece of his note below in the case that it might give you the fortitude to wade through some of my silliness.)

As I see it, there is a deep affinity between a key thought in your writings and the one concept for which I have the greatest hope to finally give us a clear view of what quantum theory is all about. I agree with you that physical law should be limited in its form by the information-processing capabilities of physical systems. The ultimate reason for this is simply that "physical law," – i.e., the equations and rules our students read in their textbooks and apply to great merit – is itself a construct of information processing systems. (In this case, the information processing system is "us," the community of physical scientists, but it need not have been so: it could have been Deep Blue for all I care.) There is the world independent of us, and then there is our attempt at describing that world in the most accurate fashion. The two things are different, and I don't see any need to make apologies for that fact. However, I think there is more to it than an ultimate "discretization" of the present physical laws that you so often allude to. This is where my views about the essence of quantum theory come into the picture. To a very great extent, I believe that your edict about the connections between information-processing limitations and physical law are already well taken into account by present-day quantum mechanics. (Don't get me wrong, though: this is certainly not to say that we don't have still a long way to go in forming a completely consistent physics ... in particular, in the manner you suggest.)

The already-discovered information-processing limitation has to do with a funny property of the world we happen to live in. It is this: *my* information gathering about a given physical system will generally disturb *your* description and predictions for that selfsame system. Nonetheless we, as communicating beings, must come to a consistent description of what we see and know about that physical system. This, I ever more firmly believe, is the essence of the quantum mechanical formalism. Quantum theory is nothing more than Bayesian-like reasoning in a world with such a funny property. (This I see as a large chunk of my personal research program: clarifying, delineating, and searching for holes in this point of view.) Perhaps ... to put it in a more amusing way ... what I am saying is that I wouldn't be surprised if the whole edifice of quantum theory couldn't be constructed from the singular fact that "quantum cryptography exists." That is to say, Hilbert spaces, the inner product rule, entanglement, and unitary evolution all from that clean, simple idea Bennett and Brassard were the first to make some currency of.

Fundamentally, I think this point of view is quite liberating. For I think it indicates that physics is far from the closed book that our quantum cosmologist friends (in their great lack of imagination and ultimate chutzpah) would have us believe. Rather than knowing all of physics, it seems more likely to me that we know almost nothing at all about it. In fact, the only thing that we've really got a firm hold of is the most basic idea of all: in learning about our world, we change its course. Now let's get to the real work of understanding and making use of the greater import of this small piece of knowledge.

Well I think that in a nutshell is where I stand in relation to you. If you've got any comments on fruitful approaches I'd like to hear them. I had planned on putting together a complete compilation of my murmurings on this subject for you. But I discovered that that's more of a task than I'm willing to undertake presently. (As you can see by the length of this note, I write a lot, probably much more than I should.) What I think I'll do instead is send you the compilations that I had already sent Carl. The first, most clear, and – I think – most interesting set of thoughts is connected to several letters I wrote to David Mermin and a few other fellows. The best part of that actually starts up on page $[X]$ and goes on to the end at page $[X + 13]$. [NOTE: This probably refers to the things between Merminitions 12 and 23 in the present volume.] The second thing in this connection is one of my long letters to Asher Peres; the most interesting stuff there comes after Asherisms 8 and 11. Finally I'll send a letter I wrote Bob Griffiths explaining what I see as the emptiness of the "consistent histories" business as anything over and above textbook quantum theory. All of these passages tackle in one way or the other the idea that physics and the world it describes can evolve in a nontrivial and interesting sense, and this all has to do with the information–disturbance tradeoff principle and information-processing limitations that you like to point out. All the documents are written in standard LATEX. If for whatever reason you have trouble getting them printed out, please let me know and I'll try to get you fixed up in another way.

Rolf's Reply

Maxwell and Boltzmann, Bohr and Einstein, sharpened up their perception via exchanges. In the day and age of phone-mail and e-mail that has largely disappeared. Most of our colleagues will only write things down for publication; and I thought I was the only one who still went beyond that (I just sent out some dozen or so copies of an e-mail note re shot noise reduction in mesoscopic samples). But you outdo me by many orders of magnitude. I am getting ready to leave the lab. for a few weeks, and giving you only a hasty first reply. I will also send you the latest version of what you call "discretization" (not an accurate label). Your views have much more in common with John Wheeler's "Participatory Universe" than with mine. No, there is no way that present day q.m., utilizing ordinary continuum math., can possibly be taking my notions into account. You believe that there is "the world independent of us", and if put into such a strong anthropocentric form, I might agree. But I do not think that in science it makes sense to discuss objects whose behavior cannot be observed or predicted. Your view that the world is doing its own thing, to the full precision allowed by our laws of physics, even though we cannot ever check that, is not exactly "wrong", but it is a matter of religion, not science.

You seem to understand "consistent histories" and I may need to get back to you on that, some day, in my non-converging attempt to catch on to what is right and wrong there. But first I need to do more homework.

28 July 1998, "Precision"

Let me comment on one of your points. I'll try to have something more intelligent to say about your thoughts after I see your new article. [You say,]

Your view that the world is doing its own thing, to the full precision allowed by our laws of physics, even though we cannot ever check that, is not exactly "wrong", but it is a matter of religion, not science.

Indeed I do think it is my view that the world is "doing its own thing" . . . but that should be qualified. For it doesn't seem to me that the "doing its own thing" before information processing units evolved and the "doing its own thing" after they evolved are precisely the same. The point of the idea of an evolutionary universe is that it evolves: if the course of that evolution were written in stone at the beginning of time, then there would hardly be any reason to invoke a notion of evolution. In any case, I don't believe that there's anything in the writings I sent you that would imply your phrase "to the full precision allowed by our laws of physics." You're fighting a battle with someone else in your mind there; it's not me. I don't see any reason to imagine a one-to-one correspondence between *our* laws and the stuff out there. You're right, my view does have some affinity with John Wheeler's: I wouldn't doubt that the stuff out there ultimately has no laws at all.

Rolf's Reply

Thanks for the clarification. I am not yet on a first name basis with evolutionary universe concepts; that barrier may take some time to disappear.

08 October 1998, "Info is Physical"

I have room in a very tight conference proceedings paper that I'm writing to make one citation to your phrase, "Information is Physical." Do you have a favorite paper that I should cite? Should I make the citation to your earliest mention of the phrase? Or should I make the citation to what you think is the clearest statement of it? Please make the decision for me and send me the correct reference. My paper is going off to the editors tomorrow, so if you could send me a quick note that'd be great!

Rolf's Reply

The concept, but not that phrase, first appeared in R. L. *IEEE Spectrum* vol. 4, issue #9, pgs. 105–109 (1967). (Like for *Physics Today*, the page numbering restarts with every issue.) The exact wording PROBABLY first showed up in my 1991 *Physics Today* paper. But it seems best to cite an early or a very recent paper. Two are on their way into print. One that is likely to appear within a few weeks: R. L. in *Feynman on Computation 2*, ed. by A. J. G. Hey, Addison Wesley, Reading (1998?).

The title of that one "Information is Inevitably Physical". About 11 days ago at a session in Helsinki I complained to my audience: I have gotten a fair amount of acceptance for that phrase, but not for the message attached to it. But thanks for checking.

P.S. I'll mail both papers, even though they will arrive too late. But if you take an instant dislike to the one you used, you can scratch it at galley proof time.

17

Letters to Hideo Mabuchi

22 November 1997, "Martha White and Her Flour"

[Note: Words supplied by Greg Comer. See notes to Asher Peres, dated 17 October 1997, Greg Comer, dated 25 November 1997, and Carlton Caves, dated 21 December 1997.]

> Martha White Theme Song
>
> (Open with a Banjo break by Earl Scruggs)
> (Next, come in with Lester Flatt singing . . .)
>
> Now you bake right,
> (other band members respond:) ah ha,
> With Martha White,
> (band response:) yes ma'am,
> Goodness gracious, good and light,
> With Martha White.
>
> Now you bake better biscuits, cakes, and pies,
> With Martha White Self-Risin' Flour,
> (band response:) that one all-purpose flour,
> With Martha White Self-Risin' Flour
> You've done all right!
>
> (Finish with an Earl Scruggs banjo solo.)

05 October 1998, "Oh Magic Eight Ball"

Democritus said, "All the world is but Atom and Void." I've decided (with Kiki's help) that I much more like the metaphor, "All the world is but Magic Eight Balls and Questions." See the resemblance? You couldn't have atom without void; but you could have void without atom. You couldn't have Questions without Magic Eight Balls; but you could have Magic Eight Balls without Questions. The sole remainder of archaic atomism is that "atoms" be "repositories." Not repositories of "properties." But repositories of answers to the questions we might ask. To keep these sentences from being contradictory, those answers can't exist

215

before we ask them (else they be properties). I'm sure all this has left you cold: but I'm in the middle of one of the fuzziest feelings I've had yet.

01 July 2000, "Destinkifiers"

Thanks for those nice compliments. I've been meaning to write you for quite some time, but there was always something: my email contact has gone down drastically in general.

Anyway congratulations on that big award. I was much impressed by your speculation on the Caltech webpage on how you might spend it. I say "here, here" for a little creativity in our field.

I'm in Greece right now, just finished with the NATO meeting. Tomorrow morning I leave for Capri (the QCMC conference), and then finally join Kiki in Munich at the end of the week. My talk was pretty successful in Mykonos; I was pretty happy with it. For Capri I'm going to make a completely new one, this time based on the stuff I did with Kurt Jacobs. I've decided the best way to say what I've been hoping to get at. Question: "If the wavefunction isn't real, then what is it that IS real about a quantum system?" Answer: "The locus of all information–information tradeoff curves that one can draw for such a system." (I've decide to stop calling it information–disturbance because it conveys bad imagery and preconceptions. The disturbance is to information, so why not just make it explicit.)

OK, gotta get up tomorrow at 4:30. Take care . . . and spend that money wisely!

18

Letters to David Mermin

18 September 1996, "Theorem II"

I encountered your "Ithaca Interpretation" paper this morning on the `quant-ph` archive ... and, I must say, I've been walking around with a nice feeling since. There are some things in it that I like very much!

Your Theorem II (or a very slight variation of it) has indeed been proven before, by Bill Wootters. I'll attach the citations below along with part of a note Bill once sent me. (Interestingly, one can also make a cut between real, complex, and quaternionic Hilbert spaces based on such a theorem.)

About your Theorem I, if I am not mistaken, it also has a history that predates the references you gave. I believe Richard Jozsa once told me that a fellow named Hadjisavvas found it well before HJW and Gisin. If you are interested you can contact Richard directly; his e-mail address is ....

As I say, I enjoyed your paper very much! (Presently) I think my only point of departure in what you say is about the issue of "objective probability." Carl Caves and I (in still another paper cited below which can be found on `quant-ph`, pp. 22–26 in particular) have advocated the view that even the probabilities of quantum theory should be interpreted in a Bayesian or subjective way. We prefer to say that it is the "indeterminism" of quantum mechanics that is "objective," and not the probabilities themselves. One always assigns probabilities based on incomplete information; it is just that in quantum physics "maximal information is not complete."

If you have any comments on our (somewhat garbled) thoughts there, I would very much appreciate hearing them. Also, you may find the various references to "objective probabilities" in it useful. (If you would like a much more extensive list; I can compile that for you ... the exercise would also be useful for me: I have an amorphous pile of things in my file cabinet at home.)

I meet Carl in Dallas Monday for a flight to Japan ... so I'll have several hours at my disposal to make him think a bit more philosophically again. I'll show him your paper then ... and that'll also give me a chance to read it still more carefully.

1. W. K. Wootters, "Quantum mechanics without probability amplitudes," *Foundations of Physics*, **16**(4), pp. 391–405, 1986.

2. W. K. Wootters, "Local accessibility of quantum states," in *Complexity, Entropy and the Physics of Information* (W. H. Zurek, ed.), (Redwood City, CA), pp. 39–46, Addison-Wesley, 1990. Santa Fe Institute Studies in the Sciences of Complexity, vol. VIII.
3. C. M. Caves and C. A. Fuchs, "Quantum information: How much information in a state vector?," to appear in *Sixty Years of EPR* (A. Mann and M. Revzen, eds.), (*Ann. Phys. Soc.*, Israel) 1996. Also `quant-ph/9601025`.

From Bill Wootters, 7 June 1996:

By the way, my favorite distinction between real, complex, and quaternionic states is hinted at in my paper "Local Accessibility of Quantum States," in *Complexity, Entropy, and the Physics of Information*. The issue there is this: given a measurement scheme that is set up to ascertain the states of the *parts* of a composite system, how well does it determine the state of the whole? For a complex space, the measurement scheme gives you just what you need to determine the state of the whole system. For a real space, it gives you less than what you need, and for a quaternionic space, it gives you more than what you need. (Ben Schumacher called this the Goldilocks principle.) Just in case you feel like verifying that statement and you actually have time to do so, the function $g(N)$ in my paper – which is the number of real parameters necessary to specify an arbitrary mixed state in N dimensions – is $(N^2 + N - 2)/2$ for a real space, $N^2 - 1$ for a complex space, and $2N^2 - N - 1$ for a quaternionic space. (You probably knew that anyway.)

13 October 1997, "Caltech and Propensities"

I've heard rumors that you'll be coming out to Caltech for a visit soon. If so, when? What are the dates? I'd like to make sure that I'm here then.

By the way, I'm much more versed in "propensity"/"objective probability" stuff than I was this summer. If you come, I'd like to talk about such things again (in connection to quantum mechanics). Plus, I still owe you that bibliography.

09 December 1997, "The Philosophy of 'Nothing-More'-ism"

... perhaps it's appropriate for a paper about Correlation Without Correlata. (If you rotate and reflect those initials, by the way, they're my old graduate advisor's.)

Anyway, did you realize you use the phrase "nothing more" 11 times in this paper. By the time you're finished, there's not a lot left to quantum theory ... but I guess that's the intention.

10 December 1997, "Deconstructing Ithaca"

I feel as if I'm caught in a tornado of deconstructionism, with your friend Derrida outside my window – pointed nose, broomstick, and all – laughing insanely. I look at your phrase "correlation without correlata" and see a story of quantum mechanics as purely knowledge-theoretic in character. You look at the same phrase and see a call to arms for objective probability (propensity). We are reading different stories, both presumably consistent, but still almost antithetical. The shocking thing is that it's the same text!

Anyway, advisable or not, I decided to read your paper for a third time! And then I decided to read the original IIQM paper again. And after that I went back to Everett's original relative state paper and Wheeler's assessment of it.

Now it is time to reach into myself: I think I really am ready to write now. So standby, prepare for goulash. Things should slowly start to trickle in through the night and into tomorrow.

10 December 1997, "Game and Format"

I don't think I told you, but the name of my laptop on our local network here is "tychism." If you look up that word in a Webster's dictionary, you'll find, "(in the philosophy of Peirce) the theory that chance has objective existence in the universe." If, on the other hand, you look in Peirce's essay, "Evolutionary Love" (*The Monist*, 1893), you'll find that he draws a fine distinction between the usage of four words in all: tychism, tychastic, tychasticism, and tychasm. The last of these refers to the actual operative principle or mechanism in the world that tychism is concerned with. If you will, it is the correlatum of the sentence, "Our world is governed by tychism."

I've been thinking about how I'm going write all these things that I'd like to write to you. When I played the game with Asher – as you saw last week – I'd highlight a passage of his paper, dub it an "Asherism #x", and then make some comment. It seems to me that I should do something no less grandiose for you. But I wouldn't want you to feel cheapened or secondary by my doing exactly the same. (You can see where this is leading.)

Here's what I'll do. I'll comment on your paper in the same way I did before, but now the quotes will each be designated "Merminition #x."

What I plan to do is send you separate installments with each Merminition as I write it. When the process is complete I'll lump them all into one long file and send it again. But these are the plans. Right now, I'm off to dinner. Then from 8:30 to 9:30 I have my favorite TV shows of the week. We'll see what happens after that.

14 December 1997, "An Idle Thought"

Merminition 1: *Meditating on your insistence that I simply have to say more about the relation of IIQM to many worlds, I'm struck by the fact that I am also told by Griffiths that I simply have to say more about its relation to consistent histories, and by Gottfried that I simply have to say more about its relation to traditional descriptions of measurement and the problematic character of "observables" as described, for example, in his book.*

If everybody (I exaggerate) from all these mutually incompatible schools thinks that I am so close to simply restating what they have known all along, then maybe I really am on to something!

It seems to me that that's probably a bad sign! I suspect that when we finally understand "what quantum mechanics is trying to tell us" it will be rather clean and decisive. If it helps, I'll be your beacon of opposition: I'm more inclined to the view that the quantum formalism has almost nothing to say about *Reality* ... only that *my* information-gathering questions/measurements disturb *your* predictability. At least that's the only firm ontological lesson that I've been able to draw from the theory. But like you ... and unlike Griffiths ... I am not 100% convinced that I'm on the right track – that's why I've been willing to study your paper. In any case, I don't see how we can reconcile our two antithetical views. So there.

Merminition 2: *But bear in mind that I am not out to convince the world that this is the answer. (I'm not convinced myself.) I only want to persuade a few smart people that this might be a useful way to think about things.*

I have born that in mind ... or at least, looking at myself through my own eyes, it looks like I have. My only point was that maybe you should go after a few of the "dumb" ones too.

17 December 1997, "Exercise in Anti-Lucidity"

The document is getting a little unwieldy, and I think I've reached a convenient breaking point. So I've decided I'll send on now what I've written so far. [See note to David Mermin, titled "Big Bowl of Anti-Lucidite," dated 10–17 December 1997.] I'd still like to say a few things about what you've written on the modal interpretations and the consistent histories stuff, your distaste of anthropocentrisms, and, most importantly, on objective probabilities. But I need a bit of a break, and it's getting hectic here: Sam Braunstein arrived Sunday. I do want to complete this project and have these things sitting in the archive of my mind. Perhaps, with a good finger crossing, I'll have some more in your mailbox with the weekend. But certainly move on and put your paper on the archive in the meantime if you wish.

The document that is about to arrive includes some of the things that I've already sent you. However, I edited those old things slightly so that they more accurately portray what I really think ... and where I have some doubts.

In all, looking back over the multitude of words, I see that I don't say nearly as much as I had wanted to. (Make sure you understand me properly on this: I say a lot of words, but those words don't say a lot!) In any case, I did put an honest effort into it; I tried very much to be unbiased and to learn from you. And I did learn a lot from this exercise ... though maybe the fruits will be slow-growing in a recalcitrant Copenhagenist. We shall see. Any reactions you have on what I've written will be most welcome.

10–17 December 1997, "Big Bowl of Anti-Lucidite"

A. My Outlook

My strongest and most basic agreement with you is this.

Merminition 3: *The IIQM does not emerge from a general view of the world out of which quantum mechanics is extracted; the strategy is rather to take the formalism of quantum mechanics as given, and to try to infer from the theory itself what quantum mechanics is trying to tell us about physical reality.*

This attitude, along with the paper's title, is what I like above all else with your efforts. The firmest thing of all in quantum mechanics is the formalism itself. It's hard to see how there can be any better starting point for a deeper understanding than that. In addition to this, however, I really, really do like the business you emphasize of focusing on correlations. I have always liked points of view that subordinate the state vector to probabilistic or correlative statements. And I think you are doing grand justice to clarifying the similarities between "measurement interactions" and general "correlating interactions." Viewing the Wootters and HJW theorems as statements of some foundational importance is a key insight, and it strikes me as a particularly fruitful strategy for getting at something real.

Because you write so clearly on all these things – and you are careful to state what you know and what you don't know – this, in my eyes, is going to be an important paper. I am glad that you're going to push this to the point of publication.

But, what is my outlook about the ultimate content here? I've said it before, I'll say it again:

I feel as if I'm caught in a tornado of deconstructionism, with your friend Derrida[1] outside my window – pointed nose, broomstick, and all – laughing insanely. I look at your phrase "correlation without correlata" and see a story of quantum mechanics that hints of being purely knowledge-theoretic in character. You look at the same phrase and see a call to arms for objective probability (propensity). We are reading different stories, both presumably consistent, but still almost antithetical. The shocking thing is that it's the same text!

I hope some of the things I say below will help elucidate this. Failing that, I hope it helps me identify how I am trying to *force* quantum mechanics to "emerge from a general view of the world out of which [it] is [to be] extracted." You do realize that I go to such lengths in writing things down, most selfishly, so that I can look at what I'm thinking. Rarely do I really and completely do it for the designated recipient of the letter! I hope you will indulge me in my self-education. You are my point of departure.

B. Teleported Goulash

I. Immediate Goulash

I'll start with the two points I tried to emphasize to you the other day. If I can make any strong suggestions for the present paper, it is these two. The remainder of my comments,

---

[1] I had promised you a reference to a decent article on the connections between Bohr and Derrida's thoughts. It is: J. Honner, "Description and Deconstruction: Niels Bohr and Modern Philosophy," in *Niels Bohr and Contemporary Philosophy*, edited by J. Faye and H. J. Folse (Kluwer, Dordrecht, 1994).

concern more the substance of the IIQM than the content of anything you'll be putting on the web next week.

Merminition 4: *The central (and possibly fatal) conceptual difficulty for the IIQM is the puzzle of what it means to insist that correlations and only correlations have physical reality.*

This point is buried on page six of a 42 page paper. To be sure, you do close the paper with a weak reiteration of the point: "Whether this is a fatal defect of the IIQM, or whether it is a manifestation of the primitive state of our thinking about objective probability, remains to be explored." But, I guess I want to take everything you've written pretty seriously (regardless of whether it is or is not on the mark) ... and I want other people to do the same.

Therefore, I think it would be immensely useful to have a summary at the end of the paper delimiting – in the most precise way that you presently know – the problems, prospects, and challenges of the Ithaca Attitude. You do after all say over and over again that the project is not complete. In what ways can a young researcher hope to make a contribution to this (possibly) final turn in our understanding quantum mechanics? And what will the completion of this project give us beside a warm fuzzy feeling that there are no longer any mysteries? Will its completion mean that we, the physics community, can finally go full-head tilt in exploring quantum cosmology? Will it possibly give us any insight into dealing with something so mundane as Kitaev's nearest-four-neighbor-interaction lattice model for a quantum computer? Of course, I don't mean for you to address these particular questions specifically, but some inspirational words would be most useful. I believe it would make this paper a more lasting contribution.

Merminition 5: *It is at this point that the IIQM comes closest to the many-worlds extravaganza. The original paper of Everett said nothing whatever about many worlds. Everett took the view that the quantum state ought to be understood only in a relational sense, but the relations he emphasized were not as straightforward as the subsystem joint distributions emphasized here.*

I think that I have a strong feeling now for how **IIQM $\neq$ Everett**, but that comes after some relatively long thought and (most importantly) after some personal contact with you. The general reader may not have the opportunity for either of these methods of clarification. About the confusion in general, though, clearly there is something amiss here: otherwise, you would not have had people both at Caltech and Montréal making the comments that they did. It seems to me that your discussion of Everett should be expanded substantially. And it needs to come somewhere in the paper *long* before page 22.

The things you have to fight hard against dispelling are the suspicious similarities between the two trains of thought. Both you and Everett purport to have *only* unitary evolution. Both you and Everett have a universe in which all possible "somethings" – whatever they be – simultaneously exist, even though their (naïve) derivatives are mutually inconsistent. In your case it is the joint probabilities that are simultaneously existent, though the conditional probabilities they seemingly give rise to cannot be. In Everett's case it is all possible relative-state decompositions of the state vector that simultaneously exist; it is the facticity of the "worlds" so-derived that cannot be imagined to fit within a single reality. (That is, unless you go so far as the extravaganza that you say.)

The main thing that you have to be on your guard against – and I'm trying to get it across that you've yet to do it sufficiently – is the loose associations that people have already drawn in their minds. Such a thing is so easy to do when there are no equations involved in a paper! Almost anyone who is going to read this paper is likely to have a long history of having read (at least cursorily) things about the many-worlds interpretation. And their dissection of your ideas is not going to be helped in any way by the many things that have already been said in the literature. Let me give you two strong examples of where the trouble is going to come from. I'll just pull two quotes out of Everett's original paper and Wheeler's assessment[2] of it (that appears immediately afterward):

Everett: Thus with each succeeding observation (or interaction), the observer state "branches" into a number of different states. Each branch represents a different outcome of the measurement and the *corresponding* eigenstate for the object–system state. All branches exist simultaneously in the superposition after any given sequence of observations.

and

Wheeler: Another way of phrasing this unique association of relative state in one subsystem to states in the remainder is to say that the states are correlated. The totality of these correlations which can arise from all possible decompositions into states and relative states is all that can be read out of the mathematical model.

The first quote comes darned close to contradicting what you say in this Merminition. The second quote (especially the last sentence thereof) comes darned close to saying, "correlation without correlata." (Or, at least to the uninitiated it does.) The main thing is to be long-winded on these points: it can't hurt.

II. Big Bowls of Goulash

Merminition 6: *According to the IIQM the only proper subjects for the physics of a system are its correlations. The physical reality of a system is entirely contained in (a) the correlations among its subsystems and (b) its correlations with other systems, viewed together with itself as subsystems of a larger system.*

With correlata excluded from the realm of physical discourse, what is it that determines the identity of these subsystems? What mechanism is left for drawing some kind of line in the sand? Can you see my worry about this? Take the four dimensional Hilbert space $\mathcal{H}_4$. There is a continuous infinity of ways of decomposing it into something of the form $\mathcal{H}_2 \otimes \mathcal{H}_2$. Presumably you want to single out one particular cut in this way as being

---

[2] By the way, I learned the most amazing thing in digging up these papers yesterday. The particular volume of *Reviews of Modern Physics* containing them displays back-to-back *eight*(!) papers from John Wheeler's group.

physically real. (You give some hint of this on the next page with your phrase, "A possible further requirement ... .") So, is that cut just built in at the outset? Physical reality consists of a Hilbert space, a fixed particular decomposition into elementary tensor product spaces, and a vector on that Hilbert space ... oops, can't say vector ... must say the complete set of all correlations. I don't suppose this is any real big deal. But it does impose one fixed absolute structure above and beyond the Hilbert space itself. If there were correlata, then the existence of the cut would be of no conceptual difficulty: that's what the world is, just a lot of separate thingies. But if you banish the correlata, why do you not banish the fixed cut while you're at it? What compels you to keep it?

Merminition 7: *The crucial formal property of a resolution into subsystems is that all* **observables** *associated with one subsystem must commute with all* **observables** *associated with any other distinct subsystem.*

Note the emphases that I've added to "observables" in this Merminition. Doesn't it bother you that so much of what you are trying to formulate relies on the old "anthropocentric"[3] concepts that you're trying to do away with in the first place? If we have no correlata, and we don't truly have any need of measurement for understanding what quantum theory is about, then what is it that singles out "observables" as the relevant concept with which to construct an interpretation?

Perhaps I haven't amassed enough evidence for this complaint yet, but every time I see something in physics or philosophy that has to strain so to get off the ground, I think of a piece of a Paul Simon tune. And it's going through my head right now, so I might as well say it:

> I don't know why I spend my time
> Writing songs I can't believe
> With words that tear and strain to rhyme.

But let me try to go into more detail on this point, instead of fixating on a diatribe.

Merminition 8: *By correlations among subsystems, then, I mean nothing more than the ... [probability distributions] ..., at any given time, ... [derivable from] ... system observables (Hermitian operators) that consist of products over subsystems of individual subsystem observables.*

Allow me to try to say more clearly what I was trying to tell you at the chalk board the other day. Consider a universe described by $\mathcal{H}_2 \otimes \mathcal{H}_2$. "Physical reality" in this universe consists, according to your prescription, of all joint probability distributions $p(\hat{a}_i, \hat{b}_j)$ that can be derived from some fixed *pure* quantum state $\hat{\rho}$ via

$$p(\hat{a}_i, \hat{b}_j) = \mathrm{tr}\big(\hat{\rho}(\hat{a}_i \otimes \hat{b}_j)\big), \tag{18.1}$$

---

[3] I put the scare quotes here because I don't really think that speaking in terms of measurement and observables is overtly anthropocentric.

where the $\hat{a}_i$ and the $\hat{b}_j$ both form complete sets of orthonormal projectors. Thus, physical reality is captured by the mathematical model of a set with a certain structure – in this case, the set of joint probability distributions derivable from the antiquated, "anthropocentric" notion of (standard, von Neumann) measurements. But why that particular structure and not another?

You shore up your argument for this by pointing out that if one knows all these probabilities, then one can automatically infer the quantum state $\hat{\rho}$. But you don't need nearly that much structure to recover the quantum state. Indeed you only need consider the joint probability distributions formed from three complementary observables on the A side and three complementary observables on the B side. You don't need the gross overkill of a continuous infinity of observables on both sides. So why not pick your six favorite local observables and call the set of nine probability distributions derivable thereof physical reality?

Well the reason you don't is that, though it doesn't take so many observables to determine a state, there are nevertheless loads of other observables that are measurable ... when we insert an external observer into the game. So you go to textbook quantum mechanics and pull out the Hermitian operators as the relevant set with which to construct your interpretation.

But why stop there? External observers can play lots of games with this toy 4-D Hilbert-space universe. In particular, they can imagine measuring completely general POVMs on the A and B sides, respectively, and build up joint distributions in that more ecumenical way. So why not take this as the starting point for an interpretation? This physical reality would be immensely larger in scope than one based on Hermitian observables.

I know you may think it silly that I bother over this issue. But mostly I want to emphasize the somewhat arbitrary feel and construction that your present "correlation without correlata" (CWC) has. How can you better justify this picture of reality that you're trying to draw? Why, if you are searching for deep reality, are you playing the role of a complemento-anti-Heisenberg? You remember the story of how Heisenberg came across the uncertainty relations: he decided to let the theory tell him what could and could not be measured. You wish – it looks like – to use the theory to tell you what is measurable and then use that in turn to tell you what is real. If that's the case, and you want to de-anthropocentrize the theory, shouldn't there be a better way to go from theory to reality?

Merminition 9: *[Wootters'] theorem follows immediately from three facts: ... (3) The algorithm that supplies observables with their mean values is linear on the algebra of observables.*

Have you thought about how the assumptions behind Gleason's theorem might be weakened to help out your cause?[4] For instance, one could/should, based upon your sought-after interpretation, consider the following.

---

[4] I know you said you spoke to Gleason recently, but it didn't sound to me like your question to him was of this flavor. What was it? Can you explain the idea a little more thoroughly?

The standard scenario behind Gleason's theorem is this:

The questions that I may ask of a physical system correspond to the set of all orthonormal bases for a given Hilbert space. The task of physical theory is to provide a probability distribution over the answers to those questions, i.e., the basis elements in each orthonormal basis. What is the most general form these probability distributions can have?

For two-dimensional Hilbert spaces, this question doesn't pin down anything interesting. For three-dimensional and above, however, as you know, one gets back the familiar Born rule *for some* particular density operator. This warrants us to say that the set of valid "physical states" for a system corresponds to the set of density operators over some Hilbert space.

Just to repeat the logic behind this: one starts with the set of valid questions (measurements), and says that it is the task of physical theory to give probabilities to the answers (outcomes). From that, one derives the structure of the state space.

What can we do in the same ilk for the set of universes crudely[5] described by $\mathcal{H}_3 \otimes \mathcal{H}_3$ ... when physical reality consists of "correlation without correlata?" Since this is a nine-dimensional Hilbert space, and $9 \geq 3$, we could snap Gleason's theorem on it like a hand tool. But that wouldn't be very interesting. For, after all, a basis consisting of vectors entangled across these two spaces corresponds – using language you don't condone – to a set of correlata with respect to a third system, one we're not allowing in our toy universe.

However, could it possibly ... wonderfully ... be the case that if we said, "it is the task of physical theory to provide a probability distribution for each orthonormal *tensor-product* basis," then that would be enough to recover the full theorem? You see it's not immediately clear that one could hope to get such a thing. In the way Gleason's theorem is proved presently, one of the crucial first steps is to show that the "frame function" must be continuous on the set of bases. But here there is no possibility of that – a probability distribution need not even be defined for a basis of entangled vectors.

This is probably just a silly thought – it seems pretty unlikely that it even stands a chance of being true – but it might be worth mulling over. In particular, if it isn't true, it seems you should ask: why is it that the old question/measurement-based foundation can go so far (through Gleason's theorem), but the CWC foundation isn't powerful enough to get back the Born rule without assuming it at the outset? That is to say, why does the CWC foundation start at the Wootters theorem and not earlier?

Merminition 10: *This is another familiar tale. The IIQM shifts the way it is told, by emphasizing that the state of a non-trivially correlated subsystem is never pure: the state of the specimen evolves continuously from a pure state through a sequence of mixed states into the "post-measurement" mixed state ... at the moment the measurement interaction is terminated. If at that stage one wishes to regard the state of the specimen as undergoing an abrupt change, it is at worst a collapse from a mixed state viewed in this fundamental way,*

---

[5] I say "crudely" here because I don't want to specify the set of physical states yet.

to the same mixed state viewed under the "ignorance interpretation". Since the internal correlations of the specimen are exactly the same regardless of which view you take, the collapse, if one chooses so to regard it, is rather ethereal.

I rather like this passage, especially the last two sentences. From my favorite point of view, the collapse is, by its very nature, pretty darned ethereal. It alone is the piece of quantum phenomena that cannot be governed by physical law. What else could you expect of it *but* to be ethereal?

But more substantially, what is it in the IIQM that warrants the *particular* transition from "fundamental" mixed state to "ignorance" mixed state that you are imagining here? That is to say, what is it in the IIQM that tells us the appropriate way to view collapse is in terms of an orthogonal mixture of vectors (the eigenvectors of the post-interaction density operator)? What about all those other decompositions of the density operator assured by the Hughston–Jozsa–Wootters theorem? Yes, the measurement interaction was set up just to elicit this possible picture of things, an ignorance-based mixture of reliably distinguishable states. But remember, you wish to have neither "real" correlata nor "real" observers out there to define a fixed set of questions of which the measurement is about.

I guess the point I'm making here isn't so different from the point (I had forgotten) that you make on the page following this Merminition. But it seems to me even worse than you had thought – this is because of all the extra density-operator decomposition freedom. Thus it seems to be that you are left with one more onus for the conscious observer that you exclude from the problem. His role is to first define a decomposition and, second, to play the measurement-problem equivalent of the separation between past and future.

Merminition 11: *There is thus no quantum measurement problem for the internal correlations of the specimen or the apparatus. . . . The measurement problem survives only in the* specimen–apparatus correlations . . . .

I still have a funny feeling about this that I'm trying to put my finger on. I suppose the problem goes all the way back to the main declaration of your paper, "Correlations have physical reality; that which they correlate, does[6] not." The problem seems to stem from my willingness to think of *entanglement* itself more as a physical "reality" than you, as something different than the sum total of correlations. Or, at least, I like to think of it as a valuable currency in its own right. You seem to only allow correlation to sit in that lofty spot.

The thing that's going on here is that "correlation" is linear in the specimen's density-operator decomposition, whereas all reasonable definitions of mixed-state entanglement that I know of, are not. For instance, consider a specimen density operator ρ^{AB} that arises

---

[6] Two points of grammar that I've been meaning to mention. (1) Shouldn't the pronoun 'that' be taken to be plural here? So you would use 'do' in the sentence? (2) If you look in a standard Webster's dictionary, you'll find 99.9% of all words starting with "non" nonhyphenated. Therefore, this has been my standard. In particular, I speak of nonorthogonal states, nondemolition measurements, etc. I'm not sure where physicists have gotten the habit of hyphenating so much. Is this an older protocol that has fallen out of favor with the new dictionaries?

from a tripartite pure state $|\Psi^{ABC}\rangle$ just as you wish: $\rho^{AB} = \mathrm{tr}_C|\Psi^{ABC}\rangle\langle\Psi^{ABC}|$. Now focus on a particular pure-state decomposition $\mathcal{E} = \{p_i, \Pi_i^{AB}\}$ for that density operator:

$$\rho^{AB} = \sum_i p_i \Pi_i^{AB}. \tag{18.2}$$

If we denote the partial traces of the states in this decomposition by $\rho_i^A = \mathrm{tr}_B \Pi_i^{AB}$, then the (average) entanglement $E(\mathcal{E})$ of the decomposition $\mathcal{E}$ is

$$E(\mathcal{E}) = \sum_i p_i S(\rho_i^A) = -\sum_i p_i \mathrm{tr}\left(\rho_i^A \log_2 \rho_i^A\right). \tag{18.3}$$

Pretty clearly, this quantity depends on the particular decomposition that is under consideration.

Aside: Quantities like this find quite some use within quantum information theory. For instance, suppose Alice, Bob, and Charlie share this tripartite pure state. Moreover suppose that, though Alice and Bob don't possess a pure state on their own, they need to use their part for quantum-state teleportation. Charlie, being their friend, decides to help as much as he can, but suppose he's quite some distance away and can't transport his system to them. With that constraint, the best Charlie can do is perform a measurement so that A and B have a pure state conditioned on his outcome (as far as he is concerned). Then, if he transmits that classical information to A and B, they can make use of it to teleport more effectively than they could otherwise.

Now, if Charlie is smart, he'll try to force the pure state between them to have as high an entanglement as possible. By the HJW theorem, for any decomposition $\mathcal{E}$, there is a measurement he can perform to generate it. Since Charlie can't generally predict the outcome of his measurements, the best thing he can do is to try to maximize his *expected* entanglement for Alice and Bob. His best expected entanglement we call the *entanglement of assistance*[7]

$$A(\rho^{AB}) = \max_{\mathcal{E}} E(\mathcal{E}). \tag{18.4}$$

The operational reason that this particular average is important is because if Alice, Bob, and Charlie actually have many, many copies of the state $|\Psi^{ABC}\rangle$, this controls the ultimate number of (perfectly entangled) singlets that Alice and Bob can get with Charlie's assistance.

Where am I going with this? The main point is that because entanglement is nonlinear, *as far as Charlie is concerned* the average entanglement between Alice and Bob will change dependent upon which measurement he performs. In this way, it is quite unlike the "correlation" between Alice and Bob that you speak of. Now because of this and your Desideratum #5, it cannot be – in your terminology – an objectively real quantity.

I think that may be fine with me: nothing to do with the quantum state is too objectively real if you ask me. But still, it feels a little funny that a quantity so close to correlation … or better yet, one that provides a measure of all-purpose correlation … isn't allowed quite

---

[7] This quantity (and what is known about it) hasn't yet made a public debut. This is something that I will eventually write up in one of six papers I told you about!

the same objectivity status as its more primitive, classically-oriented cousin. Any thoughts on this?

Merminition 12: *If we leave conscious beings out of the story and insist that physics is only about correlation, then there is no measurement problem in quantum mechanics. This is not to say that there is not a problem. But that problem is not a problem for the science of quantum mechanics but an everyday question of life: it is the problem of the nature of conscious awareness.* [8]

The scary thing is that we don't differ so much on this point! But my take on this is that the thing really being emphasized here is: quantum mechanics itself has nothing to say about the *real* measurement problem. The *real* measurement problem is precisely that quantum mechanics does not address the "mechanism" that brings about the individual outcome that we and all the fellows we talk to perceive. [9]

The other day, you and John Preskill scoffed at me when I said that I am inclined to take "measurement" as a primitive of the theory. Well, there is a sense in which that is exactly what you are doing here! Let he who has not sinned cast the first stone. For the purpose of clarifying quantum theory, you leave *this* notion of measurement unanalyzed; it is taken as a given. Albeit for the world you are trying to construct, it is an incidental primitive ... one that is not to be confused with anything inside physics proper. I, on the other hand, see it not at all as incidental: it seems to me that it is the big red flag that's telling us just exactly what quantum theory is about. The terms in the theory correspond to what we can ask about the world and what we can predict of the answers that it gives us. When there were no people (or artificially intelligent Dell laptop computers), there were no wave functions. Period.

But everyone wants an ontological lesson from the theory. I guess I am no exception.

Merminition 13: *It is a another remarkable feature of quantum mechanics (not shared with classical physics, where external correlations are always possible) that the totality of all possible internal correlations is enough to determine whether or not any non-trivial external correlations are possible.*

This statement may be the only firm ontological lesson that can be gathered from quantum mechanics. It's the one that powers the Pauli quotes at the beginning of these notes, though said in such different language. It's the principle that lies behind quantum cryptography. [10] The lesson is this: *My* information-gathering measurements (about systems that you have some knowledge of) disturb *your* predictability (over them).

This "lesson" is constructed or couched in epistemological terms, but it is ontological in nature. It is a statement about a significant *property* of the world. Confronted with the

---

[8] Grammar check: too many "buts."

[9] Please don't confuse this statement in the way that Bob Griffiths did. I am not exhibiting in it any desire for a hidden-variable theory to underlie quantum mechanics. I am perfectly happy with the indeterminism of the theory.

[10] C. A. Fuchs, "Information Gain vs. State Disturbance in Quantum Theory," to appear in *Fortschritte der Physik*. Also in LANL archive, quant-ph/9611010.

task of describing, manipulating, and studying a world with this property, one is led to ask with what consistent means such a thing can be carried out? How can one hope to describe a world in which the Kantian noumena are even further removed from observable phenomena than Kant's lofty philosophical arguments could have dreamt of?

That's the sort of question that really piques my interest. One has to wonder whether such a thing – much more carefully stated(!) – could pin down quantum theory as the only reasonable answer. Quantum theory, that is, with all its attendant silence about measurement itself and its troublesome concept of entanglement. Could it be that Bohr was not so far off the mark when he said,[11]

> They have not that instinct that it is important to learn something and that we must be prepared to learn something of very great importance. … They did not see that it [the complementarity description of quantum theory] was an objective description, and that it was the only possible objective description.

That's one thing that I'd really like to know. More than that I suppose, I'd like to know what one can do with such a wonderful piece of newfound knowledge.

But still, you ask:

Merminition 14: *Why should the scope of physics be restricted to the artificial contrivances we are forced to resort to in our efforts to probe the world? Why should a fundamental theory have to take its meaning from a notion of "measurement" external to the theory itself? Should not the meaning of "measurement" emerge from the theory, rather than [the] other way around? Should not physics be able to make statements about the unmeasured, unprepared world?*

I still see no completely satisfactory answer to these questions, but I do strongly think that so much of this is forced upon us by the "ontological lesson" above. Physics, and quantum mechanics in particular, does say something about the unmeasured, unprepared world: the level of abstraction just seems to be one above where you'd like to see it. If you ever catch yourself saying the phrase, "information is physical," like so many in the field of quantum information do, then you will have gone a long way toward buying into this. I say the mantra of "information gain vs. state disturbance" every morning with breakfast and know that I feel a lot better because of it.

So, with this, let me try to summarize my present feelings about CWC:

Merminition 15: *But to insist that physics is exclusively about measurement, is unnecessarily to relegate to an inferior ontological status the more general correlations among arbitrary subsystems.*

I must say, I completely agree with this and, still, completely disagree with it. I have no problem with your feeling that the conventional idea of measurement and more general

---

[11] I haven't yet confirmed the exact wording of this quote: it comes from Bohr's last interview (with Kuhn), but I was only able to get it second-hand.

entanglement – I am reluctant to use the phrase "general correlations" – are of the same ontological status. But that ontological status is zip, zero, triviality. Entanglement, either from a measurement interaction or a general, nonteleological interaction, *to me* signifies the information that is *potential* in one system about another. That information can be activated by the supplement of a set of correlata. How this supplementation comes about is part of the *real* measurement problem spoken of above … the part you are willing to sweep under the rug of the question of consciousness.

Correlation without correlata, as an ontological statement? Can I buy it yet? Not yet. As part of a larger story that we're still getting hints of? Yes, I think so.

C. Pause

So, where does this leave me? Strangely, perhaps too predictably, not so far from where I was on 5 December 1995, just over two years ago. On that day, I wrote my friend Greg Comer a summary of my position in the form of a little table that I'll reproduce on the next page. Expressed in a fairly poetic way, you'll find something that's not so far from your "correlation without correlata." I think it expresses fairly clearly why I've been drawn a little to your paper … and also why I've been repelled a little by your paper. Quantum theory still strikes me as a theory of "what we have the right to say" in a world where the observer is no longer detached. If anything, what you've written here has helped strengthen that impression.

But that's enough for now.

07 January 1998, "Cloud-Cuckoo-Land"

After having my first cup of coffee this morning, I took your new Section IX to that place where I do my best thinking. And I found that I liked what I was reading. A little later I looked at the new additions again and found that I still liked them. So I guess that means they're OK.

You do now say more clearly those things that needed saying the first time around. The only change that I might make if I were in your shoes would be to add a little blurb that goes something like:

Whereas the essence of IIQM can be described as "correlation without correlata," the essence of Everett can be described as "correlata without correlation."

Indeed one of the standard research projects for the Everettistas over the years has been to somehow *derive* the standard Born probability rule from the relative-state formalism itself. Everett tried his hand at it in the original paper … then Graham in his PhD thesis tried to fix that up … then Benioff in several papers tried to refine that … and then, finally, Deutsch in his papers tried to argue that the project needed to be given up and that the Born rule should be taken as a postulate for describing the *fraction* of worlds with one property or other. The main point is that in simple Everett, there are no "correlations" in the classical, statistical sense because there are no joint *probability* functions lying around with which to work. You might cash that in for something.

Also, let me comment that, in my heart, I still think you are being too soft on Consistent Histories. As far as I can tell, all the mathematical, objective-sounding coating that they put on their cake boils down to something every bit as *ad hoc* as Bohr's metaphysical ramblings on complementarity. *Why* are the single-time "events" in a history one of a set of orthogonal projectors? *Why* are the "weight functions" with which we derive the consistency requirements just exactly the ones given by the Born rule in the first place? These questions are never asked by Griffiths. If he were deadly serious about wanting to ascribe a man-independent reality to a system at every moment of time, why does he restrict himself to those structures already existent in the old "measurement-interpreted" formalism? Let me give an example. For instance, why not suppose that the elementary events can always be drawn from a set of projectors that *do not* give rise to a Kochen–Specker paradox? That is to say, why not base the notion of a "framework" on this larger set of events? For a three-dimensional Hilbert space, one might be able to get away with eight or nine *possible* events at each shot in time, only one of which is *real* from the consistent histories point of view. (I say eight or nine because I do not know what the ultimate record in the Mermin–Peres–Kernaghan small-vector-number war will be.)

What is a quantum system?	A line drawn in the sand.
What is a quantum state?	What little we can say about what's on the other side.
What is unitary evolution?	Our way of saying that once we know everything about our side, there can be no contribution to our state of knowledge about the other side without crossing the line to take a look.
What is wave function collapse?	Something not so very different in spirit than Bayes' rule for updating probability assignments in the light of new evidence.
What is a measurement model?	A description of the particular way Bayes' rule must be modified when we attempt to get some new information from the other side.
Modification to Bayes' rule?	This is another way of expressing what we have known so long about quantum phenomena. Information gathering measurements disturb quantum states.
Is the disturbance mechanical?	No more so than Bayes' rule. The space of quantum states is just as ethereal as the space of probability assignments.

What is the sand the lines are drawn in?	It is not spacetime. Spacetime apparently has nothing to do with it. The perfect case in point is the EPR thought experiment. This is why E thought it to be the final blow.
Do we really need a sand in which to draw the lines?	Probably not. Did we need an ether for the field? Did we need a caloric for heat? Did we need a phlogiston for the flame?
Ok then.	$+ + + + + + + +$
What is a quantum system?	A line drawn ...
What is a measurement outcome?	What we find when we ask a question, fully prepared to make use of our modified Bayes' rule (measurement model) to describe the system afterward.
Where does it come from?	The other side of the line.

One last thing that just dawned on me. . . .

Wait, wait, one more last thing: I defer discussion of objective probability once again. Of the lot, that's the hardest thing to talk about without writing a whole paper in the mean time.

15 January 1998, "Buttered Toast"

Merminition 16: *Could I impose on your sound literary and physical instincts once more? I've taken your advice to heart yet again, and put [. . .]*

Boy, you really know how to butter a guy up!

Anyway, I have looked at your new section, but I'm not exactly sure what I want to say about it. I am glad that you've added it, but I wish it were longer! I think that your saying "the paper is already too long" carries no force at this point. It seems to me that if a potential reader is willing to get so far into the paper as to read Section XII, then he's already demonstrated a substantial commitment ... so substantial, in fact, that he's likely to want a little more (like I did). Three or four more pages won't kill him. On the other hand, I understand that you're losing steam ... and something of the present order is likely to be the best that I can hope from you. So I am glad that you added what you did: it does help make the ending a little less lame.

A couple of points. I did find the first paragraph a little difficult to read. The sentence starting with "This shifts the terms ..." is almost Bohrian in its topology. Either that's a good sign or a bad sign, but I tend to suspect the latter. Also in the final sentence, you say, "This question ... has not ... been asked." What you mean to say is something different: as you point out above, it has been asked but the answer has always been sought in a different way. Beside that, overall, the section is a little choppy … but I guess there's no avoiding that if you want to keep the length down (while still expressing the same thoughts).

Merminition 17: *I suspect our unfathomable conscious perceptions are going to have to enter the picture, as a way of updating the correlations. To acknowledge this is not to acknowledge that "consciousness collapses the wave-packet" but it is, I suspect, to admit that quantum mechanics does not offer a picture of a clockwork world of eternally develop-ing correlation (described by "the wave-function of the universe"), but a phenomenology for investigating what kinds of correlations can coexist with each other and for updating current correlations and extrapolating them into the future.*

This paragraph intrigues me, of course. It is at that point that your opinion and mine seem to converge most closely. But along with that, I have to wonder what it is that's keeping us so far apart otherwise. This remains a mystery to me that I want to understand much better.

Your paragraphs #4 and #5 caused me a little narcissistic twinge. You actually address one of the technical points in my long note! Wow! But alas, you address it only by edict: that was a little disappointing. "Isn't it plain to everyone that 'measurement' is an anthropocentric concept? … But of course 'system' … now 'system' … no there's nothing wrong with having a objective preordained *cut* even when there are no correlata." Oh well, we do what we can. (This, by the way, induced me to carry the narcissism a little further this morning: I actually reread my long note to you. Looking back at it – this morning, at least – I decided that I sort of like it after all. Even I got something new after reading it a second time. Did you read it a second time?)

That's about it; that's all I can think to say. You've emptied me again. I really have enjoyed working through and thinking about the Ithaca point of view. Maybe in the next couple of months, while writing the paper with Caves and Schack on Bayesianism (and thinking about IIQM again), I'll be able to come to some decent perspective on it.

Oh yes, and connected to that, one final thing … one that got my dander up:

Merminition 18: *P.S. Did you know that Sidney Coleman claims to have made respectable the introduction of probabilities into Everett by his adaptation of an old argument of Hartle?*

"Claims" is the key word here. I would say that I am more than aware of the argument: all versions of it, including Coleman's (as eventually published by Gutmann) and Benioff's (which preceded Coleman by 20 years), make their ostensive progress by slipping the notion of probability in from the back door. After slipping that by, the only thing they end up doing is proving various versions of "the law of large numbers" by a long and circuitous route. Dispelling some of the "respectability" of those claims is what this paper I'm writing with Caves and Schack will be about and work very hard to make clear.

15 January 1998, "Thanks"

One last comment prompted by your last comment:

Merminition 19: *Glad to hear you think Coleman is just the law of large numbers. That's what it sounded like to me too. When he talked about it he freely admitted that probability had to be put in by hand. What seemed to be the issue was where it got put in and how natural the putting in was. There was also stuff about randomness and noncomputability that sounded different, which didn't make much sense to me.*

This is what I meant by "various versions" of the LLN. This randomness business is really just a refinement of that ... and nothing more. This remains true even when one evokes words and concepts from algorithmic information theory. Whenever you're ready for some points of departure for reading about these things, let me know: I can provide you with some citations (Benioff, van Lambalgen's review article, etc.)

15 February 1998, "Landau, Lifshitz, Correlata, and Pauli"

The other day I (accidentally!) got embroiled again in a discussion of the foundations of quantum mechanics. In particular I was making fun of the naïveté of taking state vectors as having a one-to-one correspondence with reality. My sparring partner responded with, "Well then, what would *you* take in the theory as corresponding to a reality?" In a sort of desperation I pointed to the middle part of a picture that we had already drawn on the board: two square blocks connected by a few squiggly lines. With my finger on the lines, I exclaimed "entanglement!" ... But don't you even think *"Yes!! Correlation without correlata!"* ... despite your coming to my mind in just that moment. For then I quickly followed with, "But only as a sort of shorthand for a more fundamental situation: when one part of the world tries to obtain or leave an imprint on another, both are indeterministically disturbed in the process. The apparatus of quantum mechanics appears to be the only available method for reasoning in light of that situation."

One day you'll have to respond to me why you dislike that line of thought so much (which, in any case, is not so much original in me as it is in Pauli, in his version of Copenhagen).

"What we learn about is not nature itself, but nature exposed to our methods of questioning."

(Werner Heisenberg)

"All that is solid melts into air, all that is holy is profaned, and man is at last compelled to face with sober senses his real conditions of life and his relations with his kind."

(Karl Marx)

14 March 1998, "Pauli in Ithaca"

Have you ever seen this:

Quite independently of Einstein, it appears to me that, in providing a systematic foundation for quantum mechanics, one should *start* more from the composition and separation of systems than has until now (with Dirac, e.g.) been the case. – This is indeed – as Einstein has *correctly* felt – a very fundamental point in quantum mechanics, which has, moreover, a direct connection with your reflections about the *cut* and the possibility of its being shifted to an arbitrary place.

It comes from a letter written by Pauli to Heisenberg in the wake of the EPR paper. I thought this might intrigue you with your Ithaca attitude. In case you know German better than I – I swiped this piece of the translation from another source – and you are interested in reading further, the place to look is in letters numbered 412, 413, 414, and 415 in *Wolfgang Pauli: Scientific Correspondence with Bohr, Einstein, Heisenberg a.o.*, Vol. 2, edited by Karl von Meyenn (Springer, 1985).

29 March 1998, "Ithacan Monadology"

Lest you think I've long since forgotten the Ithaca II paper and am not continually mulling it over and debating it about in my head, let me bring it up once again. I just ran across the following interesting line in an otherwise wacky article by Julian Barbour:

What distinguishes [Leibniz's philosophy] from Berkelian idealism is the most radical (and least well explained . . .) element in the Monadology – the assertion that the perceptions of any one monad are nothing more and nothing less than the relations it bears to all other monads.

What think you? Have you ever given any thought to the connection or disconnection between "C w/o C" and Mr. Leibniz?

By the way, whatever happened to the Landau/Lifshitz stuff you were going to send me? Or did you come to an opinion on it yourself? For your other question (i.e., about the renaming of Wootters' Theorem), I still don't have a strong opinion: maybe "state-vector local reconstruction theorem" or just "local reconstruction theorem" or "local representation theorem." I guess if you wanted to be really bold you could call it "The Equivalence Principle." :-)

I think to some extent I have a new objection or twist on the credence I would give this IIEP. It is this: who says – and why do they say it – that my ability to reconstruct a wave function from solely local measurements is so all-fired important? In order to give the IIEP any operational significance, as you know, we must be able to make many many local measurements on many many identically prepared systems. But why is the issue of the *efficiency* of our measurements ignored in your interpretation? That is to say, if we worry about such matters, it seems to me to become more suspect that the "correlations" (as you like to call them) are the whole story. Nonlocal measurements will in general be more efficient than solely local ones. And this is not simply due to the fact that there are nonclassical correlations (i.e., Bell inequality violations) for entangled states. I'm thinking in particular about a certain malicious example that Bennett and company and I are writing up right now. (Hopefully it'll be on the net in a couple or three weeks – the manuscript is already sitting at 21 pages.) In this case, Alice and Bob each have possession of a three-dimensional Hilbert space, and are assured that the quantum state in front of them is one of a set of nine *orthogonal* product states $|\psi_i\rangle|\phi_i\rangle$. Because the nine states are orthogonal, Alice and Bob can in principle identify the unknown quantum state with a *single* measurement. But the twist is this, the set of $|\psi_i\rangle$ and the set of $|\phi_i\rangle$ are each maliciously chosen so as not to be orthogonal sets. And in particular, they are chosen so that one can prove that a nonlocal measurement is required if one wants to completely recover the identity of the unknown quantum state. No ingeniously clever set of local measurements (including POVMs) and classical communication can take the place of a nonlocal measurement in this example. What in the Ithaca Interpretation gives warrant for ignoring the issue of efficiency of state identification in asserting the equivalence between "correlations" and "state vectors"? With the issue of efficiency taken into account, doesn't it seem like there is more to a state vector than "C w/o C"?

Anyway aside from this negative comment, I do believe that I have a much firmer understanding of what I see as important in your efforts. (I.e., I'm closer to being on the verge of identifying what it is that resonates most with my happy neo-Copenhagen predispositions.) With some luck I'll write you up a little ditty on it while I'm in Munich next week. (My wife and I are going to visit her parents, and I'm hoping to spend the time being more philosophical than usual.)

17 April 1998, "References"

Merminition 20: *Fire away. Just be polite.*

Of course, I'll be polite: I am a Texan foremost, and therefore a gentleman! ... Actually I should have used those "irony quotes" you suggested. The reference I'll make to your papers will mostly likely be in passing. The main concern of the paper I'm writing is with the Bayesian way of looking at quantum probabilities, deconstructing the Hartle–Coleman stuff, and [trampling on] the philosophical notions of objective probability (von Mises' frequency, Popper's propensity, and Lewis's chance). I'd probably like to attack your notion of *objective* probability too, but I can't figure out what your notion is!

Thanks for sending on the Landau–Lifshitz stuff. I saw the following citation in the April 1 issue of *J. Mod. Optics* the other day: "[for such and such ...] see the textbook series of Landau and Lifshitz. Admittedly, we do not know which volume should be consulted, but our Russian colleagues are positive that everything can be found in Landau and Lifshitz."

By the way, you know it has dawned on me that for all the 57 kilobytes(!) of notes that I wrote on your "correlation without correlata" you've made hardly any effort to reply to anything in detail. Either that means that you think that what I had to say was pretty foolish or ... more than likely ... you think it even more foolish than that! If I had to boil it down to one thing I would like the king to give me audience on, I would say that that would be the stuff addressed in Merminitions 7 and 8. You want to turn to the structure of the theory to give us one ontological statement, but from my perspective it doesn't look as if it gives anything unique in that regard. We can make Wootters theorem go through with a set of observables much smaller than the ones you use in your derivation in the appendix; we can make it go through with a set much larger. What in the theory gives you the right to say "Ouh, this is just right!" when you get to the complete set of (local, product) Hermitian-operator observables?

17 April 1998, "How Do I Sleep?"

I'm back again. I was thinking earlier that, among other things, I know you must ask yourself, "How can a devout Copenhagenist sleep at night?" Surely they must have awful nightmares with all the silly anti-realistic metaphysics they carry around in their heads! … Well, anyway, I've given my best shot at exposing the ontological lesson of quantum theory – the reason there's not too much worry for nightmares – and yet it seems to make no impression on my friends. So let me try once again, this time with a relatively eloquent passage that I found just today. It's from a 1957 paper of Léon Rosenfeld; I'll scan it in for you.

Sleep well.

The Curse of Positivism

According to our critics, the epistemological point of view of quantum theory undermines the sound belief in the reality of the external world, in which all physical thinking is rooted, and opens the door to the barren doctrine of positivism: we are no longer concerned with things, but only with the way to speak about things; science is degraded from a quest for truth to a verbal exercise. Dogmatic assertions that certain questions are meaningless bar the way to further inquiry; inability to understand the riddles of the quantum is hidden in the clouds of mystical renouncement.

This picture would be alarming if it were true. However, it is just another dream, a nightmare perhaps, of our critics. Obviously, it cannot result from a serious assessment of the immense broadening and enrichment of scientific thought which has resulted from the endeavor to formulate the laws of atomic phenomena. In fact, it is based on the most futile casuistics: the critics diligently excerpt from the writings in which the principles of quantum theory are discussed isolated sentences on which they put arbitrary interpretations. No wonder that they should find (as they freely confess) some difficulty in understanding Bohr: which, incidentally, does not prevent them from branding him as a positivist. There is no difficulty, at any rate, in understanding the critics' philosophy and exposing its unscientific character.

The realization of the mutual limitations imposed upon the use of classical concepts by the conditions of observation has forcefully reminded us of our own position in the world, and of the function of science in relation to this position. We are not merely contemplating the world, but acting upon it and thereby modifying its course. Accordingly, the scientific description of the phenomena is fundamentally concerned with the interaction of external agencies with the human observer; or, at least, in the narrower domain of physics, with material systems under the latter's control. The mode of description of classical physics appears, from this point of view, as a special case of wide validity, in which the quantitative effects of the interaction between observational devices and observed systems may be neglected: which does not mean, however, that the presence of these observational devices is not just as essential as in quantum theory for the very definition of the physical concepts.

It must again be stressed that there is in this view of the nature of science no arbitrary element: it is just an explicit statement of a situation which has always existed, even though it was not always so clearly recognized. Certainly, it puts an emphasis unknown to the outdated materialistic metaphysics of the nineteenth century on the active role of the observer in defining the phenomena: but in so doing, it brings the whole structure of science nearer to reality, in closer conformity with our real relationship to the external world.

There is an undeniable similarity between the epistemological conclusion drawn in such a straight-forward, unambiguous way from the peculiar character of the quantum laws, and the insistence of the early positivists on the essential part played by our sensations in determining our knowledge of the external world. This only means that, to that extent, the early positivist movement was a healthy reaction against the shallow metaphysics of mechanistic materialism. But why should scientists be made responsible for the later positivists' blundering into a metaphysics of their own? No scientist would accept the extreme positivist contention that there is nothing more in state-ments about phenomena than the conceptual expression of relations between sensations: he would maintain that such statements refer primarily to real processes of the external world; our mental representation of these processes being itself, of course, subject to definite laws depending upon our sensorium.

To point out that certain relations between classical concepts cease to be meaningful in quantum theory has none of the sinister implications fancied by our critics: it is a plain statement of fact, founded in a law of nature. The words 'renouncement' or 'resignation' often used in this context are ambiguous in their emotional connotation: renouncement may be felt as privation or as liberation. Some critics seem to take the invitation to 'renouncement' as an attempt on their personal freedom: the right to indulge in metaphysical dreams is not disputed; only, this activity is not science.

20 April 1998, "Socrates and Rosenfeld"

Merminition 21: *What do you mean, I didn't respond? I added major chunks to the paper in response to your criticisms. That kid had every right to be proud to know you. Whereof I had no answers, thereof I remained silent, if you know what I mean.*

But as for Wootters' theorem (now called the SSC theorem, by the way – I'm not going to keep changing its name every time I get a new email) I don't care if it takes less than subsystem correlations to determine the state. The crucial thing is that it doesn't take more (as it would if one were restricted to a Hilbert space over the reals).

Thank you for your extra sentence of explanation. (Oops, forgot those irony quotes again.) More to the point – and more sincerely – thank you for the three nice emails. I'm starting to feel like I really am a grain of sand in your shell. Whereof one cannot speak, thereof one should think harder.

Funny that you should find in Socrates a new poster boy for IIQM. Just the other day I wrote Asher Peres: [See note dated 5 April 1998.]

The Rosenfeld quote came from: L. Rosenfeld, "Misunderstandings about the Foundations of Quantum Theory," in *Observation and Interpretation: A Symposium of Philosophers and Physicists*, edited by S. Körner (Academic Press, New York, 1957), pp. 41–45.

21 April 1998, "C^* Hunkies"

Merminition 22: *So if you take the conventional view (which I remember from listening to C^* algebra types in ancient days) that the state is just a catalog of all possible mean values, then there's nothing to favor complex over real scalars in the underlying Hilbert space. But if you take the SSC view that the state is a catalog of the non-trivial subsystem correlations, then there is. That has to (be trying to) tell us something.*

I agree that there must be something deep going on with this. But it seems to me that the C^* people could always feel better about themselves because they had Gleason's theorem to back them up. That is to say, if they hadn't had Gleason lying around, they would have had to have given some extra justification for why they chose this particular type of catalog (i.e., the state) over another. But what about Ithaca? Take a look again at my silly idea following Merminition 9. Wouldn't you need something like that if you wanted to feel as comfortable as the C^* people?

07 May 1998, "A Rug and A Sweep"

Point 1. Didn't you use some analogy like "sweeping the problems of quantum theory under the rug of objective probability" in Ithaca I or II? If you did, I can't find the words "rug" or "sweep" anywhere in there at all now – can you fulfill my need to quote accurately by pointing me to the right place ... or telling me that I just imagined their use?

Point 2. For history's sake, let me tell you about another little discovery I made a couple of days ago. I found the main piece of the Hughston–Jozsa–Wootters thingy that you rederive in Ithaca I – namely the piece about unitary reshuffling (both if and only if) ... but not the part about creating ensembles from a distance – in an old paper by Ed Jaynes. Let me give you the citation, E. T. Jaynes, "Information Theory and Statistical Mechanics. II" *Phys. Rev.* **108**, 171–190 (1957).

Point 3. In Germany last month I read your debate with Stapp. Hate to admit, but I had a lot of trouble reading it: that could either reflect on the writing, or the state of my head but I think it was the writing this time. Anyway, I was amazed, among other things, about the trouble you went to illustrate the meaning of Bohr's phrase, "Of course there is in a case like that just considered no question of a mechanical disturbance ... But ... there is essentially the question of an influence on the very conditions which define the possible types of predictions regarding the ... behavior of the system." Amazed, maybe, because I was already pretty happy with it. Well, while amazement is running high, I just found today that Ed Jaynes also didn't think the phrase was clear at the outset. If you'd like to see his rendition in terms of Bayesian language, take a look at E. T. Jaynes, "Clearing Up

Mysteries – The Original Goal," in *Maximum Entropy and Bayesian Methods (Cambridge, England, 1988)*, edited by J. Skilling (Kluwer, Dordrecht, 1989), pp. 1–27. [To wit:]

The spooky superluminal stuff would follow from Hidden Assumption (1); but that assumption disappears as soon as we recognize, with Jeffreys and Bohr, that what is travelling faster than light is not a physical causal influence, but only a logical inference. Here is Bohr's quoted statement (italics his):

> Of course there is in a case like that just considered no question of a mechanical disturbance of the system under investigation during the last critical phase of the measuring procedure. But even at this stage there is essentially the question of *an influence on the very conditions which define the possible types of predictions regarding the future behavior of the system.*

After quoting these words, Bell added: "Indeed I have very little idea what this means." And we must admit that this is a prime example of the cryptic obscurity of Bohr's writings. So – with the benefit of some forty years of contemplating that statement in the context of his other writings – here is our attempt to translate Bohr's statement into plain English:

> The measurement at A at time t does not change the real physical situation at B; but it changes our state of knowledge about that situation, and therefore it changes the predictions we are able to make about B at some time t'. Since this is a matter of logic rather than physical causation, there is no action at a distance and no difficulty with relativity [also, it does not matter whether t' is before, equal to, or after t].

Again we see how Bohr's epistemological viewpoint corresponds to Bayesian inference, and could have been expressed precisely in Bayesian terms. However, Bohr could not bring himself to say it as we did, because for him the phrase "real physical situation" was taboo.

There's much in the article that I don't like. Jaynes, like most good Bayesians, didn't quite believe in quantum mechanics – where there is probability, there is ignorance about something objectively existent ... yuk – but he does hit the Bohr part of the discussion on the mark (as far as I am concerned).

Point 4. I had a dream about you a few nights ago. I was sitting on a park bench in Ithaca. You were jogging by and stopped for a moment to talk. The significant thing that stood out about you was that you were wearing white tennis shoes, white shorts, and now had blond hair. Ithaca – I've never seen it in real life – looked like something out of "It's a Wonderful Life." We spoke for a moment on reality and Bayesian probability, and then I awoke.

08 May 1998, "A Bug in the Soup"

Well, I'm always happy to get credit for something(!) ... but I have no idea what you are talking about. Feel free to refer away ... as long as no one gets the idea that I'm an Ithacan. (Ithacan: that species of people marked by a regressive desire to believe that the mysteries of quantum mechanics can be relieved with the formulation of an "objective" notion of probability. Compare Einstein's regressive desire to relieve the mysteries with some version of classical field theory. And compare Pauli's constant use of the word "regressive" when speaking of Einstein.)

Yeah, I hope you do take a look at that Jaynes article. Also, there's another article where he bashes on quantum mechanics. I'm sure he boobalates somewhere in there (given my religious love of the quantum and its mysteries), but I haven't yet made the effort to pinpoint these troubles. I'll do that next week with Rüdiger. Please, please, please, send me any thoughts or comments you might have. I was especially pleased to find that Jaynes's had some very vague desire to build up (something like, *but not identical to* – that's where he screws up) quantum mechanics as a system for inductive and deductive inference under the situation that information-gathering measurements cause disturbances. [For instance, he says:]

Put most briefly, Einstein held that the QM formalism is incomplete and that it is the job of theoretical physics to supply the missing parts; Bohr claimed that there are no missing parts. ... if we can understand better what Bohr was trying to say, it is possible to reconcile their positions and believe them both. Each had an important truth to tell us. ...

... When Einstein says QM is incomplete, he means it in the ontological sense; when Bohr says QM is complete he means it in the epistemological sense. ...

Needless to say, we consider all of Einstein's reasoning and conclusions correct on his level; but on the other hand we think that Bohr was equally correct on his level, in saying that the act of measurement perturbs the system being measured, and this places a limitation on the information we can acquire and therefore on the predictions we are able to make. The issue is merely whether this limitation is as great, and has the same quantitative form, as Bohr supposed. ...

Bohr had no really cogent reason for his postulate that the limitations on the ability of the QM formalism to predict are also – in complete, quantitative detail – limitations on what the experimenter can measure; this seems to us an outstanding example of the Mind Projection Fallacy.

I would have liked to believe that I was the only person to have that epiphany. But then I finally realized that that is what Bohr was talking about. I guess I don't mind making a smaller contribution – especially if it's in a form that people can understand when it's finished ... and not debate it about on the pages of *AJP*, for instance. Irony quotes, irony quotes.

04 September 1998, "The Bohrazine Shuffle"

After all these years, for some reason today I finally had the heart to wade through Bohr's reply to Einstein, Podolsky, and Rosen again. Wow, what tough going! I wonder if the man had any clue just how horribly he wrote? Not surprisingly, once again I hardly got anything from the article. *However*, there was one little thing that made my toes tingle. It's a small passage that comes just after the famous one you see quoted everywhere – I think even you quoted it recently. (I'll enter it in below for you.) What I found most intriguing about it is that it *seemed* to be talking to what I see as the most elegant – and promising for progress(!) – solution to all our quantum troubles.

Now, the word "seemed" above is where you come into this picture. Being the Bohr scholar that I know you are, I wonder if you can shed some light on what he might be saying. And whether he (Bohr) is saying anything that – vague though it is – might be of use in steering me.

Lest you have forgotten my point of view, before moving on to Bohr, let me remind you of it with a couple of summaries I wrote Adrian Kent a few days ago:

[See notes to Adrian Kent dated 30 August and 2 September 1998.]

With that in mind, let me stress once again that, for me, it is not just enough to say these things and then feel comfortable with myself. I want to see quantum mechanics come out as the end product of these vague proclamations, not as the starting point. (Then, when I'm finished with that, it'll be time to move on to some really revolutionary stuff ... but that's a different story.) Anyway, if I want to go the route of relying on much of the hard work of Gleason and Kraus, this leaves me with having to justify the following five things from the information–disturbance foundation described above:

1. Why are information-gathering measurements restricted to being orthogonal bases in a Hilbert space?
2. Why is there a noncontextuality assumption about the probabilities of overlapping basis vectors? [Gleason uses 1 and 2 to derive that all measurement probabilities can be summarized by a density operator. This gives us quantum mechanical states.]
3. Why is that Hilbert space above complex? [I know you think your "correlation without correlata" gives a good explanation of that, but I think a strict interpretation of quantum states as states of knowledge (and nothing more) gives an even better explanation – there's some chance I'll be writing this up this fall (or in a personal letter to you before then).]
4. Why must the evolution of our states of knowledge be a linear mapping (from density operators to density operators)? [Kraus needs only assumption 4 to show that unitarity can always be invoked at some sufficiently high level.]
5. Why is the Hilbert space of a composite system the tensor product of the more primitive Hilbert spaces (and not some other magical combination of the primitive spaces)?

My focus for this note is Question 1 above. In essence, the question is: How can an "information–disturbance foundation" force "complementarity" upon us? I.e., how can such a foundation force upon us the restricted (mutually exclusive) measurements described in Question 1?

OK. Now I hope you're ready for Bohr's cryptic passage.

Of course there is in a case like that just considered no question of a mechanical disturbance of the system under investigation during the last critical stage of the measuring procedure. But even at this stage there is essentially the question of *an influence on the very conditions which define the possible types of predictions regarding the future behavior of the system*. Since these conditions constitute an inherent element of the description of any phenomenon to which the term "physical reality" can be properly attached, we see that the argumentation of the mentioned authors does not justify their conclusion that quantum-mechanical description is essentially incomplete. On the contrary this description, as appears from the preceding discussion, may be characterized as a rational utilization of all possibilities of unambiguous interpretation of measurements, compatible with the finite and uncontrollable interaction between the objects and the measuring instruments in the field of quantum theory. In fact, it is only the mutual exclusion of any two experimental procedures, permitting the unambiguous definition of complementary physical quantities, which provides room for new physical laws, the coexistence of which might at first sight appear irreconcilable with the basic principles of science. It is just this entirely new situation as regards the description of physical phenomena, that the notion of *complementarity* aims at characterizing.

I'm especially interested in the part above starting with "On the contrary ...". It looks to me like he is wanting to say that: measurements causing disturbance *causes* complementarity (i.e., the mutual exclusivity of certain measurements). (Be aware however: I fully realize that he is NOT talking about something so mundane/mechanical as trying to get the Heisenberg relation from a Heisenberg microscope-like picture.) What's your take on this passage? Can you make much sense of it? What does he mean by "providing room for new physical laws?" What "basic principles of science is he talking about?" What five pages of derivation are lying behind all this business?

It nags me that Bohr often speaks as if it is clear that the structure of quantum theory is derivable from something deeper, when in fact all the while he is taking that structure as given. When did he ever approach an explanation of "Why complex Hilbert spaces?" Where did he ever lecture on why we are forced to tensor products for composite systems? It's a damned shame really: I very much like a lot of elements of what he said, but as far as I can tell all the hard work is still waiting to be done. Is he hinting above that he has a magical way of seeing where "orthogonal bases for allowed measurements" come from? Is he hinting that it comes from "information vs. disturbance"?

Any thoughts you have will be most welcome.

08 September 1998, "I'll Bohr You to Tears"

Thanks for the long reply to my query. That's the kind of email I like to get. I found myself mulling over it the whole long holiday weekend. I reread Chapters 13 and 14 in your book: now I see I can blame my frustration on an inheritance from you! I first read these chapters about 5 years ago and I guess I had forgotten them a little. I also read again the introduction to your long article in Chapter 12 and washed it all down with a rereading of Pais's Chapter 19 and Section 25c in his Bohr and Einstein biographies. [I should have chosen a better drink though: I found Pais pretty annoying.]

Did you have any luck with Plotnitsky? I'll try to dig up his book again and have a look at it sometime this week.

But, let me presently add a few comments to your comments.

Merminition 23: *You could use some interpreting yourself, you know.*

Yeah I do know that!!! That's why I like to associate with clear thinkers like you. At least as far as this aspect of my career (i.e., dabbling in the foundations) goes, I want to hone things to immaculate clarity before saying them to someone who doesn't already know me.

Merminition 24:

... my information gathering about something you know, causes you to loose some of that knowledge ...

This is an admirable way to start thinking about the problem, but there has to be a better way to put it even at this crude level. I mean my knowledge is my knowledge. I can't lose it.

You are indeed right: I apologize. I was priding myself on being pithy in that formulation, but I see I went too far. The trouble comes from my usage of the words "know" and "knowledge" ... which should be taken to be, more accurately, "can predict" and "predictability." PLEASE indulge me *two* last times, and then I promise I'll leave you alone ... that is, if you give some small evidence that you did indulge me!

1) A much better technical statement of what I'm trying to express above can be found in Section III of my paper "Information Gain vs. State Disturbance in Quantum Theory" (`quant-ph/9611010`), *Fort. der Phys.*, **46**, pp. 535–565 (1998). It is crucial to my point of view that there be at least *two* players and *two* quantum states in the game. Appreciating that will, I think, to some extent negate your questions:

Merminition 25: *Does gathering some information disturb other information? Or is it just in the nature of the information that was gathered that it no longer makes sense to contemplate the content of the other information?*

Please read the cited section in my paper, and then come back for the following comment.

I think one of the troubles in our founding fathers' discussions is that they continually focused their attention on *one* observer making measurements on a quantum system described by *one* (known) quantum state. This led them to say things – in language similar to some of the specimens in your note – like, "The gain of knowledge by means of an observation has as a necessary and natural consequence the loss of some other knowledge." (Pauli) Without at least a second player in the game, those gains and losses hardly seem to be sensible concepts to me: they can only refer to the observer's attempt to ascribe one or another classical picture to the quantum system in front of him. Since we know – from Bell's argument and the religion of locality – that it is not reasonable to assume that those classical variables (correlata) are there and existent without our prodding, it is hard to call the revelation of a measurement outcome a "gain of knowledge." What did you learn about the world that was there before your looking? Nothing. However, throw a second player into the game and that situation changes. Those random quantum outcomes now have something *existent*, some unknown truth, that they can be correlated with. The revelation of an outcome really can correspond to a "gain of knowledge," but you need at least two information processing units in the world for that to be the case.

Disturbance? The founding fathers like to say things like, "For every act of observation is an interference, of undeterminable extent ..." But again, this is a muddled concept when the discussion is restricted to *one* observer and *one* (known) quantum state. This is true for exactly the same reason as above: there just ain't no correlata already there to disturb. (Nice of me to use the word correlata just for you!) So why did they even bother with the word disturb? It's because they had semi-classical ideas in their head. But, still, trying to fix the picture by saying that wave-function collapse will do the job is no good either: if one is talking about a *single* (known) quantum state, there is no necessary collapse upon measurement. I can build a device that "measures" any observable you wish while still regenerating the known quantum state in the process. However, just as before, throw a second player and two nonorthogonal states into the game and everything changes. Now one observer (the eavesdropper) can't help but impart a disturbance to at least one of the two quantum states – that is, if she puts herself in the position to gather some information about the other observer's state of knowledge (i.e., the quantum state he prepared). Moreover, in this context, the phrase "of undeterminable extent" remains meaningless ... or superfluous at best. If Alice knows everything that Eve is doing, then the disturbance will be perfectly determinable – that's what my paper is about – but that doesn't mean the disturbance disappears. The disturbance arises because the eavesdropper's equipment necessarily becomes entangled with the system it is measuring. (Sometimes I think there is nothing to the concept of entanglement beyond this.)

But all this, everything I've just said, is derivable from quantum mechanics. It's part of the answer to your question, "What is quantum mechanics trying to tell us?" It's an example of quantum mechanics teaching us when and how to use the words "information" and "disturbance" properly. As I've told you, though, I'd like to see the tables turned. But for that I have to take on an attitude that I think is a little foreign to you. In your original Ithaca paper you speak of the minimal requirements for a quantum mechanical universe: it is, you say, two qubits – two things to have correlation without correlata. I, however, am more afraid to go that far, i.e., to some final/overarching ontological statement. Instead, the most I think I'm willing to ask is, "What are the minimal requirements for a physical *theory*?" And there, I think the answer is two "theory makers" and a physical system. (It's hard to get rid of that damned Catholic Church!) From my perspective, a *theory* never stands a chance of giving a complete description of reality – for that would require some nasty self-reference by at least one of the theory makers. However, from a theory, one can still hope to intuit *some* (safely assumed) ontology.

2) This leads to my second request for an indulgence. In a second mailing, I'll send you something I wrote Rolf Landauer along these lines. It doesn't say much more than I've already said above, but maybe it says it from a slightly different slant. Reading it may help the ideas click on ... or click off ... for you. In either case, it may help you provide me with some useful criticism. (That would be good enough small evidence that you did indulge me.)

There, that concludes my painful requests for indulgence. As far as making some real progress – not words(!) – toward actually turning those tables, that's a more difficult story. But, I don't think it is a story without hope for progress. I give an (unfortunately, fairly wimpy) example of this in Section VI of the above-said paper: it might interest you a little. I think a much more important tack, though, is to first get straight how quantum probabilities are purely subjective probabilities – for measurement outcomes, not for existent but unknown properties, not for correlata(!) – and see how that already pins some things down about the theory. This is where you and I differ most on the Ithaca call. Once that is carried out, one can come back to ask why it is that we're stuck with subjective probabilities … and that's where information vs. disturbance comes in.

So let me finally move on to some of your other comments.

Merminition 26: *The "funny property" might be intersubjective correlations, you know …*

It's funny to watch the two of us try to squeeze everything into our own pet language! In any case, I do hope you realize that despite my spitting back the words "information" and "disturbance" in each and every email to you, I do work very hard to listen to what you're saying. "How do I correlate/anti-correlate what he's saying with what I'm saying?" – this I try to do every time.

In that vein, I wanted to read your big Ithaca paper one more time over the weekend, but I was thwarted by my broken printer. It's too darned hard to read something that long on a computer screen. If I ever get back to the office, I'll try to give the final version a read and then come back to your Bohr-interpretation work again.

Merminition 27: *I haven't read Kant since I was a freshman in college. Do I have to?*

All I was trying to say here is that Kant argued that, though it must be there, we can never know the thing in itself (the noumenon). "Objective reality is forever beyond our reach." And he was even able to argue this in a classical world! Do I believe the argument? I don't know; I don't even remember it – I haven't read him since I was a freshman in college either. But, it does seem to me, if there is anything worth salvaging in his stuff, we'll find that we have it heaped even higher in a quantum world. (It's like having second quantization.) I tell you what: I'll pledge to read his Prolegomena again – it's only 163 pages – if you will. It'd probably do us both some good for party jokes. Lately I've been trying to read the neo-Kantian Schopenhauer because Pauli had great respect for him … and supposedly saw something of quantum mechanics in his philosophy … but it'd probably do me some good to have a better background before tackling him.

Merminition 28: *Who is Kraus, by the way? What should I read by or about him?*

I took the liberty to fix your spelling mistake. Karl Kraus was a mathematical physicist. He was the first to prove that a completely positive linear map on density operators can always be viewed as a unitary operation on a larger space followed by a partial trace. (Recall that a positive linear map is any map that takes positive operators to positive operators. It is completely positive if, when extended by the identity operation on any tensor product space alongside the original, it continues to preserve positivity even for entangled density operators.) This theorem is quite nice from the point of view of foundations: for it helps to show that unitarity can be viewed as a convenient, rather than essential, description of time evolutions. From the point of view of "quantum states as states of knowledge and nothing more" the criterion of complete positivity is more natural than unitarity. Why assume such a detailed thing as unitarity just to say that states of knowledge must go to states of knowledge? If one could get rid of the linearity assumption, then it would be even more natural.

Merminition 29:

2) Why is there a noncontextuality assumption about the probabilities of overlapping basis vectors?

What's that?

I like to split the premise for Gleason's theorem into two assumptions. 1) It is the task of physical theory to assign probabilities to the outcomes of all *valid* measurements on a system. The set of all valid measurements on a system corresponds to the collection of all complete sets of orthogonal one-dimensional projectors over some Hilbert space. 2) All such probabilities can be derived from a single function (a "frame function") on the set of one-dimensional projectors.

It is Assumption 2 that I am calling the "noncontextuality assumption." The probability for a given projector-as-outcome does not depend upon the particular set of projectors it is considered embedded in. Given that we know that our measurements don't simply reveal pre-existing (but unknown) properties, my question is, "Who ordered that?"

By the way, it is of some interest to note that Asher says in his paper, "An Experimental Test for Gleason's Theorem" (*Phys. Lett.* A, 163, 243–245, 1992) that *complex* Hilbert spaces are required for the proof. That is not true. Also, I think his Assumption (3) is overkill; it is rather a consequence of the theorem: one need not start out by assuming that two-dimensional projectors even correspond to valid questions. His equation (4) is a mild consequence of the definition of a frame function. I take as my bible of all things Gleason-like: Itamar Pitowsky's paper, "Infinite and Finite Gleason's Theorems and the Logic of Indeterminacy" (*J. Math. Phys.*, **39**, 218–228, 1998).

Merminition 30: *I don't see any connection between the passage and your Question 1, so I guess that's my (disappointing) answer. On the other hand (as noted above) I find the passage irritatingly obscure, so perhaps it provides room for your interpretation. [I.e.,]*

It looks to me like he is wanting to say that: measurements causing disturbance *causes* complementarity (i.e., the mutual exclusivity of certain measurements).

I guess I was taking as my cue the following rearrangement/interpretation of Bohr's words:

On the contrary this description [i.e., the quantum mechanical one] ... may be characterized as a rational utilization of all ... measurements [i.e., information gathering], compatible with the finite and uncontrollable interaction between the objects and the measuring instruments [i.e., necessary disturbance] ... . In fact, it is only the mutual exclusion of any two experimental procedures, permitting the unambiguous definition of complementary physical quantities, which provides room for new physical laws [i.e., quantum theory itself, assuming it being built upon the "quantum postulate" that he so often speaks of, and assuming the validity of equating with "finite and uncontrollable interaction"].

That's about it. I just mostly wondered if you had some insight. My attempt at understanding the passage was certainly more amateurish and self-serving than your effort. I suspect my shifting of his words came about mostly because they were so confusing: it's so much easier to impose your own reading on the scripture. Also I may have been subliminally influenced by some old readings of Pauli (which I also reread this weekend). Check out this piece of a letter from Pauli to Bohr, dated 15 February 1955:

... [I]t seems to me quite appropriate to call the conceptual description of nature in classical physics, which Einstein so emphatically wishes to retain, "the ideal of the detached observer." To put it drastically the observer has according to this ideal to disappear entirely in a discrete manner as hidden spectator, never as actor, nature being left alone in a predetermined course of events, independent of the way in which the phenomena are observed. ...

In quantum mechanics, on the contrary, an observation *hic et nunc* changes in general the "state" of the observed system in a way not contained in the mathematically formulated *laws*, which only apply to the automatical time dependence of the state of a *closed* system. I think here on the passage to a new phenomenon by observation which is technically taken into account by the so-called "reduction of the wave packets." As it is allowed to consider the instruments of observation as a kind of prolongation of the sense organs of the observer, I consider the impredictable change of the state by a single observation – in spite of the objective character of the result of every observation and notwithstanding the statistical laws for the frequencies of repeated observation under equal conditions – to be *an abandonment of the idea of the isolation (detachment) of the observer from the course of physical events outside himself.*

To put it in non-technical common language one can compare the role of the observer in quantum theory with that of a person, who by its freely chosen experimental arrangements and recordings brings forth a considerable "trouble" in nature, without being able to influence its unpredictable outcome and results which afterwards can be objectively checked by everyone.

Probably you mean by "our position as detached observers" something entirely different than I do, as for me this new relation of the observer to the course of physical events is entirely *identical* with the fact that our situation as regards objective description in "this field of experience" gave rise to the demand of a renewed vision of the foundation for "the unambiguous use of our elementary concepts," logically expressed by the notion of complementarity.

If the last passage of that is not trying to equate some notion of disturbance with complementarity, I'll eat my hat. But, of course, these are Pauli's words and not Bohr's.

Merminition 31: "In fact, it is only the mutual exclusion of any two experimental procedures, permitting the unambiguous definition of complementary physical quantities,"

Don't you think "definition" is crucial here? "Physical quantities" have no ontological status (there are no correlata); instead they are simply "defined". What defines them is an "experimental procedure"

Yes I do think "definition" is crucial here. And just for the reason you cite. That much I think we can give him credit for.

Merminition 32: *Complementarity comes first. He became very down on saying that measurements cause disturbance (after EPR). I read him as saying that complementarity opens things up (provides room) by allowing you to contemplate apparently incompatible things (irreconcilable with the basic principles of science) secure in the knowledge that only one of them can be DEFINED.*

Yeah, I agree with this too. However, remember *his* sentence, "But even at this stage there is essentially the question of *an influence on the very conditions which define the possible types of predictions regarding the future behavior of the system.*" It is exactly this notion of "disturbance" that I was invoking in my longish discussion above: a disturbance, or at the very least a *change*, in predictability ... something not so far away in flavor from Bayes' rule for updating probabilities in classical statistics. It is this notion of disturbance that I think will survive, one that he could not be so averse to.

———————

There, I think that's all the comments on comments I can muster up right now. I'm getting pretty tired. But I do want to quickly tell you about one rushed Bohr-related interpretive idea I've just had: I don't have the stamina to verify it presently, but I want to record it. Last month, when Asher was visiting, he said something that really struck me. It was something like, "I learned enough in my education as an engineer to know that one can build precision instruments from very crude tools. You don't have to have a precision instrument to build a precision instrument." His immediate application for this principle was in making fun of the word reality: Asher's view is that "reality" (certainly within quotes) is the crude tool with which we build the precision instruments of scientific concepts. ... But never mind the particular application of the aphorism. Instead, before going to bed, I want to bring the aphorism into conjunction with Bohr's article "Discussions with Einstein ... " in the Schilpp volume – the article Pais called "[Bohr's] finest exposé on complementarity." As I recall, I was pretty dissatisfied with that article. I may be wrong, but I thought all of Bohr's arguments were pretty darned circular. The game was always the same: Bohr would take a measuring device, *assume* that some uncertainty relation held for its parts, and then go on to argue that that instrument couldn't be used to perform a measurement on the quantum system more accurately than allowed by the uncertainty relations! His arguments seemed hardly the triumph that so many commentators portray them to be. But maybe the surprise, the triumph, was just in the disarming of Asher's aphorism in this context. I wonder if that's a plausible excuse for the "Bohr–Heisenberg tranquilizing philosophy"?

09 September 1998, "Unbohrlievable!"

You just won't believe this. Apparently in 1949, while reviewing his old papers, Bohr must have had trouble understanding the selfsame passage from his EPR reply that you and I have been talking about! For he writes in the Schilpp paper, just after quoting the passage,

Rereading these passages, I am deeply aware of the inefficiency of expression which must have made it very difficult to appreciate the trend of the argumentation aiming to bring out the essential ambiguity involved in a reference to physical attributes of objects when dealing with phenomena where no sharp distinction can be made between the behavior of the objects themselves and their interaction with the measuring instruments.

07 October 1998, "Carnap"

I see my last long note left you speechless …

Anyway, I ran across something this morning that reminded me of a conversation with you. It's from the notes of a 1992 conference in Helsinki. Abner Shimony says, "I was a pupil of Carnap. He said that …" Remember I once told you that I had heard Shimony say that he was Carnap's student: this seems to confirm it. As I recall, you disagreed, thinking it was another famous philosopher. Now we both know.

08 October 1998, "Oh Mighty Mermin"

Boy you are a popular guy! Take a look at this conclusion I just wrote for my contribution to the QCM conference proceedings.

CONCLUSION. What is the essence of quantum theory? What crucial features of the phenomena about us lead ineluctably to just this formalism? These are questions that have been asked since the earliest days of the theory. Each generation has its answer; ours no doubt will find part of it written in the language of quantum information. What is striking about the newest turn – the quantum information revolution – is that it provides a set of tools for this analysis from *within* quantum theory. The example of the tradeoff between information and disturbance in quantum eavesdropping is typical. Words about "measurements causing disturbance" have been with us since 1927, but those always in reference to outdated, illegitimate classical concepts. The time is ripe to consider turning the tables, to ask "What is quantum mechanics trying to tell us?" \refnote{\cite{Mermin98}} Why is the world so constituted as to allow single-bit information transfers to be disturbed by outside information-gatherers, but never *necessarily* more so than by an amount $D_{@\mathrm{MI}} \approx 0.067$? Why is the world so constituted that binary preparations can be put together in a way that the whole is more than a sum of the parts, but never more so than by $Q \approx 0.202$ bits? The answers surely cannot be that far away.

On another note about your popularity, I received a paper from Adán Cabello today titled "Quantum correlations are not local elements of reality" (to appear in *PRA*). Since you are thanked for "feedback" in the back, I presume you've seen at least a preliminary version of it. I had seen one too. In fact, I wrote him a letter about it … somewhat discouraging his attempt at publication. I don't think his point is oh so good as to warrant a paper – we should reserve those trees for something a little more substantial. Anyway, I'm pretty sure his argument didn't faze you; that it wouldn't is what I tried to express in my letter to him. Now that the cat is out, let me forward on the comments I wrote him. I was wondering whether you really do agree with the points I made to him. Not to worry, a short reply will suffice!

Cabello stuff in next message. [See note to Adán Cabello, titled "Mermin and His Correlata," dated 16 August 1998 (Other Letters).]

13 December 1998, and to several others, "Little Historical Point"

[WARNING: The present note is factually inaccurate, as David points out immediately below.]
To my friends with a little interest in the history of quantum mechanics,

Let me share a delightful little piece of information that I ran across this morning. Though, perhaps I should say that I "became cognizant" of it this morning; for I know that I have read over this passage on at least two other occasions.

I think we are all accustomed to crediting David Bohm with being the first to speak of a finite-Hilbert space version of the EPR gedankenexperiment. (At least I know that I do, and both Asher and David do in their books.) But look at this little passage from the Born–Einstein letters. It comes from a letter from Born to Einstein, dated 9 May 1948:

Let me begin with an example. A beam of light falls on to a plate of doubly refracting crystal, and is split into two beams. The direction of polarization of one of the beams is determined by a measurement: it is then possible to deduce that that of the second beam is perpendicular to the first. In this way one has been able to make a statement about a system in a certain part of space as a result of a measurement carried out on a system in another part of space. That this is possible depends on the knowledge that both beams have originated from one beam which has passed through a crystal; in the language of optics, that they are coherent. It seems to me that this case is closely related to your abstract example, which is apparently connected with collision theory. But it is simpler and shows that such things happen within the framework of ordinary optics. All quantum mechanics has done is to generalise it.

Isn't that neat?

David's Reply

I remember that passage as yet another example of Born's having completely missed the point of EPR in all his arguments with Einstein.

I don't think it has anything to do with two-particle entangled states. Nor does it have anything to do with coherence. It seems to me all Born is saying is that if you have a birefringent crystal (in a fixed orientation) then if you determine the horizontal polarization of one of the output beams then you can be quite sure that the polarization of the other will be in the perpendicular direction.

I don't think that's significantly different from saying if you have two streams of water flowing into a bathtub and you determine that one of them is the cold water then you can be quite sure the other one is hot.

Quite aside from entanglement playing no role, it also has nothing to do with what's bothering Einstein, since the two beams already have their character (ordinary and extraordinary – or horizontally and vertically polarized) before any measurement is made on either of them.

Am I missing something?

13 December 1998, "Missing Something"

Merminition 33: *Am I missing something?*

I don't know: Maybe I was missing something myself. I took the license to think that Born was not actually talking about a beam of light, but a beam of light so attenuated as to be a single photon, say. (I don't think this is too much license.) But then a birefringent crystal wouldn't do the trick, would it? We'd only have the photon coming out one way or the other ... not two entangled photons. (At best we could say that the photon is entangled with the crystal.) You don't suppose I can take so much license as to assume that Born was talking about some nice downconverting crystal, could I? Guess not. Hmm.

Well, in any case, when he returns to the abstract discussion, it seems to me that he is indeed talking about entangled states, as Einstein had attempted to steer him. He just seemed to disagree that there was anything mysterious in them as the two systems had interacted in the past. Do you agree with this? Do you have the whole passage there with you? But then why did he – being a hell of a lot better physicist than me – say the silly things that I misinterpreted above?

Merminition 34: *Quite aside from entanglement playing no role, it also has nothing to do with what's bothering Einstein, since the two beams already have their character (ordinary and extraordinary – or horizontally and vertically polarized) before any measurement is made on either of them.*

This is sort of beside the historical issue, but ... If you give me the first amount of license above, then I'm not so sure I agree with this. Each photon makes a "random" decision (though obeying the probability law) about which way to go; it's hard to say that it's polarization (as determined by the axis of the crystal) existed beforehand.

Are we making progress? Do you agree with these things?

Sorry for the sloppy scholarship! (I got carried away with the excitement of the morning.)

13 December 1998, and to several others, "I Whig-ed Out"

You may remember the Whig historians: they "analysed the events and ideas of the past from the point of view of the present rather than trying to understand the people of the past on their own terms." In that sense, I committed an utter Whigism this morning, and I apologize. In my over-enthusiasm for finding a new historical nugget, I read more into Born's passage than I should have. In particular, when my eyes saw "doubly refracting crystal", my brain said "down converting crystal." As David Mermin pointed out to me right away, that passage is just yet more evidence that Born never quite got what Einstein was talking about with his gedankenexperiment: Born's example doesn't make any sense at all as a reply to Einstein (whose previous letter was indeed about entanglement and his reasons for not believing that quantum mechanics is complete). A poor birefringent crystal wouldn't give us any entanglement between two beams of light (some fraction of the photons go here, some fraction go there): Born didn't seem to notice the difference.

Sorry for the false alarm.

27 December 1998, "Executive Summary"

I just finished skimming Kurt Gottfried's paper, "Is the Statistical Interpretation of Quantum Mechanics Implied by the Correspondence Principle?" (quant-ph/9812042). I'm not sure I know what to make of it; I'm not sure it added anything to the compulsion I already feel for the "statistical interpretation" of the theory (where by that I mean "quantum states as states of knowledge" much in analogy to the role I see for classical Liouville distributions). But I did notice in the end that you were thanked for "asking several pointed questions." If you have a moment, could you give me an executive summary of what you think of the paper and maybe ask some of those pointed questions again.

I hope you're having a nice holiday, and not spending too much time working on your fences again (as I recall you were last year at this time). My in-laws are visiting us in California; we're having a wonderful time doing basically nothing but eating and drinking.

David's Reply

I can't offer you much to go on beyond what Kurt says in the paper itself. He's posing the question whether, if you just gave a classical physicist the Schrödinger equation, he could figure out the probability interpretation without any further help.

My primary "pointed question" of his first draft was simply to say, over and over again, whenever he claimed to have extracted a probability, "probability of WHAT?". The ultimate result of this prodding was that he retreated from a version in which a probability density in configuration space emerged naturally, to the one you see, in which only probabilities for internal degrees of freedom are, in some sense, natural.

He's driven by some long conversations he had with John Bell a couple of months before Bell dropped dead, where Bell claimed that in some deep way QM lacked a naturalness that all classical theories possessed – that there was a qualitative difference in the "interpretation problem" – that indeed, there either was no "interpretation problem" for classical physics or that it was trivial, while for QM something profound was missing. Kurt never felt this in his bones (the way you and I do) and this paper is (a) an attempt to articulate why he feels this way but also (b) an acknowledgment that winning his argument with Bell is not going to be as straightforward as he once thought.

I'm not sure he would agree with this "executive summary" but that's more or less how I see it.

We celebrated the first anniversary of the half-mile fence a few days ago. No deer have got onto the premises and the rhododendra have been liberated from their cages and remain uneaten.

Hope your digestion survives the holidays.

11 January 1999, "Chewin' the Fat"

Merminition 35: *My primary "pointed question" of his [Gottfried's] first draft was simply to say, over and over again, whenever he claimed to have extracted a probability, "probability of WHAT?".*

I never did reply to you on this. That's a good pointed question. I presume you've noticed that it's the same one you ask – in one form or another – to me (and other Copenhagenists like me) over and over. Probabilities of the ANSWERS to the QUESTIONS we can ask, I say. Who could want more from a physical theory? Yeah, yeah, I know you want more ... but I suspect you can't get it.

In this connection, by the way, I read a very nice article by Anton Zeilinger the other day, "On the Interpretation and Philosophical Foundation of Quantum Mechanics," that lays out what he thinks is the very most important feature of quantum mechanics. You can find the article at: http://info.uibk.ac.at/c/c7/c704/qo/philosop.html. It looks like his hopes and dreams aren't so different from mine.

11 January 1999, "Chewin' the Cud"

Merminition 36: *Thanks for the Zeilinger reference. I hadn't seen it. At a glance it looks very reasonable, but I will have to look further before pronouncing Anton a closet Ithacan.*

I'm having a hard time believing you said this!!!! Is my understanding of the Ithaca Attitude so far off the mark? Or, is this just a sign that Anton should be in politics rather than physics? Can he make us both believe we have an ally?

Seriously, I read that article on three separate occasions; the one I mentioned to you this afternoon was only the last of the lot. I was so taken by his speculation that quantum mechanics might be a hint of something even bigger – via the recognition that quantum observers are not "detached" (or at least their instruments cannot be) – that I had little daydreams of trying to secure a postdoc with him in case all else fails in my job search. (I hear he's going to Vienna this year; sounds like an awfully nice place to take a postdoc.)

If you find upon another reading that you still think Zeilinger "reasonable", please write me again. I want to know more details about what you like about that article.

17 January 1999, "My Homework"

I took your assignment over the weekend and read Schrödinger's *Nature and the Greeks and Science and Humanism* (all in one Canto volume) and also his little book *Mind and Matter*. You were right: it was quite enjoyable reading. I guess I'll say that even though in my diligent search for Ithacan ideas I found so little. I presume you were most happy about pages 54 and 92–98. This issue, by the way, he calls the "principle of objectivation" and devotes a chapter to it in the latter book above. Have you read it? I think it may be more directly relevant to your quest than *Nature and the Greeks*.

If I could say I liked one idea in N & G a lot, it would be the one expressed on pages 72 and 73: the real world is by definition that which can be distilled from our common experiences (that and nothing more). But, of course, my liking of that is self serving.

How's that reading of Kant's Prolegomena coming along? (Cf. my note dated 8 September 98, "I'll Bohr You to Tears.")

Laugh at you? Never! I submit this note as evidence.

21 January 1999, "Schrödingerization"

As you could tell from my last note, you helped spur my interest in Schrödinger. . . . Well, that's not completely true I guess. My interest in him was mostly spurred by reading his paper "The Present Situation in Quantum Mechanics" (1935) and the discussion of entanglement therein which I think is particularly deep (aside from his silly remarks about relativistic stuff at the end). But you did give me another point of contact for his thought. Let me now give you another point of contact.

The book is: William T. Scott, *Erwin Schrödinger: An Introduction to His Writings* (University of Massachusetts Press, Amherst, 1967). It's not beautiful and elegant in its composition like Schrödinger's writings, but it does give an introduction to lots of things Mr. S wrote and lots of good references. In particular, you might be interested to know that Michael Polanyi (in apparently several books) has also discussed the thing S calls the "principle of objectivation."

21 January 1999, "More Stupendous Literature Man"

Let me send you a little more in connection with your (probably minor) interest in Schrödinger. This is a letter I sent off to Caves and Schack for a completely different purpose. [See note to Rüdiger Schack, dated 21 January 1999, titled "Stupendous Literature Man."] But I started to wonder what you might think of Schrödinger's "dangerous" Wheeler-like tendencies: recall Wheeler's big U with an eye atop it to represent the universe. Have a look at the Schrödinger quote in the Schack letter.

No need to reply if you're not moved.

28 January 1999, "The Propaganda Cocktail"

On another subject: your prodding me into reading *Nature and the Greeks* sent me into a wild Schrödinger spiral. I am surprised I shunned him all these years; I now think he actually came up with quite a bit of thought provoking material. Have you studied all of his 1935 papers on entanglement carefully? That in conjunction with his apparent belief in indeterminism before quantum mechanics (1916–1922) makes for a really interesting mix.

(In that connection have you read Paul Hanle's article, "Indeterminacy Before Heisenberg: The Case of Franz Exner and Erwin Schrödinger"?)

31 January 1999, "Schrödinger Note"

Thanks for the encouragement and the advice.

Merminition 37: *I don't know if I've seen all the 1935 Schrödinger papers on entanglement, but the ones I did look at were very impressive.*

The one that I think is particularly elegant and insightful is this: J. D. Trimmer, *Proc. Am. Philos. Soc.* **124**, 323–338 (1980). For some reason it is listed under the translator's name rather than the original author. (It can also be found in the Wheeler–Zurek volume on quantum measurement.) The paper is advertised as the "Schrödinger cat" paper, but it's really about entanglement through and through.

21 May 1999, "A Bohr, a Door, and a . . . "

Let me give you the coordinates to that Honner article again: it's listed below along with a sampling from it in a letter I wrote Greg Comer almost two years ago. [See letter to Greg Comer dated 09 June 1997, "Dictionaries and Their Problems."] (How long have I known you now?) I read the article again yesterday, and didn't like it nearly as much as I had thought I had. But I am very happy I picked up that book again: it's a marvelous collection of papers trying to make sense of the old man. This time around, I was especially impressed with Henry Folse's article (of course because I think it reminds me of myself). And I'm slowly going to try to make my way through some of the other articles that defend Bohr as a peculiar form of realist (instead of as an idealist, positivist, or instrumentalist, etc.). I think that depiction of him is what captures best the yearning of my own heart.

08 June 1999, "I'm Not Alone"

I thought you might enjoy the little passage I scanned in today. Pay particular attention to the last paragraph where Pauli expresses what he means by complementarity. [Letter from Pauli to Bohr, dated 15 February 1955]

In quantum mechanics, on the contrary, an observation *hic et nunc* changes in general the "state" of the observed system in a way not contained in the mathematically formulated *laws*, which only apply to the automatic time dependence of the state of a *closed* system. I think here on the passage to a new phenomenon by observation which is technically taken into account by the so-called "reduction of the wave packets." As it is allowed to consider the instruments of observation as a kind of prolongation of the sense organs of the observer, I consider the unpredictable change of the state by a single observation – in spite of the objective character of the result of every observation and notwithstanding the statistical laws for the frequencies of repeated observation under equal conditions – to be *an abandonment of the idea of the isolation (detachment) of the observer from the course of physical events outside himself.*

To put it in non-technical common language one can compare the role of the observer in quantum theory with that of a person, who by his freely chosen experimental arrangements and recordings brings forth a considerable "trouble" in nature, without being able to influence its unpredictable outcome and results which afterwards can be objectively checked by everyone.

Probably you mean by "our position as detached observers" something entirely different than I do, as for me this new relation of the observer to the course of physical events is entirely *identical* with the fact that our situation as regards objective description in "this field of experience" gives rise to the demand of a renewed revision of the foundation for "the unambiguous use of our elementary concepts," logically expressed by the notion of complementarity.

13 June 1999, "Crackpots in Quantum Land"

You're speaking to one! I've just discovered something this morning that I have to make a confession to. Recall the little Pauli passage I sent you the other day? Well, I had already sent it to you! (Which means of course that I had already read it and scanned it into the computer at least once before!) What a crackpot! (I'm sure you'd already caught me . . . right? . . . you just didn't want to embarrass me . . . right?) Anyway, if you've got nothing better to do, have a look at the old note [dated 8 September 1998, a long one titled "I'll Bohr You To Tears"]. It builds up the context of the passage much better than the last rendition.

15 October 1999, "In Transit"

How did your quantum information course go? I was amused by your remarks on the spelling "qubit." Carl Caves and I once put the following footnote in one of our papers:

Qubit is a shorthand for the minimal quantum system, a two-state quantum system, that can carry a bit of information. Logically, if one wishes to give a special name to the minimal physical system that can carry a bit, one should do so for both classical and quantum two-state systems, calling them perhaps c-bits and q-bits. We are reluctant to use the neologism "qubit," because it has no standard English pronunciation, the "qu" being pronounced as in "cue ball," instead of as in "queasy." We prefer "q-bit," but acquiesce in the use of "qubit," which has attained a degree of general acceptance.

15 October 1999, "More In Transit"

Merminition 38: *Your samizdat package just [arrived] and I look forward to perusing it (and to anything else you have to tell me, especially if it is surprising).*

Thank you for teaching me the word samizdat! If you have any thoughts on my wacky train of thought, I would like to hear. The goal is, of course, some new equations and a lot less words. So the word samizdat fits perfectly. I'd especially look forward to hearing any of your thoughts on the later things I wrote Howard Barnum and John Preskill. I felt that

I was especially lucid in explaining the consistency of the state-vector-is-knowledge point of view there.

Merminition 39: *Hey, if you guys also think it should be q-bit or qbit or Qbit (I thought I was alone) it's not too late to fight back. (The precedent, by the way, was set by Qtips.)*

No, it's too late. We wrote that in 1996 and the quantum info world has certainly gotten a lot larger since then. Everyone uses qubit, including the press.

Merminition 40: *Will you be at Gilles's meeting in Montréal in December? (I'm trying to decide whether to go myself.)*

Please do come. I keep giving Gilles subliminal messages that we should organize a small "Quantum Foundations in the Light of Quantum Information" meeting. (He's interested.) You can help reinforce that. I will be there from (late) Dec 7 through the end (Dec 11 or 12, I can't remember). I'm showing up late because the week before that, I'll be at Naples at a "Chance in Physics" conference. In case you're interested in that, it may not be too late. And there should be some interesting characters there. Check out http://www.mathematik.uni-muenchen.de/~bohmmech/chance.htm. You won't by chance be at the APS meeting in Waterville, MA on Nov 5 and 6, will you? I'll be there too.

I guess I'll give a talk on our quantum de Finetti theorem at Gilles' meeting. It's about how the notion of an "unknown quantum state" (a phrase you see everywhere in quantum information) is probably only meaningful in complex Hilbert spaces, not reals. That might intrigue you. In part, it's sort of a more malicious version of the phenomenon Bill uncovered.

16 December 1999, "Peierls and Commutivity"

Asher and I finally came to some kind of agreement on our *Physics Today* piece ... what a relief!! Did I tell you the other day that I had scanned in that Peierls article that we were talking about? I scanned it in because Asher had wanted a copy for something to do with the *PT* article I just mentioned.

Anyway, I thought you might like a copy too. It's placed below. [See note to Asher Peres, titled "Miracles of Office Technology," dated 21 November 1999. (In it, for Asher, I had placed some commentary about how Peierls wasn't being consistent with his language, etc.)] I just reread the part about commuting density operators. I'm still not quite sure what he is saying, but there may be some connection with one of my better technical results. Have a look at: H. Barnum, C. M. Caves, C. A. Fuchs, R. Jozsa, and B. Schumacher, "Noncommuting Mixed States Cannot Be Broadcast," *Phys. Rev. Lett.* **76**(15), 2818–2821 (1996). (There's a slightly extended/smoother discussion in my PhD thesis, which you can find on quant-ph. Also Göran Lindblad rederived the result from some more standard C^* algebra results in *Lett. Math. Phys.* last year – but honestly, I couldn't understand his proof.)

Anyway the result is a generalization of the no-cloning theorem that explores when an automatic device can create a (correlated) copy of a density operator: it can if and only if the set of possible density operators are restricted to a commuting set. Perhaps Peierls had that kind of thing in mind. I.e., thinking about people's knowledge as being represented by density operators, and our ability to share information one with the other having to do with attempting to "broadcast" those density operators into other people's heads.

Tell me what you think about this interpretation.

17 December 1999, "Reality of the Symbol"

Merminition 41: *The Eco quote, for what it is worth, is from Umberto Eco,* Kant and the Platypus, *Chapter 1 ("On Being"), section I.I (Semiotics and the Something) which begins: [no scanner]*

Why should semiotics deal with this something? Because one of the problems of semiotics is to say whether and how we use signs to refer to something, and a great deal has been written on this. But I do not think that semiotics can avoid another problem: What is that something that induces us to produce signs?

It's the part after the colon that I like, though I've got to admit, seeing it above in black and white, I guess it could be grinding almost anybody's axe. But for me the sign is the quantum state. Haven't yet read Eco's answer to his question. He can get pretty difficult, though.

Thanks for sending the Eco quote. Now it's safely in my quantum archive. I did a search on the word "semiotics" in my samizdat, but couldn't find it anywhere: so you must have gotten a misimpression from something else I wrote. I do have a couple of interesting tidbits on the "reality of the symbol" that you might be interested in. One is the note titled "A Fleck of Fleck" [see 14 September 1999 note to Herb Bernstein]; the other is a note titled "Echoes" [see 20 August 1999 note to Bill Wootters].

I think in the next note, I'll send you a copy of the final version of the thingy I wrote with Asher. It's changed quite a bit since what you last saw. I would like to get your comment as to what you see is deficient in the point of view. I'm toying with the idea of writing a much extended version of it for the proceedings of this Bohmian meeting that I just returned from. The title would be "Knowledge About What?" and I think I would be much less timid about making statements about the "what" than we were in the *Physics Today* article: the great lesson of the quantum is that the world is so "sensitive to the touch" that we cannot (conceptually) distill a free-standing "reality" from our experiences. What a wonderful property for the world to have! Anyway, in that regard, I wonder if you wouldn't mind looking at one other thing I put together: it's a note titled "Penrose Tiles" [see 29 August 1999 note to Rüdiger Schack]. In it I express the kernel of the new language I want to push for: forget "measurements" and "outcomes," quantum mechanics is about "acts" and "consequences." Those ideas too would go into the Naples proceedings paper.

I want to continue thinking about the Peierls thing you brought to my attention. The more I keep thinking about it, the more it all resonates with some other stuff I've been thinking about for the last month. I'm searching for a notion of "macroscopic" within my "information gain vs. information disturbance" (don't forget the adjectives "information") foundation for quantum mechanics. Here's the direction. Suppose I describe some system by either ρ_0 or ρ_1 but you just don't know which. So you perform a measurement on the system to try to gain some of the same information I have, i.e., you try to guess which quantum state I'm using. If the two states are pure and noncommuting, then we know that there will be some inevitable disturbance to my description because of your intrusion – the minimal disturbance can even be quantified purely as a function of your information gain. But what happens if the two states are very, very mixed (almost the identity matrix)? I suspect your information gathering will cause little disturbance to my description: it's just a question of quantifying this now. If it works out as I suspect, then I think I know what "macroscopic" means. Take the tree outside my window. I know some things about it, but in the grand scale of things I know very, very little. If I were to assign a quantum state to it, it would be a very mixed quantum state. Suppose you've narrowed down my assignment (just by a pure guess, no measurement) to two possibilities, and now you intrude on the system to gather some further information about my precise choice. Your information gathering won't cause much disturbance to my description. And that is what means "macroscopic": when I know very little about a system, so little that your information gathering *need not* disturb my description, then the system is macroscopic *with respect to me*. That's the idea. If the system is also macroscopic with respect to you, then we can cross-check our information, and the system (with respect to both of us) will be no worse for the wear. I should work to try to quantify this.

26 January 2000, "Note #2 – Foundations in Montréal"

Merminition 42: *P.P.S. Have you seen a little book by Schrödinger called "My View of the World"? It contains two extended essays on the nature of objective reality, one in 1925 (before his Equation!) and one in 1960. They're surprisingly similar. They turn upside down the notion that objective reality is a valid inference from the fact of intersubjective agreement. He argues instead (I think) that intersubjective agreement is a manifestation of the unity of all consciousness, which is also what creates [the illusion of] objective reality. A kind of global mass-solipsism. But very beautifully put. Makes me think I should read Spinoza, which I never have done. So the IIQM is being tugged in one direction by QIP and quite another by goddamed mysticism. It's really delicious that Schrödinger is the hero of those who believe Bohr abandoned hard-headed rationality.*

Regarding the big S, I have indeed read that book. After you suggested reading *Nature and the Greeks*, I read everything by him that I could get my hands on. Now that you mention it though, with such an extreme form of idealism working in the background of his thoughts, I can't quite remember why he still had so many troubles with Copenhagenish kinds of views about quantum mechanics. Do you have any insight?

20 March 2000, "Invitation Coming Finally"

Gilles has asked me to ask you about one further possibility: Jeffrey Bub. Do you know him? Do you have an opinion? What's he like personally? I have never met him. I am a little against him in that his book gives him the appearance of having a more strongly set opinion on the foundations than I perceive from the rest of us. The point of this meeting is that I don't think the set opinions existing presently are very adequate. Gilles, on the other hand, is for him because he is Canadian (and that sort of thing may be important for the source of our funds), but also because he has had dinner with him once and enjoyed his company. And of course there is the issue that he does represent a school of thought in the foundations that is not directly reflected in any of the other potential participants. It is because of this that I could easily be swayed to plop him into the invitation list.[12]

Your candid opinion will be useful to the both of us.

By the way I enjoyed your *Physics Today* column this month. If the issue with Bub hadn't come up, I would have written you a note titled "Ginger or Mary Ann?" If you've ever watched *Gilligan's Island*, then you'll know what I'm talking about. I was always a fan of Mary Ann; my best friend was always a fan of Ginger. I just can't believe that this issue of beauty and elegance in science goes any deeper than that. And I get really annoyed by the opinion-Nazis who tell me otherwise. Strangely Richard Jozsa and I had a heated debate on just this issue last week in Tokyo.

23 March 2000, "Can You Complete This?"

Merminition 43: *Rereading your article with Asher I was struck by your (not entirely convincing) assertion that it's OK for Cat's density matrix to be either fruit or cake while Erwin's still is a superposition.*

That's because you're not entirely convinced that the wavefunction is an epistemological beast. You don't yet really believe that it has no ontological content. Too bad.

If it helps, I believe I explained this "striking assertion" a little more clearly in some notes to Howard Barnum. Read "It's All about Schmoz" and "New Schmoz Cola" in this compilation. If not more clearly, at least I can claim to have said things more vividly. The Fierz article that is referred to can be found in the note to Rüdiger Schack, titled "Penrose Tiles," dated 29 August 1999.

I'm sorry I myself don't have time to think about your question (or even read it in detail actually) until late next week. I'm in a bit of a panic getting ready for my lectures in Scotland. But I will get back to you soon thereafter.

---

[12] I simply could not resist including this very personal paragraph from the note: It more than anything else in the book demonstrates the dangers of the close-mindedness I am predisposed to (and that I battle every day). See Chapter 9 "Letters to Jeffrey Bub" to understand this remark!

03 April 2000, "Gleason, For the Record"

I'm flying back to ABQ from Boston in some of the roughest wind I've ever been in! Yeaks.

Anyway, I just wanted to write this memory down for my record, and I thought you might enjoy a little bit of it. It turns out that Andrew Gleason was at the AMS meeting that I just attended. I told Mary Beth Ruskai how I would like to catch a glimpse of him, but that I was too shy to meet him ... not having anything to say. Well, she introduced me nevertheless. At first all I could think to say was, "I don't really have anything to say, other than that I really love your theorem. It was a real pleasure to work through its proof." He said, "Which theorem?" So I said the 1957 one on quantum mechanical probability measures. Then somehow my automatic mouth kicked in and I started talking. I told him about how David Meyer had shown that it doesn't hold for vector spaces over rational number fields. And then I asked him his opinion about whether it might not still hold when the field is only weakened from the full set of reals down to the algebraic numbers. Somehow this led to a discussion of Kochen–Specker and I told him about the minimal known examples in R^3. I explained the idea of the whole thing, and he said(!), "Is that what that is about? That you just can't make that kind of coloring? I never realized that that's all they were talking about. Can you give me a reference where I can find some of these minimal examples?" So I gave him a reference to Peres' book, and we parted.

He struck me as such a gracious sort of person. It really made my day. Rarely am I in awe, like some people in the presence of a movie star, but this time I really felt it.

I hope you're working on your problem set for FILQI.

05 April 2000, "Commutivity"

I finally had a chance to sit down and reread the Peierls passage you recommended and also your accompanying note. When I had last looked at his piece in *Physics World*, I too thought he was up to something interesting. But now, after looking at his *More Surprises* passage and thinking a little more carefully, I understand why you're having such difficulty sussing it all out: Peierls is just trivially wrong here. People ascribe noncommuting density operators to the same system all the time – it's called quantum cryptography. Just consider the B92 protocol. Alice ascribes one of two possible nonorthogonal (pure) states to the system she just prepared. Eve, if she is aware of Alice and Bob's protocol, can do no better than ascribe a density operator that is the simple mixture of the two possibilities. This mixed-state density operator will commute with neither of Alice's possible preparations. Moreover, this will generally remain true even after Eve has done some eavesdropping ... and even after Alice has taken into account that Eve has done some eavesdropping.

I general I think Peierls just has the whole idea backward (from a Paulian point of view, that is). In his *Physics World* article, he writes: "... there are limitations to the extent to which their knowledge may differ. This is imposed by the uncertainty principle."

It seems to me, people can believe whatever they want – I like the word "belief" better than "knowledge." There are no limitations to that. The unique thing about the quantum world is that it never lets people's opinions get so aligned that they can jump to the conclusion of a free-standing reality. This is captured by the fact that the best that can be said of almost all measurements can only be couched in statistical language: there are no certainties for almost all observables even when one has maximal knowledge (i.e., a pure state).

I'll put some further thoughts on this below, in the form of a draft that I'm presently writing on information and disturbance. If you've got the stamina to get all the way to the paragraph starting with "It is at this point ..." then you might see what I'm talking about when I say that Peierls has it backward. (By the way, be merciful on my English: this is only a draft, and I'm not even very far into it yet.) [See C. A. Fuchs and K. Jacobs, "Information Tradeoff Relations for Finite-Strength Quantum Measurements," ArXiv:quant-ph/0009101.]

David's Reply

P.S. Peierls is sometimes wrong, but never trivially wrong. (He is my Bill Wootters, you know.) I'll think more about your remarks when peace returns.

05 April 2000, "Pauli Understood Entanglement"

Yes, by FILQI I did mean the Montréal meeting.
Also I forgot to reply a little more to this:

Merminition 44: *Rereading your article with Asher I was struck by your (not entirely convincing) assertion that it's OK for Cat's density matrix to be either fruit or cake while Erwin's still is a superposition.*

Authority certainly won't help with you, but I was struck the other night in reading *Niels Bohr Vol. 6* – I've never been able to get hold of that volume before – by another letter from Pauli to Bohr. This one was dated 17 October 1927! Among other things, it says the following:

"Further down on page 5 where the 'proper reduction of the spatial extension of the fields' is mentioned, one could perhaps still add something for the sake of clarity. This is of course just a point which was not quite satisfactory in the Heisenberg paper; there the 'reduction of the wavepacket' seemed a bit mysterious. Now it should of course be emphasized that such reductions first of all are not necessary when all the measuring instruments are *included* in the system. In order to be able to describe the results of observation theoretically at all, one must ask what can be said about one *part* of the total system on its own. And then one sees as a matter of course the complete solution – that the omission of the instruments of observation in many cases (not always, of course) may formally be replaced by such discontinuous reductions."

Isn't that cute?

06 April 2000, "Mist of the Future"

Merminition 45: *I would like nothing better than a convincing demonstration that it's all about knowledge and only knowledge, but the ways in which the knowledge of different people can and cannot be interrelated ... It seems to me that you, being it's-only-about-knowledge-ish, have to say a little more about this. There remains work to be done.*

The first sentence in this quote shocks me actually. How can I forget desideratum #1 in your original Ithaca paper? But of the second two sentences, you should know that you have my acute sympathy. I think my main role in life lately – second only to being a father – has been to drum up support and turn our community to just this issue. There remains not only work to be done, but A LOT of work to be done. And it's a gut feeling admittedly, but the fruits will be great.

20 July 2000, "Zing!"

I promised you a report on Plotnitsky, and now maybe I finally have some time to do it. As you predicted, I did find him extremely enjoyable to talk to. He does know his Bohr, and is certainly my philosophical cousin. I spent a lot of time in Mykonos with him, on the beach and at dinners, and here and there otherwise. I challenged him to write an essay comparing and contrasting Pauli and Bohr's thoughts on QM: he took the challenge! So I'm looking forward to seeing that, or at least talking about that with him.

Thanks for sending me your Firing Line article. It didn't irritate me at all; I enjoyed it. However, it does reveal that you are far more Platonic than I. (Or is it, me? I must admit that this is one rule of grammar I never remember.) Bill Wootters, Richard Jozsa, Peter Shor and I went to dinner one evening in Capri and the conversation ended up degenerating into one about the foundations of quantum mechanics and the foundations of mathematics. It seems that Peter wouldn't admit the real existence of the number continuum. Richard protested (in his usual way) that that was just silly, and started to try to build up a hierarchy of ideas. "Without saying, the natural numbers exist." Peter and Bill both agreed quickly; I didn't. Richard was incensed! "Well, then are you going to tell me that the number three doesn't exist!?!? What about the three glasses in front of you?" I just replied, "There's no sense in which those objects are the same without our talking about them. If I weren't here to make the judgment that they're *roughly* the same, what principle of nature would do it for me?" He found me boring (in his usual way) and left the subject.

Let me give you yet another suggestion for reading (that'll you'll probably ignore 2/3 of ... you told me you only read about 1/3 of what I suggest). I really enjoyed this one on my flight home the other day.

H. J. Folse, "Niels Bohr's Concept of Reality," in *Symposium on the Foundations of Modern Physics 1987: The Copenhagen Interpretation 60 Years after the Como Lecture*, edited by P. Lahti and P. Mittelstaedt (World Scientific, Singapore, 1987), pp. 161–179.

As always, I almost surely liked it because it was saying something I wanted to hear.

Somehow I feel that I had an epiphany in Mykonos. Do you remember the parable of "Genesis and the Quantum" from my Montréal problem set? [See note to Gilles Brassard, titled "Problem Set," dated 15 May 2000.] And do you remember my slide of an empty black box with two overlays. The first overlay was of a big $|\psi\rangle$ (hand drawn in blue ink of course). I put the slide of the box up first, and said "This is a quantum system; it's what's there in the world independent of us." Then I put the first overlay on it and say, "And this symbol stands for nothing more than what we know of it. Take us away and the symbol goes away too." I then remove the $|\psi\rangle$. "But that doesn't mean that the system, this black box, goes away." Finally I put back up the $|\psi\rangle$ over the box, and the final overlay. This one says: "Information/knowledge about what? The consequences of our experimental interventions into the course of Nature."

Well, now I've made another overlay for my black box slide. At the top it asks, "So what is real about a quantum system?" In the center, so that it ends up actually in the box, is a very stylistic version of the word "Zing!" And at the bottom it answers, "The locus of all information–disturbance tradeoff curves for the system." In words, I (plan to) say, "It is that zing of the system, that sensitivity to the touch, that keeps us from ever saying more than $|\psi\rangle$ of it. This is the thing that is real about the system. It is our task to give better expression to that idea, and start to appreciate the doors it opens for us to shape and manipulate the world." What is it that makes quantum cryptography go? Very explicitly, the zing in the system. What is it that makes quantum computing go? The zing in its components!

Anyway, I'm quite taken by this idea that's getting so close to being a technical one – i.e., well formed enough that one might check whether there is something to it. What is real of the system is the locus of information–disturbance (perhaps it would be better to say "information–information") tradeoff curves. The thing to do now is to show that Hilbert space comes about as a compact description of that collection, and that it's not the other way around. As I've preached to you for over two years now, this idea (though it was in less refined form before now) strikes me as a purely ontological one ... even though it takes inserting an Alice, Bob, and Eve into the picture to give it adequate expression. That is, it takes a little epistemology before we can get to an ontological statement.

I looked back at your original Ithaca Interpretation paper, and I'll be bold enough to say that this idea satisfies all your most important desiderata: (1), (2 suitably modified), (3), and (5).

Part of this, by the way, is why I liked so much Folse's paper. Also, believe it or not, for a moment while reading it I thought I could finally "SEE" correlation without correlata. (Not lying.) But then I thought I liked the phrase "Interaction without Interactoids" even more. My wife just thought I was being silly. Maybe you will too.

PS. I did take the job at Bell Labs. I wonder if they really know what they've gotten themselves into?

20 July 2000, "Schrödinger from a Different Angle"

Me again. Let me see if I can troll your good memory for a reference I'm looking for. It's something Schrödinger said, but where? (I would be able to find it if my library hadn't burned up.)

This is a passage from a paper by Ed Jaynes. He writes:

[The Copenhagen interpretation] denies the existence of an "objectively real" world.

But surely, the existence of that world is the primary experimental fact of all, without which there would be no point to physics or any other science; and for which we all receive new evidence every waking minute of our lives. This direct evidence of our senses is vastly more cogent than are any of the deviously indirect experiments that are cited as evidence for the Copenhagen interpretation.

Perhaps our concern should be not with hidden variables, but hidden assumptions; not only about the theory, but about what we are measuring in those experiments. Consider a cascade decay experiment. As soon as we say something like "In this experiment we observe two photons emitted from the same atom," we have already assumed the correctness of a great deal of the theory that the experiment was supposed to test. This initial stacking of the cards then affects how we analyze the data. ...

Do we need more hidden variables? Perhaps eventually, but maybe our immediate problem is the opposite; first we need to get rid of some. To define the state of a classical particle we must specify three coordinates and three velocities. Quantum theory, while denying that even these degrees of freedom are meaningful and claiming to "remove unobservables," replaces them with an infinite number of degrees of freedom defining a continuous wave field. To specify a classical wave field, we need one complex amplitude for each mode; for a quantized field we need an infinity of complex amplitudes for each mode.

So perhaps quantum theory, far from removing unobservables, has introduced infinitely more mathematical degrees of freedom than are actually needed to represent physical phenomena. If so, it would not be surprising if a few infinities leak out into our calculations.

The part in particular that I'm interested in is the last two paragraphs. I remember Schrödinger saying something so similar somewhere. Where was it?

Thanks for any help you can give!

23 July 2000, "Great Garrett Quote"

AUTHOR'S NOTE: In my *Physics Today* article "Quantum Theory Needs No 'Interpretation'" with Asher Peres, there is a passage that says the following:

Contrary to those desires, quantum theory does *not* describe physical reality. What it does is provide an algorithm for computing *probabilities* for the macroscopic events ("detector clicks") that are the consequences of our experimental interventions. ...

This passage (usually taken out of context) has been found distasteful by several readers. At least one of those readers was Alvaro Carvalho, and at least one reader of Carvalho was the iconic many-worlds activist David Deutsch.[13] I know this because Michael Nielsen forwarded to me Carvalho's posting of 12 July 2000 and Deutsch's reply of 16 July 2000 from the electronic bulletin board Fabric-of-Reality@egroups.com. Carvalho writes:

> For those who have not read the Letter: "Quantum theory needs no 'interpretation'" by C. Fuchs and Asher Peres (*Physics Today* –March 2000), here are some short excerpts (with comments...): [...]
> "Contrary to those desires, quantum theory does not describe reality ..." I wonder what it can possibly describe. Is there anything else beyond reality?

While Deutsch responds:

> No, but that's not what they think. They think it describes our observations, but that we are not entitled to regard this as telling us anything about a reality beyond our observations. Why? Just for the Bohring old reason that they don't like the look of the reality that it would describe, if it did describe reality. Why? – I have many speculations, but basically I don't know. I don't understand why.
> It's sad enough when cranks churn out this tawdry old excuse for refusing to contemplate the implications of science, but when highly competent physicists – quantum physicists – dust it off and proudly repeat it, it's a crying shame.

In bemusement, I forwarded these quotes to David Mermin, who responded with:

Merminition 46: *Funny, there was this English Bayesian, Tony Garrett, who said more or less the same thing about anybody who had given up the search for hidden variables. And I suppose there's a sense in which many worlds, insofar as it can be made coherent, is the ultimate hidden variables theory.*

This comment spurred the following note on my part.

Indeed I know about Garrett. I have a great quote from him in the passage below. In fact, I wrote this passage to combat people just like him. I thought you might enjoy it as whole (along with the quote, that is).

When am I ever going to publish all this crap I've written?

———————

The first opposition we must tackle is not the physicist but the average Bayesian himself. Bayesians generally tend to be "naïve" philosophical realists. Their frame of mind goes something like this: "There is a real world out there. We use probability to quantify our ignorance of it. When we have maximal information about it, we have no ignorance. Consequently, all probability assignments we might make in that situation should be of the trivial sort, simple Kronecker deltas – 0 if false, 1 if true. In contradistinction, when there is ignorance, we should seek to alleviate it. By definition, when we have ignorance our information is nonmaximal, and therefore stands to be completed."[14]

---

[13] It probably goes without saying, but perhaps I should have said "many-worlds activist and Star Trek fan." See http://www. qubit.org/people/david/David.html for details.
[14] In this connection, it is probably worth noting a widespread misconception of non-Bayesians. It is sometimes said that Bayesians have dangerous tendencies toward philosophical idealism. Nothing could be farther from the truth: this passage ▷

Because of this, most Bayesians have trouble coming to grips with the fundamental indeterminism posed by quantum mechanics.[15] It makes no sense within *their* framework. Perhaps this point of view has been expressed most clearly by A. J. M. Garrett [4]:

> The nondeterministic character of quantum measurement can, and should, be taken to imply a deeper 'hidden variable' description of a system, which reproduces quantum theory when the unknown values of the variables are marginalised over. Differences in measurements on identically prepared systems then represent differences in the hidden variables of the systems. Not to seek the hidden variables, as the Copenhagen interpretation of quantum mechanics arbitrarily instructs, is to give up all hope of improvement in advance, and is contrary to the purposes of science.

But then again, most Bayesians are relatively far removed from the countless attempts that have been made to find a *reasonable* hidden-variable model underneath quantum mechanics [5,6]. The indifference of most physicists to yet further attempts is far from arbitrary. The strong constraints placed by the Bell and Kochen–Specker theorems are quite compelling in this respect [7,Chaps. 5–7]. What practicing physicist would really want to bother with a hidden variable theory of the Bohmian type, which is nonlocal to its core [8]? Its defining feature appears to be no more than this: more difficult mathematics than the standard theory, while at the same time making no further predictions that can actually be tested. Hardly a bargain. All this is done for the sole sake of an easily pointed to "reality" underneath the statistical predictions – a "reality" so far removed from common experience that it is practically unrecognizable [9,10].

This turn is not for us. It need not be. The open-eyed Bayesian has no use for hidden variables in the quantum context. Our strength actually lies in adopting the selfsame language that Einstein has already introduced. The quantum mechanical description – the quantum state – is in fact not a *complete* description of the quantum system. *However,* we temper this by adding that, at the same time, we staunchly reject Einstein's assessment of the situation [11]:

> There exists [...] a simple psychological reason for the fact that this most nearly obvious interpretation is being shunned. For if the statistical quantum theory does not pretend to describe the individual system (and its development in time) completely, it appears unavoidable to look elsewhere for a complete description of the individual system; in doing so it would be clear from the very beginning that the elements of such a description are not contained within the conceptual scheme of the statistical quantum theory. With this one would admit that, in principle, this scheme could not serve as the basis of theoretical physics.

The theory prescribes that no matter how much we know about a quantum system – even when we have *maximal* information about it – there will always be a statistical residue. There will always be questions that we can ask of a system for which we cannot predict the outcomes. *In quantum theory, maximal information is not complete and cannot be completed.*

---

tries to show that. The confusion seems to come about precisely because Bayesians support a *subjective* notion of probability rather than an objective one. Somehow, this is seen instead as an endorsement of a "subjective reality." The Bayesian is simply careful to point out that there is the world and then there is what we know about it. The two categories should not be confused. In particular, the Bayesian insists that probability should be recognized as inhabiting the latter category only; this says nothing negative about the existence of the former.

[15] Hints of this are peppered throughout many of the papers in Ref. [1]. See also, in particular, Refs. [45,46].

To think that this is an arbitrary rejection of the principles of science is to work within a prejudice imposed by classical physics. This statement is actually a physical postulate, perhaps not so different in character than that the speed of light is constant in all reference frames. It breaks an old tradition, but that need not be a blemish. The clarity the "c = const" assumption brought to the already-existent Lorentz transformations cannot be questioned [12].

1. R. D. Levine and M. Tribus, eds., *The Maximum Entropy Formalism*, (MIT Press, Cambridge, MA, 1979); C. R. Smith and W. T. Grandy, Jr., eds., *Maximum-Entropy and Bayesian Methods in Inverse Problems*, (D. Reidel, Dordrecht, 1985); J. H. Justice, ed., *Maximum Entropy and Bayesian Methods in Applied Statistics*, (Cambridge University Press, Cambridge, 1986); J. Skilling, ed., *Maximum Entropy and Bayesian Methods: Cambridge, England, 1988*, (Kluwer, Dordrecht, 1989); P. F. Fougère, ed., *Maximum Entropy and Bayesian Methods: Dartmouth, U.S.A., 1989*, (Kluwer, Dordrecht, 1990); B. Buck and V. A. Macaulay, eds., *Maximum Entropy in Action: A Collection of Expository Essays*, (Clarendon Press, Oxford, 1991); W. T. Grandy, Jr. and L. H. Schick, eds., *Maximum Entropy and Bayesian Methods: Laramie, Wyoming, 1990*, (Kluwer, Dordrecht, 1991); C. R. Smith, G. J. Erickson, and P. O. Neudorfer, *Maximum Entropy and Bayesian Methods: Seattle, 1991*, (Kluwer, Dordrecht, 1992); A. Mohammad-Djafari and G. Demoment, eds., *Maximum Entropy and Bayesian Methods: Paris, France, 1992*, (Kluwer, Dordrecht, 1993); G. R. Heidbreder, *Maximum Entropy and Bayesian Methods: Santa Barbara, California, U.S.A., 1993*, (Kluwer, Dordrecht, 1996); J. Skilling and S. Sibisi, *Maximum Entropy and Bayesian Methods: Cambridge, England, 1994*, (Kluwer, Dordrecht, 1996); K. M. Hanson and R. N. Silver, *Maximum Entropy and Bayesian Methods: Santa Fe, New Mexico, U.S.A., 1995*, (Kluwer, Dordrecht, 1996).
2. E. T. Jaynes, "Clearing Up Mysteries – The Original Goal," in *Maximum Entropy and Bayesian Methods (Cambridge, England, 1988)*, edited by J. Skilling (Kluwer, Dordrecht, 1989), pp. 1–27.
3. E. T. Jaynes, "Probability in Quantum Theory," in *Complexity, Entropy and the Physics of Information*, edited by W. H. Zurek (Addison-Wesley, Redwood City, CA, 1990), pp. 381–403.
4. A. J. M. Garrett, "Making Sense of Quantum Mechanics: Why You Should Believe in Hidden Variables," in *Maximum Entropy and Bayesian Methods (Paris, France, 1992)*, edited by A. Mohammad-Djafari and G. Demoment (Kluwer, Dordrecht, 1993), pp. 79–83.
5. F. J. Belinfante, *A Survey of Hidden-Variables Theories*, (Pergamon Press, Oxford, 1973).
6. M. Jammer, *The Philosophy of Quantum Mechanics: The Interpretations of Quantum Mechanics in Historical Perspective*, (John Wiley, New York, 1974).
7. A. Peres, *Quantum Theory: Concepts and Methods*, (Kluwer, Dordrecht, 1993).
8. D. Bohm and B. J. Hiley, *The Undivided Universe: An Ontological Interpretation of Quantum Theory*, (Routledge, London, 1993); J. T. Cushing, *Quantum Mechanics: Historical Contingency and the Copenhagen Hegemony*, (U. of Chicago Press, Chicago, 1994).
9. J. T. Cushing, A. Fine, and S. Goldstein, eds., *Bohmian Mechanics and Quantum Theory: An Appraisal*, (Kluwer, Dordrecht, 1996).
10. M. O. Scully, "Do Bohm Trajectories Always Provide a Trustworthy Physical Picture of Particle Motion?," *Phys. Scripta* **T76**, 41–46 (1998).

11. A. Einstein, "Remarks Concerning the Essays Brought Together in this Co-Operative Volume," in *Albert Einstein: Philosopher-Scientist*, third edition, edited by P. A. Schilpp (Open Court, La Salle, IL, 1970), pp. 665–688.
12. C. Rovelli, "Relational Quantum Mechanics," *Int. J. Theor. Phys.* **35**, 1637–1678 (1996).

05 October 2000, "Bad Gleason Mornings"

Merminition 47:

As far as Q foundations are concerned, some pretty interesting things are starting to happen again. (Though you know I've got this pathetic backlog of things to write up – one of these days somebody needs to write that stuff up, de Finetti, Dutch book, etc.) Thinking about a POVM version of Gleason's theorem has been immensely useful. (Remember it was one of the problems in my Montréal problem set?) In particular, with POVMs taken as the base for a notion of frame function, one can even make Gleason work for two dimensions! We're writing this one up immediately, making PRL the target.[16]

I'm eager to see whatever you've come up with. I seem to remember something about Gleason's theorem and POVMs by somebody named Busch some time within the last year. Am I making that up?

Boy, you know how to spoil a morning. If anybody, it's Paul Busch. I'm gonna have to do a search on it when I get into work. The result is simply this: if we think of the basic elements in quantum measurement theory to be the POVMs (i.e., resolutions of the identity into positive operators), then there is essentially only one way to assign probabilities to the measurement outcomes in a noncontextual way. That is to say, if I suppose a function f from positive operators to the real interval [0, 1] such that $\sum_i f(E_i) = 1$ whenever the E_i form a POVM, then the theorem says there must exist a density operator ρ such that $f(E_i) = tr\rho E_i$. What is cute about this is that it works even in 2-D, whereas Gleason's standard theorem doesn't.

Oh, I hope Paul didn't do it already!

12 October 2000, "Turning Heads ... At Least One"

Thanks for bringing my attention to Paul Busch's paper. Yep, it pretty much looks like he did it. We have some extra stuff, but he's taken most of the wind from our sails. It seems like I'm always a little late.

But look at this below. It's a note from Jeffrey Bub, and it thrills me to no end. [See "Jeff's Reply" to my note "Commit the Bit!," dated 30 September 2000.] I think I take particular indulgence in this because the guy has pretty much spent his life on the opposite side of the fence from me – that is, trying to find a free-standing reality within quantum

---

[16] This describes work of Joseph Renes, Kiran Manne, Carlton Caves, and myself.

mechanics. He even went so far as to write a book on the subject. So to turn his head some strikes me as an immense triumph for the power and reasonableness of the Paulian idea. I used to tell Carl Caves that if we could just convert David Mermin and John Preskill (two very rational sceptics) to our way of thinking, that would be a great victory. But maybe this is even better.

03 November 2000, "Do You Know?"

Do you know who first used the terms "separable," "nonseparable," and "inseparable" to describe quantum systems? I have a faint memory that it might have been d'Espagnat, but I'm not sure why I have that faint memory.[17]

23 April 2001, "New Posting"

I just made the next posting.

Since my last writing to you, the thing has mainly changed by my beefing up the Bennett chapter. In particular, there are two notes added there mostly for the purpose of saying two sentences (embedded within them). [See notes to Charlie Bennett titled "Emma Jane Fuchs" and "Emma Jane," dated 13 January 1999 and 5 June 1999 respectively.] Recontemplating them while taking a shower, I was led to recontemplate the main (and enduring) confusion about my point of view on quantum mechanics. There seems to be a great tendency to think that when one says "the wave function is about knowledge, only knowledge" one is placing the observer outside of the physical world. But it is exactly the opposite of that! And I just can't see why no one can see it. We are forced to the quantum *description* precisely because there is nothing special about us (beyond the specialness of the world itself). In stark contrast, it is the ideal of classical physics that makes the observer irrevocably special.

Now, I must work on "The Quantum State of a Laser Field" for a while to convey the illusion that I am a physicist. I'll get back on the samizdat this afternoon.

David's First Reply

Congratulations, you're almost to Schrödinger! All you have to do is change the last sentence to

> We are forced to the quantum *description* precisely because there is nothing special about the world (beyond the specialness of us).

David's Second Reply

Funny you adding the note you sent me last week about the exact opposite of placing the observer outside of the world.

---

[17] Anyone who knows the answer, please do write me.

You never replied to my comment on it: that you were getting close to Schrödingerian mysticism. I suspect the only difference (if it is a difference) is that you say the observer is in the world while Schrödinger – this is a cartoon version of what he says but is hard to resist in the current context – says that the world is in the observer. Both of you say (with Pauli) that they cannot be separated.

But setting my growing taste for mysticism aside, do you really want to add a letter that comments on the preparation of the text itself?

26 April 2001, "Fire of the Soul"

Merminition 48: *You never replied to my comment on it: that you were getting close to Schrödingerian mysticism. I suspect the only difference (if it is a difference) is that you say the observer is in the world while Schrödinger – this is a cartoon version of what he says but is hard to resist in the current context – says that the world is in the observer. Both of you say (with Pauli) that they cannot be separated.*

God, I loved that! Can I put it in the samizdat??!?

19

Letters to David Meyer

23 February 1999, "From Gleason to Pitowsky"

You'll note that I've forwarded you three emails from my correspondence with Itamar. [See note to Pitowsky dated 13 October 1998 (Other Letters).] I'd be keen to hear any further thoughts you might have on Gleason or Kochen and Specker. My email is always an open door when it comes to better understanding the quantum ...

11 May 1999, "SD and KS" and "Even More Relevant"

How about we plan on my dropping by sometime during the week of May 24–28. You guys choose the exact dates (either one or two nights, as you said). I'll probably bring my wife and little girl down with me. They'll drop me off to play with you, and go off to play a little on their own.

I have a question about your rational-point anti-KS result. I'm starting to wonder how robust the idea is. The way you pose the problem is to consider the points on the sphere with rational cartesian coordinates. But are you sure that nothing changes under rather natural coordinate changes? For instance, imagine working in spherical coordinates for vectors on the unit sphere in R^3. If we restrict ourselves to rational angles θ and ϕ, the cartesian coordinates so generated will almost always be irrational. And this discrete set of vectors is not just a rigid rotation of the ones you consider. Can we find subsets of vectors within this set that cannot be consistently colored? You write, "Surely the meaning of quantum mechanics should not rest upon such non-experimental entities. But, at least in the three dimensional arena for the Kochen–Specker theorem it does ... " But the notion of a rational point for the experimentalist has to be specified with respect to something natural within the experimental procedure: who is to say it is not the rational angles that are accessible to the experimentalist instead of the rational cartesian coordinates on the sphere?

Am I overlooking something simple? (Likely.) Anyway, regardless of the philosophizing, it would be interesting to know if there is a fundamental distinction between these two distinct sets of "rational points" with respect to colorability.

Maybe an even more relevant question might have been not to consider the rational angles, but instead those that are given by rational fractions of π. What happens then?

This indeed seems much more to me like a reasonable constraint to put on the gedanken-experimentalist.

13 November 1999, "What Did vN Not Know?"

I quote from page 836 of the original Birkhoff and von Neumann paper:

One conclusion which can be drawn from the preceding algebraic considerations, is that one can construct many different models for a propositional calculus in quantum mechanics, which cannot be differentiated by known criteria. More precisely, one can take any field F having an involutory anti-isomorphism satisfying $Q4$ (such fields include the real, complex, and quaternion number systems[1]), introduce suitable notions of linear dependence and complementarity, and then construct for every dimension-number n a model $P_n(F)$, having all the properties of the propositional calculus suggested by quantum mechanics.

Stephen Adler in the pages of his book that I copied long ago, writes "...the subject was further explored in an important article by Finkelstein, Jauch, and Speiser (1959)." Unfortunately I don't have a more complete reference than that. I'll talk to Adler about all this when I meet him in Naples a couple of weeks from now.

13 November 1999, "A Couple More Things"

Jauch [*Synthese* **29** (1974) 131–154] writes, "It is known that every proposition system $\mathcal{L}$ admits a representation in a linear vector space with coefficients from the real, complex, or quaternion fields. ... Under some mild additional restrictions one can show that the coefficients of the Hilbert space are the complex number field C." He cites C. Piron, "Axiomatique Quantique," *Helv. Phys. Acta* **37** (1964) 439–467, for that. But he also writes in a footnote, "The question of the number field remained for a long time beyond empirical test. The recent work by Gudder and Piron ['Observables and the Field in Quantum Mechanics,' *J. Math. Phys.* **12** (1971) 1583–1588] is the best one can do."

---

[1] ... Conversely, A. Kolmogorov ... has shown that any projective geometry whose k-dimensional elements have a *locally compact topology* relative to which the lattice operations are continuous, must be over the real, the complex, or the quaternionic field.

20

Letters to Jeff Nicholson

06 June 1996, "Cyclic Never-Endings"

Yes, yes, I like Montréal! Does that finally answer your question? The only thing I really miss about the states is having a swarm of good physicists around me.

El Jeffy 1: *You know, I really miss listen to you babble about quantum mechanics. Nobody around any more to tell me that the moon isn't there when I'm not looking; you even had me believing for a while.*

Damn, if I had only stayed there a bit longer ... perhaps you would have remained a true believer!

22 August 1996, "Laws of Thought"

[Referring to the phrase "Quantum Mechanics is a Law of Thought.":]

El Jeffy 2: *I was just reading about the mathematician George Boole, the guy who invented Boolean Algebra. In 1854 he published a book called* An Investigation of the Laws of Thought, on which are founded the mathematical theories of logic and probabilities. *Is that where your lovely little phrase comes from?*

Yes indeed, that's exactly where it comes from. Quantum Mechanics is a Law of Thought. God I like the ring of that.

22 August 1996, "Plagiarists"

El Jeffy 3: *I am disappointed in you. I never realized that you plagiarized that from poor ol' Boole. Can't think up your own catch phrase, eh?*
I think the phrase bothers me far less when applied to something like Boolean algebra. It still gives me the heebies when you refer to real live physical systems with such terminology.

Hey, what right do you have calling me a plagiarist?!?! I can hardly believe that Boole knew anything of quantum mechanics in 1854! Please reevaluate!

And, sir, by the way, I am not referring to "real live physical systems with such terminology." I am referring to the mathematical structure of quantum mechanics. Some of the best examples of real live physical systems I've ever seen have been in your lab ... and – I promise you – try as I might each time I would visit you there, I never could see a Hilbert space sitting on your optics table. I looked and looked; never could find the damned thing. Never saw a state vector either. Just silly little arrangements of mirrors and cavities.

09 October 1997, "Pickin', Scratchin', and Grinnin'"

Thanks for the long note. I promised you a long one back, didn't I? Oops, I lied. But maybe I'll say a couple of things while I have my first quarter cup of coffee. Things are fine out here in Lava Lamp Land. Kiki is busy, frustrated, and happy with her job: she now has a class of 24 little kindergartners. Me, I, myself have been a little apathetic lately – don't ignore the "a" – but I still dream of finding the ultimate philosophy. It has something to do with "information" you know. (Lucky for you, my coffee's too hot to drink.) I've been thinking about the similarities between "introspection" and the quantum mechanical measurement situation lately – they're really a lot alike. The maximum information I can attain about what I am thinking can never be complete information ... that is, in the sense of allowing me complete predictability over my next action. The reason is that complete information about one's own present thought requires self-reference and an infinite regress. "OK, I am thinking about whether to stay in bed or get up. Wait, I'm thinking about my thinking of it. Doh! I can't stop this. Will I get up or won't I?" At some point the action just happens – the causal mechanism that brought it about is usually filled in slightly later, after the action has actually occurred. Anyway, one has to wonder whether the mathematics of quantum mechanics – I didn't say the "physics of"! – might be useful in getting at a precise description of the process.

23 November 1997, "Long Note on a Sunday Morning"

This morning I caught myself reading Jean Baudrillard (in the five hours I couldn't think of anything to say in). He's one of those weird Frenchies who – even though he knows no quantum mechanics – thinks nothing is real.

10 March 1999, "Schnockerization"

I tell you what, I'll do a little schnockerizing with you tonight. Tell me how *my* grammar fares!

The biggest fear I had to deal with when I was alone was exactly the feeling of permanence that came with it. If somehow, someway I could have heard a whisper "You will meet the girl of your dreams, your wife-to-be in two years." If I had better understood the finiteness of my time alone, I could have coped so much better – in fact, I don't think it would have been a problem at all. It was the fear of the permanence of the emptiness that really set me off. I guess I've always thought it is a load of crap that one has to "be for themselves" before they can "be for someone else." My view of man has always been quite overtly trinitarian: man, woman, (potential) child. We make each other, define each other, and give each other meaning.

In that vein, let me tell you one of my favorite acknowledgements ever in a scientific paper. It was Schrödinger in his 1935 paper "The Present Situation in Quantum Mechanics": "My warmest thanks to Imperial Chemical Industries, London, for the leisure to write this article." Schrödinger could not have been Schrödinger without the support of the community. And so the same for us all.

21

Letters to Michael Nielsen

22 February 1996, "Goo-Losh"

Mmm, goulash ... is that how you spell it? Thanks for the tip. I haven't heard the reaction of Bennett and company on this business yet, but when I do, I'll send a report your way.

I've been thinking about Kochen–Specker theorems lately; they're very nice you know ... especially the ones with small numbers of vectors. Implications for "quantum information"? There must be! In particular, why can't one go much further in this effort to prove the theorem with smaller and smaller numbers of vectors? Why is there no proof of this property (i.e., no noncontextual hidden variables for the outcomes of standard von Neumann measurements), the very second you admit a measurement that is nonorthogonal to the others? Just food for late night thoughts.

18 March 1996, "The Commitments"

It's a good movie; did you ever see it?

Suppose you need some cash and so you go to your Automated Teller Machine (ATM). You insert your bank card and the machine asks you for your Personal Identification Number (PIN). It does this for the obvious reason that it wants to know that you are indeed who you say you are before it hands over the money. That's a good safeguard. However suppose, as really did happen a year or two ago, someone had set up a fake ATM in place of the real one. By inserting your bank card into it, you have revealed your account number. By typing in your PIN, you have revealed how to make transactions on that account. The (dishonest) maker of this machine is now capable of withdrawing money from a real ATM machine anytime he wishes.

Is there a way of circumventing this sort of thing? Yes, in fact there are several ways – but they are all based on a problem that is computationally hard (classically). For instance one scheme is based on the user choosing two large prime numbers instead of a PIN. The user only reveals the product of the primes to the bank. Then it can be shown that there is a way the user can interact with the ATM so that he reveals no information about his primes and yet the ATM can verify that the user is indeed who he says he is. This is a very handy little protocol ... so long as large numbers cannot be factored effectively. (By the way, the rubric for these sorts of protocol is "zero knowledge proof.")

Well it turns out that as long as you have a secure "bit commitment" scheme you can implement a zero knowledge proof. The point is that bit commitment has loads of (practical) applications. Now it would be very nice to build an analogous scheme for which the security of it relied on the validity of quantum physics ... instead of some mathematically hard problem. That is the idea behind trying to build a quantum bit commitment scheme.

Whether such a thing is really possible, I do not know. Dominic Mayers is now running around here saying that they're not ... but I have not been able to understand his reasoning yet.

I think philosophically the existence of such things would be very nice. I used to think that the existence of quantum key distribution was important to Prof. Wheeler's mumblings on quantum mechanics. He likes to say that it is the "*community* of communicators" that is at the base of what is real in the world. This is different from "idealism" and "solipsism" because it relies crucially on the existence of many individuals, all communicating with each other. The world is not just a big thought. Well what good would it be to say that the world consists of individuals if there were nothing in physics to allow them completely private communication? If everyone could, in principle, listen in to everyone else's conversation, then the world could hardly be said to be composed of individuals at a basic level. This sort of thought excited me about quantum key distribution. A similar set of thoughts excites me about quantum bit commitment. What good would it be to say that the world consists of individuals if there were no means to protect one's identity from emulation by others? (Of course this idea could also be implemented with QKD, but dissimilar examples are good.)

So I think settling the question of quantum bit commitment is not only interesting from the practical side but also from the wispier side.

Entropy measurements on individual systems? I need a Latin phrase similar to "non sequitur" to answer that.

What am I doing researchwise? Nothing that I'm supposed to be doing! I still haven't been able to really bring myself back to the dissertation stuff ... but every day I look at it a little. I've written a couple of very nice referee reports. I worked an inordinate amount of time trying to turn 6 into 1 ... to no avail. The last couple of weeks have been filled with giving lectures on inference–disturbance tradeoff. (Lectures go extremely slowly here; the first two were essentially a mini-course on quantum mechanics.) Also I've spent a lot of time trying to work things out further for two-state systems, i.e., I'm trying to fill in the gaps that were only filled with computational work in the paper with Asher.

Philosophically, I've been thinking about Kant again. (Ask Jeff Nicholson the story of Kant Cola.) I'm about half way through a book by C. F. von Weizsäcker on Kant and Quantum. Quantum Theory as a law of thought – nothing excites me more! Before that I read a book on G. E. Moore and the Higher Sodomy; didn't glean much from it though. Let's see, after this book, the plan is to go back to Russell's writings on "logical atomism."

Kiki continues on her rampage to culture me. You should see, we have a very "cultured" apartment this time round. She very much likes it up here. I do too actually. The only thing I miss is a good used bookstore.

01 April 1996, "Mo' Better Sophistry"

I thought it was time to unload some thoughts. I hope you don't mind being chosen "designated driver." Please just make sure I make it home safely.

Oh, by the way, before I get started, could you do me a favor and enquire about the meaning of the term "three-lock box." It comes from an old Sammy Hagar tune. A good source is probably either Chris or Amy. (Actually, taking the mean of the answers would be even better.)

I'm just back from a two week foray into the netherworld of cryptology; God save my soul! Maybe I did learn something and didn't waste my time, but that still remains to be seen: I'm not sure. In the meantime, however, this may give some fodder to the old question of quantum measurement.

The idea of a "bit commitment" is, in the end, a very simple one. Alice would like to give Bob some evidence that she is committed to a bit, 0 or 1, and that she will not change her mind, but she would like to do so in such a way that Bob can gain no information about the bit she actually chose. A simple classical way to enact such an idea is for Alice to write the bit on a piece of paper, place it in a safe, and give the safe to Bob without the key. When Alice would like to reveal that she was actually committed to the bit, she gives Bob the key. He opens the safe and sees the bit. As long as there is no intricate mechanism for transferring a last-minute change of mind from the key to the paper, this can be safely deemed "evidence" that Alice did indeed commit. Moreover, as long as Bob cannot break into the safe, he can gain no information about Alice's bit.

The way such a thing is done in practice is that Alice hides her bit in a mathematically hard problem, working under the assumption that Bob does not have the computational power to break it. The problem with this, along with the simple example of above, is that one has to work under the assumption that Bob does not have the power to break into the container that Alice puts her bit in.

Does quantum theory give a way to get around this difficulty? At first sight it might appear that quantum mechanical thingies are a great place to hide information. If Alice prepares a quantum system in one of two nonorthogonal states and gives it to Bob, then there is no way he can recover the full bit – that's something of an example. As another (pseudo-)example, Alice can dump a bit into EPR correlations by way of teleportation. This is very much something that looks like an unbreakable safe: without the device that Alice used to prepare the state that was teleported, or the two classical bits arising from measurement of the Bell basis, there is nothing that Bob can do to recover the information from his element of the EPR pair.

However, the idea of Quantum Bit Commitment calls for something even stronger. Alice *must not* be able to reverse her commitment, and yet Bob *must not* be able to recover any information whatsoever about the bit she committed to. These two requirements appear to be mutually exclusive in the quantum world – *IF* I believe Dominic Mayers general attack (and I think I do, though I'm still not 100% convinced, only 99.8%). I suppose my purpose for writing this note is to say, *viewed in this way*, the impossibility of quantum bit commitment is something we – as physicists – should have known long ago ... and if the problem hadn't been mired in cryptological notation, we would have discounted it at the start.

"No quantum phenomenon is a phenomenon until it is an observed phenomenon, brought to close by an irreversible act of amplification."

(*J. A. Wheeler*)

Well, it's not exactly that, but it's very much of the same flavor. Alice and Bob play with two systems A and B and the game is the following. Alice "commits" to a bit by preparing the total system AB in one of two distinct states. She may not have had initial control over B, but that seems to be irrelevant. At the end of the day, after the "commit phase" has been carried out, Alice can only diddle with system A, Bob can only diddle with system B. The requirements for a bit commitment are this. After the commit phase there must be no unitary evolution that Alice can perform on A alone that will change the two states into each other. And there can be no measurement that Bob can perform on B alone that reveals any information about the commitment. These two requirements are mutually exclusive. If there is no tangible evidence on Bob's side after the "commitment" (i.e., evidence capable of revealing at least to some extent which bit was actually committed to), then there is no real coupling between Alice and Bob and she can reverse without in any way coming into contact with B again. That's what it boils down to as I see it.[1]

04 September 1996, "One Mo' Book"

Here's another book that I just heard about that may also be of interest to you. Apparently it anticipated the many-worlds interpretation in 1931 (according to Chris Isham). It's *Many Dimensions* by Charles Williams (Faber, London, 1968).

16 May 1999, "Purifications"

How's this? If you look at H. Araki and E. H. Lieb, "Entropy Inequalities," *Commun. Math. Phys.* **18**, 160–170 (1970), you'll find in Lemma 4, the following statement:

Let ρ^1 be a density matrix on H^1. Then there exists a Hilbert space H^2 and a pure state density matrix ρ^{12} on $H^1 \otimes H^2$ such that $Tr^2 \rho^{12} = \rho^1$.

---

[1] Looking back over this paragraph for the first time in five years, I have to wonder what on earth I was thinking? Still it intrigues me that I *might* have been thinking something that – with a sufficient amount of modification – *might* even be sensible. I have no qualms with intellectual alchemy!

After that follows the proof without any further attribution. Figuring that these guys were probably pretty well-established by then, this might well be the first appearance of a purification.

Wait ... no I'm probably wrong about that. I bet the first place was: D. Bures, "An Extension of Kakutani's Theorem on Infinite Product Measures to the Tensor Product of Semifinite w^*-Algebras," *Trans. Am. Math. Soc.* **135**, 199–212 (1969).

I'd momentarily forgotten about that one. Well, anyway, maybe somewhere between the two of these papers lies the truth. (Is that a bad pun?)

23 August 1999, "A Convergence of Sorts"

Friday morning as I was walking in to the office, I caught myself singing an old Paul Simon song that you may not know. It's called Kathy's Song. You should look up the words on the internet; I hope you'll take a slow moment to read them.

My singing actually had something to do with you in that it set my mind to remembering a conversation you and I had had in one of the Cambridge pubs. I was thinking of the one concerning what is most important in life, and, in particular, the contrast of your thoughts with those of David Deutsch's. This led me to think about the things that Deutsch is likely to be willing to define as "truth," in contrast to the things that Paul Simon would. What does Simon take to be the bedrock of our world, and what does Deutsch? And to which of these camps am I closer at this time in my life?

Then I remembered something I had written Carl and Rüdiger just a couple of days earlier. [See note to Rüdiger Schack dated 18 August 1999.]

I held that thought, mulling it over and over trying to refine it, and then – by happy circumstance – happened to read two wonderful articles by Silvan Schweber over the weekend. The first was: S. S. Schweber, "Physics, Community and the Crisis in Physical Theory," in *Physics, Philosophy, and the Scientific Community*, edited by K. Gavroglu, J. Stachel, and M. W. Wartofsky (Kluwer, Dordrecht, 1995), pages 125–152. The second was: S. S. Schweber, "The Metaphysics of Science at the End of a Heroic Age," in *Experimental Metaphysics: Quantum Mechanical Studies for Abner Shimony, Vol. 1*, edited by R. S. Cohen, M. Horne, and J. Stachel (Kluwer, Dordrecht, 1997), pages 171–198. (If you want to copy them, I'll have them in my office for the next couple of days ... that is, once I come in again.) Let me pluck a piece out of the first one that I thought was significant [pages 142–144].

For Einstein physics became a retreat from the world, and he found a monastery at the Institute for Advanced Study in Princeton. Meaning was given by his faith in the comprehensibility and the comprehension of nature. More specifically, Einstein believed "in the simplicity, i.e., the intelligibility of nature."

Although Bohr did appreciate Einstein's position, for him science could not be an escape from the world. In fact for him there is no escape from the world. There is no way to step outside of it and obtain a God's eye view of it. We are always on the inside and whatever we know of the world we have mediated and constructed. He believed that all scientific understanding – all understanding – required that one test and risk one's convictions and prejudgments in and through encounters with others and in particular in encounters with what is radically new and alien. For Bohr the practice of science exhibited a commitment to an underlying moral order. Scientists form an ideal society that relies on the virtues of honesty, tolerance, trust, truthfulness, and cooperation.

Without their moral commitment to be truthful and trustworthy members of this society that guarantees tolerance for the views of others and the verification of their claims, no important statements of fact could see the light of day and be stabilized. It was always self-evident to Bohr that science is a human activity, and constitutes a collaborative and cooperative enterprise. In fact, Bohr had a vision that the scientific community could serve as a paradigm for a possible future international order. Einstein would have mostly agreed – but he also believed that the communal aspect of science emphasized by Bohr could be transcended by individual acts of transcendence.

Ultimately, the difference between the two is the following: deep down – although a child of the Enlightenment, Einstein was really a gnostic priest. He believed that he could read the mind of his Spinozan God and discover the laws that govern the objective physical world out there. He believed in Truth with a capital T. For Bohr science does not determine the truth, truth with a capital T – rather there is convergence toward truth and the understanding of physical reality. On the basis of their experiments – their queries to nature – scientists constantly reorder and reinterpret external reality, and these imaginative reorderings must be presented to the members of the community for their critical assessment by means of language. And because language is a uniquely human activity – our distinctive (though imprecise) means to communicate with one another and convince one another – and a unique means to stabilize communities without recourse to physical violence – language is a central and constant concern in Bohr's philosophical gropings.

Bohr's conceptions of community, as well as those of truth and reality have close affinity to the views of Peirce. Peirce's notion of reality combines a social conception, that "the real is the idea in which the community ultimately settles down" with the notion that the real is what it is regardless of what anyone may think of it. For Peirce, the scientific enterprise is the model for how a society of free inquiring individuals can generate discourse and cooperation without chaos. What gives the actions of individuals their coherence is their decision as agents to aim toward a definite goal. We are all at work building and maintaining a scaffolding that will give stability and meaning to our lives and to the communities in which we share our lives. Science is part of that scaffolding; just as are literature, music, the visual arts, and the other activities we engage in. Science is a linkage between individualism and coordinated public choices and actions.

So too for Bohr. Bohr's conception of scientific activities in fact nurtures a wider vision of human interactions. Understanding the origin of the universe, its possible demise, unravelling the constitution of matter in its simplest manifestations and comprehending the rich and complex diversity matter can assume, understanding the limitations of mathematical theorizing and gaining insights into why it has been so effective in describing nature – are all part of our collective scaffolding – irrespective of the kind of scientific practitioners we are. Equally importantly, science is a piece of the scaffolding that makes for bridges to the members of other civilizations and cultures. Science, in fact, is that piece of what may be called the collective mind of the human race that is independent of place or individuality. And only in building together and in maintaining together the scaffolding can we guarantee stability.

So what? Nothing really, just that I'd probably buy you a beer at the Athenaeum if you're seeing the same sort of convergence that I am.

25 August 1999, "Late Night Threads"

Your spelling was impeccable. The running thread (the convergence) was meant to be: 1) There is no truth with a capital T within science. 2) Truth (with a small t) is actually grounded in the fuzzy rather than the precise. "Having been trained as an engineer, I see nothing mysterious about the fact that I can use very clumsy imprecise tools to manufacture delicate, high precision instruments."

About the cat, a side point. One thing that I was wanting to get at, but couldn't think of some good words for at the time is this. The "property" of being alive versus being dead is not a property intrinsic to the cat alone. It never was. Life is now "restored" routinely in ERs across the nation. In the ultimate limit of "maximal knowledge" of the patient (i.e., a pure quantum state for its description), the doctor would always be able to restore life.

I personally feel the luckiest when I have a little fuzzy truth. I wonder if Paul Simon still thinks that of his time with Kathy?

14 July 2000, "Quantum Theory Does Need an Interpretation"

Thanks for sending the evidence of Mr. Carvalho's short attention span. He writes [regarding my *Physics Today* article with Asher Peres],

> "... We set up this or that experiment to see how natures reacts. We have learned something new when we can distill from the accumulated data a compact description of all that was seen and an indication of which further experiments will corroborate that description ..."

This reminds me of Bertrand Russell's chicken. They also learned a lot from experiment. According to this description of the purpose of experiment, where does explanation fit in? I guess inductivism is not dead after all ...

But if he would have just read the remainder of the paragraph starting one sentence further down, he might have had some pause to think. Why would a follower of pure inductivism write such a thing as this:

If from such a description we can *further* distill a model of a free-standing "reality" independent of our interventions, then so much the better. Classical physics is the ultimate example in that regard. However, there is no logical necessity that this worldview always be obtainable. If the world is such that we can never identify a reality independent of our experimental activity, then we must be prepared for that too.

Like I wrote Asher just yesterday, as we were working on these replies for *Physics Today* (Mr. Carvalho got that wrong too):

I am not pushed to the rejection of a free-standing reality in the quantum world out of a predilection for positivism. I am pushed there because that is the overwhelming message quantum THEORY is trying to tell me.

But then, you probably already knew that of me ...

09 August 2000, "Big Red Xs"

Let me tell you a good one. When I was in Greece last month, I started out my talk by saying: "The last couple of days I've noticed that most talks start out at a pretty heuristic level, with lots of pictures and motivation, etc., and slowly become more and more technical. But, by the end of the talk all of the speakers are really pressed for time and the pace becomes phenomenal. It's kind of an uphill battle as equation after equation flashes across the screen. So, today, I've decided to give you a downhill slide: I'll start out with some technical results and derivations, slowly using more and more pictures as I go, and by the end – if I have time – I'm hoping the talk will degenerate into a discussion on the foundations of quantum mechanics."

I got a good laugh. And true to form, at the beginning while I was talking about things like Wojciech's "predictability sieve," Ben's "distinguishability sieve," and my "mutual information sieve" all the young guys were taking notes pretty thoroughly and all the old guys were nodding off. But by the end of the talk, when I changed to honest-to-god foundation stuff, all the young guys put down their pencils, and all the old guys became livid. It was quite a scene. I guess I riled them with my first slide (for that portion of the talk). It had at the top "Decoherence Studies" and at the bottom "Quantum Foundations" and it had an arrow running down from the top phrase to the bottom phrase. And I said, "Everyone tells me that decoherence has something to do with quantum foundations. But I have to admit, try as I might, and as much as people tell me that they're saying something convincing, I just don't see it. And I have tried to understand. So either I'm really thick, or my colleagues just aren't being as convincing as they think they are. I think it's the latter." Then I put up an overlay that had a big red X. "I think decoherence studies like the ones I've just showed you, important as they are, have nothing to do with the foundations of quantum mechanics." You should have seen the explosion! (But I'm used to that sort of thing: my house burnt down.)

22

Letters to Asher Peres

I am just returning from the library, where – completely by accident – I ran into and read an entertaining little article: A. Peres, "Einstein, Gödel, Bohr," *Foundations of Physics* **15**(2), 1985, 201–205. I was especially entertained by your closing line, "Any attempt to inject realism in physical theory is bound to lead to inconsistencies." Do you have any other writings (beside the *AJP* paper with Zurek) where you expand upon the ideas presented in this?

In connection to this, I have another question. If I recall correctly, you once told me that you finally came to grips ("interpretationally," that is) with quantum mechanics in 1986. (Perhaps I have the year wrong, but that's part of the question.) Would you be willing to expand upon that point? In particular, was your arrived-at comfort founded upon the line I quote above?

I suppose I ask these questions for several reasons, perhaps the strongest of which is a renewed dilettante interest in the foundations of quantum theory. Among other things, I am trying to plan (or dream up!) a talk for the Hull conference in September to be titled something like: "Quantum Information Theory's Implications for the Foundations of the Quantum as a Whole."

My own pet "idea for an idea" (as Wheeler would say) is that the quantum theoretical description – along with its Hilbert-space structure – is forced upon us when the information we have about a physical situation/system is of a certain variety. In this way, quantum theory is not so different from Ed Jaynes' molding of classical statistical mechanics and the maximum entropy principle. It just so happens in the latter case that the prior information is in the form of an expectation value for some physical quantity that can be assumed objective and independent of measurement: this prior information along with a few small desiderata leads uniquely to the canonical probability assignment. In the case of uniquely quantum problems, we start with a different kind of prior information. The *hope* is that, upon pinpointing the nature of that information (and again a few small desiderata), one would see that a "wavefunction assignment" is the natural (and unique) way of summarizing what we know and what we can predict.

From these remarks, perhaps you can see my attraction to your little article. Please fill me in!

16 October 1997, "Royal Order of Recalcitrant Positivists"

I'm perfectly happy with the comment you'd like to make. So use it, please. (In fact it's a little flattering almost.)

I was intrigued by one of the sentences in the Bub quote that you sent me. Namely his,

The disturbance terminology itself suggests the existence of determinate values for observables, prior to measurement, that are disturbed . . .

I thought I had seen something similar before, so this morning I dug and dug until I found it. It was Braunstein and Caves in the paper "Wringing out Better Bell Inequalities"; I like it much better than Bub's:

In quantum mechanics the "disturbance" (quotes because one should think twice before speaking of a disturbance if there are no objective properties to disturb) comes from using amplitude logic instead of probability logic (wave-function collapse instead of Bayes's theorem).

Is Bub's book any good? As I recall, Bub did work in Bohm-like hidden variable theories, but also with things to do with "quantum logics" (which have always struck me as overly-elaborate empiricist or Copenhagen-type frameworks). Both ends of the spectrum, so to speak. Reading him was a hard go – so I gave up a few times. What's the title of this new book? Maybe I should wait for your review to be finished, and you can send me that.

17 October 1997, "Energy ... Really"

Asherism 1: *Not energy but ENTROPY! (when observed separately by Alice and Bob).*

No, energy ... really. Really. Energy is the "tangible" in classical thermodynamics; entropy is the "intangible." I realize the analogy may not be so strong that I might take it outside the poetic license of a personal letter – that's one of the reasons I qualified the letter as philosophical-sounding and left it indented – but there may be just a little to it. Energy is a sort of "Martha White's All-Purpose Flour" for physics. Just as flour may be thought of as a potential bread, biscuit or cake, energy is a common "ingredient" that, if we wish, can be transformed into inertial mass, electromagnetic field intensity, or motion itself. It was in this way that I was thinking of entanglement. "Entanglement" is not "non-classical correlation" itself, but rather the *potential for correlation*. By choosing their measurements appropriately and adjusting their actions, Alice and Bob can build up – within limits of course – just the right type of correlation that they need. Quantum cryptography is one example of this: Alice and Bob would like to have a perfectly correlated binary string. However, teleportation is another example: there the "unknown state" preparer – Charlie say – would like future measurement outcome statistics on Bob's particle to be as he pleases. I think the things Bennett, Smolin, and I call "entanglement-enhanced classical communication" may show this even more strikingly.

For whatever it's worth, that's the sort of thing I was really thinking.

20 October 1997, "A Little More Reality"

Thank you for giving me the opportunity to compile this set of notes: even if they turn out to be of little use to you, going through this exercise has been of great use to me! It has helped me re-gel my opinions on the subject, so to speak. And that, if nothing else, will be of use to me for a project – tentatively titled "A Bayesian Reconstruction of Quantum Probabilities" – that Rüdiger Schack and I are presently involved in.

First off, let me express just how much I really have enjoyed reading and contemplating your manuscript. You and I are clearly on the same wavelength when it comes to "quantum philosophy." Perhaps there's no greater pleasure than to read something that you yourself think you are thinking!

Harsh Criticisms

But, now on to the scathing criticisms (… though I promise not to belabor them too much). My biggest worry about your draft is that, *maybe to too large an extent*, it is more a systematic account of your views on the interpretation of quantum mechanics than a review of Bub's book. Ultimately you're the one to decide on that – you're the writer – but that is the impression I get. This, for instance, is indicated in your curt dismissal of Bub's dominant program (i.e., attempting to translate various quantum interpretations into statements about lattices and sublattices):

Asherism 2: … *Bub … endeavors to assign numerical values to observables, or truth values to lattices of propositions. These are notions that have been borrowed from the classical world, and I don't see why quantum reality, whatever it is, has to be described in terms of observables or lattices of propositions.*

Indeed what you say is true. However, given that you, in the end, say:

Asherism 3: *In conclusion, there is much to learn from Bub's book, and I recommend it without hesitation to all those who are interested in the foundations of quantum theory.*

you really don't give the reader much to base the validity of that opinion on. Why is this book recommendable if so very much of it is devoted to something that can be dismissed in a single sentence? (I do realize you spent some time saying good things about it, but I think the things you praised were things *he* likely saw as tangential for building his main argument.)

My main disappointment in this respect is the very little attention you pay to what he clearly thinks of as the backbone of the book: the uniqueness theorem for the "no collapse" interpretations. What is this theorem? Couldn't you at least give the reader of the review a little of the flavor of it? Even just a small extra paragraph might make up for this. Also, I think it would be very useful for you to give an opinion – one way or the other – as to what you think of his attempt to view Bohr's interpretation as one of these "no collapse interpretations" that his theorem covers. It appears to me at least that, though he may not go into great detail about Bohr's rebuttal of EPR, he does spend quite some effort in Chapter 7 trying to make sense of Bohr in his language. In particular, I think he does make the appearance of actually agreeing with your remark:

Asherism 4: *Bohr also was sometimes quite elusive. He never explicitly treated a measurement as an interaction between two quantum systems, and he could thereby completely elude the measurement problem, as formulated by Bub.*

You can see this in the last full paragraph on p. 195 in addition to the thought being basically spread here and there throughout the chapter.

I don't want to give you the impression that I understand this book better than I actually do: I've only really given it the very slightest of skimmings. For the most part, I was spurred to notice these things by testing your manuscript. Nevertheless, with hindsight, I can say that I – as a reader – would have liked to see you touch on this stuff at least a little. (Plus, who knows, going into that might have helped out the things you would like to say about Bohr.) Ultimately, I think – assuming that all the theorems in the book are on the up and up – the main use of this work for the conceptualist may be in giving an effective way of lumping the main features of these "mutations" of quantum mechanics (as you call them) into a single description. This may be of use in the relativistic domain where one might hope to find something testable (in the spirit of a Bell inequality) for discrediting these modifications. Bohm's, for instance, might be a good test case.

Typos and Englos

But that's enough negative criticism. Let me now get on to some more positive comments. First a point of English. In the last sentence of the last full paragraph on the first page, you write "Bub's book gave me an opportunity of understanding why." That's nonstandard. I would say "for" instead of "of", or I would say "to understand" instead of "of understanding." Second, a typographical error: you forgot the word "problem" after "observational" in the long Bohr quote on page 8.

Also – upon some reflection – I think your "recalcitrant positivists" quote might convey something of a dangerous flavor. See discussion in the next section.

Foods for Thought [sic]

Finally let me comment on some of the many sentences in your paper that caused me to think a little. These remarks may not be of direct relevance for improving your exposition, but like I've already told you, that wasn't my only reason for reading the manuscript.

Asherism 5: *To interpret quantum theory, to explain its meaning, to make it understandable, the way has been shown to us: we have to translate the abstract mathematical formalism in such a way that "we can tell others what we have done and what we have learned and [this] must be expressed in unambiguous language with the terminology of classical physics."*

I've always liked this very much, but in certain ways it has always haunted me a little. To show you the ghost, let me quote a little of a letter from Schrödinger to Bohr, dated 13 October 1935.

As a matter of fact I did not really want to talk about this point, at least not with the purpose that you should reply, but rather about something else. You have repeatedly expressed your definite conviction that measurements must be described in terms of classical concepts. For example, on p. 61 of the volume published by Springer in 1931: "It lies in the nature of physical observation, that all experience must ultimately be expressed in terms of classical concepts, neglecting the quantum of action." And ibid. p. 74 "the invocation of classical ideas, necessitated by the very nature of measurement." And once again you talk about "the indispensable use of classical concepts in the interpretation of all measurements." True enough, shortly thereafter you say: "The removal of any incompleteness in the present methods of atomic physics ... might indeed only be effected by a still more radical departure from the method of description of classical physics, involving the consideration of the atomic constitution of all measuring instruments, which it has hitherto been possible to disregard in quantum mechanics."

This might sound as if what was earlier characterized as inherent in the very nature of any physical observation as an "indispensable necessity," would on the other hand after all just be a, fortunately still permissible, convenient way of conveying information, a way we presumably sometime will be forced to give up. If this were your opinion, then I would gladly agree. However, the subsequent stringent and clear comparison with the theory of relativity makes me doubt whether, in what I just said, I have understood your views correctly. ...

However that may be, there must be clear and definite reasons which cause you repeatedly to declare that we *must* interpret observations in classical terms, according to their very nature. Whenever you say that, you state it so definitely and clearly, in the indicative, without any reservations like "probably," or "it might be," or "we must be prepared for," as if this were the uttermost certainty in the world. It must be among your firmest convictions – and I cannot understand what it is based upon.

It could not be just this point (about which you talked so insistently to me already in 1926): that our traditional language and inherited concepts were completely unsuited to describe the phenomena with which we are confronted now. Because, in the course of the development of our science (and mathematics), from its earliest beginnings to the situation at the end of the nineteenth century, this was certainly the case over and over again. If the break with the old traditions seems greater now than ever before, then we should take into account that a particular time perspective is responsible for forming the impression that *that* development in which we ourselves take part, stands out as being more important and more essential than earlier ones, which we cite only from history, and whose stages we get to know mostly in reverse order. In fact, it is often difficult for us to imagine *earlier* ways of thinking. And although the difficulty of such a historical step *back* actually speaks most eloquently of *how* significant [the step] must have seemed to the pioneers at their first advances, still now and then we cannot avert the feeling: "Incredible that, up to then, people were so narrow-minded!" Here, the underestimation of the time perspective shows itself most clearly.

Thus I think that the fact that we have not adapted our thinking and our means of expression to the new theory cannot possibly be the reason for the conviction that experiments must always be described in the classical manner, thus neglecting the essential characteristics of the new theory. It may be a childish example [but I use it] only to say briefly what I mean: after the elastic light theory was replaced by the electromagnetic one, one did not say that the experimental findings should be expressed – just as before – in terms of the elasticity and density of the ether, of displacements, states of deformation, velocities and angular velocities of the ether particles.

Forgive my long-windedness. What I mean to say is: whether you couldn't make this point completely clear in the more detailed paper you announce in your *Phys. Rev.* note: Why do I [Bohr] emphasize again and again that according to the very nature of a measurement, it can only be interpreted classically? And above all: Is this a temporary resignation, or can we somehow recognize that we will never get beyond it?

Bohr's reply of 26 October 1935 gave very little help in that respect: it simply repeated over again the same *seemingly* "purely logical" argument.

On this point I must confess that I cannot share your doubts. My emphasis of the point that the classical description of experiments is unavoidable amounts merely to the *seemingly obvious fact that the description of any measuring arrangement must, in an essential manner, involve the arrangement of the instruments in space and their functioning in time, if we shall be able to state anything at all about the phenomena.* The argument here is of course first and foremost that in order to serve as measuring instruments, they cannot be included in the realm of application proper to quantum mechanics.

(By the way, these quotes are coming from *Niels Bohr: Collected Works*, vol. 7.) I think Schrödinger has a point that just can't be ignored. The gist I would like to give this, I think, is that the term "classical description" is an awfully loaded term: we have grown accustomed to thinking of it in a very technical sense that precludes Aristotelian mechanics, for instance. What Bohr was more particularly trying to get at is that all measurements must be describable with respect to the presumption of some stable, predictable, repeatable phenomena. That's the important aspect of his edict: it is exactly that that allows us to communicate the experimental set-up to our fellow observers. It does not depend in particular on the exact equational/conceptual form of our "classical physics." It's that the language has to be based upon a certain primitive form of stability. It so happens that the present form of classical mechanics is our best description of such stable phenomena, but it need not have been so for the formulation of the edict. For instance, General Relativity which succeeds Newtonian Mechanics could have been discovered in 1964 and it would not have made a bit of difference to Bohr's declaration of 1935.

Asherism 6: *Bohr was very careful and never claimed that there were in nature two different types of physical systems. All he said was that we had to use two different (classical or quantum) languages in order to describe different parts of the world.*

Indeed you are right about this as far as I can tell from my various readings of Bohr (and some of the other Copenhagenists). So, I'm very glad you say this so clearly! So many are apt to screw it up in just the way you hint ... and it generally annoys me! For your entertainment I'll attach some quotes from the "new orthodoxy" in that regard.

Zurek wrote in *Physics Today* (October 1991, p. 36):

The first widely accepted explanation of how a single outcome emerges from the many possibilities was the Copenhagen interpretation, proposed by Niels Bohr, who insisted that a classical apparatus is necessary to carry out measurements. Thus quantum theory was not to be universal. The key feature of the Copenhagen interpretation was the dividing line between quantum and classical. Bohr emphasized that the border must be mobile, so that even the "ultimate apparatus" – the human nervous system – can be measured and analyzed as a quantum object, provided that a suitable classical device is available to carry out the task.

In the absence of a crisp criterion to distinguish between quantum and classical, an identification of the "classical" with the "macroscopic" has often been tentatively accepted. The inadequacy of this approach has become apparent as a result of relatively recent developments: a cryogenic Weber bar must be treated as a quantum harmonic oscillator even though it can weigh a ton. Nonclassical squeezed states ...

Gell-Mann and Hartle in their article "Quantum Mechanics in the Light of Quantum Cosmology" (in *Complexity, Entropy and the Physics of Information*, edited W. H. Zurek, 1990) write:

The familiar rule of the "Copenhagen" interpretations described above is external to the framework of wavefunction and Schrödinger equation. Characteristically these interpretations, in one way or another, assumed as fundamental the existence of the classical domain we see all about us.

And Hartle in his article "Excess Baggage" (in *Elementary Particles and the Universe*, edited by John H. Schwarz, 1991) wrote:

All singled out the measurement process for a special role in the theory. In various ways, these authors were taking as fundamental the manifest existence of the classical world that we see all about us. That is, they were taking as fundamental the existence of objects which do have histories obeying classical probability rules and, except for the occasional intervention of the quantum world as in a measurement, obey deterministic classical equations of motion. This existence of a classical world, however it was phrased, was an important part of the Copenhagen interpretations for it was the contact with the classical world which mandated the "reduction of the wave packet."

The Copenhagen pictures do not permit the application of quantum mechanics to the universe as a whole. In the Copenhagen interpretations the universe is always divided into two parts: to one part quantum mechanical rules apply. To the other part classical rules apply.

These quotes may not completely pin the tail on the donkey, but definitely they're written in such a way as to be apt to cause a belief such as you described.

Asherism 7: *I wonder whether it is my idea that the apparatus must accept both quantum and classical descriptions, or someone else said that before me. Do you know?*

Consider the following passage in David Cassidy's biography of Heisenberg, *Uncertainty: The Life and Science of Werner Heisenberg* (page 261).

Heisenberg made the same point differently in his unpublished manuscript and published lecture. [Cassidy is here talking about Bohr's reply to EPR.] Any experiment entails two types of laws: those applying to the laboratory measuring apparatus, which, as Bohr tirelessly argued, are thoroughly classical in nature, and those pertaining to the atomic phenomena to be studied, which are thoroughly quantum mechanical in nature. Each set of laws holds in its own domain and is precise and nonstatistical. Statistics enters only with the researchers measuring experiment. Since every experiment attempts to bridge the gap between the classical laboratory domain and the quantum world of the atom, the experimental apparatus introduces a "cut" or interface between the two domains. Because physical laws are precise on either side of the cut, the statistical characteristics of quantum mechanics enter with and along the cut; that is, on the performance of the measurement. A deterministic completion of quantum mechanics must occur along the cut. But if new determining variables are introduced along the cut and the cut is subsequently moved, "then ... a contradiction between the lawlike consequences of the new properties and the [precise] relationships of quantum theory will be unavoidable." A deterministic completion of quantum mechanics is not possible, after all.

The argument, whatever the original really was, is probably stated somewhat loosely here – for instance the author completely botches the EPR argument a few pages before this – but the passage may still convey the essentials. Simply substitute "description" or "language" for "law" and it starts to sound pretty nice. The main point in my quoting this is that certainly by 1935, the issue of making the domain of classical description fluid or mobile was well understood by the Copenhagen circle.

The idea of the mismatch in, but necessary use of, two languages and the lack of a translating dictionary may be all yours. However, it sounds awfully a lot like the passage above. Heisenberg uses "cut"; you use "apparatus." In any case, it's likely that you've given it a more precise or modern spin than the founding fathers could have.

Asherism 8: *Bub wants to get rid of these fictitious observers, in spite of the fact that they appear to be quite harmless. (They are like the ubiquitous observers who send and receive signals in textbooks on special relativity.)*

I'm not so sure I agree with this. It's not so obvious to me that they are completely as "harmless" as their special relativistic cousins. In fact, they form part of the excitement of the quantum world for me. The difference as I see it is this. The observers in special relativity serve the role of defining various perspectives on a single underlying reality, the spacetime manifold. Their existence serves to delimit all the possible points of view that a real observer might take for splitting space and time into separate concepts.

On the other hand, the quantum observers – with their free choice over various incompatible measurements – serve to define just what it is that can be taken to be real. That is to say, they define the referent of the measurement record (as you say pretty clearly on page 3). Without these observers, or measurement records, there is no underlying reality. There is no analogue of the spacetime manifold in this case.

I'm not ashamed to admit that I'm perfectly willing to subscribe to (at least mild forms of) Wheeler's observer-participancy. It's the puckishness of these new kind of "observers" that Wheeler was talking about when he said, as quoted in Bub p. 191 (my emphasis),

In broader terms, we find that nature at the quantum level is not a machine that goes its inexorable way. Instead, what answer we get depends on the question we put, the experiment we arrange, the registering device we choose. We are inescapably involved in bringing out *that which appears to be happening*.

This is not idealism or solipsism – it is simply a recognition of what one must deal with because of the "no-go theorems" (modulo nonlocal theories). Pauli said it like this in his essay "Matter,"

... This takes into account the observer, including the apparatus used by him, differently from the way it was done in classical physics, both in Newtonian mechanics and in Maxwell-Einsteinian field theories. In the new pattern of thought we do not assume any longer the *detached observer*, occurring in the idealizations of this classical type of theory, but an observer who by his indeterminable effects creates a new situation, theoretically described as a new state of the observed system. In this way every observation is a singling out of a particular factual result, here and now, from the theoretical possibilities, thereby making obvious the discontinuous aspect of the physical phenomena.

Nevertheless, there remains still in the new kind of theory an *objective reality*, inasmuch as these theories deny any possibility for the observer to influence the results of a measurement, once the experimental arrangement is chosen.

and like this in his essay "Albert Einstein and the Development of Physics,"

We often discussed these questions together, and I invariably profited very greatly even when I could not agree with Einstein's views. "Physics is after all the description of reality," he said to me, continuing, with a sarcastic glance in my direction, "or should I perhaps say physics is the description of what one merely imagines?" This question clearly shows Einstein's concern that the objective character of physics might be lost through a theory of the type of quantum mechanics, in that as a consequence of its wider conception of objectivity of an explanation of nature the difference between physical reality and dream or hallucination becomes blurred.

The objectivity of physics is however fully ensured in quantum mechanics in the following sense. Although in principle, according to the theory, it is in general only the statistics of a series of experiments that is determined by laws, the observer is unable, even in the unpredictable single case, to influence the result of his observation – as for example the response of a counter at a particular instant of time. Further, personal qualities of the observer do not come into the theory in any way – the observation can be made by objective registering apparatus, the results of which are objectively available for anyone's inspection. Just as in the theory of relativity a group of mathematical transformations connects all possible coordinate systems, so in quantum mechanics a group of mathematical transformations connects the possible experimental arrangements.

Einstein however advocated a narrower form of the reality concept, which assumes a complete separation of an objectively existing physical state from any mode of its observation. Agreement was unfortunately never reached.

Why did I say this forms part of the excitement of the quantum world for me? For whatever it's worth, let me include a little essay I wrote to my old friend Greg Comer on 15 April 1996 (titled "Old Twenty Q's").

> C: (While buttering the toast . . .) I really can't see why you contend that the world, in its fundamentals, is independent of us, what we ask of it, and what we do with it.
>
> K: It's just ridiculous to think that we're that God-like. The world is what it is. It was here long before we were, and it'll be just what it is long after we're gone. This is just nonsense.
>
> C: But what about this toaster? It is most certainly a free creation of man. It is quite real. (Ouch!) You can't possibly tell me that it is something innate in the world independent of man. If you can't believe this for a toaster, why on earth do you have so much trouble with quantum mechanics?
>
> K: Bring your coffee; let's eat.

My friend,

I gained a little insight the other day, something new. Now it's time to tell it to my alter ego. Wheeler, like so many others, has always said that his understanding of the quantum came straight from the mouth of Bohr. But now I don't think so.

Recall the Game of Twenty Questions (surprise version). Wheeler walked into the room and started probing the other participants for a word. First asking one yes/no question, then another, and another, until there were 20 in all. Each person in the audience – playing a dirty trick on our friend – could answer completely at random so long as their answer was consistent with the answers to the previous questions. At the end of the questions, Wheeler himself answered with a relatively random choice: a concept, a word, the only one he could think of consistent with the 20 previous questions and answers . . . "cloud." The audience thought and thought, each individual checking his mind for consistency. Finally it emerged that "cloud" was OK! The audience rolled in laughter, and so the "cloud" came into existence.

Wheeler quite literally took this game as a basis for an ontology. Nature plays the part of the audience; the experimenter (communicator) plays the part of the poor sucker that was the butt of the joke. Each quantum measurement is a little act of creation – a little bit of it is on the part of nature with its seemingly random answer; a little bit of it is on the part of man in freely choosing the particular question to ask. If all of the (random) creation were on the part of nature, one would hardly expect the nice world we have. If all the creation were on the part of man, everything we hold dear would vanish into the ethereal mist of a dream.

That's the idea. I still like it very much and, in fact, I like it more and more every day. However, contrast this with Bohr. This is something I know that I've read before, but it never quite made the same impression on me.

> [At the Solvay meeting in 1927] an interesting discussion arose about how to speak of the appearance of phenomena for which only predictions of statistical character can be made. The question was whether, as to the occurrence of individual effects, we should adopt a terminology proposed by Dirac, that we were concerned with a choice on the part of "nature," or, as suggested by Heisenberg, we should say that we have to do

with a choice on the part of the "observer" constructing the measuring instruments and reading their recording. Any such terminology would, however, appear dubious since, on the one hand, it is hardly reasonable to endow nature with volition in the ordinary sense, while, on the other hand, it is certainly not possible for the observer to influence the events which may appear under the conditions he has arranged. To my mind, there is no other alternative than to admit that, in this field of experience, we are dealing with individual phenomena and that our possibilities of handling the measuring instruments allow us only to make a choice between the different complementary types of phenomena we want to study.

I snatched this quote from Aage Petersen's book *Quantum Physics and the Philosophic Tradition* (which is worth reading), but it also appears in the Schilpp volume on Einstein. There you have it in any case. Concerning the "free choice" in quantum measurement: Dirac voted "nature," Heisenberg voted "man," Bohr voted "neither," and Wheeler voted "both." The last point of view is, as I've written you before, quite similar to things in Pauli's writings.

Asherism 9: *The fundamental dilemma in the quantum formalism is not there [i.e., in assigning truth values to lattices of propositions]. It is that quantum mechanics permits the occurrence of all possible events (with definite probabilities), but in our consciousness there is only one world.*

Do you really believe there is a dilemma, i.e., a problem that seems to defy a satisfactory solution? This seems to go against the grain of the way you've presented so much else in this essay. Or are you using this word in order to make easy connection to Bub's words and points of view?

Asherism 10: *... a completely different problem is whether we can consider just a few collective degrees of freedom of the universe, such as its radius, mean density, total baryon number, etc., and apply quantum theory to these degrees of freedom, which do not include the observer and other insignificant details. This seems legitimate: this is not essentially different from quantizing the magnetic flux and the electric current in a SQUID, and ignoring the atomic details. You may object that there is only one universe, but likewise there is only one SQUID in my laboratory.*

I really like this passage a lot! Where on earth the community got the idea that it must "include the observer in the wave function" (as Wheeler would say) in order to do any sensible quantum mechanical calculations for cosmology, I don't know. It's very nice the way you pinpoint the problem here.

Also I like very much the essentially Bayesian viewpoint you take here! "There is only one SQUID in my laboratory." One does not need large ensembles of repetitive preparations to make sense of quantum probabilities.

Asherism 11: *For the opposite opinion of some recalcitrant positivists, see Fuchs and Peres (1996).*

I liked your earlier, less committal (but still funny) quote better. The reason for my backing out on this is that I'm not sure how to read the "opposite" in this sentence – I think it is a little ambiguous and I'm not sure I want to be too associated with a possibly outlandish interpretation.

Would you consider either restoring the original (i.e., simply without the word "opposite") or trying something like: "To see how two recalcitrant positivists actually define 'disturbance' in quantum mechanics see Fuchs and Peres (1996)."?

Of course the important thing is that the "disturbance" is not to any "dynamical variable" as Bub would like to say, but rather to Alice's predictability of Bob's measurement outcomes. That's the whole content of it.

You know, looking at a few more passages of Bohr, like

The essential lesson of the analysis of measurements in quantum theory is thus the emphasis on the necessity, in the account of the phenomena, of taking the whole experimental arrangement into consideration, in complete conformity with the fact that all unambiguous interpretation of the quantum mechanical formalism involves the fixation of the external conditions, defining the initial state of the atomic system concerned and the character of the possible predictions as regards subsequent observable properties of that system. Any measurement in quantum theory can in fact only refer either to a fixation of the initial state or to the test of such predictions, and *it is just the combination of measurements of both kinds which constitutes a well-defined phenomenon.*

and

As a more appropriate way of expression I advocated the application of the word *phenomenon* exclusively to refer to the observations obtained under specified circumstances, including an account of the whole experimental arrangement. In such terminology, the observational problem is free of any special intricacy since, in actual experiments, all observations are expressed by unambiguous statements referring, for instance, to the registration of the point at which an electron arrives at a photographic plate. Moreover, speaking in such a way is just suited to emphasize that the appropriate physical interpretation of the symbolic quantum-mechanical formalism amounts only to predictions, of determinate or statistical character, pertaining to individual phenomena appearing under conditions defined by classical physical concepts.

it dawns on me that quantum cryptography (the BB84 setup, say) might just provide a wonderful vehicle for explaining and making clear Bohr's notion of "quantum phenomenon." And it, of course, has the added advantage of having questions of "information" and "disturbance" built into the very reason for defining the game in the first place.

I don't know if you recall, but I once told you that I would really like to write something for the *American Journal of Physics* that explains well what is going on in the question of information gain versus disturbance in quantum theory. It now strikes me that one could incorporate so much of this discussion (from your paper and these notes) into a relatively nice exposition of both Bohr and information–disturbance (an article that has "something old, something new"). Would you be interested in seeing if we could tackle this in an effective way?

Asherism 12: *Bub's goal is to liberate the quantum world from its dependence on observers. ... The tacit assumption made by Bub ... is that the wave function is a* *genuine physical entity, not just an intellectual tool invented for the purpose of computing probabilities.*

Finally let me leave you on a happy note. Recall how I scolded you above for your "curt dismissal" of Bub's lattices? Well the truth of the matter is that he wasn't any more sympathetic, explanatory, or observant of our views! Just look at pages 38–39:

The only argument for introducing a special mode of state transition characterizing measurement processes is that "experience only makes statements of this type: an observer has made a certain (subjective) description; and never any like this: a physical quantity has a certain value." This echoes Bohr's remark … that the task of physics is not "to find out how nature *is*," but rather "what we can *say* about nature." But where is the argument that quantum mechanics, as opposed to classical mechanics, is a theory about subjective descriptions or what we can say about nature? After all, quantum mechanics is, in effect, a non-Boolean generalization of classical mechanics, and classical mechanics manages quite nicely to be a theory of how nature *is*, in principle, by ascribing values to physical quantities.

Perhaps after-all your article is a deserved "touché"!

22 October 1997, "Honesty"

Asherism 13: *The idea of the "cut" is not due to Heisenberg, but already appears in von Neumann's book (1932). However, this is not quite the same as my way of formulating this issue. von Neumann, and others after him, considered a sequence of apparatuses, described by quantum mechanics up to some point, and the following apparatuses were treated classically. My contention is that the SAME apparatus must get both descriptions, at different stages. Was it really my personal chutzpah to have this idea?*

I understand your contention (or, at least, think I do). However, please read the passage I sent you again: there is more being said there than what von Neumann had to say. It may not be – and probably *is not* – exactly what you have to say, but it is distinct from my understanding of von Neumann (which mimics what you said above). Cassidy says, pretty explicitly, "the experimental apparatus introduces a 'cut' or interface between the two domains [i.e., the quantum and the classical]." I am, in particular, thinking of "interface" in the computer science sense as something like Windows95 or Sun OpenWin that is half computer program, half scratch pad for my thoughts. If I am not misled by the larger context of Cassidy's remarks, I think that is the sense in which Mr. Heisenberg was using it too. He speaks of the "cut" as if it were a receptacle in which one might place hidden variables, but because of its mobility this would cause a contradiction.

In any case, other than these loosely related things, I don't think I have heard your idea before. (That is, other than reading it in your book two years ago!) I've certainly never seen it posed in a quantitative way by anyone else.

23 October 1997, "Tyche or Moira?"

It must have been an interesting combination of the headache I had today, curiosity, and just plain luck that led me to the right place in the library. I ran across the following in an article by Shimony (*AJP*, vol. 31, p. 755, 1963):

Bohr is saying that from one point of view the apparatus is described classically and from another, mutually exclusive point of view, it is described quantum mechanically. In other words, he is applying the principle of complementarity, which was originally formulated for microphysical phenomena, to a macroscopic piece of apparatus. This application of complementarity on a new level provides an answer to the difficulty regarding macrophysical objects which confronted "positivism of higher order": the macro-physical object has objective existence and intrinsic properties in one set of circumstances (e.g., when used for the purpose of measuring) and has properties relative to the observer in another set of circumstances, thereby evading the dilemma of choosing between realism and idealism. Two important conclusions follow from this discussion. ... The second conclusion is that Bohr's extension of the principle of complementarity beyond its original function of reconciling apparently contradictory microphysical phenomena is not gratuitous, as critics have often claimed. The internal logic of his position requires the application of complementarity to macroscopic measuring instruments.

It still doesn't capture what you said, namely,

The peculiar property of the quantum measuring process is that we have to use *both* descriptions for the *same* object: namely, the measuring apparatus obeys quantum dynamics while it interacts with the quantum system under study, and at a later stage the same apparatus is considered as a classical object, when it becomes the permanent depository of information. This dichotomy is the root of the quantum measurement dilemma: there can be no unambiguous classical-quantum dictionary.

but it comes closer to the flavor. In particular, the passage captures there being no classical-quantum dictionary. However it does not completely address the dual use of the measuring apparatus as such – it only says "in another set of circumstances." Maybe that is not explicit enough. In any case, though, I am coming to understand you a little better.

With that, I sign off on this issue.

15 December 1997, "Hay–Peres Comments"

I think it may be harder to comment on your paper than I had thought it would be. I take it that your largest concern is how a referee might react to the section on "von Neumann's cut" and also the concluding section. Even after reading it over twice now, I find it difficult to estimate. Very likely it will be the toss of a coin as to whether a referee will like it or not. In any case, if the calculation is sound, I don't see how a referee can deny that it was worthwhile doing. If you get unlucky enough to get a holier-than-thou follower of the many-worlds idea, then it seems likely that you'll get a hand-spanking for your "antiquated" ideas and some suggested modifications, but it'll still be accepted. (Do you actually ever get papers rejected?)

I am concerned by the growing abandonment of standard (Copenhagen) quantum mechanics that I see all around me. Actually, I mostly see that in the theorists; the experimentalists around here seem OK with the idea that the state vector describes what one knows about their lab table and nothing more. Everywhere I turn, I see theorists climbing onto the many-worlds bandwagon for no well-defined scientific reason. It just seems to give them a happier, stable feeling – at least that's all I can see. I worry about how to combat this in an effective way, but I haven't found one yet. Talking to you, Mermin, and Preskill has been helpful in that respect, but I have a long way to go I think. (Notice that I purposely placed you three in just that order: you being on the far-left, and John being on the far-right.)

29 January 1998, "Strange Surprise"

Have you looked at Mermin's latest tractatus on quant-ph? His aspirations for an (objective) "interpretation" of quantum mechanics are a little annoying, but if you can put a distaste for that aside, there are a couple of nice points in it (relying on Wootters' theorem). Aside from that – in a way that he hadn't intended, of course – I think the paper makes a strong case for the standard interpretation that you, for instance, wrote about in your last book review. Somehow, we've got to get out of all these interpretational quandaries and look for that aspect of the standard interpretation that is ripe for building upon. So that we can move on to much bigger, much better physics. I wish I could find the key for convincing good guys like Mermin of that point of view.

05 April 1998, "Other Things"

Asherism 14: *After two months, they sent the paper to a decoherentist, who demanded to mention decoherence. It's surely better than a many-worldist, or a Bohm-trajectorist, etc.*

First off, congratulations to you and Ori Hay on the new paper! You have, of course, told Ori about the coveted Einstein Number that he now possesses?! Are you so sure, though, that the decoherentists are more desirable than the Everettistas and the Bohmians? I find them all equally unpalatable ... though each for a different specific reason. My blanket complaint is that they all hope to find ontology where there is none. I, by the way, have been working on a new way to express what I find most positive about my neo-Copenhagenist views. Namely, we find a lesson in quantum theory not so different from the one taught by the great Socrates: we are only in a position to learn when we realize that we are fully and completely ignorant. The version of the theory with the least ontological commitment is certainly the one poised for the greatest, most revolutionary progress next century. (That is, once the next Einstein ... (perhaps a future son or daughter of mine?) ... is born!)

Speaking of numbers and lineage, etc., there's a little story that I haven't told you. When I won the Michelson prize I started to look for a good way to open my colloquium and it dawned on me that I had once heard John Wheeler (my academic great-grandfather) describe his Ph.D. advisor as "the great optical physicist ..." though I couldn't remember the precise name. So I hoped very hard that whoever he was, he might actually be Michelson's student. Alas, it wasn't to be! But I ended up learning something in the process of my search that was really interesting to me. Wheeler's advisor was Karl Herzfeld, who first taught in Munich (where he and Sommerfeld gave birth to Heitler), then at Johns Hopkins (where he gave birth to Wheeler), and finally at the Catholic University of America. The thing that I love to know is that, as best one can trace, Herzfeld may have been the first person to introduce the term "quantum mechanics" into the literature!! Born thought that he had done it himself, but his first usage of it was in a paper of 1924. Lorentz preceded that in 1923 with "the mechanics of quanta." However, Herzfeld had a paper of 1921 titled with the German equivalent of "Quantum Mechanics of Atomic Models." (These last details come from a footnote in Mehra and Rechenberg, vol 4.) Herzfeld by the way obtained his Ph.D. in Vienna in 1914 but I haven't been able to trace who his advisor was.

01 December 1998, "Here Comes the Judge"

Asherism 15: *The measuring process is an external intervention in the dynamics of a quantum system.*

I know that I've already expressed what I'm about to say several times, but let me just set things off to a good start again: I really like this turn of phrase! For years now, I have believed exactly what you say, namely,

Asherism 16: *[I]t is preferable not to use the word "measurement" which suggests that there exists in the real world some unknown property that we are measuring.*

but I had never invented such a clever phrase to express it. For a long time instead I made use of the word "creation" in my tract-like emails on the subject. In some ways, I still think – taking my cue from John Wheeler – that that word captures a certain central truth of quantum theory, but your word is more encompassing. I think it captures better the idea that quantum measurement is a double-edged sword, learning *and* creating.

But I wouldn't be doing my scholarly duty, if I didn't remind you of some passages of Niels Bohr, whose words you seem to have so much respect for. In Bohr's 1949 article "Discussion with Einstein on Epistemological Problems in Atomic Physics," he wrote:

[At the Solvay meeting in 1927] an interesting discussion arose about how to speak of the appearance of phenomena for which only predictions of statistical character can be made. The question was whether, as to the occurrence of individual effects, we should adopt a terminology proposed by Dirac, that we were concerned with a choice on the part of "nature," or, as suggested by Heisenberg, we should say that we have to do with a choice on the part of the "observer" constructing the measuring instruments and reading their recording. Any such terminology would, however, appear dubious since, on the one hand, it is hardly reasonable to endow nature with volition in the ordinary sense, while, on the other hand, it is certainly not possible for the observer to influence the events which may appear under the conditions he has arranged. To my mind, there is no other alternative than to admit that, in this field of experience, we are dealing with individual phenomena and that our possibilities of handling the measuring instruments allow us only to make a choice between the different complementary types of phenomena we want to study.

And, more to the point, in his 1958 article "Quantum Physics and Philosophy: Causality and Complementarity," he wrote:

In the treatment of atomic problems, actual calculations are most conveniently carried out with the help of a Schrödinger state function, from which the statistical laws governing observations obtainable under specified conditions can be deduced by definite mathematical operations. It must be recognized, however, that we are here dealing with a purely symbolic procedure, the unambiguous physical interpretation of which in the last resort requires a reference to a complete experimental arrangement. Disregard of this point has sometimes led to confusion, and in particular the use of phrases like "disturbance of phenomena by observation" or "creation of physical attributes of objects by measurements" is hardly compatible with common language and practical definition.

In this connection, the question has even been raised whether recourse to multivalued logics is needed for a more appropriate representation of the situation. From the preceding argumentation it will appear, however, that all departures from common language and ordinary logic are entirely avoided by reserving the word "phenomenon" solely for reference to unambiguously communicable information, in the account of which the word "measurement" is used in its plain meaning of standardized comparison.

Do you see these passages as meshing well with your new terminology? I don't really, but then again I don't feel that Bohr was all that right when he came to these points. I think it might be useful for your ultimate reader to add some words about your opinion on this.

Asherism 17: *As a concrete example, consider the quantum teleportation scenario. The first intervention is performed by Alice: she has two spin-$\frac{1}{2}$ particles and she tests in which Bell state these particles are.*

In which state they *are*?!?! There's got to be a better way of putting this … especially one more consistent with your whole view of the measurement process.

Asherism 18: *I do not want to use the word "histories," which has acquired a different meaning in the writings of some quantum theorists.*

What happened to your nice word "chronicle"? I liked it a lot. Can't you find some way to reinstate it? I guess I liked it because it always reminded me of a passage that I like to quote from Pauli. In a letter to Markus Fierz in 1947, he wrote:

Something only really happens when an observation is being made …. . Between the observations nothing at all happens, only time has, "in the interval," irreversibly progressed on the mathematical papers!

Asherism 19: *Quantum mechanics is fundamentally statistical, and any experiment has to be repeated many times (in a theoretical discussion, we shall imagine many replicas of our gedankenexperiment).*

Asherism 20: *Each one of these records has a definite probability, which is experimentally observed as its relative frequency among all the records that were obtained.*

Asherism 21: *Note that the "detector clicks" are the only real thing we have to consider. Their probabilities are objective numbers and are Lorentz invariant.*

As a disciple of the Reverend Bayes, you should know that I strongly dislike all these expressions. A good Bayesian would say that probability quantifies a state of knowledge. It is that and that alone, meeting an operational verification *only* through a subject's betting behavior. In particular, probability has no *a priori* connection to the frequency of outcomes in a repeated experiment. It matters not whether that repetition is real or, instead, imaginary and virtual.

Let me just try to drive this home with two of the simplest possible examples.

(1) Consider a coin for which you have no reason to believe that a head will occur over a tail in a toss. Now imagine that you flip that coin an infinite number of times, tabulating the number of heads and tails. Do you really believe that a frequency of precisely 1/2 *must* occur in the infinite limit? Answer this question to yourself very honestly. What is to bar you from getting a head as the outcome in literally every single toss? What is there in the coin to favor a random-looking sequence to a nonrandom one? Nothing. To say something like, "Well the nonrandom-looking sequences have probability zero" is just to beg the question. For one thing, you have to invoke the concept of probability again to even say it. For another, even if you allow yourself that, it still carries no force: just take the set of nonrandom-looking sequences and add to it any sequence that you would consider a valid random one (i.e., one that you believe could be generated by a repeated coin toss). That set still has probability-measure zero, but now you would have to say that that means the random-looking sequence could not be generated by your imaginary coin. The point: probability has no direct connection to frequency.

(2) Consider now the case where I toss a coin repeatedly for you. I assure you strongly that the coin is weighted 80–20 heads vs. tails *or* 20–80, but I steadfastly refuse to tell you which way it goes. What probability would you ascribe to the outcome of a head upon my first toss? Fifty-percent of course. The point is, my probability ascription is not your probability ascription. As I toss the coin ever more, if you are rational, your probability ascription will *very likely* approach mine, but there are no absolute assurances. The point: again probability has no direct connection to frequency. But also, two perfectly rational people can make different probability assignments to the same experiment without being inconsistent with each other. There is nothing objective about a probability assignment per se.

If you'd like to understand better this point of view, then I'd strongly suggest the wonderful collection of essays, "Studies in Subjective Probability," edited by Henry E. Kyburg, Jr., and Howard E. Smokler (Wiley, NY, 1964). Also there is a later edition of the book with a few other essays. For a much more in-depth study I couldn't recommend anything more than Ed Jaynes' book, "Probability Theory: The Logic of Science." It unfortunately may never be published properly, but preprints of the parts that were written can be found at the "Probability Theory as Extended Logic" web page, http://bayes.wustl.edu/.

Asherism 22: *[P]robabilities are objective numbers .... On the other hand, wave functions and operators are nothing more than abstract symbols. They are convenient mathematical concepts, useful for computing quantum probabilities, but they have no real existence in Nature.*

I think it is really cute the way you cite Stapp here!

But again I want to come back to the last point above. A good Bayesian already knows that probabilities cannot be taken to be objective numbers, existent "out there" in nature. And a good quantum physicist similarly knows that the wave function is *not* something "out there" existent in nature. The similarity of these two points of view is not an accident: it had to be the case! For, from one point of view, the wave function is nothing more than a compendium of probabilities. It may be the case that "unperformed measurements have no outcomes," but it is not the case that unperformed measurements have no probabilities! If we specify the probabilities of the outcomes for every *potential* measurement, then we have specified the wave function. There cannot be a difference in the objectivity levels of these two mathematical objects: they are both abstract symbols that summarize our states of knowledge. Their empirical meaning comes about precisely in how a rational being would behave in the light of that knowledge.

There is a famous quote in Bayesian probability theory due to Bruno de Finetti. It starts out the two volumes of his book.

My thesis, paradoxically, and a little provocatively, but nonetheless genuinely, is simply this:

PROBABILITY DOES NOT EXIST.

The abandonment of superstitious beliefs about the existence of Phlogiston, the Cosmic Ether, Absolute Space and Time, ..., or Fairies and Witches, was an essential step along the road to scientific thinking. Probability, too, if regarded as something endowed with some kind of objective existence, is no less a misleading conception, an illusory attempt to exteriorize or materialize our true probabilistic beliefs.

In contrast, my paper "Bayesian Probability in Quantum Mechanics" with Caves and Schack, opens up with the following lines:

My thesis, paradoxically, and a little provocatively, but nonetheless genuinely, is simply this:

QUANTUM STATES DO NOT EXIST.

The abandonment of superstitious beliefs about the existence of Phlogiston, the Cosmic Ether, Absolute Space and Time, ..., or Fairies and Witches, was an essential step along the road to scientific thinking. The quantum state, too, if regarded as something endowed with some kind of objective existence, is no less a misleading conception, an illusory attempt to exteriorize or materialize the information we possess.

(the ghost of Bruno de Finetti)

I don't know of any better way to express it than this.

Asherism 23: *The physical evolution that leads to Eq. (1) is the following. The intervener receives a quantum system in state ρ and adjoins to it an auxiliary system (an ancilla) in a known state ρ_a. The composite system, in state $\rho \otimes \rho_a$, is subjected to a unitary transformation . . .*

In my present way of speaking, I like to make this sound much less absolute. I usually say that this is only one *representation* of a superoperator. It is the superoperator that is the only thing that need be given; anything else is just one way or another of thinking about it. I guess the point is I see no reason to give unitary evolutions and von Neumann measurements a special status in the axiomatics of the theory. POVMs and superoperators are perfectly good (and much less specific) starting points for the theory. You can certainly do as you please, but it seems to me that there is a lot to be gained from this point of view. For instance, it seems to me to de-emphasize the point of view that Charlie Bennett labels "the Church of the Larger Hilbert Space" (which is, as he admits, a euphemism for the many-worlds interpretation).

Asherism 24: *If we wish to consider only the states just before and after the intervention, without entering into the detailed dynamics of the intervention, the result appears as a discontinuous jump of the wave function (often called a "collapse"). Clearly, this jump is not a dynamical process. It results from the introduction of an ancilla, followed by its deletion, or that of another subsystem. The jump is solely due to abrupt changes in our way of delimiting what we consider as the quantum system under study.*

I agree with you to some extent here, but not the whole way. The abrupt change that comes about when we conceptually delete a subsystem from a larger composite system corresponds to a *partial trace*. Where does the *random* jump come from that corresponds to the measurement outcome label k? I've heard you say things like this many times, i.e., that the statistics in quantum mechanics comes about because of a mismatch in two languages, the quantum and the classical – your talk at QCM was just one example – but how does one see that you're not just talking about a partial trace here? I am perfectly willing to take measurement/intervention as a primitive of the theory – it's the very thing that gives the theory meaning – but you seem to want to go further, to have it fall out of something more primitive, namely the act of drawing a conceptual line in an overall unitary dynamics. I guess I still don't see it.

Asherism 25: *Summing over them is like saying that peas and peanuts contain on the average 42% of water, instead of saying that peas have 78% and peanuts 6% [25].*

I loved the citation here! You don't think there might be a connection between Stapp and the USDA, do you?

Asherism 26: *Between these two events, there is a "free" (that is, deterministic) evolution of the state of the quantum system. What distinguishes such a free evolution from an intervention is that the latter has unpredictable output parameters, for example which one* *of the detectors "clicks," thereby starting a new chapter in the history of the quantum system. As long as there is no such a branching, the evolution will be called free, even though it may depend on external classical fields, that are specified by the classical parameters of the preceding interventions.*

I like this distinction, but I cannot completely agree with it. Take any trace-preserving completely positive map that is not unitary, i.e., most any quantum channel like the amplitude damping channel or the depolarizing channel will do. It gives rise to no "branching" changes in the quantum state, but I think one would be hard pressed to call it a "free" evolution.

I overheard Herb Bernstein once say something like, "Of course there are two kinds of evolution in quantum mechanics; there are the times when we learn something and then there are the times when we learn nothing." The line struck me. I know that this is something like what you're trying to get at, but the cut isn't between unitary and nonunitary then.

Asherism 27: *Quantum mechanics asserts that during a free evolution the quantum state undergoes a unitary transformation.*

This needs cleaning up exactly because of the point above. By the way, note that there is a typo in the sentence following Eq. (7).

Asherism 28: *Note that* $\mathrm{Tr}(\rho_f) = \mathrm{Tr}(\rho'_f)$ *is the joint probability of occurrence of the records k and μ. This is the only observable quantity in this experiment. … Einstein's principle of relativity asserts that both descriptions given above are equally valid. Formally, the states ρ_f (at time t_f) and the state ρ'_f (at time t'_f) have to be Lorentz transforms of each other.*

I am afraid that this might be construed to mean that the only way the equality of the traces can be satisfied is if ρ_f and ρ'_f are "Lorentz transforms" of each other. Surely that's not true and that's not what you mean. Maybe you could clarify this a bit.

Asherism 29: *As a further simplification, let all the U and V operators be unit matrices, so that the two particles are really free, except at the intervention events.*

What do you mean by the phrase "really free" here? Is it a joke?

Asherism 30: *We thus have to accept that unit matrices of any order may also be legitimate Lorentz transforms of each other.*

So I guess what you're meaning is that you want this problem to partially define what is meant by the very term "Lorentz transform?" I'm looking forward to understanding the business about Green's functions much better. After that I'll come back for some more substantial comments on everything in this section following Eq. (10).

Asherism 31: *Quantum nonlocality has led some authors to suggest the possibility of superluminal communication.*

Notice the correction to *your* spelling mistake ... ahem. Grin. Anyway, in this connection let me point you to a very nice article by a philosopher: J. B. Kennedy, "On the Empirical Foundations of the Quantum No-Signalling Proofs," *Phil. Sci.* **62**, 543–560 (1995). The point he makes – and I think validly so – is that any so-called "proof" that quantum mechanics cannot support superluminal signalling by using entanglement is essentially circular. It's not that the "proofs" are invalid, but essentially that there was no use in doing them in the first place: no-signalling is much more of an *axiom* than a result of the standard structure of quantum mechanics. Kennedy argued by way of digging into the historical references, finding for instance von Neumann's original motivation for introducing the tensor product rule for combining Hilbert spaces – it was essentially to block the possibility of superluminal signalling! You might enjoy reading the paper.

Asherism 32: *The classical–quantum analogy becomes complete if we use statistical mechanics for treating the classical case. The distribution of the bomb fragments is given by a Liouville function in phase space. When Alice measures $\mathbf{J}_1$, the Liouville function for $\mathbf{J}_2$ is instantly altered, however far Bob is from Alice. No one would find this surprising, since it is universally agreed that a Liouville function is only a mathematical tool representing our statistical knowledge. Likewise, the wave function ψ, or the Wigner function, which is the quantum analogue of a Liouville function, are only mathematical tools for computing probabilities. It is only when they are considered as physical objects that superluminal paradoxes arise.*

Indeed you have captured the point here. Part of this, i.e., that the proper comparison is between state vectors and Liouville distributions, is what Carl and I were striving to convey in our paper for Nathan Rosen's festschrift – so I'm already very sympathetic to this. (Of course I know that you already made this point long ago in your paper "What is a State Vector?") More particularly, though, this is one thing that Steven van Enk and I had planned to write in a silly little paper – titled "Entanglement is Super ... but not Superluminal!" – for our contribution to a book on superluminal signaling! I hope you won't mind the overlap.

The thing that holds so many up from immediately grasping this point of view is the *difference* between the quantum and classical states of knowledge (i.e., those things that are changing instantaneously as you say). Classically, the probability is attached to an existent property. "The particle has a position; I just don't know it. I capture what little bit I do know with a probability assignment." Quantumly, the probability is attached to *potential* measurement outcomes and nothing more. How is it that my probability assignment for some potential measurement outcome can change from 50–50 to complete certainty when there is no sense in which that measurement outcome can already be said to be "out there" without the performance of a measurement? That's the thing that has people perplexed and, I think, is the source of their confusion about using entanglement for communication.

Asherism 33: *In our approach, there is no "delayed choice" paradox. It is indeed quite surprising that JAW, who is a hard core positivist, introduced this "delayed choice" idea.*

Sometimes it's pretty hard to figure out what precisely it is that John is trying to say. Perhaps you shouldn't be overly harsh on him. In some of his discussions of "delayed choice" he makes it very clear that he is not talking about affecting the "past" itself (whatever that might mean) but instead "what we have the right to say about it." Combine that with one of the phrases he is fond of saying, i.e., "The past exists only insofar as it is recorded in the present," and I think you will have to position him closer than not to the positivist you used to think of him as.

16 December 1998, "Nonlinear CPMs"

I think the question you are starting to ask is a good one. In fact, one of the big things on my (thinking) wish list for the coming year is to get a better handle on the *physical* assumptions underlying complete positivity. Weinberg's evolution operator is trivially not a completely positive map because – in the standard usage of the term – completely positive maps can only be *linear*. There is a good open question about the proper way to define nonlinear generalizations of complete positivity. I had a conversation about this with Holevo at QCM and he said it would be a mess, the difficulties arising somehow because the tensor producting operation is explicitly a linear operation. I didn't understand him very well, but I have wanted to come back to think about it.

In a certain sense, complete positivity is relatively natural from the Bayesian viewpoint of quantum mechanics. (The Bayesian viewpoint, by the way, is what I'm starting to call my way of thinking about quantum mechanics: the term will make more sense to you once you read the opus that Caves and Schack and I are writing on it.) Whatever physical evolution is, it had at the very least better take states of knowledge to states of knowledge. That *plus* linearity gives complete positivity. And, once one has that one is also warranted in thinking about unitarity in the Church of the Larger Hilbert space. (Unitarity is a theorem, not a starting point!) But what compels linearity within the Bayesian view? That I don't understand yet. The standard story of the Karl Kraus types is that linearity comes from the reasonable requirement that the evolution of a mixture be a mixture of the evolutions of the independent states. But that is not so motivated within the Bayesian view where the quantum state is not solely a recipe (as you like to call it), but rather a state of knowledge full stop. (A recipe is a special case of a state of knowledge.) Things to ponder over ... though I suspect the issues won't make too much sense to you until you read our paper.

There is perhaps one thing interesting to read presently in the literature on this. A paper by Czachor and Kuna, "Complete positivity of nonlinear evolution: A case study," *PRA* **58**, July 1998, 128–134. I think their point is that the one nonlinear generalization to CPMs that the mathematicians have proposed is not a physically reasonable generalization. But I have not read the paper yet. Part of lectures in Torino will be on this subject. (I am lecturing on quantum channels, i.e., the details of CPMs.)

26 December 1998, "The Deep Intervention"

Well I have finally had some time to read through your new draft of RQM completely. I am not sure, however, what I should compose as comment yet – I think I will chew upon it for a while, before saying anything more. In the meantime, let me share a passage I just entered into my computer by way of my "voice recognition system." This came from a recent paper on quant-ph by Doug Bilodeau, "Why Quantum Mechanics is Hard to Understand." In his introduction he says, "In this paper I present the conceptual system which I think gives the clearest understanding of what it is we are doing when we use quantum mechanics." I think he fails at that, as I cannot figure out what he is talking about for much of the paper. *But*, there were some parts of the paper that struck me as particularly deep. The passage below was one of them. It concerns another use of the term "intervention," one that I am particularly fond of (but is maybe a little more risky than the meaning you introduce in your paper). I hope you enjoy it as much as I did.

A thing is historical insofar as it is objective (can be observed and treated as an object). It then enters into the realm of recordable objective occurrences which can be ordered in historical space and time. It is dynamical insofar as it is defined as an abstract element of the dynamical theory which explains causal relationships between objects. ...

Imagine that we could see the universe as omniscient external observers, all space and time at once, and that what we "see" is a tangle of intersecting particle world-lines (cf. Ch. 1 of Misner, Thorne, and Wheeler). We might detect some patterns which would constitute physical rules or laws in some sense, but it would be quite difficult or impossible to know whether we had found all the important patterns, or to distinguish significant relationships from accidental ones. Even more difficult would be to translate this omniscient description into the kinds of relationships and laws which would be observed by the huge clumsy bunches of world-lines which constitute ourselves.

When we set out to investigate Nature, we are not like that external omniscient observer at all. We look for relationships and patterns in the behavior of objects we know. We want to find out – does this kind of object always behave this way under these circumstances? The phrases "this kind," "this way," and "these circumstances" imply the ability to abstract relevant or significant features from what are really unique events. They also imply that we can find or (even better) set up many instances of these typical situations. The result is that the concepts we develop to describe physical phenomena depend not only on what we can observe, but also on what we can do.

To say that A affects or causes or influences or interacts with B implies a counterfactual: if A had been different, B would have been different, too. The most convincing way to establish a connection is to "wiggle" some parameter in A more or less randomly and then observe the same odd pattern showing up in some property of B. If I want to know whether a wall switch controls a certain light, I can flip the switch on and off and observe whether the light follows my actions. There is always the possibility that the light is being controlled by someone else or goes on and off spontaneously; but if I put the switch through a very irregular and spontaneous sequence of changes and the light still follows along, then the probability of a causal connection is very high (barring a conspiracy to deceive the experimenter).

Physical theory is possible because we *are* immersed and included in the whole process – because we can act on objects around the us. Our ability to intervene in nature clarifies even the motion of the planets around the sun – masses so great and distances so vast that our roles as participants seems insignificant. Newton was able to transform Kepler's kinematical description of the solar system into a far more powerful dynamical theory because he added concepts from Galileo's experimental methods – force, mass, momentum, and gravitation. The truly external observer will only get as far as Kepler. Dynamical concepts are formulated on the basis of what we can set up, control, and measure.

08 April 1999, "Delayed Entanglement"

When I first entered the University of Texas – coming out for the first time from my bucolic upbringing in the small town of Cuero, TX – one of the things that impressed me most was the level of intellectualism I found in the graffiti on the bathroom walls. A popular graffito at the time was: "God is love. Love is blind. But Ray Charles is blind. So Ray Charles is God?" (Ray Charles, in case you don't know, is a famous blues singer and piano player.) Despite the miscarriage of the syllogism, this was still light years ahead of what I normally read on the bathroom walls in my old home town!

Today while I was having a look at your new paper, I couldn't help but think about a new bathroom syllogism (i.e., an equally miscarried one) that I've been playing with for the last few weeks. "Knowledge is power. Power is a resource. But entanglement is a resource. So entanglement is knowledge?" OK, as advertised, there's no syllogism here. But the lesson is sound: "Entanglement is a kind of knowledge and that knowledge can be a resource."

The overall lesson of course is that knowledge is a resource. Actual entanglement is only surrogate to that. And that is what your example brings out nice and clearly. Nonclassical correlations can arise when one has a *nonclassical* state of knowledge, be that knowledge of present entanglement or instead delayed entanglement.

Let me just make a few almost trivial comments.

I don't really like the sentence you end Section 2 with. You say "Alice and Bob find experimentally that each one of the four subsets consisted of maximally entangled pairs." They don't really find that; they find post-selected measurement statistics consistent with measurement statistics that could have arisen from entangled pairs. I think there is an important point that shouldn't be misrepresented here. There is never entanglement in the more common sense, and yet there is a Bell inequality violation. "Nonlocality without Entanglement" – where've we heard that before? It's a completely differently flavored example, but the slogan still applies here (a little at least).

That's about it. As a general comment, I didn't find the example very surprising. But I've been your student for a long time. You've already taught me to "clearly understand quantum mechanics and firmly believe in its correctness." Nevertheless I think the example will still turn some heads – Richard Jozsa (who doesn't in his heart really believe QM) might be one of them. It's crisp and clean, and therefore not easily forgotten. It's good pedagogy.

10 April 1999, "My Turn"

A few weeks ago you pointed out a typo in my CV. Now it is my turn to return the favor. In the publication list you sent me, you list the following article: A. Peres, "Schrödinger's Immortal Cat," *Found. Phys.* **17** (1988) 57. That should be Vol. 18 instead.

The title intrigues me, but I have no clue what the article is about. The way I came across it was that I read your article "On Quantum-Mechanical Automata" (commenting on a paper by David Albert), and was intrigued by your phrase, "There are, however, some distinguished theorists who claim that irreversibility ... is not fundamental ..., in brief, that human ignorance or weakness has no role in physics. Although I may disagree on this point [15] I shall tentatively adopt this approach just for the sake of argument ..." That sent me to your CV to see what the mysterious Ref. [15] was. It was an article titled "The Physicist's Role in Physical Laws." And I thought, while I'm in the library, I might as well see what other (previously unseen) Peres tidbits I can find. Hence my finding that I could not find the article mentioned above.

In any case, I liked your anti-Albert article very much. It seems to me that it (along with your 1982 paper with Zurek) contains a point that is in need of much more exploration. (Particularly the point you make in the last three paragraphs of the paper.) In my heart I feel that you are on the right track with that, but you have only scratched the surface. In that connection, by the way, I have given your new appendix to RQM a first reading. But I want to withhold my comments until I can read it again more carefully, and have some time to construct a proper commentary. There is something about it that troubles me, something that is incongruous with the anti-Albert article I just mentioned. I will need some time to articulate that.

About the mysterious Ref. [15], the title itself almost sent me into Pavlovian salivation. So I dropped everything to find it. I made a copy of it for my files, but still I came back from the library with my head hung low. It was that – certainly through no fault of your own – the paper just didn't say what I had wanted it to say!! I can't have all my wishes, can I? Of course I very much like part of the sentiment you express at the beginning, "The physicist's role as an observer ... appear[s] ... in the formulation of the physical laws themselves." A large part of the foundation of my research program (as you know by now) is to give some substance to just this thought. But my tack on it is that the structure of quantum mechanics itself is just this sort of thing:

"In the beginning God created the heaven and the earth. And the earth was without form, and void ... [Day 1], [Day 2], [Day 3], [Day 4], [Day 5] ... And God saw everything that he had made, and behold, it was very good. And there was evening and there was morning, a sixth day. Thus the heavens and the earth were finished, and all the host of them." But in all that, there was still no science. For science must of its nature be a creation of man – it is the expression of man's attempt to be less surprised by this God-given world with each succeeding day. So the society of man slowly but surely set out to discover and form physical laws. Eventually a wonderful fact came to light: information gathering about the world is not without a cost. When I learn something about an object, you are forced to revise (toward the direction of more ignorance) what you could have said of it. And yet from our comparing our notes (and our comparing our notes with those of the larger community) we are still in a position of constructing a scientific theory of the things we observe. The world is volatile to our information gathering, but not so volatile that we have not been able to construct a successful theory of it. The *speculation* is that quantum theory is the unique expression of that happy circumstance: it is the best we can say in a world where information gathering and disturbance go hand in hand.

But I must get back to more serious things: my own *JMO* article needs finishing.

10 May 1999, "Eat My Words"

Also I found one very interesting piece of history that I guess I had forgotten. Karl Kraus in his paper, "General State Changes in Quantum Theory" [*Ann. Phys.* **64**, 311–335 (1971)], starts the introduction with:

The state of quantum systems may be changed by external interventions.[1] Haag and Kastler[2] have discussed a particular case of such state changes in algebraic quantum theory. They called them operations. The same notion has been used by Pool[3] in a lattice-theoretic framework.

You'll note in particular that he used the word "intervention" too. I really like that word much more than "measurement." (I hope also you've reinstated the word "chronicle" into your latest version.)

19 May 1999, "I Finally Decohere"

Asherism 34: *In the next section of this article, the notion of measurement will be extended to a more general one: an intervention. This is the interface of the classical (Boolean) world and the quantum world.*

Do you really mean to say that the notion of intervention is "more general" than measurement? Or, instead that it is what we have always had, but now stated in a crisper language? The word "measurement" is one that fits more within the classical world view: it seems to me, part of your great contribution is to point out that limitation in a nice manner.

On a different point, what should the cryptic use of the word "Boolean" be taken to mean in the reader's mind? I like its insertion there, but I'm not sure what to make of it.

Asherism 35: *Bell pointed out that the notion of measurement, or observation, was logically inconsistent in a genuine quantum world.*

---

[1] Familiar examples are the state change due to an ideal measurement ("reduction of the wave packet"), or the state change produced by the action of an external field. In the Heisenberg picture used here, the state is constant in the absence of such external interventions.

[2] R. Haag and D. Kastler, *J. Math. Phys.* **5** (1964), 848.

[3] J. C. T. Pool, *Comm. Math. Phys.* **9** (1968), 118.

I don't like this phraseology. It is my working assumption that the world really is *genuinely quantum*. But by that I mean that the best description I can make of the world's workings must make use of the quantum formalism. What you mean however (I think) is that Bell was considering a kind of world where the wave function is reified so to speak: a world in which there is an overall wave function controlled by unitary dynamics that is a state of nature, not a state of knowledge.

I think a simple substitution of a different adjective than "genuine" would do the trick for me.

Asherism 36: *The probabilities of the various outcomes of an intervention can be predicted by using a suitable theory, such as quantum theory.*

These "outcomes" are the "creations" that you speak of later? That is probably a good way to put it. David Mermin once asked something like: "Where there is subjective probability, there is ignorance. What are quantum probabilities to signify our ignorance of if the ignorance cannot be of objective properties?" The outcomes of our interventions, I say!

Asherism 37: *Note that the "detector clicks" are the only real thing we have to consider. Their probabilities are objective numbers and are Lorentz invariant. On the other hand, wave functions and operators are nothing more than abstract symbols. They are convenient mathematical concepts, useful for computing quantum probabilities, but they have no real existence in Nature.*

I very much agree with the first sentence of this. But I don't like (or agree with) the distinction you draw thereafter. If wave functions or density operators are not objective entities, then neither are the probabilities we calculate from them through Born's rule. This is the Bayesian point of view: probabilities are subjective numbers to aid in our decision making and gambling (in the case of physics, gambling with the world). Period. It is inconsistent to take one as objective but not the other. And there are perfectly adequate arguments to show that neither is objective, i.e., agent independent.

Asherism 38: *Our partial ignorance is not a sign of weakness. It is fundamental. If everything were known, acquisition of information would be a meaningless concept.*

I like this point very much. In fact I had wanted to have a discussion along these lines with Rolf Landauer before he fell ill. "What can you mean by your phrase 'Information is Physical'? Was information physical before there were physicists doing physics? I get the feeling that that is not what you mean. But then why call what you are talking about 'information'?" I never got the chance to have the discussion.

Asherism 39: *Clearly, quantum jumps are not dynamical processes. They result from the introduction of an apparatus, followed by its deletion or that of another subsystem. The jump in the quantum state occurs because there are abrupt changes in our way of delimiting the object we consider as the quantum system under study.*

Regardless of the formalism you've introduced and the discussion of decoherence you've given up to this point, I still don't believe you've given a convincing argument for this point. From my point of view, the random jump that we find upon measurement is simply not contained within the quantum formalism. This is not a blemish to the theory: it is just a fact.

Asherism 40: *Consider the evolution of the quantum state in the Lorentz frame where intervention* A *is the first one to occur and has outcome* μ, *and* B *is the second one, with outcome* ν. *Between these two events, nothing actually happens in the real world. It is only in our mathematical calculations that there is a deterministic evolution of the state of the quantum system. This evolution is not a physical process.*

I like this very much. It reminds me of a quote of Pauli that I often repeat. In a 1947 letter to Fierz, he wrote:

Something only really happens when an observation is being made [...]. Between the observations nothing at all happens, only time has, 'in the interval,' irreversibly progressed on the mathematical papers!

Also it reminds me of quote of Heisenberg that I just read recently. Henry Stapp in a recent paper quotes him as saying:

Since through the observation our knowledge of the system has changed discontinuously, its mathematical representation also has undergone the discontinuous change, and we speak of a "quantum jump".

A real difficulty in understanding the interpretation occurs when one asks the famous question: But what happens "really" in an atomic event?

If we want to describe what happens in an atomic event, we have to realize that the word "happens" can apply only to the observation, not to the state of affairs between the two observations.

Asherism 41: *What distinguishes the intermediate evolution between interventions from that due to the intervention itself is that the latter has unpredictable output parameters, for example which one of the detectors "clicks," thereby starting a new chapter in the history of the quantum system.*

Did you want to say "history" or "record" here? I like history better – it has a nicer flow – but you had earlier said that you wouldn't use the word history.

02 June 1999, "9906006"

In all, I'm not sure what to make of Meyer's result. Strictly speaking – without placing an interpretation on the mathematics – it is that all the orthonormal triads in $\mathbf{R}^3$ with rational Cartesian coordinates can be consistently colored. One could ask a similar question about rational-angled spherical coordinates, and maybe the result changes drastically, but Meyer didn't do that. Or one might consider sets for which the θ and ϕ for the spherical coordinates are exclusively rational fractions of π, but Meyer didn't do that either. Instead, he wants to derive a *physical* conclusion from the result based on the fact that the rationals are dense in the reals: thus the issue of experimental verifiability.

But I'm not too sure what it even means to speak of an experimental test of the Kochen–Specker theorem. We can never measure all these overlapping triads simultaneously. (Didn't Cabello and Garcia-Alcaíne once address the issue of the "experimentalization" of KS? – that memory just came back to me.)

In my mind, if there's anything interesting to be learned here it is that the character of quantum mechanics might change drastically if we were to imagine the Hilbert spaces of the theory to be vector spaces over the field of (complex) rational numbers, rather than the full set of complex numbers. Several in the past (Adler, Wootters, Stueckelberg, etc.) have pointed out oddities in the theory if the base field were taken to be the reals instead of the complex numbers. But I think this – Meyer's paper – may be the first to consider the oddities of the theory if posed over the rationals.

26 August 1999, "A Contextual Thought"

The following thought just popped to mind; let me pass it by you. Would you call our "nonlocality without entanglement" result a "contextuality" result in the tradition of the BKS idea? We didn't say that in the paper, but it seems that it has a bit of the flavor. Charlie instead wrote the following in the paper:

In what sense is a locally immeasurable set of states "nonlocal?" Surely not in the usual sense of exhibiting phenomena inexplicable by any local hidden variable (LHV) model. Because the ψ_i are all product states, it suffices to take the local states α_i and β_i, on Alice's and Bob's side respectively, as the local hidden variables. The standard laws of quantum mechanics (e.g. Malus' law), applied separately to Alice's and Bob's subsystems, can then explain any local measurement statistics that may be observed. However, an essential feature of classical mechanics, not usually mentioned in LHV discussions, is the fact that variables corresponding to real physical properties are *not* hidden, but in principle measurable. In other words, classical mechanical systems admit a description in terms of local *unhidden* variables. The locally immeasurable sets of quantum states we describe here are nonlocal in the sense that, if we believe quantum mechanics, there is no local unhidden variable model of their behavior. Thus a measurement of the whole can reveal more information about the system's state than any sequence of classically coordinated measurements of the parts.

It seems that one can explain nonlocality without entanglement with a local hidden variable model ... but only via hidden variables that are necessarily disturbed by the measurement process. That's something like a contextual hidden variable theory, isn't it? I wonder how the Clifton–Kent model fares with respect to this?

Do you think there's anything to this?

23 September 1999, "End of the Day"

Asherism 42: *Yesterday in my mail I got two related items. One was an e-mail from the PRA editor, asking my opinion on proposed changes in PACS: 03.65.Bz, which* *includes now too many papers, would be split into 10 subsections. One of them is "Foundations and interpretations of QM". I objected to an explicit mention of "interpretations" and Bernd Crasemann agreed with me. The other item was Physics Today of August, with a paper by our consistent historian friends Bob and Roland (I have not yet read it).*

This follows a couple of papers on Bohmian trajectories, and led me to wonder whether the PT editors have lost their mind and I (or we) should try to educate our colleagues and explain that quantum mechanics needs no interpretation.

You've heard the story that Rudolf Peierls refused to use the words "Copenhagen interpretation." For according to him, "the Copenhagen interpretation IS quantum mechanics."

I haven't read the Griffiths/Omnès article yet. But you do know that this subject has been getting under my skin lately. Let me read it over the weekend and see what I think ... though I'm almost sure what my conclusions will be. What about we perhaps write a letter to the editor of *PT* concerning it ... if we turn out to have similar opinions. Or were you thinking of a larger project?

Now to something serious. I know you know Wigner's marvelous theorem on symmetries; I think you described it in your book. If we have a bijection from unit vectors to unit vectors on a Hilbert space that preserves all inner products, then that mapping must either be a unitary linear map or an antiunitary antilinear map. If we further suppose that mappings are continuously connected to the identity, then we are left with the unitaries.

Here is my question. Can something of the same flavor be proven within classical physics? What I am thinking of is this. Start with a phase space, and consider the space of all normalized probability densities over that space. (These functions all live within the positive cone of the vector space of ALL square-integrable functions over the phase space.) Let us suppose that whatever dynamics is, it must be a bijection on the probability densities that preserves overlaps. Will that be enough to give a Wigner-like theorem for classical physics: i.e., that all allowed evolutions must satisfy a Liouville equation.

Have you seen this proven anywhere rigorously? If so, can you give me a reference?

24 September 1999, "More Peierls"

Asherism 43: *Have you a reference for that phrase? I asked Ady Mann who was Peierls's post-doc, and he promised to try to find.*

That's a good question: what is the origin of that phrase? I know that I have seen it in print, but presently I don't recall from where. The first time I used it was in a letter to my friend Greg Comer on 26 July 1993. I know that because I did a quick look-up yesterday: I was thinking for a moment that Rosenfeld had said it, and I wanted to make sure I was correct. To my surprise I found Peierls. (And by the way, I just discovered that I misspelled Peierls yesterday; I'm probably the fault of your doing the same.)

26 September 1999, "Wigner"

Asherism 44: *I have no immediate answer about Wigner's theorem for Liouville functions. The proof given by Wigner is for discrete Hilbert spaces, and Liouville's equation has a continuous spectrum. I assume that you are familiar with Koopman's theorem (pp. 317–319 in my book). I'll think about it, but I don't see the physical motivation.*

Essentially what I am asking for is a converse to Koopman's theorem. The physical motivation is simple: it would mean that Newton's laws are almost content free ... that they almost couldn't have been anything else once one has established that what classical physics is about is the evolution of probability distributions over phase space (x, p). To say that one knows a system is isolated and evolving according to some intrinsic dynamical law is to say that one is neither losing information about the system nor gaining information about the system. A way of mathematizing that would seem to be simply the statement that the overlaps of any two unknown probability distributions should not change with time.

Maybe this sort of question is tackled in Arnold's book on classical mechanics? Or perhaps in some of Prigogine's work?

Anyway, that's the idea.

27 September 1999, "contra-Koopman"

Many thanks for sending the further references about the Liouvillian.

Asherism 45: *Anyway, your idea of an anti-Koopman theorem is probably correct from the point of view of mathematics. I am not yet convinced by the physical motivation, so that you should make it clearer.*

I don't presently know how to say it much clearer than I already have. I will work on that. In the mean time, though, let me get Asher Peres to say a few words about the subject:

... a quantum state is *not* the analog of a point in the classical phase space. The classical analog of a quantum state is a *Liouville probability density*. [p. 347, *Quantum Theory: Concepts and Methods*]

and

Since there are analogies between classical and quantum mechanics, why not try to use quantum methods for solving classical problems. [p. 317, *Quantum Theory: Concepts and Methods*]

It seems to me that we just have potentially (if all works out) a new way of thinking about Hamiltonian dynamics. It comes not so much from the "mysterious" (to me at least)

Principle of Least Action, but rather from the requirement that statistical distinguishability never increase.

While I'm quoting from great men, let me also call Max Born to my aid until I can say things better myself.

In his lecture "In Memory of Einstein" he writes:

Einstein, however, stuck to his opinion. ... It was in fact a matter of a fundamental difference in our views of nature. ... For years afterwards I kept turning over in my mind the philosophy which was behind my theory, and then gave a very brief summary of it in the Festschrift in honor of Heisenberg's 60th birthday. What it boils down to is that scientific forecasts do not refer directly to 'reality,' but to our knowledge of reality. This means that the so-called 'laws of nature' allow us to draw conclusions from our limited approximate knowledge at the moment on a future situation which, of course, can also be only approximately described.

And more pertinently, in his essay "Is Classical Mechanics In Fact Deterministic?" he writes:

To summarize, we may say that it is not the introduction of the indeterministic statistical description which places quantum mechanics apart from classical mechanics, but other features, above all the concept of the probability density as the square of a probability amplitude $P = |\Psi|^2$; the phenomenon of probability interference results from this, and therefore it is impossible to apply without modification the idea of an "object" to the mass particles of physics: the concept of physical reality must be revised. This however is beyond the scope of these elementary considerations.

What I'm thinking is that a Wigner-like theorem is part of the "indeterministic statistical description" that is common to both classical and quantum mechanics. It is just another way of helping to clarify where the two theories diverge.

18 December 1999, "KS Crypto"

Asherism 46: *We are working on qutrit cryptography – it's not completely silly and quite amusing, because related to KS theorem.*

That sounds quite interesting actually. I look forward to seeing your results. I have often wondered whether there is an interesting "phase transition" in the information vs. disturbance curves when one goes from doing q-crypto with a colorable set of bases to doing q-crypto with a noncolorable set. Just as a first guess, one might suspect that a noncolorable set of vectors would be intrinsically more secure: though with so many bases in the protocol the data rates would be quite atrocious.

19 December 1999, "State of Confusion"

Many thanks for the PostScript file and the note. However, now I'm really confused. I don't understand the motivation for considering tri-colorings, etc., etc. Yesterday when I wrote you, I thought you were talking about the standard KS colorings. No need to bother with me any more: I'll just wait until your paper appears on quant-ph or if you send a review copy to me beforehand.

27 December 1999, "Fuchs, but not Laflamme"

About Steven van Enk's comment on David Meyer: it is about his paper on quantum games. Steven's comment, as you will read, is pretty much on the mark. As I recall, he made three points. 1) Meyer restricts himself to two-level systems, so in principle there is nothing quantum mechanical about his example. Simple two-level systems can always be simulated by an underlying hidden-variable theory. 2) Meyer's example could have just as well been constructed using classical wave mechanics. In quantum computing, of course, there is the extra issue of scaling – classical wave computers have an exponential scaling of resources (energy, space) that quantum computers don't – but for Meyer's quantum game that is not an issue. Meyer claims to have shown something interesting at the simple qubit level. 3) This one, I think, is the most devastating and at the same time the most obvious. In Meyer's game, one player must abide by the rules of classical physics while the other one has the full freedom of quantum mechanics. It is no wonder that one player can win all the time! There is nothing here that is not in ANY two-player game (regardless of physics): if one player obeys the rules, while the other CHEATS then of course the latter will always win. van Enk shows an explicit example of this in the classical context: it is precisely Meyer's game, but with one player having the extra power to stand a coin on its edge, rather than just flipping it. That player, of course, can always win ... and there is nothing quantum mechanical about this example.

One interesting tidbit about van Enk's comment and Meyer's rebuttal. Meyer thanks both Raymond Laflamme and me for discussions, while van Enk simply thanks me. Steven, for a while, toyed with the idea of writing, "I thank Chris Fuchs, but not Raymond Laflamme, for discussions." (That made me chuckle, though he certainly didn't mean it as an insult. Steven doesn't even know Laflamme.) Anyway, while they both thank me, I am on Steven's side with this one.

09 January 2000, "KS Cryptoland"

I had a chance to give your paper with Helle a brief "once over." Let me just make a couple of comments.

1) I very much liked the mutually unbiased basis example. It has so much pretty symmetry, it makes one wonder if it might not be (easily) amenable to a full analysis for the information–disturbance tradeoff as we did in *FGGNP* (and I believe Dagmar did for the three-basis protocol).

2) However I was a little dissatisfied with your so-called KS analogue. I think it would be much more interesting to look at how well a real KS example fares. By that, I mean something like the 33 vector, 16 basis example that you found, or the 31 vector, 17 basis example that Conway and Kochen found. In those cases, there can be no underlying noncontextual hidden variable carrying the key information between Alice and Bob. What is the connection between that and

the information–disturbance tradeoff that quantum mechanics supplies? Is there any connection? I think these questions are quite interesting from a fundamental point of view.

Would you be interested in collaborating on questions of this sort during my visit to Haifa? Perhaps also with Helle or whoever else is interested?

Tomorrow I go to Albuquerque for my weekly collaboration with Caves. Tuesday, however, I plan to start searching for tickets for my visit to you.

09 March 2000, "Petra"

Asherism 47: *Curiously, there is another student who is going to work with me by remote control. Her name is Petra Scudo, and she just got a MSc from Pavia. She has family in Israel and came to visit me just before your visit (she came with her mother and little sister). Chiara likes her and promised to help guiding her until she arrives here in the fall. The project I have in mind for her involves data analysis, probably by MaxEnt methods. Where can she read about Bayesian statistics? Once you mentioned to me a good book but I don't remember which one.*

The very best reference is:

E. T. Jaynes, *Probability Theory: The Logic of Science*. This already-huge book was, very unfortunately, never finished before Prof. Jaynes' death. An incomplete draft dated 13 May 1996 is available on the World Wide Web at http://bayes.wustl.edu/.

Beside that though, she can find a lot of information in the following series of books:

R. D. Levine and M. Tribus, eds., *The Maximum Entropy Formalism*, (MIT Press, Cambridge, MA, 1979); C. R. Smith and W. T. Grandy, Jr., eds., *Maximum-Entropy and Bayesian Methods in Inverse Problems*, (D. Reidel, Dordrecht, 1985); J. H. Justice, ed., *Maximum Entropy and Bayesian Methods in Applied Statistics*, (Cambridge University Press, Cambridge, 1986); J. Skilling, ed., *Maximum Entropy and Bayesian Methods: Cambridge, England, 1988*, (Kluwer, Dordrecht, 1989); P. F. Fougère, ed., *Maximum Entropy and Bayesian Methods: Dartmouth, U.S.A., 1989*, (Kluwer, Dordrecht, 1990); B. Buck and V. A. Macaulay, eds., *Maximum Entropy in Action: A Collection of Expository Essays*, (Clarendon Press, Oxford, 1991); W. T. Grandy, Jr. and L. H. Schick, eds., *Maximum Entropy and Bayesian Methods: Laramie, Wyoming, 1990*, (Kluwer, Dordrecht, 1991); C. R. Smith, G. J. Erickson, and P. O. Neudorfer, *Maximum Entropy and Bayesian Methods: Seattle, 1991*, (Kluwer, Dordrecht, 1992); A. Mohammad-Djafari and G. Demoment, eds., *Maximum Entropy and Bayesian Methods: Paris, France, 1992*, (Kluwer, Dordrecht, 1993); G. R. Heidbreder, *Maximum Entropy and Bayesian Methods: Santa Barbara, California, U.S.A., 1993*, (Kluwer, Dordrecht, 1996); J. Skilling and S. Sibisi, *Maximum Entropy and Bayesian Methods: Cambridge, England, 1994*, (Kluwer, Dordrecht, 1996); K. M. Hanson and R. N. Silver, *Maximum Entropy and Bayesian Methods: Santa Fe, New Mexico, U.S.A., 1995*, (Kluwer, Dordrecht, 1996).

07 April 2000, "Why Fidelity?"

I finally(!) had a chance to read your note "why fidelity?" Has Dagmar responded to you? I would like to hear her thoughts.

Asherism 48: *It is customary to evaluate the quality of a quantum transmission, or cloning, or other set of operations by the "fidelity" of the result. What is the justification for that, other than the simplicity of the formula? It seems to me that the use of fidelity as a criterion is completely arbitrary.*

I think you are on the mark here: it's pretty much arbitrary as far as I can tell too. It seems to me that distinguishability measures or closeness measures should always be motivated by the context of the problem when one can do that. If a problem (like optimal cloning) doesn't lead uniquely to one measure over another, then perhaps one should reexamine one's motivation for looking at the problem in the first place.

In our FP96 paper, we considered mutual information as a measure for quantifying what Eve wanted and fidelity for quantifying how well she was hidden. I think the latter of these wasn't too bad. The fidelity in this case corresponds exactly to the probability of Alice and Bob detecting Eve. On the other hand, mutual information may have been a little ad hoc on our part. What if Eve had wanted to optimize her ability to guess the identity of the signal and not just sharpen the posterior probabilities as much as possible? Then she would have used Helstrom's techniques instead of Holevo's. What should she really do for the somewhat related cryptographic problem. Probably still something different. (I'm sure Tal Mor has two or three different answers for that!)

Asherism 49: *In general, I'll want to have a classical criterion for any process, even if quantum notions are used at intermediate steps. As Niels Bohr told us long ago, the account of any experimental procedure has to be done in a classical language, because we should tell "what we have done and what we have learned".*

I like this philosophy. But even classically there is no unique solution independent of the context. And thus I don't see the following as cut and dried as you do:

Asherism 50: *Thus, if the input is a discrete set of signals, the natural measure for quality of transmission (or of identification) is the resulting mutual information.*

There has been some work in the classical information theory literature trying to (rigorously) justify the mutual information as an interesting quantity outside of the channel capacity scenario. But none has been completely successful in my eyes. Here are a couple of references.

1. D. V. Lindley, "On a measure of the information provided by an experiment," *The Annals of Mathematical Statistics*, vol. 27, pp. 986–1005, 1956.
2. J. L. Kelly, Jr., "A new interpretation of Information Rate," *The Bell System Technical Journal*, p. 917, July 1956.

22 April 2000, "Cats and Kitties"

Recently you wrote me:

Asherism 51: *Another subject about which I had been thinking is our old friend the Cat. Of course there is no pure |living⟩ state. The initial state is represented by a density matrix, or rather a class of density matrices (if we remove one atom from the cat, it still is a living cat). The "complementary" set of matrices – whatever this means – represents dead cats. How can we clone a cat? We don't need to reproduce all the details. Only a restricted set of characteristics are important. I don't know how to quantify that.*

Let me just ramble a bit about this. I think the answer to your question, at least partially, lies in our old "no-broadcasting" paper [*PRL* **76**, 2818 (1996), also see Lindblad, *Lett. Math. Phys.* **47**, 189 (1999)]. There the question was, when can an "unknown" density operator be cloned in a suitably general way – i.e., the kind of way one usually thinks of when one is thinking of a classical record being made. The answer was – perhaps not surprisingly (though it was really hard to work out) – if and only if the "unknown" density operator is drawn from a known commuting set.

Once that is in hand, it seems to me, the main question becomes this. Why is it that most states that we are willing to ascribe to a class of living beings are commuting, or at least, awfully close to commuting? Wojciech – I'm sure – would like to think that it has something to do with some kind of environmentally induced superselection rules (and thus depends crucially on the characteristics of the Hamiltonia around us). But I suspect it has almost nothing to do with that. Instead I would be willing to bet it's this. Whenever we have a system so large and so complex that we would be willing to call it "living," we will most surely know so little about it that our density operator assignment will be very close to being proportional to the identity operator. All such states are very nearly commuting, and therefore, very nearly broadcastable.

That is the direction I would like to flesh out more fully at some point in the future.

By the way, let me give you another tip for a paper that has something to say which I perceive as (very) loosely connected to this issue. P. Busch and J. Singh, "Lüders Theorem for Unsharp Quantum Measurements," *Phys. Lett.* A 249, 10 (1998). It's kind of a trivial result, but you might enjoy it in a way. (I did.)

Finally, let me send you the little review I wrote on Wigner's thoughts on reproduction and cloning. It is published at Dan Gottesman's Quick Reviews in Quantum Computation and Information webpage: http://quickreviews.org/qc/. [See note to Daniel Gottesman, dated 3 November 1998 (Other Letters).]

24 April 2000, "Fuzzy Cats"

Thanks for the long note. I will have a thorough look at it tomorrow. Most of today I will be preparing and then giving a lecture at LANL. Let me just make a quick comment to an early paragraph.

Asherism 52: *You wrote that living things would be represented by commuting density operators, and mentioned environmentally induced superselection rules (what's that?).*

I mentioned that Wojciech Zurek would say that it had something to do with environmentally induced superselection rules. I certainly didn't endorse that, as I made clear. So the two ideas shouldn't be lumped back together. In any case, you ask what is EISR about? Only Wojciech's main research program in quantum foundations for the last 20 years! (See *PRD* **26**, 1862 (1982) for the phrase, and *Phil. Trans. R. Soc. Lond.* A **356**, 1793 (1998) for its latest mutation.)

Asherism 53: *Here I must disagree: take a living cat, and translate it by 0.1 nanometer. It's still a living cat and is not orthogonal to the preceding one.*

I said "commuting or almost commuting" (the latter phrase being something awaiting a suitable quantification). I never said anything about the states being orthogonal.

But I'll be back for a more thorough read of your note later.

12 July 2000, "Two Answers"

You've asked me at least two questions in my backlog of email.

Asherism 54: *Were the conferences you attended interesting? Any new discovery?*

Actually both were fairly interesting. At Mykonos, I met one person of real note, Arkady Plotnitsky. David Mermin had told me that I would enjoy his company and it was quite true. He was a real oddity at this meeting. He is a Professor of English at Purdue University but is absolutely enamored with Bohr's writings. I felt that his understanding of Bohr was quite deep. In case you have seen him in print (he wrote a book titled *Complementarity*), I know that he doesn't look so good there. I couldn't understand a word of his book and gave up very quickly; but, I know him enough now to know that that is a misimpression of his real intellect. I challenged him to write an essay comparing and contrasting Bohr's and Pauli's ideas on the foundations of quantum mechanics, and he has accepted the challenge. So I am enthusiastic to see the result of that.

Another person I took the time to get to know better in Mykonos was Roland Omnès. I think that was quite fruitful too, though he remains a little of an enigma to me. His ideas seem strangely closer to Copenhagen than his *Physics Today* article with Griffiths reveals.

At Capri, I think the most exciting thing was a talk by Nicolas Gisin. He seems to have found an interesting classical "analogue" to entanglement, and even to bound entanglement. I think his paper with Stefan Wolf (which I believe is on quant-ph) is well worth studying.

24 August 2000, "Expenses, Reimbursements, and Memories"

Thank you so for sending me the Enz article on Pauli. The fire has not only caused me to retabulate my old properties. It has also made me painfully aware of how fragile all my thoughts are, and how like anything associated with a computing machine they should be backed up. To that end, I have also been working very hard to recall everything in my various file cabinets. One of those cabinets was devoted to what I call the Paulian idea. I've done a good bit of a reconstruction job on it, recollecting papers and the like, but this time I'm making sure to make it more electronically accessible. In the next mail, I'll send you the fruits of that. I deem that the project is about 1/4 of the way done. When it is done completely, I will look for a more permanent repository.

01 November 2000, "Catching Up"

Thank you for your long letter. I'm glad to hear that you have family in Morristown; it never ceases to amaze me how small the world is.

I have a funny anecdote to tell in that regard. One morning during my last week at Los Alamos, I was getting out of my car when I saw that Stirling Colgate was getting out of his just three further down the row. I knew who Stirling was, but I had never met him. We both started to walk toward the Theory Building. His nose was buried in a scientific paper, and he was walking quite slowly. I sped up to overtake him, but strangely he sped up too; he never looked up from the paper. Now walking beside him, I sped up once again. And so did he! Finally, after an uncomfortable moment, he said, "If I walk beside you, I can read a little more easily. You steady me." I said, "Oh, like a seeing-eye dog." We then walked further, but I was uncomfortable with the situation. Finally, I couldn't help but break the silence. I said, "Funny, I was just in a town last week where seeing-eye dogs are trained." Stirling said, "Morristown, New Jersey." I said, "Yeah, how did you know?!?!" He said, "My dad started that service." In shock, I said "Really?" He said, "Well, my dad and Mrs. Eustice." So it is a small world.

Thank you again for the offer of sending some books. You will find a willing recipient in me! I'll place both my addresses below; you can send them to either one you wish.

One of these days I will have to reply to Todd. There is nowhere *anywhere* in the Cathy–Erwin example where a *single* system is given two distinct pure state assignments. He is confused on this point for the simple reason that he hasn't absorbed our main point: Quantum states are states of knowledge, not states of nature. One of these days I will break down and formulate a reply.

But now I've got to get to more serious things.

23

Diary of a Carefully Worded Paper: More Letters to Asher Peres

"Partners in crime again?"
CAF to AP
29 September 1999

Introduction

Irving John Good once titled a paper, "46,656 Varieties of Bayesians." With little stretch of the imagination, one can believe that a similar paper on the interpretations of quantum mechanics exists somewhere out there too. And that may be true even though 46,656 is greater than the whole membership of the American Physical Society! It is no wonder that taking a stand on quantum foundations is treacherous business. Nevertheless, in the March 2000 issue of *Physics Today*, Asher Peres and I flouted the usual rewards of a calm life and did just that. Our article was titled, "Quantum Theory Needs No 'Interpretation'."

This chapter is about the evolution of that article and a later "reply to critics." When I was an undergraduate at the University of Texas, Bryce DeWitt made a great impression on me by saying something I have not forgotten: "We learn mathematics so that we don't have to *think* when we do physics." My paper with Asher contained no mathematics. We had to think, and think very hard. The fruits of that labor, for my part, were in a greatly enhanced keenness for seeing through various foundational issues. How I wanted to share that newfound keenness with the reader! But it was difficult, if not impossible, to convey such an aspect of the paper-writing process in the very paper itself. I could not give the reader as much as I would have liked.

I ask, "What substitute can I give for the lucky gamble I put myself through: that is, having had enough (unfounded) faith to go down one path and then overcome enough trials to build an overpowering confidence that the path taken was the *right* path?" There is no full-scale substitute, but here I wish to further the point of view Asher and I put forward in those articles by opening the curtains on the process of writing itself. After a reprint of the final article and the "reply to critics," I give a play-by-play account of (predominantly)

326

my side of things as the two articles were hammered out. In this chapter, I have opted to be unusually open, revealing many letters that are only tangential to the actual words in the articles: they set the scene, and thus form part of the reality behind the science that Herb Bernstein so often emphasizes. The interest in this, however, is not only for the historian of science, but for the physicist with the nagging feeling that perhaps all is well with quantum mechanics after all: look at the trials we faced, and see the way we hammered. The deeper physics will be found by taking quantum mechanics at face value, rather than resorting to the tools of science fiction to slay an imagined beast.[1]

A. *Quantum Theory Needs No 'Interpretation'*

Quantum Theory Needs No 'Interpretation'
Christopher A. Fuchs and Asher Peres
Physics Today **53**(3), 70–71 (2000)

Recently there has been a spate of articles, reviews, and letters in PHYSICS TODAY promoting various "interpretations" of quantum theory (see March 1998, page 42; April 1998, page 38; February 1999, page 11; July 1999, page 51; and August 1999, page 26). Their running theme is that from the time of quantum theory's emergence until the discovery of a particular interpretation, the theory was in a crisis because its foundations were unsatisfactory or even inconsistent. We are seriously concerned that the airing of these opinions may lead some readers to a distorted view of the validity of standard quantum mechanics. If quantum theory had been in a crisis, experimenters would have informed us long ago!

Our purpose here is to explain the internal consistency of an "interpretation without interpretation" for quantum mechanics. Nothing more is needed for using the theory and understanding its nature. To begin, let us examine the role of experiment in science. An experiment is an active intervention into the course of Nature: We set up this or that experiment to see how Nature reacts. We have learned something new when we can distill from the accumulated data a compact description of all that was seen and an indication of which further experiments will corroborate that description. This is what science is about. If, from such a description, we can *further* distill a model of a free-standing "reality" independent of our interventions, then so much the better. Classical physics is the ultimate example of such a model. However, there is no logical necessity for a realistic worldview to always be obtainable. If the world is such that we can never identify a reality independent of our experimental activity, then we must be prepared for that, too.

The thread common to all the nonstandard "interpretations" is the desire to create a new theory with features that correspond to some reality independent of our potential experiments. But, trying to fulfill a classical worldview by encumbering quantum mechanics with hidden variables, multiple worlds, consistency rules, or spontaneous collapse, without any improvement in its predictive power, only gives the illusion of a better understanding. Contrary to those desires, quantum theory does *not* describe physical reality. What it does is provide an algorithm for computing *probabilities* for the macroscopic events ("detector clicks") that are the consequences of our experimental interventions.

---

[1] See note to David Mermin, dated 23 July 2000.

This strict definition of the scope of quantum theory is the only interpretation ever needed, whether by experimenters or theorists.

Quantum probabilities, like all probabilities, are computed by using any available information. This can include, but is not limited to information about a system's preparation. The mathematical instrument for turning the information into statistical predictions is the probability rule postulated by Max Born [1]. The conclusiveness of Born's rule is known today to follow from a theorem due to Andrew Gleason [2]. It is enough to assume that yes–no tests on a physical system are represented by projection operators P, and that probabilities are additive over orthogonal projectors. Then there exists a density matrix ρ describing the system such that the probability of a "yes" answer is $\text{tr}(\rho P)$. The compendium of probabilities represented by the "quantum state" ρ captures everything that can meaningfully be said about a physical system.

Here, it is essential to understand that the validity of the statistical nature of quantum theory is not restricted to situations where there are a large number of similar systems. Statistical predictions do apply to single events. When we are told that the probability of precipitation tomorrow is 35%, there is only one tomorrow. This tells us that it is advisable to carry an umbrella. Probability theory is simply the quantitative formulation of how to make rational decisions in the face of uncertainty.

We do not deny the possible existence of an objective reality independent of what observers perceive. In particular, there is an "effective" reality in the limiting case of macroscopic phenomena like detector clicks or planetary motion: any observer who happens to be present would acknowledge the objective occurrence of these events. However, such a macroscopic description ignores most degrees of freedom of the system and is necessarily incomplete. Can there also be a "microscopic reality" where every detail is completely described? No description of that kind can be given by quantum theory, nor by any other reasonable theory. John Bell formally showed [3] that any objective theory giving experimental predictions identical to those of quantum theory would necessarily be nonlocal. It would eventually have to encompass everything in the universe, including ourselves, and lead to bizarre self-referential logical paradoxes. The latter are not in the realm of physics; experimental physicists never need bother with them.

We have experimental evidence that quantum theory is successful in the range from 10^{-10} to 10^{15} atomic radii; we have no evidence that it is universally valid. Yet, it is legitimate to attempt to extrapolate the theory beyond its present range, for instance, when we probe particle interactions at superhigh energies, or in astrophysical systems, including the entire universe. Indeed, a common question is whether the universe has a wavefunction. There are two ways to understand this. If this "wavefunction of the universe" has to give a complete description of everything, including ourselves, we again get the same meaningless paradoxes. On the other hand, if we consider just a few collective degrees of freedom, such as the radius of the universe, its mean density, total baryon number, and so on, we can apply quantum theory only to these degrees of freedom, which do not include ourselves and other insignificant details. This is not essentially different from quantizing the magnetic flux and the electric current in a SQUID while ignoring the atomic details. For sure, we can manipulate a SQUID more easily than we can manipulate the radius of the universe, but there is no difference in principle.

Does quantum mechanics apply to the observer? Why would it not? To be quantum mechanical is simply to be amenable to a quantum description. Nothing in principle prevents us from quantizing a colleague, say. Let us examine a concrete example: The observer is Cathy (an experimental physicist) who enters her laboratory and sends a photon through a beam splitter. If one of her detectors is activated, it opens a box containing a piece of cake; the other detector opens a box with a piece of fruit. Cathy's friend Erwin (a theorist) stays outside the laboratory and computes Cathy's wavefunction. According to him, she is in a 50/50 superposition of states with some cake or some fruit in her stomach. There is nothing wrong with that; this only represents his knowledge of Cathy. She knows better. As soon as one detector was activated, her wavefunction collapsed. Of course, nothing dramatic happened to her. She just acquired the knowledge of the kind of food she could eat. Some time later, Erwin peeks into the laboratory: Thereby he acquires new knowledge, and the wavefunction he uses to describe Cathy changes. From this example, it is clear that a wavefunction is only a mathematical expression for evaluating probabilities and depends on the knowledge of whoever is doing the computing.

Cathy's story inevitably raises the issue of reversibility; after all, quantum dynamics is time-symmetric. Can Erwin undo the process if he has *not yet* observed Cathy? In principle he can, because the only information Erwin possesses is about the consequences of his potential experiments, not about what is "really there." If Erwin has performed no observation, then there is no reason he cannot reverse Cathy's digestion and memories. Of course, for that he would need complete control of all the microscopic degrees of freedom of Cathy and her laboratory, but that is a practical problem, not a fundamental one.

The peculiar nature of a quantum state as representing information is strikingly illustrated by the quantum teleportation process [4]. In order to teleport a quantum state from one photon to another, the sender (Alice) and the receiver (Bob) need to divide between them a pair of photons in a standard entangled state. The experiment begins when Alice receives another photon whose polarization state is unknown to her but known to a third-party preparer. She performs a measurement on her two photons – one from the original, entangled pair and the other in a state unknown to her – and then sends Bob a classical message of only two bits, instructing him how to reproduce that unknown state on his photon. This economy of transmission appears remarkable, because to completely specify the state of a photon, namely one point in the Poincaré sphere, we need an infinity of bits. However, this complete specification is not what is transferred. The two bits of classical information serve only to convert the preparer's information, from a description of the original photon to a description of the one in Bob's possession. The communication resource used up for doing that is the correlated pair that was shared by Alice and Bob.

It is curious that some well-intentioned theorists are willing to abandon the objective nature of physical "observables," and yet wish to retain the abstract quantum state as a surrogate reality. There is a temptation to believe that every quantum system has a wavefunction, even if the wavefunction is not explicitly known. Apparently, the root of this temptation is that in classical mechanics *phase space* points correspond to objective data, whereas in quantum mechanics *Hilbert space* points correspond to quantum states. This analogy is misleading: attributing reality to quantum states leads to a host of "quantum paradoxes." These are due solely to an incorrect interpretation of quantum theory. When correctly used, quantum theory never yields two contradictory answers to a well-posed question. In particular, no wavefunction exists either before or after we conduct an experiment. Just as classical cosmologists got used to the idea that there is no "time" before the big bang or after the big crunch, so too must we be careful about using "before" and "after" in the quantum context.

Quantum theory has been accused of incompleteness because it cannot answer some questions that appear reasonable from the classical point of view. For example, there is no way to ascertain whether a single system is in a pure state or is part of an entangled composite system. Furthermore, there is no dynamical description for the "collapse" of the wavefunction. In both cases the theory gives no answer because the wavefunction is not an objective entity. Collapse is something that happens in our description of the system, not to the system itself. Likewise, the time dependence of the wavefunction does not represent the evolution of a physical system. It only gives the evolution of our probabilities for the outcomes of potential experiments on that system. This is the only meaning of the wavefunction.

All this said, we would be the last to claim that the foundations of quantum theory are not worth further scrutiny. For instance, it is interesting to search for minimal sets of *physical* assumptions that give rise to the theory. Also, it is not yet understood how to combine quantum mechanics with gravitation, and there may well be important insight to be gleaned there. However, to make quantum mechanics a useful guide to the phenomena around us, we need nothing more than the fully consistent theory we already have. Quantum theory needs no "interpretation."

References

1. M. Born, *Zeits. Phys.* **37**, 863 (1926); **38**, 803 (1926).
2. A. M. Gleason, *J. Math. Mech.* **6**, 885 (1957).
3. J. S. Bell, *Physics* **1**, 195 (1964).
4. C. H. Bennett, G. Brassard, C. Crépeau, R. Jozsa, A. Peres, and W. K. Wootters, *Phys. Rev. Letters* **70**, 1895 (1993).

B. Quantum Theory – Interpretation, Formulation, Inspiration

Quantum Theory – Interpretation, Formulation, Inspiration: Fuchs and Peres Reply
Christopher A. Fuchs and Asher Peres
Physics Today **53**(9), 14, 90 (2000)

Like Paul Harris, we had an old friend in Alexandria. His name was Euclid. When we asked him whether his famous books *Elements* needed an interpretation, he answered categorically: "Absolutely not! Geometry is an abstract formalism and all you can demand of it is internal consistency. However, you may seek material objects whose behavior mimics the theorems of geometry, and that involves interpretation. For example, light rays might be considered analogous to straight lines." More than 2000 years later, we met a much younger friend in Göttingen and asked him eagerly, "Bernhard, have you really found a new, improved interpretation of Euclidean geometry?" His answer was no less categorical: "Riemannian geometry is not at all a new way of presenting Euclid's work. It is a broader formalism, having Euclidean geometry as a limiting case. Just reinterpreting Euclidean geometry without introducing radically new features would have been an illusion of progress. If some day people find that light rays do not behave as Euclid's straight lines, my geodesics may turn out to be a good description of them."

We do not claim that contemporary quantum theory is the final word for a description of nature; this should be clear from the last paragraph of our "Opinion" essay. There may some day be indications that our description of nature by means of vectors in a complex linear space is insufficient to represent *experimental* evidence. A more general theory may then be needed to extend the present formalism. However, as Daniel Styer correctly points out, the various "interpretations" of quantum mechanics cannot be distinguished through *experiments*. If they could, they would not be new interpretations, but proposals for new theories.

Styer compares these "interpretations" with the various formulations of classical mechanics: Lagrangian, Hamiltonian, Liouvillian, and others. The analogy is not correct. Quantum mechanics also has different formulations, such as those due to Heisenberg (close in spirit to Hamilton), Schrödinger (close to Liouville), and Feynman (which relies on a classical Lagrangian). All these formulations are mathematically equivalent and the choice of one of them for solving a particular problem is a matter of convenience. On the other hand, the so-called interpretations of quantum theory introduce new concepts, such as an infinity of parallel worlds, without any experimental support nor any benefit to the theorist who performs calculations to be compared with experiment. These gratuitous additions to quantum theory are the true analogs to Ptolemy's epicycles in Harris's story.

Stanley Sobottka apparently wishes to retain some objective status for the quantum wavefunction. He writes that interference phenomena always suggest that physical waves are interfering, and that "our understanding is greatly enhanced if we assume that the pattern is caused by actual, physical waves." However, Sobottka himself has reservations about this assumption, which is untenable for higher-order interference effects involving two or more particles. This is a problem Wendall Holladay runs into immediately when he says, "The wavefunction of the four outer electrons in the ground state of the carbon atom produces a tetrahedral structure in Euclidean three-dimensional space that undergirds the observed tetrahedral structure of the diamond crystal. This is an objective fact about the physical world ...." The truth is that the wavefunction of the four outer electrons lives in a 12-dimensional space, while our tangible physical world has only three dimensions. This example (contrary to its intended purpose) is an excellent one for showing that the wavefunction is a mathematical tool, not a physical object.

Holladay asks rhetorically, "How could a theory that does not describe physical reality give such accurate results for the magnetic moment of the electron and correctly predict the existence of antiparticles?" Similarly we could ask: How can ordinary probability theory give such reliable results in the gambling house, knowing as it does nothing whatsoever about the dynamics of the roulette wheel? The probabilistic predictions we make in gambling would work just as well whether the ultimate underlying physics were deterministic or, instead, indeterministic [1]. The point is that a theory need make no direct reference to reality in order to be successful or to be perfectly accurate in some of its predictions. Probability theory is a prime example of that because it is a theory of how to reason best in light of the information we have, regardless of the origin of that information. Quantum theory shares more of this flavor than any other physical theory. Significant pieces of its structure could just as well be called "laws of thought" as "laws of physics." However, this does not preclude quantum theory from making *some* predictions with absolute certainty. Among these predictions are the quantitative relationships between physical constants such as energy levels, cross sections, and transition rates that Holladay mentions.

The fallacy in Holladay's presumptive question is common to people not accustomed to quantum lines of thought. He makes no distinction between nature, which we try to understand, and the description of our experimental interventions into it. Accepting a distinction between these concepts requires only that we humble ourselves before nature, something our present scientific community is often reluctant to do. On the other hand, attempting to identify the two concepts discloses nothing more than a prejudice for a method that, in the past, seemed to work well in the classical world.

When experimenters have similar information, they should make – on the rules of quantum mechanics – similar predictions and draw similar conclusions. By this account, quantum mechanics is a scientific theory without rival. We feel comfortable in saying that diamonds have tetrahedral symmetry because, on any number of occasions, various experimenters have checked this with great accuracy. They could do that because this aspect of carbon is part of the "effective reality" quantum theory produces in some regimes of our experience. Indeed, this "effective reality" forms the ground for all our other quantum predictions simply because it is the part of nature that is effectively detached from the effect of our experimental interventions. But, if one tries to push this special circumstance further and identify an overarching "reality" completely independent of our interventions, then this is where the trouble begins and one finds the *raison d'être* of the various "interpretations."

Todd Brun and Robert Griffiths point out that "physical theories have always had as much to do with providing a coherent picture of reality as they have with predicting the results of experiment." Indeed, *have always had*. This statement was true in the past, but it is untenable in the present (and likely to be untenable in the future). Some people may deplore this situation, but we were not led to reject a free-standing reality in the quantum world out of a predilection for positivism. We were led there because this is the overwhelming message quantum theory is trying to tell us.

The main point of disagreement we have with Brun and Griffiths is about the existence of a wavefunction of the universe that would include *all* its degrees of freedom, even those in our brains. We assert that this would lead to absurd self-referential paradoxes. Therefore, it is necessary to restrict the discussion to a (reasonably small) subset of the dynamical variables. Brun and Griffiths ask, "Can we only describe the Big Bang, or an exploding supernova, in terms of the light that reaches our telescopes?" We never demanded such a restriction. We did not claim that only what is directly observed exists. There is much more to say about astrophysical phenomena than just describing the light that originates from them. Yet, their description cannot be so detailed as to include every particle involved in their observation, such as those in the retina of the observer, in the optic nerves, in the brain cells, etc. A limit must be put somewhere between the object of our description and the agent that performs that description. Quantum theory can describe *anything*, but a quantum description cannot include *everything* [2].

We agree with Brun and Griffiths that the violation of Bell's inequality by quantum theory is not a proof of its nonlocality. Quantum theory is essentially local. Bell's discovery was that any *realistic* theory that could mimic quantum mechanics would necessarily be nonlocal. Near the end of his life, Bell was indeed inclined to seek such a theory, bearing traces of realism and nonlocality. We do not rule out that such an extension of quantum theory may some day be produced, but no one so far has achieved this goal in a useful fashion, nor is an extension required for a clear understanding of the quantum phenomena about us.

We surely agree with Brun and Griffiths that "in science, one cannot rule out alternatives by fiat; one must evaluate them on their merits." We do not find any merit in the various alternatives that were proposed to the straightforward interpretation of quantum theory: It is a set of rules for calculating probabilities for macroscopic detection events, upon taking into account any previous experimental information. Brun and Griffiths may think this a "straitjacket," but it prevents the endless conundrums that arise solely from shunning quantum theory's greatest lesson – that the notion of experiment plays an irreducible role in the world we are trying to describe.

References

1. P. A. Hanle, Indeterminacy before Heisenberg: The Case of Franz Exner and Erwin Schrödinger, *Hist. Stud. Phys. Sci.* **10**, 225 (1979).
2. A. Peres and W. H. Zurek, Is quantum theory universally valid? *Am. J. Phys.* **50**, 807 (1982).

29 September 1999, "The Next Move"

Asherism 55: *You just heard from Eugen Merzbacher. What shall we do next? How busy will you be in the near future? I shall be "normally" busy, teaching my course on foundations of QT, and doing research as usual.*

I would classify myself as "normally" busy too. Carl and I are working on a National Science Foundation proposal, but I should be able to rise above that before my move to Los Alamos. Also there's the quantum de Finetti theorem paper to be finished, but I'm quite sure I can multi-task between you and Carl. Perhaps the biggest thing to get in our way, is that I am likely to be significantly out of commission from October 7–18. I must drive our dogs to New Mexico (Kiki and Emma will fly), then I will be spending several days helping Kiki get all the furniture arranged, our bank accounts and other affairs opened, and build a fence in the yard to contain the dogs.

I think the *Physics Today* article is a nice opportunity and we should run with it if we get a chance. Too many of our close colleagues are going by the wayside: our efforts may not change their opinions, but we may help suppress the snowballing effect with younger students. And that has to be worth it. I'm very glad you had the idea of writing to Eugen. (BTW, I also received a very nice personal letter from him this morning; it appears that he has been watching my career a bit from afar … and has asked me to give a colloquium at his university.)

Asherism 56: *My idea, for starting this opus, was to collect excerpts of my book and articles, with nice slogans ("Unperformed experiments have no results"), and let you add your own bibliographical sources – surely you have much more, if you include writing from third parties – and then let you try to sort all that raw material into definite subjects to work on. Or have you a better idea?*

No, I think this is an excellent idea and a good starting point. I myself would like to find a way of (gently) inserting my provocative slogan, "Quantum states do not exist," and also I would like to cover a little bit of the in-principle reversibility of measurement in much the same way in the letter I wrote to Howard Barnum (that I forwarded to you): I was very pleased that that discussion finally turned his head a bit. [See notes to Howard Barnum, dated 30 August 1999 and 5 September 1999.] I'm sure we can come to some agreement about substitutes for the words "know" and "knowledge" that you found disagreeable in that exposition. Also if we could figure out a way of saying a word or two about quantum information (much like Eugen did in his letter to Benka), I think that it would help to give an air of legitimacy and forward-looking feel to all that we have to say.

Asherism 57: *Also, though Eugen was very helpful, we should have ASAP direct contact with PT, before we start writing anything more than an outline (a kind of abstract). Then we must ask them whether they want any particular format, about the deadline, what they are willing to reveal about the other "feature" and so on.*

That is a good idea too. Since you are the senior among us – and anything an editor would say to you would likely be more binding – would you take the task of initiating that? I think it would be most useful to understand who the authors of the other "feature" are. That could also help us gauge what to say, to understand whether it will ever materialize, and, if we know them, well perhaps we could contact them directly for further information. (An off the cuff guess would be that Anton Zeilinger might be writing something.)

Partners in crime again?

01 October 1999, "One More Round"

Thank you for getting the *PT* process started by writing such a nice detailed letter: I think the article's already starting to shape up in my mind. Below I'll place the modifications I made; you should have a look at them before we send the letter off to Eugen. I think you'll find most of the modifications minor, only adding a little more detail to what you already said. In a couple of cases I Americanized the English a little bit. I also removed the "alternative medicine" remark that Eugen started in order to keep in line with his comment to Benka that we would not write a polemic.

The largest change you'll notice is that I deleted the original item #3. As you know, I am Bayesian through and through now, and Bayesians are pretty allergic to ensembles. "The world is given to us only once." I decided to delete the passage rather than attempting to modify it now because I don't think the deletion will change the flavor of the letter enough that Benka will even notice its absence. In the meantime, it will buy us some time to come to agreement among ourselves: I think we can do that and will probably both learn a little bit in the process.

Tomorrow afternoon I'll try to write you a more detailed explanation of why I disagreed with item #3.

If you are OK with the letter as it is presently, feel free to send it off to Eugen yourself. Otherwise, I will send it off first thing tomorrow morning or review any further modifications you make over the evening.

01 October 1999, "We Are Converging"

I made only a couple more minor style changes (that I would have done last night if I were completely awake), so I went ahead and sent it to Eugen.

Asherism 58: *I think that we completely agree that QM applies to a single system, such as the Universe or the SQUID mentioned in #6.*

Absolutely. About the semantics, I'll try to come back to that this afternoon or sometime over the weekend.

05 October 1999, "Sorry So Late"

I'm sorry to have been out of contact the last couple of days. For some reason my dial-up networking decided to quit on me this weekend, and things just got too busy for me to get in to the office. I made all the changes you suggested except:

Asherism 59: *#6: delete "But," in line 2 and the comma at the end of line 6.*

I left the "But" but added "standard" in the paragraph:

Many feel the need to go to conceptual extremes with quantum mechanics because they want to do quantum cosmology. But, standard quantum theory places no obstacle to that. Is there a "wave function of the universe?" If we consider just a few collective degrees of freedom, such as the radius of the universe, its mean density, total baryon number, etc., we can apply quantum theory to these degrees of freedom, which do not include ourselves and other insignificant details. This is just as when we quantize the magnetic flux in a SQUID and ignore the atomic details. One may object that there is only one universe, but likewise there is only one SQUID in our laboratory. For sure, we can manipulate that SQUID more easily than we can manipulate the radius of the universe. Still, that SQUID is unique.

The reason I did that was to try to set some contrast. I have noticed that most people who go to the extreme of the Everett interpretation do so because they say that one cannot do quantum cosmology without it. Taking the "But" changed the meaning of the thought, I believe. In any case, this is only a proposal letter: so I hope you will not mind my not bouncing off you once more. We can reserve endless iterations about meaning for the real thing!

I think we can have some fun writing an article like this: I'm looking forward to it.

Asherism 60: *Even a joint publication is conceivable, depending of what he intends to write.*

I think we will have a stronger position for our view if we're able to get it represented by two articles in a single issue. So I'm a bit reluctant to join forces outside of a mutual proofreading and encouragement, etc. Certainly, though, I think we should contact Anton before we're ready to set down to write something.

Ok, now I must get all the final things packed up before the moving people come tomorrow. Tomorrow is reserved for their packing everything in boxes. The truck comes to be loaded the following day. So I may be out of email commission for a few days: at least until Sunday or Monday.

03 November 1999, "Explanation of My Absence"

Below is a partial explanation of why you haven't heard from me. I feel pretty weak right now ... so I'll leave it at that. I say we start up our efforts on the opinion piece next

Tuesday. Have you looked at Bub's criticism – recently on `quant-ph` – of your (our) view. Studying that (mildly) may help us fend off similar criticisms to our opinion piece.

07 November 1999, "A Sunday the Way I Like to Spend It"

Today I had the chance to do a little work around the house, take Emma and the dogs to the park, listen to the morning jazz show on the radio, and, in between, to read two papers. Now this is the way I like to spend a Sunday!

The first paper was your "Karl Popper and the Copenhagen Interpretation." Of course, I agree on most counts and liked it a lot. (I spotted one typo: the first sentence of the last paragraph of page 3 in the `quant-ph` version. You're missing an "it" near the beginning.) I especially found comfort in discovering that you and I (and Bohr) seem to agree on the following point:

Asherism 61: *Note that Bohr did not contest the validity of counterfactual reasoning. He wrote:*

Our freedom of handling the measuring instruments is characteristic of the very idea of experiment ... we have a completely free choice whether we want to determine the one or the other of these quantities ...

Thus, Bohr found it perfectly legitimate to consider counterfactual alternatives. He had no doubt that the observer had free will and could arbitrarily choose his experiments. However, each experimental setup must be considered separately.

I say I found comfort in that because recently Howard Barnum, Charlie Bennett, and John Preskill all have told me (on separate occasions) that I'm being mystical when I say things like that. They seem to see me as somehow trying to buck the scientific worldview, which angers me. I simply say that I can't understand what experimental science (or even science at all) is without that simple presupposition – that we actually do meaningful experiments, one or another or another ... that we actually learn things by actively pursuing to learn them. I think I replied most forcefully to John Preskill. Let me paste that in, in case there's a phrase or two in it that we might want to use in our *PT* article. [See letter to John Preskill, dated 8 September 1999, titled "Smug Note."]

The other paper I read today was Jeffrey Bub's "Quantum Mechanics as a Principle Theory" (`quant-ph/9910096`). That wasn't a waste of time: in learning his motives, I think I come to a better understanding of our position. He devotes the last section of the paper (titled "Instrumentalism") to criticizing you and van Kampen (but mostly you). The only thing he has going for him is a question, "Knowledge of what?" My answer would be terse: what we are speaking of is our probabilistic knowledge of the (experimental) consequences of our (experimental) *interventions* into the course of Nature. It was only sloppy thinking about the nature of classical physics that ever led us to (foolishly) believe that we could have more. Perhaps I shall compose a letter to Bub and send it directly.

Tomorrow morning early I drive to Albuquerque for my standard Monday collaboration with Carl Caves. Tuesday, I should be ready to do something useful with respect to our *PT* project. Let me give one word of opinion now though: I think we should lay a little low on talking about Bohr and Copenhagen directly. As you make it clear in your paper that I just read, there's not even a consensus on what Bohr actually said: two people can have two completely different readings of his words. It has been my experience that sometimes when people hear the words Bohr or Copenhagen they just get riled and refuse to listen to anything further. I think we should stick with the factual situation that quantum mechanics presents us with – much in the style of the letter we sent to Benka – and let the simple force of the proper way of thinking carry itself. But I probably don't need to express this: you're probably already ahead of me on this.

Let me tell you a cute anecdote in that connection: I'll just cut and paste from a letter I wrote Carl following the Baltimore conference. (Notes within notes within notes ... kind of like Ezekiel with his wheels.)

In general, by my reckoning, the talk went very, very well. This time, I focused more on the overall program "quantum states as states of knowledge" than on the technicalities of the proof ... and it really seemed to pay off in this audience. I was absolutely surprised at the number of heads that seemed to turn to these ideas. Even when people disagreed (like Nicolas Gisin and Philip Pearle did), they seemed to come away as if they had heard a fresh idea. As I wrote Herb Bernstein earlier today, [See note to Herb Bernstein date 17 August 1999, titled "A Bernstein Off the Earth?"]

OK, I go to bed now: it is a long drive to ABQ and back tomorrow. I hope you are feeling better (and that this letter didn't exhaust you). Please give Aviva my best wishes. Tell her that the two miniature rugs she gave us have found a temporary home above our fireplace. (They will have to be moved to a more permanent and safe place before we actually build a fire in it!)

12 November 1999, "Embryo.tex"

What a dreadful couple of days I've had! The troubles began yesterday morning when we discovered [...]

I have created a file titled "Embryo.tex" containing the embryo of the article you sent me and will hopefully start some modifications to it today. Now, though, I must go to the Lab for the Friday Quantum Lunch. Wish me more luck than I've had in the last two days.

13 November 1999, "Putting the Finger"

Well I procrastinated a bit from the proper part of our project today, but maybe I came out the better for it. I finally sat down and read all the relevant things in *Physics Today* that I could find: the Goldstein article, the Griffiths and Omnès articles, that article by Mara Beller on the Sokal hoax [*PT*, Sept. 98, pp. 29–34] in which she makes a strong stand for Bohmianism, and the letters replying to Goldstein [*PT*, Feb. 99, pp. 11–13, 15, 89–90, 92].

The only sensible thing written in all this lot was Anton Zeilinger's letter contra Goldstein. Goldstein's own writings were, by far, the worst representatives. As far as I could tell, his 10 pages worth of article said almost nothing ... other than perhaps weakly expressing what he *wants* of quantum theory! And this guy is just way over confident with himself. Listen to what he writes on the first page of his first article:

Many physicists pay lip service to the Copenhagen interpretation, and in particular to the notion that quantum mechanics is about observation or results of measurement. But hardly anybody truly believes this anymore – and it is hard for me to believe that anyone really ever did.

I dread meeting him two weeks from now in Naples.

The Griffiths and Omnès article also only helped confirm the conclusions I've already drawn on consistent histories. Despite the overtures RBG and RO (and JBH and MGM) make, they haven't in any way taken the observer out of quantum mechanics, nor extended the theory beyond its old form. Let me put my finger on it in the most rhetorical way I can. To do this, just step back for a moment to standard quantum mechanics and fix a specific von Neumann measurement on some system. When it is performed, the observer will find one and only one of the possible outcomes (the outcomes, of course, are taken to correspond to a fixed set of orthogonal vectors). As long as that measurement is fixed and we never speak (counterfactually) about another, we can play AS IF one of those outcomes actually "was the case" all along, independent of our having performed the measurement. In such a simplistic situation, one can act AS IF "unperformed measurements have out-comes." Or, another way to say this is that as long as a single measurement is fixed, we can imagine that the quantum state has nothing deeper to say than a classical probability distribution would have had for a fixed set of outcomes in some probabilistic trial. Quantum mechanics only starts to make its mark when we *contemplate* simultaneously various noncommuting von Neumann measurements.

All the Consistent Historians do is extend the simple case above to a family of "his-tories" – strictly speaking, a history is nothing more than a projector on a tensor product of many copies of the original Hilbert space. When the histories all commute, and we refuse to speak (counterfactually) about any other noncommuting set of histories, then we can always act AS IF one of the histories is truly the case (i.e., existent, real, objective, observer-independent). But this is no surprise and nothing deep. What can be said about two noncommuting sets of histories from their point of view? "Nothing," RBG and RO say, "such an inquiry is meaningless." But that is just as in standard quantum mechanics, where we never speak of measuring simultaneously position and momentum. AND THAT IS ALL THERE IS TO CONSISTENT HISTORIES. There's nothing new here that wasn't in the old theory. In particular, if our friends would just have the mental nerve to broach the questions you and I do on a regular basis – about distinct noncommuting sets of observables – then they would be stuck in the same lot as us ... i.e., being forced to the conclusion that unperformed measurements have no outcomes.

But I rant and rave. Tomorrow I have the feeling I will be able to turn some of this energy to creative purposes. Sundays are generally good for me in that way. The other day you wrote:

Asherism 62: *If the title of our essay is "QM needs no interpretation," our task is to explain why indeed no interpretation is needed. Not that the various interpretations (Bohm, Everett, consistent histories) are wrong – they probably are just complicated reformulations of the theory – they simply are redundant.*

I agree with you completely on this. (Don't let my words above lead you to think that I'll come out swinging a battle axe for RBG and RO. My head will be much more sensible tomorrow.)

15 November 1999, "Have a Good Trip"

In the next note, I'll send a little bit of progress to the manuscript. Please note that it is titled "Embryo.tex" … so everything I have written is certainly still in its formative stages. It is not too late to scrap it all and start again if you find that you do not agree with what I have written. There are so many similarities in our thoughts, I am sure we will in one way or other be able to find a middle ground if need be.

The main changes/additions of mine can be found in the first five paragraphs and in the final paragraph. I spent most of my time today with the beginning, trying to set up a general point of view. What I plan to do next is tie in, in a more seamless fashion, what you have already written (both in the manuscript and in your supplementary notes).

One point that we may need some private side discussion on before we set it in stone, is captured among other places in your sentence:

Asherism 63: *The notion "state" refers to a method of preparation, it is not an intrinsic property of a physical system.*

In general I have noticed in this manuscript that you lean more heavily on the word "preparation" than we did in our letter to Benka. (In fact, I can't find any mention at all of the word "preparation" in that letter.) Unless I misunderstand your usage of the word, it may actually be a little too anthropocentric even for my tastes. The problem is this: consider what you wrote in the paragraph about the wave function of the universe. It seems hard to me to imagine the wave function of those degrees of freedom which we describe quantum mechanically as corresponding to a "preparation." Who was the preparer?

It is for this reason that Carl Caves and I prefer to associate a quantum state (either pure or mixed) solely with the compendium of probabilities it generates, via the Born rule, for the outcomes of all potential measurements. And then we leave it at that. Knowing the preparation of a system (or the equivalence class to which it belongs) is one way of getting at a set of such probabilities. But there are other ways which surely have almost nothing to do with a preparation. An example comes about in quantum statistical mechanics: when the expected energy of a system is the only thing known, the principle of maximum entropy is invoked in order to assign a density operator to the system. There may be someone beside me in the background who knows the precise preparation of the system, but that does not matter as far as I am concerned – my compendium of probabilities for the outcomes of all measurements is still calculated from the MaxEnt density operator.

To help ensure that I was not jumping to conclusions on your usage of the term, I reread today your paper "What is a state vector?" [*AJP* **52** (1984) 644–650]. There was a time when I agreed with everything you wrote there (in fact, I think it was the first paper with which I got to know you). But as of today at least, I think a more neutral language as in our letter to Benka is more appropriate.

Please let me know your thoughts when you have time. If you prefer we can wait until you return from your conference. If, on the other hand, you have no strong disagreement with me, I will continue working on the manuscript Tuesday morning and perhaps we will have a first draft by Wednesday.

Please have a safe trip and give my regards to Anton.

17 November 1999, "Holding Pattern"

By the way, I decided to wait upon your return to write any more on our project. Once I hear your opinions, I'll reengage. I hope you are enjoying being 2000 light years from home.

18 November 1999, "Gamete Actually"

Thanks for "deprosing" me a little! I know that I sometimes have a tendency to go overboard on my first drafts – luckily I usually trim them down myself on second and third readings. For the most part, I am pleased with your results. I'll place the draft below with a few yet further changes and the voicing of a little disagreement here and there. I won't go too deeply tonight as I haven't seen the other things you have in mind (and it is late for me given that I got up this morning at 2:30 and really never fell back asleep). I'll use the same system you did, adding footnotes when relevant.

Parsing the Paper

Quote from Draft: Recently there has been a spate of articles, letters, and reviews in *Physics Today* promoting various "interpretations" of quantum theory. The running theme in these is that from the time of quantum theory's inception until the emergence of a particular interpretation, the theory was in a crisis because its foundation was ????.

Comment: Looking back at what I wrote here before, i.e., "mystical muddle of words," I see that it was a bit flowery. However, I don't think the phrase "logically inconsistent" captures the right notion. For instance I know that Omnès would say that he is only extending the Copenhagen interpretation as it is "incomplete." The main criticism of most these guys seems to me to be that Copenhagen is just a lot of words without substance and that exalting measurement as the primary process of quantum theory verges on the mystical. I think we should work a little harder for a better phrase here, but one does not come to me right now.

Quote from Draft: An experiment is in its essence an active intervention into the course of Nature: we set up this or that experimental situation so as to understand the consequences of these interventions.

Comment: I agree with you that, "We should have a defensive strategy: avoid loose statements that can easily be attacked, if they are not essential to the argument." What I was shooting for though – and I think it is essential – was a kind of "positivism without positivism." I am afraid that the simple word "understand" can be construed in too many ways, and it may be attacked on that account. In fact, I remember well a conversation with David Mermin in which he took the point of view that "to understand a phenomenon" is to pinpoint the *reality* that gives rise to it. So I think it kind of important to further emphasize the distinction between positivistic/operational understanding and the classical worldview. Can you agree to the version of the sentence that I reinstated?

Quote from Draft: We do not deny the existence of an objective reality independent of what observers may perceive. However, this reality, whatever it may be, is not described by quantum theory, nor by any other theory known to us. John Bell formally showed that any objective theory giving experimental predictions identical to those of quantum theory would necessarily be nonlocal. It would have to encompass all the degrees of freedom of the Universe. As Bell eloquently puts it: "Separate parts of the world would be deeply and conspiratorially entangled, and our apparent free will would be entangled with them."

Comment: I am a little worried that these three sentences about Bell will open up a can of worms. The Bohmians should agree, but I am afraid that the Consistent Historians, the Everettistas, and GRWians will all come out of the woodwork for the attack. They'll cry foul, and say that we've just never taken the time to understand their theory. So I'm a little afraid to make these statements. Are they essential?

Quote from Draft: Indeed, a natural question is whether the universe has a wave function. There are two ways of understanding this. If this "wave function of the universe" has to give a complete description of everything, including our own brains, we encounter bizarre self-referential paradoxes. These paradoxes are not in the realm of physics. Experimental physicists never have to bother about them. On the other hand, if we consider just a few collective degrees of freedom, such as the radius of the universe, its mean density, total baryon number, etc., we can apply quantum theory to these degrees of freedom which do not include ourselves and other insignificant details. This is just as when we quantize the magnetic flux in a SQUID and ignore the atomic details. One may object that there is only one universe, but likewise there is only one SQUID in our laboratory. For sure, we can manipulate that SQUID more easily than we can manipulate the radius of the universe. Still, that SQUID is unique.

Comment: Before that, we have to explain why it is legitimate to apply statistical reasoning to a single system. You're starting to sound awfully Bayesian!! Congratulations.

19 November 1999, "The Nut House"

Asherism 64: *I am sending now all that I wrote. It's a hodge-podge of paragraphs in random order. I have no time to reread that, because we now have dinner with my granddaughter and her parents. Please forgive any insanities in that text. It's just raw material, as you know.*

If that's temporary insanity, it ain't bad! I quickly skimmed the first three pages, and am very much starting to like the flow. Hopefully I will get a chance to do some minor things on the draft tonight. But tomorrow I should be able to do some more serious work with it. (Today I am working with Carl on our project.)

Have fun with your granddaughter.

19 November 1999, "End o' Day"

I give up. I was hoping to have a real look at your draft before the day was out, but it doesn't look like I'm going to be able to do it. Kiki (and Emma) keep saying "Come downstairs, it's Friday night!" (What is your secret?!?!)

20 November 1999, "Almost Fetal"

Well I put in a solid day's work on our project today. I was hoping to completely finish it by the evening, but it has just become too difficult and now Kiki is calling me to dinner. Tomorrow morning I will rise early and continue the pace. I've made many modifications (almost all minor), tweaking this and that and working toward getting it under our size limitation (max 1800 words) without compromising the integrity. Please give me another day, and then I will send you the next iteration so that it can be your turn again.

I think its really starting to take shape.

21 November 1999, "Miracles of Office Technology"

That article by Peierls is indeed quite good. Plaga was right. In some places, Peierls' writing could have just been ours (perhaps more accurately I should have said it the other way around). It dawned on me that it wouldn't take any time to scan it in for you. So I'll paste it below. [Relevant parts only reproduced here.] There may be some nonsense in there: Optical Character Recognition isn't perfect yet and I didn't reread it completely. Still, here and there I have placed some footnotes expressing my reactions to what he said.

In Defence of "Measurement"
Rudolf Peierls
Physics World, January 1991, pp. 19–20

In my view the most fundamental statement of quantum mechanics is that the wavefunction, or more generally the density matrix, represents our *knowledge* of the system we are trying to describe. I shall return later to the question "whose knowledge?". It is well known that we have to use a wavefunction if we have a "pure state" i.e., if our knowledge of the system is complete, in the sense that any further knowledge is barred by the uncertainty principle.[2] Failing such complete knowledge we must use a density matrix, which therefore contains both quantum and classical ignorance. The wavefunction is a special case of a density matrix, and I shall here talk about "density matrix" when I mean "wavefunction or density matrix".

More precisely, while the time variation of the density matrix is given by Schrödinger's equation, the initial values represent knowledge usually obtained from observations. (There are not always measurements; for example, if an atom has been for a reasonable time in free space we know it must be in its ground state.)

Our knowledge is not fixed, but may increase or decrease. It increases if further observations are made; it decreases if the system is disturbed by external factors which we cannot control. There is nothing new in this. In classical physics our knowledge may increase and decrease in the same way. The only difference is that in quantum mechanics we have to be specific about what we know, because our possible knowledge is confined by the uncertainty principle. In classical physics there is no reason in principle why we cannot know everything about the system and we usually argue as if we did. But in a practical situation our knowledge may increase or decrease as indicated.

In quantum mechanics any increase in our knowledge is usually accompanied by a decrease in some other respect, because of the uncertainty principle.[3] This applies particularly when we are concerned with a "pure state". Then we can gain no new information (other than confirming what we know already) without losing some of the existing information.

Once this significance of the density matrix is understood, it is clear that upon a change in our knowledge the density matrix must change. This is not a physical process, and we certainly cannot expect it to follow from the Schrödinger equation. It is just the fact that our knowledge has changed, and thus must be represented by a new density matrix.

---

[2] I like this point of view about the pure state except for the loose term "uncertainty principle." I think instead our information–disturbance relations hint at the right sort of rigor that one would need for putting some flesh on this sentence. But that defines the large part of my research program. Also Caves and I would advocate the term "maximal knowledge" here in place of "complete knowledge" (preferring to say that when one is forced to probabilistic predictions it is always because of incomplete knowledge).

[3] Again I think the term "uncertainty principle" causes trouble here because it makes one think that the knowledge is about a *preexisting* x or p independent of our interventions. See footnote toward the end of the article.

When I refer to "observation," this term has its common-sense meaning. The observation usually (but not necessarily) involves an apparatus which interacts with the system in question, and which produces a signal (visible, audible, or other) which we can recognise, and which is correlated with the variables of the system. Bell quoted the view of Landau and Lifshitz (and therefore of Bohr) that the apparatus must necessarily obey classical physics. In my view this is not correct. It is of course true that our senses are macroscopic, and that the instruments we find convenient are also macroscopic and in practice classical. But this is a practical point, not one of principle. The sensitivity of the human eye is almost sufficient to detect a single photon. If some experimentalist has sufficient vision to see one photon, the observation of that photon might perfectly well serve as a measurement.

The apparatus usually consists of a chain of correlated events. I have elsewhere (Peierls 1979, 1985) discussed as an example the observation of a spin component of a spin-atom by a Stern–Gerlach magnet. The first step, the passage through the inhomogeneous magnetic field, sets up a correlation between the spin component and the position of the atom. It is not yet a measurement; we have not yet gathered any information. This requires determining the position of the particle, i.e., in which part of the split beam it travels. To find this out, we may use a counter, but again this conveys no information – and nothing collapses! – until we find out whether the counter has been activated. We can obviously pursue this chain: the counter will be part of an electrical circuit, the circuit will operate a digital recorder, we may read this recorder by means of the light it reflects into our eye, etc. Each step is correlated with the preceding ones and therefore with the spin component of the particle. Each step keeps both options open until we "see" the result, and then we revise our density matrix.

Because of the uncertainty principle we cannot acquire knowledge of, say, the z component of the spin without losing what information we had previously about, say, the x component. Is this happening in the first step, the passage through the magnet? At first sight this looks likely, because information about s_x is contained in the phase relation between the components of the wavefunction belonging to $s_z = +\frac{1}{2}$ and $s_z = -\frac{1}{2}$. Since the beams corresponding to the two s_z values are now split, they do not overlap and do not interfere, so their phase relationship is not observable. However, the information is not irretrievably lost. By arranging a further magnet we could recombine the two beams and observe their phase relation (thereby foregoing the possibility of observing s_z). We do finally lose the "forbidden" information when we "see" the atom in one of the beams. We then have to replace our density matrix by one containing only the one s_z value, so there is no interference.

As long as we do not "see" the atom in the beam, the reconstruction of the seemingly lost information is troublesome, but easy to visualise. At the next stage, i.e., after the counter, it becomes much more involved. Since the density matrix now contains the variables of the counter, interference requires not only that the two atomic beams be made to overlap, but in addition that there be an overlap between the density matrices for the activated and unactivated states of the counter. The observation of the phase relation therefore requires an operator capable of deactivating the counter coherently. This is possible in principle, but in practice prohibitively difficult. As we go further down the chain of connections involved in our "measurement," this difficulty gets worse.

This is the origin of the belief that the apparatus makes the off-diagonal matrix elements of the density matrix disappear. In most cases that is true "for all practical purposes," but not in principle. The off-diagonal matrix elements disappear only when we know the result of the measurement.

The "system" to which we apply our description can be as large as we like, including the whole world if we want. However, if we make the system too large, the amount of information we can obtain is relatively small, so that the density matrix is made up mostly of parts proportional to the unit matrix (which denotes complete ignorance) and it becomes hard to do any useful physics. In any case the "system" cannot include the mind of the observer and his knowledge, because present physics is not able to describe mind and knowledge (and it is not obvious that this is a proper subject for physics).[4]

The objection is sometimes made: "How can one apply quantum mechanics to the early Universe, when there were no observers around?" The answer is that the observer does not have to be contemporaneous with the event. We can, from present evidence, draw conclusions about the early Universe, the classical example being the cosmic microwave background. In this sense we are observers. If there is a part of the Universe, or a period in its history, which is not capable of influencing present-day events directly or indirectly, then indeed there would be no sense in applying quantum mechanics to it.

That leaves the question: whose knowledge should be represented in the density matrix? In general there will be many who may have some information about the state of a physical system.[5] Each of them has to use his or her density matrix. These may differ, as the nature and amount of knowledge may differ. People may have observed the system by different methods, with more or less accuracy; they may have seen part of the results of another physicist. However, there are limitations to the extent to which their knowledge may differ. This is imposed by the uncertainty principle. For example if one observer has knowledge of s_z, of our Stern–Gerlach atom, another may not know s_x, since the measurement of s_x would have destroyed the other person's knowledge of s_z and vice versa. This limitation can be compactly and conveniently expressed by the condition that the density matrices used by the two observers must commute with each other.

1. J. S. Bell, 1990, Against "measurement", *Physics World*, August, 33-40.
2. R. Peierls, 1979, *Surprises in Theoretical Physics*, Princeton section 1.6 (Some of the points made in this article will also be discussed in a forthcoming volume, *More Surprises in Theoretical Physics*, Princeton).
3. R. Peierls, 1985, Observations in Quantum Mechanics and the "Collapse of the Wave Function", in *Symposium on the Foundations of Modern Physics*, World Scientific.

21 November 1999, "Kathie and Erwin"

I understand Erwin, but why Kathie? Is that one of your granddaughters?

22 November 1999, "A Baby Kicks?"

I bet you're already awake and have already looked at your email (and wondered, "Where is that Fuchs?"). I'll send you two files momentarily. The first is the latest draft including detailed footnotes explaining why I did what I did, etc. The second is the same draft with

---

[4] Sound familiar?

[5] Here is an example where he is being a bit sloppy. "Information about the state of the system" is not a well defined concept as we know. His problem comes from slipping back into objectivist language: he is here acting as if a "state" is an objective property of the system. In my view we ought to always be strict: Knowledge of what? Information of what? *Probabilistic knowledge of how a system will react to our external interventions!*

all the footnotes stripped out so that you may have an easier time editing if you decide to pick up where I left off (for the most part).

It's starting to sound pretty good to me, but that could simply be from the weariness of reading it over and over. By my editor's count we have presently 1849 words. But the editor gets a little confused by TEX commands, etc., so the real count should be somewhat below that (maybe 1750).

I think the transition between the penultimate and ultimate paragraphs is still a little abrupt, but I couldn't think of a slicker way to pull it off. Maybe you'll have better luck.

Also, I would still really like to say just a little something about Erwin's in principle possibility of reversing Kathie's superposition eating. The reason is I've seen too many people think that our point of view precludes such a thing. But I realize space limitations are tight and I don't see much else that I would like to cut out.

Parsing the Paper

Quote from Draft: To start, let us assess the goals of experimental science. An experiment is an active intervention into the course of Nature: we set up this or that experiment to see how Nature reacts to our prods.

Comment: I started thinking that I had overused the word "intervention." And besides I think this captures what I had wanted to say better. It adds the right amount of emphasis and twist away from "realist" thinking.

Quote from Draft: Understanding is achieved when we can distill from the accumulated data a compact description of all that was seen and an indication of which further experiments will help corroborate that description. This is what science is about. If from our understanding we can *further* distill a model of a free-standing "reality" independent of our interventions, then so much the better. Classical physics is the ultimate example in that regard. However, there is no logical necessity that such a further step always be obtainable. If our world is so "sensitive to the touch" that we can never identify a reality independent of our experimental intrusions, then we must be prepared for that too.

Comment: I decided I liked the word "sensitive" better than "ticklish." Also I did a little more "intervention" trimming.

Quote from Draft: Contrary to these aspirations, quantum theory does not describe physical reality. It only provides an algorithm for computing probabilities for the macroscopic events ("detector clicks") we choose to associate with a measurement process.

Comment: Predicting → computing. It's more Bayesian that way. I chose to use the word "choose" because of a very nice paper I once read by some guys named Hay and Peres.

Quote from Draft: These clicks may be viewed as the consequences of our experimental interventions.

Comment: I threw this sentence in so that it would be clear how "macroscopic events" makes a connection to our previous discussion. I think such a thing is necessary.

Quote from Draft: This strict definition of the scope of quantum theory is the only interpretation ever needed by experimenters and theorists alike.

Comment: I didn't want us theorists left out.

Quote from Draft: The probabilities spoken of here are computed by means of *any* information that may have been gathered previously, including, but not limited to, information about a system's preparation.

Comment: I think it is really necessary that we get away from the notion of "a state *defined* purely as a preparation or recipe" if we want to talk about a very general situation like the wavefunction of the radius of the universe, etc., later in the text. That is to say, we need to be a little more Peierls-ian in our statement. Also if we want to be able to say "this strict definition is the only interpretation ever needed by experimenters" then we'll most certainly have to do something like this. Just think of an engineer who must worry about the thermodynamics or statistical mechanics of some very fine microdevice. As a first characterization of his design, he will most likely assign a quantum state based solely on some knowledge of the running temperature or mean energy of the system. In the latter case, he will most likely use the principle of maximum entropy for his quantum-state assignment. He generally won't have the luxury of associating a recipe or preparation with his quantum state. If a technician later comes along and presents the engineer with a detailed analysis of the energy flow into the device from the reservoir in which it sits, then so much the better. The engineer will just update his state assignment to be in accord with his new information and forget about his crude earlier methodology.

Quote from Draft: The precise latitude with which such information can be incorporated into quantum statistical predictions finds its formal expression in the probability rule postulated by Max Born. The conclusiveness of this rule is made sharp by a remarkable theorem due to Andrew Gleason.

Comment: I wanted to stress that Gleason really puts a cap on things. And by the way, the theorem is indeed remarkable. I have walked through every step of its proof, and it was no easy task!

Quote from Draft: So long as one assumes that yes–no tests on a system are represented by projection operators P on a Hilbert space and that probabilities are additive over orthogonal projectors, then there exists a density matrix ρ so that the probability of a "yes" answer is $\text{tr}(\rho P)$.

Comment: I tried to make this sentence a complete statement of the theorem, skipping only that dimensionality greater than two is required (which doesn't seem so essential for an "opinion" article). And another "by the way": in your book, p. 190, you write, "The premises needed to prove that theorem are ..., supplemented by reasonable continuity arguments." That is not true. One of the things that makes the *mathematics* that Gleason did so remarkable is that he *did not* assume continuity! He proved it! Now any physicist (except our friend Kent) would think that continuity is perfectly reasonable. But it still shocks me that Hilbert space is such a nice place that continuity need not even be assumed.

Quote from Draft: We do not deny the existence of an objective reality independent of what observers perceive. However, this reality, whatever it may be, is not described by quantum theory nor by any reasonable theory known to us.

Comment: I changed "other" to "reasonable" because most of our antagonists claim to have just that.

Quote from Draft: John Bell formally showed that any objective theory giving experimental predictions identical to those of quantum theory would necessarily be nonlocal. It would have to encompass all the degrees of freedom of the Universe, and as Bell put it, "separate parts of the world would be deeply and conspiratorially entangled, and our apparent free will would be entangled with them."

Comment: I struck the phrase "it would be beyond human understanding" and replaced it with parts of what you had written before. Bohm's theory, for instance, may be ugly and add features extraneous to quantum mechanics, but I don't see that it makes the world beyond human comprehension.

Quote from Draft: We can, on the other hand, have a description of an "effective" objective reality in the limiting case of macroscopic phenomena such as planetary motion, but such a description ignores most degrees of freedom for the system and is necessarily incomplete.

Comment: I rearranged this sentence somewhat so that it read better for me. But I am not sure that I completely agree with what you are thinking here, so let me withhold complete judgement until a later draft. The main problem is this. Take the information–disturbance relations from our 1996 *PRA*. It seems to me that they capture a gross aspect of reality (suspending for the moment our phrase "whatever reality is"). Really I just mean that all observers will agree that the information–disturbance phenomenon exists. But that phenomenon has nothing to do with an incomplete description. Instead it seems more an integral part of the reason that we cannot get at a world with "a free-standing reality independent of our interventions." From this point of view a phenomenon is "macroscopic" precisely when the tradeoff is weak between information gain and state disturbance for various observers trying to come to agreement on the predictions they can make about potential further interventions on their system. That in general can/will happen when we are looking only at certain gross aspects of a system, for instance by ignoring most degrees of freedom. I understand that this is too advanced a topic for this essay. But I want to make sure that I don't sin by writing something I don't believe. So I wonder if there might be some way of adding two or three extra words that would make me feel comfortable?

Quote from Draft: Here it is essential to understand that the statistical nature of quantum theory does not restrict its validity to situations where there is a large number of similar systems. Statistical predictions do apply to single events. When we are told that the probability of precipitation tomorrow is 35%, there is only one tomorrow. This tells us that it is advisable to carry an umbrella.

Comment: I removed: "Statistical predictions such as the above one are prepared by using available information, which is incomplete, and considering various hypotheses for the missing data. We thus create a conceptual ensemble of initial conditions. For each one, we model the future dynamical evolution, and thereby obtain a probabilistic forecast, such as a weather forecast."

Quote from Draft: Probability theory is the formal quantification of how to make rational decisions in the face of uncertainty; this carries over as much to quantum phenomena as it does to anything else. When one makes a probability statement concerning a quantum measurement outcome, one is essentially making a *bet* about what will be seen. As long as the probability calculus is used, an adversarial gambler can never force the bettor to a sure loss.

Comment: If, on the other hand, it is ignored, there exists a bet leading to a sure loss *even* in the single trial. This I believe is the best, clearest, and most famous "operational" definition for probability within the Bayesian approach, and is why I always say to you that one need never deal even with "conceptual ensembles" to define probability. (Besides, the conceptual ensemble approach is inconsistent, as I tried to explain to you once before.) The formal argument I am alluding to in the text is called the "Dutch book argument" and can be found in *Studies in Subjective Probability*, Second Edition, edited by H. E. Kyburg and H. E. Smokler (Krieger Publishing, Huntington, NY, 1980). If you wish, I can send you a summary of the argument that Rüdiger Schack wrote up in LaTeX; I also have some notes in LaTeX that Carl Caves wrote up. Alternatively, if you wish to read Bruno de Finetti's famous article in French (where the argument was first promoted), have a look at *Annales de l'Institut Henri Poincaré* 7 (1937), 1–68.

Quote from Draft: The peculiar nature of a quantum state is best illustrated by the fact that although an unknown quantum state cannot be duplicated because of the no-cloning theorem it can nevertheless be faithfully teleported.

Comment: I really like the idea of saying some things about teleportation and extracting something deep or expository from it, but I had a hard time trying to figure out what you were getting at here. Moreover, I have to admit of being a little doubtful that we yet understand what it is that is peculiarly quantum mechanical here. On the one hand, take two nonorthogonal Liouville distributions; by Koopman's theorem there is a good sense in which they cannot be cloned (this issue is what spurred my no-broadcasting paper. On the other hand take the paper by Cerf, Gisin, and Massar, quant-ph/9906105. If instead of entanglement, Alice and Bob are correlated via some local hidden variable theory, they can still teleport an unknown Liouville distribution for those hidden variables for only 2.19 bits on average. It's not quantum teleportation for sure, but what's the difference of principle that would make quantum teleportation illustrative of "the peculiar nature of the quantum state?" Though I'll be trimming this passage out of the concise version of the draft I'm also sending you, I did make some changes in this paragraph. So I'll go ahead and leave it all in this footnoted version.

Quote from Draft: It may seem paradoxical that in order to completely specify the polarization of a photon, we need three real numbers (represented by one point in the Poincaré sphere), but to teleport that state from one photon to another with the help of a previously shared entangled pair, only two bits of classical information have to be transferred.

Comment: Sentences combined. I didn't understand what you were wanting to express by entanglement "contains no information at all."

Quote from Draft: The difference is that if the state of the photon is explicitly known, we can produce an arbitrarily large number of photons with that state.

Comment: Just for one last emphasis: suppose a pure state were a pretty good approximation to what we could say about all the observables associated with the radius of the universe. How would we produce arbitrarily large numbers of that state?

Quote from Draft: On the other hand, the unknown state that is teleported disappears from the particle to which it "belonged," and still remains unknown, even though we are sure (in an ideal experiment) that another particle acquires that state.

Comment: I put quotes around "belonged" because that was a pretty blatant slip into objectivist language. Do you see what I mean?

I hope that you now agree that the following passage was superseded by all the previous discussions. "It thus appears that a 'state' is like a recipe. Once it is defined, we can prepare as many systems as we wish according to that state. Indeed, the simplest way to understand the notion of state is to relate it to a preparation protocol. However, it is not always possible to do so. When we observe photons that originated in a distant star, no one was there to prepare them millions of years ago. How shall we define such a naturally prepared state? A more general (though more abstract) way of defining a state is to specify the probabilities of the outcomes of all potential observation processes. These probabilities determine the best strategy for our future actions."

Quote from Draft: It is curious that some people are willing to abandon the objective nature of the world we see around us with its meaningful experiments and detector clicks, and yet work so hard to retain the abstract quantum state as something real. There is a temptation to believe that each quantum system has a wavefunction even if no one knows it.

Comment: Stripped out "The latter may not be known to any physicist; if its value is needed for further calculations, one would have to make reasonable assumptions about it, just as in classical statistical physics. However, conceptually, the state vector of any physical system would have a well defined, objective value."

Quote from Draft: Most likely, the root of that temptation is that in classical physics one has a "state space," the points of which can be assumed to correspond to an objective reality, and in quantum mechanics one again has a "state space" (only this time a Hilbert space).

Comment: The roots of realism. I suspect that if one were to do a thorough study of the Aristotelians, one would find that they had no such compulsion.

23 November 1999, "Airport Time"

I'm waiting at the Albuquerque airport for my mom and stepfather to arrive. They'll be staying with us until Friday when I depart for Naples and they depart back to Texas. This is only my mom's fourth flight ever, and she dreads flying like the dentist. So I keep my fingers crossed that they'll emerge from the plane confident and comfortable. If all is on time, they'll be here in 15 minutes.

So chances are, I won't be able to touch our draft again until tomorrow evening. So I hope you can be patient with me! But I would like to have it all tidied up before my European trip.

Asherism 65: *I restored the teleportation paragraph (in a more appropriate location) because I think it's important.*

I very much endorse trying to find some way of making quantum information relevant to understanding quantum mechanics in its entirety. So I'm not put off by this as I tried to express. The main hurdle is to get me to understand the significance of what you wrote. Perhaps after looking at your new version, I'll feel more comfortable. Do you have a rejoinder for why the "classical teleportation" of Gisin doesn't make the same point? (I'm too weary to open up your bigger files right now; it's easier to just babble my own thoughts.) Of course, maybe classical teleportation does make the same point (only quantitatively different): probability distributions are states of knowledge without objective reality just as quantum states are. I will think about this more.

Asherism 66: *I think that Gleason's proof necessitates some continuity assumptions.*

No, Gleason really does NOT need to assume continuity, that is what makes his theorem so remarkable! As I said, I've worked through every step of it (in Pitowsky's simplified version). Have a look at:

1. R. Cooke, M. Keane, and W. Moran, "An Elementary Proof of Gleason's Theorem,"
   Math. Proc. Camb. Phil. Soc. **98**, 117–128 (1981).
2. I. Pitowsky, "Infinite and Finite Gleason's Theorems and the Logic of Indeterminacy,"
   J. Math. Phys. **39**, 218–228 (1998).

Even Gleason's original paper did not assume continuity of the frame functions: it proved it. The most difficult part of the theorem is proving continuity: it turns out to just be a wonderful property of R^3 (the reals in 3-D). Meyer's example on *rational* vector spaces does not contradict this. On rational spaces, one cannot make the same set of moves (and build up the same graph structure) as Gleason did.

My mom's plane is here.

25 November 1999, "Apologies"

Thank you for the Thanksgiving day wishes. I apologize for my hiatus. I had hoped to get work done with my mother here, but it just hasn't happened. I hope you won't mind waiting too much for one more day. I will send you the next draft from the Chicago airport tomorrow afternoon. I must run now.

26 November 1999, "Too Late to Abort?"

I'm in the Chicago airport waiting for my flight to Naples. In the next note I will send you the latest draft. I feel like we are converging: I didn't make so many changes this time. The draft contains a hybrid notation for the changes: as you, I used ¶, ¶¶, and ¶¶¶, but for me the number of ¶'s represented simply (my subjective assessment of) the magnitude of the change. Also I put a few footnotes.

I have no idea whether I will be able to make email contact from the conference. If you make no major changes, perhaps it would be best for you to submit to Benka and get the process started with *PT*. You and I can always agree upon the minor tweakings later. If you do make significant changes though, perhaps we should risk it that I will have some email contact and I will be able to review what you have done.

By the way, I have no problem if we ask our colleagues for their opinions.

The Naples conference is a resort hotel on the island of Ischia. Next Saturday through Tuesday morning I go to Rome to visit Francesco De Martini. Tuesday evening through Dec. 12 I will be in Montréal. So the only time I could possibly be out of complete email contact is my time in Naples.

Parsing the Paper

Quote from Draft: To start, let us assess the goals of experimental science. An experiment is an active intervention into the course of Nature: we set up this or that experiment to see how Nature reacts. We have learned something new when we can distill from the accumulated data a compact description of all that was seen and an indication of which further experiments will corroborate that description. This is what science is about. If from such a description we can *further* distill a model of a free-standing "reality" independent of our interventions, then so much the better. Classical physics is the ultimate example in that regard. However, there is no logical necessity that this worldview always be obtainable. If the world is such that we can never identify a reality independent of our experimental interventions, then we must be prepared for that too.

Comment: I thought there were too many "interventions" so I went to "intrusions" even though that didn't seem wholly satisfactory (I couldn't find another good synonym). But I don't like "tests" in this position either: to me (and I'm guessing others) the word "test" conveys a sense of revealing a truth value – just precisely the sort of thing we want to get away from here. When we test our students, we ascertain whether they have studied or not. So I reinstated "interventions", as presently it doesn't seem like the word has been overused and I like the sentence better now.

Quote from Draft: The thread common to all the nonstandard "interpretations" – which actually are alternative theories – is the desire to create a theory with features corresponding to some reality independent of our experiments.¶ As we see it, trying to fulfill a yearning for a classical worldview by inserting into the theory extra hidden variables, extra worlds, extra kinds of time evolution, or extra consistency rules, while not at the same time improving its predictive power, is utterly counterproductive.

Comment: This gives a stronger feeling that these extra things are foreign and should be viewed as such.

Quote from Draft: These probabilities are computed by using any available information. This includes, but is not limited to, information about a system's preparation. The apparatus for turning that information into statistical predictions is formally expressed by the probability rule postulated by Max Born at the time quantum mechanics was invented.

Comment: I didn't like "how to convert" in this place because it seemed far too specific. Generally there is no clean cut algorithm for turning all the gathered information into a quantum state assignment; that is still something of an art. This is what I tried to capture the last time when I used the phrase "the latitude with which such information can be incorporated." The quantum rule only sets the limits; the rest is the experimentalist's art.

Quote from Draft: We do not deny the possible existence of an objective reality independent of what observers perceive.

Comment: I cannot agree with either of your two new phrases: "in particular, the 'detector clicks' are definitely objective" and "they remain real even if no observer is there." The problem is that, as we say later, "detector clicks" come about when we leave something out of the picture, when our information is incomplete. If we leave the observer out of the picture, I don't know what a detector click is. We are on safer ground if we leave "objective reality independent of our interventions" alone. Part of what you are hoping to get at might be better described as an "intersubjective reality" … but still I think we should leave it all alone. I rearranged the part of your new stuff that I could agree with to near the bottom of the paragraph.

Quote from Draft: The peculiar nature of a quantum state as a "state of knowledge" can be illustrated by the teleportation process.

Comment: If we say it like this, I think I can accept it. The Gisin and company "classical teleportation" was pretty powerful for me. Also inserting this phrase helps make connection to the previous paragraph.

Quote from Draft: In order to completely specify the polarization state of a photon, we need three real numbers (one point in the Poincaré sphere). Yet to teleport that state from one photon to another – with the help of a previously shared entangled pair – only *two bits* of classical information have to be transferred. The difference is that when a preparer knows the state of a photon, he can produce arbitrarily many further photons with the same state.

Comment: You wrote, "**ANSWER** It's just a matter of money. As Archimedes answered to the king: give me a firm support, and I shall move the Earth." *I didn't find that answer adequate!!*

Quote from Draft: On the other hand, the unknown state that is teleported disappears from the original photon. It remains unknown to the teleporter even though, with respect to the original preparer, another photon acquires precisely that state.

Comment: I thought it important that we made more explicit that there are two points of view here. Hence I was explicit about what came from what point of view.

Quote from Draft: It is curious that some well intentioned theorists are willing to abandon the objective nature of "observables," and yet wish to retain the abstract quantum state as a surrogate reality.

Comment: Why don't we just try it without the phrase "(that is, potentially observable properties)?" The thing I'm having trouble with is the word "properties" – I no longer think of quantum systems as having properties independent of our interventions. You made the mistake of inventing a word that I could finally feel comfortable with (i.e., interventions), and now you have to take the consequences!!

Quote from Draft: Of course, all this is *not* to say that there are no fundamental questions still to be explored within quantum theory. One can ask for better clarification of the features in our world that compel us to this theory instead of another or any number of other questions. However, to make quantum theory a useful guide to the phenomena about us and a self-contained theoretical edifice, we need nothing more than we already have. Quantum theory needs no "interpretation."¶¶¶

07 December 1999, "Took Care of Bureaucracy"

You'll notice in one of my notes that I did the final paper work for your symposium next year. You had previously asked for titles and abstracts for my talks; you will find them within that message.

I feel that I *should* have been working on our paper today while flying, since I was able to retrieve the draft yesterday. However, I found that it was easier to get my head around the many, many bureaucratic matters that had piled up in my email absence; so I spent my energy in that direction.

My interaction with De Martini was quite fruitful and I am glad that I went. I got to see many nice Alice and Bobs on the experimental tables! And I was able to propose various experiments that he might perform related to work that I had done with Charlie Bennett and John Smolin. Also there is an experiment that one can imagine for quantum state tomography related to this quantum de Finetti theorem of ours. I was really pleased to find that both Enrico Fermi and Bruno de Finetti had actually been on faculty at that university.

However, the meeting in Naples might have been a waste of the taxpayer's money! (Tax money from the European Union, that is, not the US! My funding was completely covered by the organizers.) The meeting was advertised as science, but as far as I could tell most of the attendees might have just as well belonged to the priestly set and worn white robes. The two main contingents were the followers of GRW and the Bohmians. (The Bohmians were the larger.) In all their talks and discussions, it seemed to me that they said nothing and listened to nothing. I was the lone hold-out at the conference for a Copenhagen-like sensibility. I gave a talk trying to show that quantum information theory only deepens our understanding in that regard, and, in fact, may be regarded as a much more effective way of exploring the foundations of quantum mechanics. I first gave a little of a philosophic spiel, then I focused on three significant examples (teleportation, our information–disturbance relations, and the BDFMRSSW nonlocality without entanglement). My message fell on deaf ears! And the thing that really surprised me is how so few of them were even interested in the physical phenomena themselves despite foundational issues. A particular thorn in my side was David Albert. I found him most annoying when he branded our point of view about quantum mechanics – precisely the point of view in our *Physics Today* article – as unscientific! (I wanted to say, "Then why is it that I am employed as a physicist, but you are employed as a philosopher?")

But there were some small highlights to the meeting. I had many chances to talk to Adrian Kent, Jeremy Butterfield, Orly Shenker, Chris Dewdney (who is a Bohmian, but with a more open mind), and Jos Uffink (I especially enjoyed his input).

I now continue flying for several hours. (I am stuck in the Chicago airport for over four hours I believe.) I will try to make some progress on our project if I can fight the exhaustion. In any case, you should expect a revised draft from me no later than Thursday.

11 December 1999, "Productivity"

This conference has turned out to be quite a productive affair. It's so good to be back with real scientists. I've been making changes here and there to our draft when I've had a chance. I think we're getting into the final leg. However, I am sorry to keep you waiting so long. Tomorrow I will be in planes and airports most of the day (Montréal, Chicago, Dallas, Albuquerque); I should be able to put all my final touches on the draft during that time.

Do you have any plans to go to the Benasque meeting this summer? It falls straight on top of QCM. (I have also heard that Sandu is organizing a meeting during that same time, but I don't know where.)

Today, I meet with the mathematician Mary Beth Ruskai. (Her career was made by proving with Lieb the property of quantum entropy known as strong subadditivity. It had been an open question for 10 years before that.)

Last night it snowed here. Everything is beautiful. I'm starting to get so homesick for Kiki and Emma. (That always happens to me when I see a beautiful scene and she is many miles away.)

12 December 1999, "Thinking It's a Toddler"

Now I'm starting to think its a toddler. Let us see what you think. I just did a precise count on the words and it came to 1790 words in the text alone (i.e., not counting title, biographies, or references). That's cutting the limit very close. But I think the draft is starting to be very good now, something we will both be proud of some time from now.

The biggest changes you will notice are in the teleportation paragraph and the closing paragraph. I have thoroughly documented all my thoughts and reasons for making changes in footnotes.

Let me bring up one more change that I have not yet inserted. You wrote:

Asherism 67: *Incidentally, Vaidman also wrote to me that Cathy's digestion was not an irreversible process, and I privately agreed with him on this point, but explained that we didn't want to open a second front, and we had constraints on the length of the text.*

I understand that you don't want to open up a new front, but I think Vaidman's reaction will be the general one. I have in the past heard Howard Barnum, John Preskill, and David DiVincenzo all say precisely the same thing. I think it is much more likely that by not saying anything about the issue we will by default open up a new front. We could of course defend ourselves when we write our reply to critics, but don't you think it would be so much better if we expressed the complete consistency of our point of view in the article proper? I think the article would be better for it, and demonstrate too that we have a well thought out position.

I'm sure we could fix the problem with just two sentences. For instance, something like:

If Erwin had chosen not to take a look that would be fine too. With so much information about Cathy and her digestion, he could in principle reverse the whole chain, wiping Cathy's stomach and mind of all memory of her tasty treat.

The problem is that that's 41 more words. But surely we could try to push the limits a bit. We could just wait and see what happens after the editors start wielding their knife. I am inclined to do that. In my mind, the manuscript has now reached a certain stage of completeness that I wouldn't want to see it toyed with (at least voluntarily on our sides). I like the way it reads now.

Please do think about this. (I've been pushing for this point since we started the project: Lev's criticism has only reactivated my worry.)

OK, now I must act like a good husband and spend some time with Kiki. I'll be in touch with a report of Mermin's further thoughts sometime tomorrow.

Parsing the Paper

Quote from Draft: The thread common to all the nonstandard "interpretations" – some would say alternative theories – is the desire to create a theory with features corresponding to some reality independent of any potential experiments.

Comment: I thought "look like" was too wishy-washy. You might remember that I had once deleted this whole phrase because I thought it would cause trouble, but then you reinstated it. After you did that, I actually grew to like it a lot. It is simply a mistake to think that the Many Worlds Interpretation is the same theory as the quantum mechanics you and I use; it is a contentless structure, whereas quantum theory is a meaningful guide to our experiments and engineering. What could it mean to say I "choose" this or that experiment in MWI? Why not give up on life and just let the universal wavefunction carry us along? I have no great trouble with creating controversy with this article. We should be brave and say what needs to be said. If you are not averse to reinstating your old stronger statement, I am not; I certainly prefer your old formulation over my present one. Damn the torpedoes, I say, full speed ahead! As you said before, we can come out kicking again in our inevitable "reply to critics."

Quote from Draft: As we see it, trying to fulfill a yearning for a classical worldview by inserting into the theory extra hidden variables, extra worlds, extra kinds of time evolution, or extra consistency rules, while not at the same time improving its predictive power, is utterly counterproductive.

Comment: Now, here's where I would weaken the grip. It is our *opinion* that it is *utterly counterproductive* to attempt to insert a classical worldview. (I know from last week in Naples that Dürr and Goldstein, for instance, think it is extremely productive: at the very least it gets them enough funding to hold a conference every two years.) I know you don't like the phrase "as we see it," but I think we need something here that carries the same flavor. Also notice that a sentence starts with "however" just four sentences above – it conveys a little too much a see-sawing feeling to the read to me.

Quote from Draft: Quantum probabilities, like all probabilities, are computed by using any available information. In the quantum case this includes, but is not limited to, information about a system's preparation. The mathematical instrument for turning that information into statistical predictions is the probability rule postulated by Max Born at the invention of quantum mechanics.

Comment: Good point. So I went one step further and made it clear that we are speaking of a "mathematical instrument." Also I shortened the sentence just slightly, by rearranging the ending phrase to flow a little easier.

Quote from Draft: The conclusiveness of Born's rule is known today to follow from an important theorem due to Andrew Gleason. So long as one assumes that yes–no tests on a physical system are represented by projection operators P, and that probabilities are additive over orthogonal projectors, then there exists a matrix ρ such that the probability of a "yes" answer is $\text{tr}(\rho P)$. The compendium of probabilities represented by the "quantum state" ρ captures everything that can meaningfully be said about the physical properties of the system.

Comment: I still don't know what the phrase "depending solely on that system (not on the apparatus that probes it)" can possibly mean. If one of the author's can't understand the phrase, then the innocent readers will surely have trouble. The quantum state cannot depend on the system at all, as I understand the meaning of the word "depend." This is because, as we have already agreed, the quantum state is not an *objective* property of the system itself. Its assignment depends only on the information that has been gathered (by whatever means). So then what does it mean to say that the quantum state depends on the system? Well certainly it does, if by that one means the dimensionality of the density matrix: presumably the dimensionality of the system of interest is an "objective" property. But I can't see anything beyond that. Please explain what you are trying to get at in more detail, or let's just drop the phrase. The present wording seems perfectly adequate to me and less likely to create confusion.

Quote from Draft: The peculiar nature of a quantum state as representing information is strikingly illustrated by the quantum teleportation process. In order to teleport a quantum state from one photon to another, the sender (Alice) and the receiver (Bob) prepare a pair of photons in a standard entangled state. The experiment starts when Alice receives another photon whose state is unknown to her. She performs a measurement on her two photons, and then sends Bob a *two bit* classical message instructing him how to exactly reproduce the unknown state on his photon. This economy of transmission is remarkable, because to completely specify the polarization state of a photon, we need three real numbers (one point in the Poincaré sphere), that is, an infinity of bits. The difference is that as far as Alice and Bob are concerned, nothing whatsoever changes about Bob's system: they start off maximally ignorant about it and remain maximally ignorant about it when the process is complete. The two bits of classical information transfer only signify the cost of transferring the preparer's information about one of Alice's systems – i.e., the "unknown" state – to Bob's previously correlated system. Nothing physical is transferred.

Comment: I understand that Eugen endorsed this paragraph with the words "much clearer now," but I have to admit that I still do not get the point you are trying to make here. (It should always be the job of the authors to be more critical of the words than the casual reader.) For instance, I really don't see what it has to do with the remarkableness of teleportation that if we have a system in a pure state, we might be able to make many *other* copies of that state. It just doesn't click for me – and that's a minimal requirement. So let me try to rewrite the passage so that I can understand the import of it. However, I'll keep the old words here so that we can reuse parts of them if need be: "The difference is that when we know the state of a photon, we can produce arbitrarily many photons with that state. On the other hand, the unknown quantum state that is teleported disappears from Alice's laboratory. It is also unknown to Bob who receives it. Yet, the original preparer who knows that state can verify that it is correctly transferred to Bob's photon."

Quote from Draft: It only gives the evolution of our probabilities for the outcomes of potential experiments on the system.

Comment: In agreement with Eugen, I think it is best to minimize the usage of the loaded term "measurement." Hence I put "experiment" into this sentence since we had defined it precisely previously. Note also that I deleted the last sentence (that I had added in the previous draft). I decided that it didn't have such a transition value after all.

Quote from Draft: All this said, we would be the last to say that the foundations of quantum theory were not worth further scrutiny. For instance, it may be quite useful to search for various minimal sets of physical assumptions that give rise to the theory. Since it is not yet understood how to combine quantum mechanics with relativity and gravitation, there may well be insight to be gleaned there. However, to make quantum mechanics a useful guide to the phenomena about us and a self-consistent theory, we need nothing more than we already have. Quantum theory needs no "interpretation."

Comment: OLD PARAGRAPH: All this is *not* to say that there are no fundamental questions still to be explored within quantum theory. How to combine it with relativity and gravitation is not yet satisfactorily understood. Another issue is whether any quantum state or operator that we know to write can be realized, in principle, experimentally. However, to make quantum mechanics, as we know it, a useful guide to the phenomena about us and a self-contained theoretical edifice, we need nothing more than we already have. Quantum theory needs no "interpretation."

Comment: How about a compromise here? The thing I didn't like about your examples is that both seemed to be more about difficulties in *application* of quantum theory, rather than its interpretation. I think it is important that the reader understand that we are not doomsayers about the interpretation – that is to say, we are not completely rigid in the way that Bohr and Rosenfeld were. A substantial number of Peres' papers in the last 15 years have been precisely about clarifying the foundations. It has been a fruitful pursuit precisely because the foundations are not yet a closed book. Flesh and bones still need to be added even to Copenhagen-like ideas. So I made what I was trying to get at a little more concrete by changing the wording and adding two references of serious works. Then I gave an even more concrete reason for pursuing that: namely precisely your point about relativity and gravitation. Now I like the ending even better; I hope you will agree.

14 December 1999, "Fear of Numbers"

Please don't be frightened by the Annotated Toddler. Despite the number (and size) of footnotes, I think we are reaching rapid convergence. I think essentially everything is now stable (at the very least readily negotiable) except the paragraphs on teleportation and reversibility. I expect though that we'll also have all that cleared up soon, probably tomorrow. I've given them my best shot again. Let's see what you think about this iteration.

This afternoon, Kiki, Emma, and I trudged through the snow to find a Christmas tree. For $10, Los Alamos County gives a permit that lets you cut anything on the mountain that is not over 10 feet tall. It was quite fun, and I can see that it will turn into a family tradition. We bundled up Emma quite well (in a miniature ski suit) and she seemed to take in every scene. The only problem now is that tonight she has developed a deep cough (though it is probably unrelated). We are a little worried, however, because whooping cough is going around; we just read in the paper this morning that 42 cases have been reported in the last month. Kiki will take Emma to the doctor tomorrow.

Thanks, by the way, for the stories about your grandchildren. I sometimes do silly things like rub a rubber coaster over Emma's hair and then watch the hair rise from the static. Kiki says, "She's not a science experiment!"

Parsing the Paper

Quote from Draft: The thread common to all the nonstandard "interpretations" is the desire to create a new theory with features corresponding to some reality independent of our potential experiments.

Comment: "Any" → "our."

Quote from Draft: These attempts to fulfill a classical worldview by inserting into quantum mechanics extra hidden variables, extra worlds, extra kinds of time evolution, or extra consistency rules, while not at the same time improving its predictive power, only yield the illusion of a better understanding.

Comment: I took out the phrase "a yearning for." The sentence seems simpler this way without changing its meaning. Also I replaced "into the theory" with "into quantum mechanics" because I didn't want there to be any ambiguities about which theory we are referring to. Also, by the way, I think this now boils down to a much subtler way of saying that the other "interpretations" are not quantum mechanics – so we get our cake and eat it too.

Quote from Draft: The compendium of probabilities represented by the "quantum state" ρ captures everything that can meaningfully be said about a physical system.

Comment: I am ashamed to admit that the phrase "physical properties of the system" has made it past my sieve in all these iterations. This is precisely the sort of objectivist language that we are trying to move past with this article: unperformed measurements have no outcomes, systems have no "properties." There are only the macroscopic consequences of our potential interventions. We need to be consistent on this point.

Quote from Draft: Cathy's story inevitably raises the issue of irreversibility. As quantum dynamics is time-symmetric, can Erwin undo the process? Again, why not? For this, Erwin would need complete control of all the degrees of freedom in Cathy's laboratory, but that is a difficulty of practice, not of principle. The main point is that the information Erwin possesses has to do with the potential consequences of his experiments. The information is not about what is "really there." If Erwin performs no experiments, there is no reason he cannot reverse Cathy's evolution. If, on the other hand, he intervenes in her unitary evolution, then he forfeits the capability: he too will have become entangled with the laboratory and all its contents.

Comment: OLD SENTENCES: "This is obviously far beyond Erwin's ability. He has no choice but to consider Cathy as a macroscopic object that is part of an "effective" reality, as defined above."

Comment: Boy, do we diverge in opinion on this one!! The problem is not at all one about Erwin using a less than complete description (the kind of thing that creates the macroscopic reality of detector clicks). In fact that would be inconsistent with our previous explanation about Cathy being in a 50/50 superposition. There is nothing to stop us from imagining that Erwin has absolutely maximal information about Cathy's laboratory, i.e., that he describes it all via some large entangled PURE state. The problem instead stems only from a holdover of the objectivist training of most physicists. The knowledge Erwin possesses has to do with the potential consequences of his experimental interventions on Cathy's laboratory. It doesn't mean anything more than that. In particular, it has nothing to do with what is "really there," whatever that might mean (i.e., like Cathy really eating the cake, or really eating the fruit). That is the only consistent position. If Erwin doesn't intervene on Cathy's laboratory with a "measurement", then there is no reason that he cannot reverse the evolution of everything in her laboratory. What was your Archimedean quote? (Now I'm being facetious.) If on the other hand, he does intervene on the laboratory with a measurement then he has forfeited all rights to reverse her evolution: he has now become entangled with that system (to use a concept from the Church of the Larger Hilbert Space). We can only completely control the evolutions of systems to which we are completely external.

I propose that we condense one or two sentences from what I just wrote above in this footnote to fill the rest of the paragraph. Let me try my hand at it. I will do this in a tentative way, that you will likely want to make clearer. The main point is that we need to get the argument right.

Anyway, for extra reference, let me record here how I put it to Howard Barnum when he had the same problem as Lev: [See the note to Howard Barnum, titled "It's All About Schmoz," dated 30 August 1999.]

Quote from Draft: The peculiar nature of a quantum state as representing information is strikingly illustrated by the quantum teleportation process. . . . The conundrum is solved by realizing that as far as Alice and Bob are concerned nothing changes about Bob's photon: they end up describing it by the same density matrix they started with, the completely depolarized state. The two bits of classical information serve only to transfer the preparer's information to be from being *about* the original photon to being *about* the one in Bob's possession. This happens because of the previously established correlation between Alice and Bob's photons.

Comment: I have trouble agreeing with the sentences you wrote here. You write "this complete specification is not transferred." But it IS, for the preparer! It is just a question of with respect to whom that the specification is transferred. That is the thing that needs to be made clear to give the example of teleportation any relevance to our quest. In any case, I don't think the unprepared reader will be able to pick this up from the description you wrote. So let me try rewriting this again – somewhat along the lines of my proposal yesterday – but this time I'll be careful not to introduce any technical jargon that the whole physics community does not know. You were right that my last attempt was not so good; I hope to do better this time. OLD SENTENCE: "However, this complete specification is not transferred: the unknown quantum state that is teleported disappears from Alice's laboratory, and it is also unknown to Bob who receives it. Yet, the original preparer who knows that state can verify that it is correctly transferred to Bob's photon."

14 December 1999, "Lost Toddlers"

I see from your pile of notes to me that somehow you didn't receive my toddlers until long after I had sent them. It is strange how email can disappear for so many hours. You just have to wonder where it is sitting (or bouncing) all that time.

Emma coughed a deep cough all night. Looking back at the note I wrote you last night, I realized that I didn't mention that the 42 cases were all in this small county in the mountains. I think it is unlikely that she has whooping cough (it sounds to be more a problem of chest congestion), but still I worry. The curse of new parenting?

I forgot to answer your question about the abstract of the abstract. What you have written is just fine. I gave a slightly more technical version of that talk in Montréal Saturday. I was the second-to-last talk of the conference, so I didn't really expect people's full attention, but still it flowed quite smoothly. I was left believing a counterfactual: if the audience had not been exhausted, then the talk would have been received well! Actually, I got some nice questions from Dan Gottesman, Hoi-Kwong Lo, and Michael Nielsen. The main thing that made me think the talk wasn't a waste was that David DiVincenzo came up to me afterward and said that he now understood quite clearly what the problem was all about. That was a triumph!

Aha, I just saw you have sent me some more notes. I will go to them now.

14 December 1999, "Training Pants?"

Another day, another draft. I again follow my usual protocol: one draft with footnotes, one draft without. The draft with footnotes gives all the details of what I was thinking as I made changes. I keep my fingers crossed that we will reach complete agreement soon. The main thing that makes me feel giddy is that with *Physics Today*, our potential readership may be on the order of 50,000 people or more! You know that John Wheeler had this idea that things become more real as they become acknowledged by more and more members of the "community of communicators." I would hate one of the things we said to become real if later we decided we didn't like it!

Once we finish this project, I hope to have a look at the Brukner/Zeilinger paper that you didn't recommend. I contacted Anton once by email saying that I thoroughly enjoyed his web article "On the Interpretation and Philosophical Foundation of Quantum Mechanics." (You can find it on his homepage.) He wrote me back that he had something much more serious in the works, something that would rigorize the point of view he expressed there. It's an article that appeared in *Foundations of Physics* last year (or maybe this year); I suspect the *PRL* you didn't recommend is a follow-on to that. Well anyway it was not very good, and I couldn't believe that the same man had written such an empty paper. A curious fellow he must be.

Tonight Kiki and I decorate the Christmas tree we cut yesterday. We've already placed it in the stand: so far, it looks marvelous. Neither of us is religious of course, but we still have faith in Santa Claus.

Parsing the Paper

Quote from Draft: Our purpose here is to explain the internal consistency of an "interpretation without interpretation" for quantum mechanics. Nothing more is needed for using the theory and understanding its nature. To start, let us examine the role of experiment in science.

Comment: The word "define" seemed too haughty to me in this context: I thought it conveyed a bit of the flavor that we view ourselves as the arbiters of what science should be. Anyway, the more I thought about this sentence, the more talking about the "goals" didn't seem right either. So I reconstructed the sentence completely. I think this one hits the mark a lot better. By the way, completely as an aside, you might enjoy the historical article (I did): M. Jammer, "The Experiment in Classical and in Quantum Physics," in

Proceedings of the International Symposium: Foundations of Quantum Mechanics in the Light of New Technology, edited by S. Kamefuchi (Physical Society of Japan, Tokyo, 1984), pp. 265–276.

Quote from Draft: The thread common to all the nonstandard "interpretations" is the desire to create a new theory with features corresponding to some reality independent of our potential experiments. But, trying to fulfill a classical worldview by inserting hidden variables, parallel worlds, nonlinear evolutions, or consistency rules into quantum mechanics, while not at the same time improving its predictive power, only yields the illusion of a better understanding.

Comment: At first I put a "however" here, but the word seemed overused. Then I tried "but" and it seemed so much better: it seems to give the sentence a much more accusatory feel. And I like it this way. My only worry was that you would not like a sentence starting with "but." So I looked in my *American Heritage*, and lo and behold it said the following: "*But* may be used to begin a sentence at all levels of style." I'm keeping my fingers crossed that you'll like it too.

Quote from Draft: Yet, it is legitimate to attempt to extrapolate the theory beyond its present range, for instance when we probe particle interactions at superhigh energies, or in astrophysical systems, including the entire Universe. Indeed, a common question is whether the universe has a wavefunction. There are two ways to understand this. If this "wavefunction of the universe" has to give a complete description of everything, including ourselves, we get the same meaningless paradoxes as above.

Comment: Somehow the word "same" seems essential for emphasis to me. Take Vaidman as an example: he rejects Bohmianism as a solution to the quantum measurement "problem," but embraces MWI. He sees one solution as better than the other, and somehow doesn't see that they are both ridiculous.

Quote from Draft: Does quantum mechanics apply to the observer? Why would it not? To be quantum mechanical is simply to be amenable to a quantum description. Nothing in principle prevents us from quantizing a colleague, say.

Comment: This small point may only have to do with our difference in our culture or ages or the literature we read, etc., but adding a little extra phrase like "say" or "for instance" to me somehow gives this sentence the flavor of being tongue-in-cheek. Without it, the joke seems to fall a little flat.

Quote from Draft: Cathy's story inevitably raises the issue of reversibility since quantum dynamics is time-symmetric. Can Erwin undo the process if he has not yet observed Cathy? A superficial reading of the story might lead some to think not. That misconception arises from forgetting that the information Erwin possesses has only to do with the consequences of his potential experiments. The information has nothing to do with what is "really there." If Erwin has performed no observation, then there is no reason he cannot reverse Cathy's digestion and memories. Of course, he would need complete control of all the degrees of freedom in Cathy's laboratory, but that is a practical problem, not a fundamental one.

Comment: YOUR LAST VERSION: For this, he needs complete control of all the degrees of freedom in Cathy's laboratory. His difficulty is practical, not fundamental. In practice, Erwin has of course only incomplete control and he must consider Cathy as a macroscopic object that is part of an "effective" reality, as defined above. (If Erwin observes Cathy, he forfeits the reversal capability: he becomes entangled with her and only a "superobserver" can reverse the process.)

Comment: OK, it looks as if we still have a disagreement on some points: some practical and some presentational. Let me try to separate those points and be as clear as I can be. Your point about what Nathan Rosen said is well taken. So, I hope we can come to some agreement on this (and thus label it physics): I think our readers will definitely benefit if we do it right.

1) There is a sense in which this paragraph is completely redundant. We have already said quite clearly that the wavefunction is information about the consequences of interventions, not about what is "really there." Therefore no one should have the right to think that our example about Cathy and Erwin implies that quantum mechanics should become irreversible at some sufficiently high level. I believe this may have been the reason you were reluctant to include this subject in the first place. However the plain fact is that people do seem to misunderstand the point. Conversations and email with LV, HB, JP, DPDiV, and possibly CHB, attest to that. The point is that people don't listen so readily when they already have some preconceived notion in their head, no matter how sincere they are. So as I deem it, the main goal of this paragraph is to reiterate the point of "knowledge about what" in a slightly new context, that of reversibility.

2) You wrote in your accompanying note: "We don't disagree at all, but were concerned by different issues. It is quite obvious that after Erwin performs his measurement, there is no return to Cathy's $|hungry\rangle$ state. The question is whether this is still possible before Erwin measures." Actually, your second sentence was not the foremost issue on my mind: I added a discussion of that because other people (like Barnum) often speak sloppily of the observer reversing his own measurement – that phrase is an oxymoron. The foremost issue, as I see it, is to push people away from thinking that our story implies an in-principle *irreversibility* before Erwin measures, i.e., the same concern as you. The only reason people think that what we are talking about is irreversible is (in my experience) because they have not yet fully absorbed the lesson that quantum mechanics is not about some free-standing reality.

3) I cannot agree with your sentence "In practice, Erwin has of course only incomplete control and he must consider Cathy as a macroscopic object that is part of an 'effective' reality, as defined above." Previously we defined "effective reality" as having to do with ignoring most degrees of freedom and considering a description that is necessarily incomplete. Of course, both of those things could enter into our discussion of Cathy and Erwin's experiment, but that is not the essential point as I see it. We did not introduce an incomplete

description previously (i.e., in the first paragraph about Cathy and Erwin) and I think it would strongly cloud the relevant issues if we do it now. If we had done it previously we would have had no right to talk of 50/50 superpositions – instead our whole discussion would have had to be in terms of significantly mixed density operators. The solution to Vaidman's question does not come about because of Erwin's technical inability. We should stick to our guns: the theory is not about our knowledge of a free-standing reality. Therein lies the solution to Vaidman's conundrum.

4) I did not like your introducing the phrase "and only a 'superobserver' can reverse the process." Vaidman would like to think that we edge upon acceptance of MWI, and I think that would only fuel his opinion. I know that I mentioned in my last note to you something about the Church of the Larger Hilbert Space, but I would like to keep that short-hand private. I don't like the Church; the last I want to do is give it a public endorsement by acting as if there were always a superobserver available. If so, then why not speak of the superobserver of the whole universe and his wavefunction for it?

5) Given all that, let me try to take another shot at closing the paragraph. If we still can't come to agreement (after perhaps your next iteration), then maybe we should follow Rosen's dictum.

Quote from Draft: The peculiar nature of a quantum state as representing information is strikingly illustrated by the quantum teleportation process. In order to teleport a quantum state from one photon to another, the sender (Alice) and the receiver (Bob) prepare a pair of photons in a standard entangled state. The experiment starts when Alice receives another photon whose polarization state is unknown to her. She performs a measurement on her two photons, and then sends Bob a classical message of only two bits, instructing him how to reproduce the unknown state on his photon. This economy of transmission appears remarkable because to completely specify the state of a photon, namely one point in the Poincaré sphere, we need an infinity of bits. However, the conundrum is only apparent. As far as Alice and Bob are concerned nothing changes about Bob's photon; they describe it by the same density matrix throughout the whole process. The two bits of classical information serve only to transfer the preparer's information to be from being *about* the original photon to being *about* the one in Bob's possession. This is not magical: it comes about by the previously established correlation between Alice and Bob.

Comment: WORDS OF THE LAST DRAFT: "However, quantum teleportation does not involve such an explicit specification. Neither Alice nor Bob can know the state that is teleported. Yet, the original preparer who knows that state can verify that it was correctly transferred."

Comment: Believe it or not, I probably spent over an hour or an hour and a half composing the sentences that I put in this place last time. That, of course, does not mean that they should not be discarded if they don't convey the proper information! But I did think hard about what it would take to make the significance of the example clear to myself, and then how to translate that to the layman. I just can't believe that the uninitiated reader will understand what you are really getting at if the point is made in such a terse manner: I use myself as an example, I didn't understand it at first.

You write, "quantum teleportation does not involve such an explicit specification." But, again, it does for the original preparer; it only doesn't for Alice and Bob. That point needs to be made clear. You write, "Neither Alice nor Bob can know the state that is teleported." True, but what does that have to do with "strikingly illustrating the peculiar nature of a quantum state as representing information" in the context of teleportation?

Finally, if one does not emphasize that the effect comes about because of the previously established correlation – the resource the preparer uses to update his knowledge of Bob's system – then teleportation does indeed look like some mysterious superluminal transportation of the state. And that annuls the point of the example.

Please give my words from yesterday (modified a fair amount now) one more shot and see if you cannot find a kernel of a good explanation within them.

15 December 1999, "Insanity and Insomnia"

I just quickly skimmed the notes you sent me.

Asherism 68: *When you wrote "extra kinds of time evolution", I thought of the two-time formalism of Aharonov and Vaidman, who claim they have a better way of understanding QM. Now you wrote instead "nonlinear evolutions".*

This was insanity. Actually, all along, I had been thinking of stochastic collapses of the GRW type (since Goldstein had mentioned them in his article). Why I wrote nonlinear evolutions yesterday to describe that, I don't know. I like your new sentence. Should we though keep connection with the *PT* articles we are criticizing by substituting "stochastic evolutions" for "multiple times"? I don't believe the Aharonov–Vaidman ideas were mentioned in any of the *PT* articles. In any case, I get the impression that the GRW wackiness is more "mainstream" than the AV wackiness.

I'll look into the draft in more detail later (after I wake up).

Emma's cough has subsided a bit, but it's still here. The doctor yesterday gave her a thorough examination and said that there are no signs of whooping cough, only deep chest congestion. Strange though that whooping cough is running so rampant in Los Alamos since all children are required to be immunized against it. It suggests to me that perhaps there is a new strain that is resistant to the old immunization.

Now I go back to bed (since after all I am asleep).

15 December 1999, "One Last Stumble"

Well, it does look like we're narrowing down. The only stumbling block for me is still the teleportation paragraph. As the last two sentences are presently worded, I still don't see how the reader will come to the intended realization, i.e., your changes seem like much more than MSC [Minor Style Changes] to me. Perhaps I'm just being stubborn and somehow have blinded myself to a significantly simpler formulation – that is a real possibility, I'm not discounting it. I wonder what you think about the following. Might we see if Eugen would be willing to make a comment on the two opposing versions? He has thought a little about the issue, but is still much more likely than us to be representative of the average readers of *PT*. I could send him a note today, and if he responds by tomorrow we wouldn't be any worse off. If he does not respond quickly, we can just do something with whatever we have – I'll volunteer now to let you have the last word on the subject – and send it to Benka.

I don't want to be completely brick-headed. I realize that in this writing I have been quite stubborn about getting my way … and you probably don't deserve so much heartburn. I don't want you in the end wishing that you had just written the comment yourself and been done with it. I just want to feel comfortable with the clarity of our exposition to the extent to which it is in my power.

I'll go ahead and send this note off now, in case you are nearing bedtime. In the next note I'll write a detailed commentary of why I don't think your teleportation changes are just MSC.

15 December 1999, "Detailed Commentary"

Let me try to expose where I see differences of substance between what you wrote and what I write. After that I will propose one last version of the paragraph, to see if we can both agree upon it.

I can tell that in the end we are in agreement about the physics and *our* understandings of the phenomenon. Our only difference may be in what we wish the reader to take away from the paragraph. In fact we only differ in the description of the closing part of the paragraph.

Here are your words:

However, the two bits of classical information serve only to transfer the quantum state from being that of the original photon to being that of Bob's photon, while neither Alice nor Bob knows that state. Only the original preparer who knows it can verify that it was correctly transferred.

Here were mine:

However, the conundrum is only apparent. As far as Alice and Bob are concerned nothing changes about Bob's photon; they describe it by the same density matrix throughout the whole process. The two bits of classical information serve only to transfer the preparer's information to be from being *about* the original photon to being *about* the one in Bob's possession. This is not magical: it comes about by the previously established correlation between Alice and Bob.

———

1) In the first sentence you say, "the two bits … serve to transfer THE quantum state." The word "the" implies that there is one and only one state for that system. But that is not the case. This, in my eyes, is why "the peculiar nature of a quantum state as representing information is strikingly illustrated by quantum teleportation." There is the state Charlie,

the preparer, assigns to his original system, $|\psi\rangle$. Then there is the state that Alice and Bob assign to that system. If they have no prior information about Charlie's preparation, they will assign the completely depolarized state, $\rho = \frac{1}{2}I$. One is mixed and one is pure, to be sure, but from our point of view – as endorsed by Gleason's theorem – they are two equally valid states.

Then there is the state that Charlie originally assigns Bob's system. Because he knows that Alice and Bob share a maximally entangled state, his assignment is the completely depolarized state. Alice and Bob too assign the completely depolarized state to Bob's system. So for this aspect of the experiment, there is one state.

After the teleportation process is complete, Charlie updates his assignment to Bob's system to be $|\psi\rangle$. However, Alice and Bob continue to describe that system via $\rho = \frac{1}{2}I$. Thus at the beginning and ending stages of the game, there is not one quantum state in the picture, but rather two.

To present the idea in the way you did, it seems to me, only endorses the bad mode of thought that has caused us to write this article in the first place. What right do Alice and Bob have to say that there is a quantum state there and they just don't know it? For all they know, Charlie could have given them half of an entangled pair to teleport. Or maybe there is no Charlie at all and the photon has come from the three-degree microwave background radiation.

As useful as the idea of the "unknown quantum state" is, the phraseology is a holdover from the idea that systems have "states" independent of the scientists speaking about them. That is the point of view we are trying to fight in this paper. I realize that we can't rectify the whole situation in a simple opinion piece, but there is no reason to fuel the fire either.

2) In the second sentence you say, "Only the original preparer who knows it can verify that it was correctly transferred." Again I don't feel that this sufficiently emphasizes that it was ONLY the preparer's description that was transferred and nothing physical beyond that. The wording you used is much more of the feel that there is a state THERE, and Charlie can verify it if he wishes, but nevertheless the raw fact is that there is a state there. Again, why help fuel the fire that is consuming our community?

It is just a question of getting away from bad language. Remember my footnote (#4) about Peierls, i.e., the last one inserted into the paper I scanned for you? The major drawback about his little piece was that every now and then he would revert to objectivist language: he wasn't as consistent as he could have been. Sometimes he would say that the quantum state stands for what we know, and then sometimes he would speak about having information about THE state of a system. This is what I'm trying to get at: I want us to do better than that.

If we are going to be consistent, and educate the community a bit, we should not speak of "it being correctly transferred" in any way that hints of it being done independently of the preparer who knows it.

3) You don't like "conundrum" and "magical." That's fine: I can change that, they weren't essential. The main point to me is that after all our effort to de-objectify the

quantum state, your language in the teleportation example only seems to re-objectify it. I want us to be careful about that. For instance, I know for a fact that Richard Jozsa takes that language very seriously: he thinks that our way of looking at the quantum state is just hogwash (he told me again last week in Montréal). He, like his old graduate advisor Roger Penrose, likes to think that systems have objective states and that sometimes they evolve unitarily and sometimes they collapse. Most importantly for the point at hand though, he thinks that quantum teleportation endorses this view – he told me so. This, I believe, comes about partially because people have been fixated by the inappropriate phrase "unknown quantum state."

4) You didn't like my phrase "As far as Alice and Bob are concerned, nothing changes about Bob's photon." But that is the only position one can take, unless one believes that there is more that we can know about a system than the quantum state one ASSIGNS to it (based on any previously gathered information). If you really want to take that position, then we would have to retract our earlier statement, "The compendium of probabilities represented by the "quantum state" ρ captures everything that can meaningfully be said about a physical system." I for one, however, don't want to retract that statement. Now, maybe there is an issue that perhaps I shouldn't have inserted the "nothing changes" phrase into the actual text. But you must see the force of the point: with respect to Alice and Bob, nothing changes about Bob's system ... even if they perform a unitary operation on it. Their compendium of probabilities for all measurements remains the same, and that is all that can be meaningfully said about THAT system. It is true that their state assignment for the Charlie–Alice–Bob system changes in the teleportation process, but that change refers to a composite system, not the singular system in Bob's possession. We have to have the courage of our convictions.

5) There is nowhere in your discussion an emphasis on why Charlie's state can be transferred from Alice to Bob after only one of four potential unitary transformations has been performed. This is what I was trying to get at with my sentence, "It comes about by the previously established correlation between Alice and Bob." If Charlie hadn't "known" something about Bob's system through the previously established entanglement, then he would have never been able to transfer his "state of knowledge" (quantum state) onto that system.

6) You didn't like my usage of the word "about" ... but that's not overly essential. In our discussion of Gleason's theorem, we used the word "describe" (after your convincing me to put it there) so it doesn't seem out of place to me to use it in this context.

7) Putting this all together, let me now give the passage one more shot.

———

I'm sorry, I wasn't able to complete this note this morning. I got tied up in meetings at the lab. In the meantime I've noticed that you have sent me a few more notes. I will study them and then try to incorporate your further comments into the redraft of the paragraph I was going to write.

———

NEW PROPOSAL FOR TELEPORTATION PARAGRAPH:

The peculiar nature of a quantum state as representing information is strikingly illustrated by the quantum teleportation process. In order to teleport a quantum state from one photon to another, the sender (Alice) and the receiver (Bob) need a pair of photons in a standard entangled state. The experiment starts when Alice receives another photon whose polarization state is unknown to her, though known to some preparer in the background. She performs a measurement on her two photons, and then sends Bob a classical message of only two bits, instructing him how to reproduce the unknown state on his photon. This economy of transmission appears remarkable because to completely specify the state of a photon, namely one point in the Poincaré sphere, we need an infinity of bits. However, the disparity is merely apparent. The two bits of classical information serve only to transfer the preparer's information, i.e., his *state*, to be from describing the original photon to describing the one in Bob's possession. This can happen precisely because of the previously established correlation between Alice and Bob.

———

OK, so what do you think of that? I took away conundrum and magical. Also I stopped mentioning that "as far as Alice and Bob are concerned, nothing changes about Bob's system" in a way that I feel will not compromise my fidelity. True as I believe it is, I will back down that it is essential for making a point here. A few other things to notice. 1) I introduced the preparer in the same sentence that "unknown state" is introduced. I still don't like that phrase (unknown state), but now is probably not the time to buck the vernacular too hard. (I will do that in the colloquium at Technion.) By introducing the knower at precisely the same spot, I could stomach it a bit better. 2) I reiterated that state and information are precisely the same thing. 3) I stuck to my guns that the emphasis should be that it is the preparer's information that is transferred from one system to the other. Saying the word "state" over and over in this context without reminding the reader that it is information we are talking about would have been defeatist. 4) I emphasized that teleportation can be done for the economical rate of only two bits because of the previously established correlation between Alice and Bob's photons.

All these points, I think, are essential to me. And looking back at all that you wrote today, I don't think that you will be in any essential disagreement with all this. Aside from the "nothing changes" stuff, it was really just a question of emphasis that kept us apart.

Chances are I'll wake up with insomnia and read your notes early tomorrow. Thanks for having so much patience with me.

16 December 1999, "Our Baby Leaves Home"

Well, we did it! I've just sent the manuscript to Benka and carbon copied that letter to you. I didn't make any further changes to the manuscript; so it is the one you sent me this morning.

Looking back at the manuscript one last time, I thought, "Wow, that's pretty good." And we didn't once mention Copenhagen!

For the Naples conference proceedings, I'm going to write a much extended version of our little piece titled "Knowledge About What?" The plan is to address some of the issues where we (rightly) feared to tread in the *Physics Today* article. In particular, I'm hoping to hammer out some of my ideas that what quantum mechanics is about is the best intersubjective agreement we can come to in a world that is "sensitive to our touch." As it takes shape, I will consult with you about your opinion.

I never did tell you what David Mermin told me: he said the copy editors at *PT* will be ruthless about changing things in our paper in nonsensical ways. So, despite what Benka told us, we should be on our toes. On a scientific note, David said that in the last year he has become much more inclined to our point of view about quantum mechanics. Now that's progress!

22 January 2000, "Big Old Jet Airliner"

Now I am on my way home. The conference was quite fun. It started when, just as I arrived, Bill Wootters arrived too. We went to dinner together, and had a nice time trying to derive the Hamilton equations in classical physics from the principle that statistical distinguishability of phase space distributions remain constant. It wasn't so hard! I think we succeeded in broad outline. When I have a chance, I will try to make it completely rigorous. Then there were several very good talks at this meeting: Daniel Gottesman, Gilles Brassard, Charlie Bennett (he gave a talk that I had never heard before), Bill's talk (it was just excellent), Ben Schumacher, Nolan Wallach. One sad thing: Paul Benioff became ill this morning (I don't know with what) and had to go to the hospital. His talk was cancelled.

You asked about our anti-paper. It is funny: in Schumacher's talk he had a slide titled "Fuchs's Interesting Non-Property" in which he discussed a certain aspect of the Holevo bound. Anyway, I haven't heard anything from Benka. Do you think I should send a cordial note to him?

28 January 2000, "Benka"

I've now written Benka. I do hope he sends the proofs soon so that the article will actually appear in the March issue.

Many thanks for Emma's birthday wishes: I've already passed them on to her. (She woke up in such a good mood this morning.)

01 February 2000, "Us vs. Them"

By the time you read this, you will no doubt have seen the fax that Benka sent us. Below I'll tabulate the changes to our manuscript that I was able to catch. (I suspect I missed some: there were so many despite the sugar-coating he put in the letter!) Accompanied with the tabulation, I'll also put down my opinion of the change.

Here we go:

Us: To start, let us examine the role of experiment in science.
Them: To begin, let us examine the role of experiment in science.
Opinion: None.

Us: An experiment is an active intervention into the course of Nature: we set up this or that experiment to see how Nature reacts.
Them: An experiment is an active intervention into the course of Nature: we set up this or that experiment to see how Nature reacts.
Opinion: You'll notice no difference between these two passages. However, let me take this opportunity to point out that in the remainder of the paper they always capitalized the first word following a colon. I don't care about the capitalization. But at least we were consistent: They should be too!!

Us: Classical physics is the ultimate example in that regard. However, there is no logical necessity that this worldview always be obtainable.
Them: Classical physics is the ultimate example of such a model. However, there is no logical necessity for a worldview to always be obtainable.
Opinion: I don't like what they've done! Our wording was on the mark: Classical physics is a "compact description" AND, in that case, one CAN distill a model of a free-standing reality. But even there, there is no necessity that has to distill such a model: this was even pointed out by Immanuel Kant so long ago. The second sentence, as they rewrote it, is just nonsense.

Us: If the world is such that we can never identify a reality independent of our experimental activity, then we must be prepared for that too.
Them: If the world is such that we can never identify a reality independent of our experimental activity, then we must be prepared for that, too.
Opinion: Don't know why they inserted a comma. Looks funny to me ... but maybe we have to pick our battles.

Us: It only provides an algorithm for computing *probabilities* for the macroscopic events ("detector clicks") that are the consequences of our experimental interventions.
Them: What it does is provide an algorithm for computing *probabilities* for those macroscopic events ("detector clicks") that are the consequences of our experimental interventions.
Opinion: I don't like their construction at the beginning of the sentence: It's like listening to fingernails scratching on a chalkboard. Why did they change "the macroscopic events" to "those"? More than likely that's their philosophy spilling over into our paper. They want to

imagine that there are macroscopic events independent of our experimental interventions: That's why they changed the qualifier.

Us: This strict definition of the scope of quantum theory is the only interpretation ever needed both by experimenters and theorists.
Them: This strict definition of the scope of quantum theory is the only interpretation ever needed, whether by experimenters or theorists.
Opinion: None.

Us: Quantum probabilities, like all probabilities, are computed by using any available information. This includes, but is not limited to, information about a system's preparation.
Them: Quantum probabilities, like all probabilities, are computed by using any available information, not just information about a system's preparation.
Opinion: Can't stand it. It blurs two ideas that were distinct. While reinstating our sentence, I would however like to take the opportunity to modify it to the following: "This can include, but is not limited to, information about a system's preparation." Again the issue is the old one that we debated so much: Some systems have no preparations (in the usual anthropic sense).

Us: The mathematical instrument for turning that information into statistical predictions is the probability rule postulated by Max Born.
Them: The mathematical instrument for turning the information into statistical predictions is the probability rule postulated by Max Born.
Opinion: None.

Us: Here it is essential to understand that the statistical nature of quantum theory does not restrict its validity to situations where there is a large number of similar systems.
Them: Here, it is essential to understand that the validity of the statistical nature of quantum theory is not restricted to situations where there are a large number of similar systems.
Opinion: None.

Us: In particular, there is an "effective" reality in the limiting case of macroscopic phenomena like detector clicks or planetary motion: their objective occurrence would be acknowledged by any observer who happens to be present.
Them: In particular, there is an "effective" reality in the limiting case of macroscopic phenomena like detector clicks or planetary motion: Any observer who happens to be present would acknowledge the objective occurrence of these events.
Opinion: None.

Us: John Bell formally showed that any objective theory giving experimental predictions identical to those of quantum theory would necessarily be nonlocal.

Them: John Bell formally showed that any objective theory giving experimental predictions identical to those of quantum theory would necessarily be non-local.

Opinion: How many times do I have to complain about this to people in editorial positions, i.e., people who are supposed to know the English language better than I!!! Look in any modern dictionary and count the number of words that start with "non" AND are hyphenated. The fraction is minuscule (unfortunately it is not zero). Why do people in the sciences want to hyphenate when no one else does?

Us: It would eventually have to encompass everything in the Universe, including ourselves, and lead to bizarre self-referential logical paradoxes.

Them: It would eventually have to encompass everything in the universe, including ourselves, and lead to bizarre self-referential logical paradoxes.

Opinion: So they don't have the same reverence for the Universe that we do ... probably another symptom of being a journal editor! In any case, we weren't completely consistent on our capitalization of this (as exhibited in the very next paragraph).

Us: Yet, it is legitimate to attempt to extrapolate the theory beyond its present range, for instance when we probe particle interactions at superhigh energies, or in astrophysical systems, including the entire Universe.

Them: Yet, it is legitimate to attempt to extrapolate the theory beyond its present range, for instance, when we probe particle interactions at superhigh energies, or in astrophysical systems, including the entire universe.

Opinion: None.

Us: On the other hand, if we consider just a few collective degrees of freedom, such as the radius of the universe, its mean density, total baryon number, etc., we can apply quantum theory to these degrees of freedom which do not include ourselves and other insignificant details.

Them: On the other hand, if we consider just a few collective degrees of freedom, such as the radius of the universe, its mean density, total baryon number, and so on, we can apply quantum theory to these degrees of freedom that do not include ourselves and other insignificant details.

Opinion: I think I like their rendition better in this case.

Us: According to him, she is in a 50/50 superposition of states with some cake or fruit in her stomach.

Them: According to him, she is in a 50/50 superposition of states with some cake or some fruit in her stomach.

Opinion: None.

Us: From this example, it is clear that a wavefunction is only a mathematical expression for evaluating probabilities and depends on the knowledge of whoever is computing them.

Them: From this example, it is clear that a wavefunction is only a mathematical expression for evaluating probabilities and depends on the knowledge of whoever is doing the computing.

Opinion: I prefer their formulation in this case too.

Us: Cathy's story inevitably raises the issue of reversibility since quantum dynamics is time-symmetric.

Them: Cathy's story inevitably raises the issue of reversibility; after all, quantum dynamics is time-symmetric.

Opinion: I suppose I'm fine with this one.

Us: Of course, he would need for that complete control of all the microscopic degrees of freedom of Cathy and her laboratory, but that is a practical problem, not a fundamental one.

Them: Of course, for that he would need complete control of all the microscopic degrees of freedom of Cathy and her laboratory, but that is a practical problem, not a fundamental one.

Opinion: None.

Us: In order to teleport a quantum state from one photon to another, the sender (Alice) and the receiver (Bob) need a pair of photons in a standard entangled state.

Them: In order to teleport a quantum state from one photon to another, the sender (Alice) and the receiver (Bob) need to divide between them a pair of photons in a standard entangled state.

Opinion: Darn, wouldn't you know that after we spent so very much time coming to agreement on this paragraph, they would come in and demolish it!!! I suppose I'm OK with this sentence presently (unless you propose something better).

Us: The experiment starts when Alice receives another photon whose polarization state is unknown to her, but known to some external preparer.

Them: The experiment begins when Alice receives another photon whose polarization state is unknown to her, but known to some external **[or third-party? It's not Bob, is it?]** preparer.

Opinion: To make them happy, why don't we change the sentence to this? "The experiment begins when Alice receives a photon whose polarization state is unknown to her, but known to its external preparer – a third party in the protocol."

Us: She performs a measurement on her two photons, and then sends Bob a classical message of only two bits, instructing him how to reproduce that state on his photon.

Them: She performs a measurement on her two photons – one from the original, entangled pair and the other newly received – and then sends Bob a classical message of only two bits, instructing him how to reproduce that state **[Which state? ...]** on his photon.

Opinion: Why don't we try this? "She performs a measurement on her two photons – that is, on one from the entangled pair and the photon that was just received – and then sends Bob a classical message of only two bits, instructing him how to reproduce the unknown state on his photon."

Us: The two bits of classical information serve only to convert the preparer's information, from describing the original photon to describing the one in Bob's possession.
Them: The two bits of classical information serve only to convert the preparer's information, from a description of the original photon to a description of the one in Bob's possession.
Opinion: I'm OK with this.

Us: It is curious that some well intentioned theorists are willing to abandon the objective nature of physical "observables," and yet wish to retain the abstract quantum state as a surrogate reality.
Them: It is curious that some well-intentioned theorists are willing to abandon the objective nature of physical "observables," and yet wish to retain the abstract quantum state as a surrogate reality.
Opinion: They got us on this one. Both my *American Heritage* and my *Webster's* dictionaries have "well-intentioned" hyphenated.

Us: There is a temptation to believe that each quantum system has a wavefunction even if the latter is not known explicitly.
Them: There is a temptation to believe that every quantum system has a wavefunction, even if the wavefunction is not explicitly known.
Opinion: OK.

Us: This analogy is misleading. On the contrary, attributing reality to quantum states leads to a host of "quantum paradoxes."
Them: This analogy is misleading: Attributing reality to quantum states leads to a host of "quantum paradoxes."
Opinion: ????

Us: The latter, when correctly used, never yields two contradictory answers to a well-posed question.
Them: When correctly used, quantum theory never yields two contradictory answers to a well-posed question.
Opinion: None.

Us: In particular, no wavefunction exists before we start an experiment, nor after its completion.
Them: In particular, no wavefunction exists either before or after we conduct an experiment.
Opinion: ????

Us: Also, the theory has no dynamical description for the "collapse" of the wavefunction.
Them: Again, the theory has no dynamical description for the "collapse" of the wavefunction.
Opinion: This was a dumb change. Makes no sense; why would we say "again"? Why don't we propose: "Furthermore, the theory has no dynamical description for the "collapse" of the wavefunction."?

Us: Its collapse is something that happens in our description of the system, not to the system itself.
Them: Collapse is something that happens in our description of the system, not to the system itself.
Opinion: None.

Also they changed our biographies a bit. I think I will ask for the capitalization in "director-funded" back as that is actually the job title. As I say, I probably missed some of their changes. You'll probably miss some too. Let's just hope that they're not big enough to make a difference.

Tomorrow morning I go to ABQ for my weekly collaboration with Caves (I usually go on Mondays, but this week I was recruited to give their journal club talk and so I traded days). Nevertheless I may get a chance to work on your next iteration, or compose a letter to Benka, while I'm there.

Oh, almost forgot. My preference for the cartoon is strongly toward the first one. I tried them out on a few people here (Leslie Weaver our secretary, Kurt Jacobs a postdoc, and Tanmoy Bhattacharya a staff member) and they all voted likewise.

02 February 2000, "Cartoon URGENT"

Yes, please do send a fax of the cartoon when you can. I will compose a letter to Benka this evening (gently) asking if he would be willing to consider a cartoon from your source.

Now we have our lunchtime meeting.

03 February 2000, "Sorry"

Sorry I haven't gotten back in touch with you. It turned out that I got in quite late last night. I will conglomerate our comments in some acceptable way and send them to Benka before the day is out.

04 February 2000, "Running in Circles"

It has turned out that Kiki needs me to help her in Albuquerque today (she's going to pick up her parents but also getting some small furniture). So I will drop by the lab and pick up the fax before we depart. Then on the ride to ABQ I will compose the letter to Benka and send it off when I arrive.

I believe I was confused about the cartoon you were sending. There was one sitting on my desk yesterday: "Then a Miracle Occurs." With that one, one of the most interesting coincidences occurred. The boy in the office next to mine, walked in wearing a shirt with just that cartoon emblazoned on it! I told him (mistakenly) that you had wanted that cartoon for our article. He said, "Maybe that's not a good choice: everyone knows this cartoon." (His shirt made that quite believable.)

So hopefully I can retrieve the fax when I go in to the office in a few minutes.

04 February 2000, "Oops"

I'm sorry I forgot to carbon copy you the letter I wrote to Benka! Here it is below. I took some liberties to modify slightly some of your suggestions – in all cases, I think your opinion won't be too much different than mine (so there shouldn't be much to worry about).

Let's now keep our fingers crossed that they will abide by our wishes, and that this project will be done with.

Hi Steve,

Thanks for the great job in editing you guys did. The paper is indeed generally clearer now. Of the 31 changes we detected, we only have an issue with a few of them. I will detail these below. The protocol will be this. I will first list the complete sentence or sentences you have in the present draft that we disagree with, then I will list what we would like it or them changed to, and then finally I will give an explanatory remark if necessary.

You: CHRIS FUCHS, previously the Lee DuBridge Prize Postdoctoral Fellow at Caltech, is now a director-funded fellow at Los Alamos National Laboratory.
Us: CHRIS FUCHS, previously the Lee DuBridge Prize Postdoctoral Fellow at Caltech, is now a Director-Funded Fellow at Los Alamos National Laboratory.
Remark: As Asher wrote me (and which is actually true), "'Director-Funded Fellow' must have capitals, just as the other titles in our biographies. This denotes a prestigious award, not a budgetary item."

You: An experiment is an active intervention into the course of Nature: we set up this or that experiment to see how Nature reacts.
Us: An experiment is an active intervention into the course of Nature: We set up this or that experiment to see how Nature reacts.
Remark: The first letter after the colon should be capitalized to be consistent with all the other instances of the same in the text. (Thanks for teaching us this rule!)

You: Classical physics is the ultimate example of such a model. However, there is no logical necessity for a worldview to always be obtainable.

Us: Classical physics is the ultimate example in that regard. However, there is no logical necessity that this worldview always be obtainable.

Remark: Please reinstate the original wording here. It was chosen quite carefully to be exactly what it is. The way we see it is this: Classical physics is, on first pass, nothing more than a compact description. We are lucky that we can further distill a MODEL of a free-standing reality from it IF WE WISH TO GO TO SUCH A METAPHYSICAL LEVEL. So in changing the words, you have changed the meaning of what we wanted to convey. Also you'll note that in your second sentence in this cluster, there is clearly a necessary qualifier missing from "worldview."

You: What it does is provide an algorithm for computing *probabilities* for those macroscopic events ("detector clicks") that are the consequences of our experimental interventions.

Us: What it does is provide an algorithm for computing *probabilities* for the macroscopic events ("detector clicks") that are the consequences of our experimental interventions.

Remark: We wrote "It only provides" – actually quantum theory does other things too, so we understand the motivation for your first change. However, please change back the "those" to a "the." Asher finds this construction more palatable, and I think it better conveys my overall belief: "those" conveys a flavor that there are things that qualify as "macroscopic" beyond the consequences of our experimental interventions. That does not reflect what I want conveyed.

You: Quantum probabilities, like all probabilities, are computed by using any available information, not just information about a system's preparation.

Us: Quantum probabilities, like all probabilities, are computed by using any available information. This can include, but is not limited to, information about a system's preparation.

Remark: This is an issue that we had much debate about as we constructed our draft. Please instate the wording above (which is very slightly different from our original wording). It conveys the idea that some systems have no preparations (in the usual anthropic sense). Nevertheless, we make do with whatever information we have in making our predictions. That set of thoughts is absent in your formulation.

You: John Bell formally showed that any objective theory giving experimental predictions identical to those of quantum theory would necessarily be non-local.

Us: John Bell formally showed that any objective theory giving experimental predictions identical to those of quantum theory would necessarily be nonlocal.

Remark: We've never seen a dictionary (English or American) that hyphenates nonlocal. In fact the vast majority of all words starting with "non" have no hyphenation; it only appears to be a habit that physicists have gotten into.

You: On the other hand, if we consider just a few collective degrees of freedom, such as the radius of the universe, its mean density, total baryon number, and so on, we can apply quantum theory to those degrees of freedom that do not include ourselves and other insignificant details.

Us: On the other hand, if we consider just a few collective degrees of freedom, such as the radius of the universe, its mean density, total baryon number, and so on, we can apply quantum theory to these degrees of freedom, which do not include ourselves and other insignificant details.

Remark: As Asher wrote, "They replaced 'these degrees of freedom which' by 'those degrees of freedom that'. The original meaning, as I understood it, was 'these degrees of freedom, which ...' (that is, the degrees of freedom that were just mentioned and which have the following property ...). The new text may be incorrectly interpreted as including ALL the degrees of freedom that do not include ourselves. Please restore 'these degrees of freedom, which ...' with a comma before 'which' so that the text is unambiguous."

You: Let us examine a concrete example: The observer is Cathy (an experimental physicist) who enters her laboratory and sends a photon through a beam splitter.

Us: Let us examine a concrete example: The observer is Cathy (an experimental physicist) who enters her laboratory and sends a photon through a beamsplitter.

Remark: "Beamsplitter" is usually written as a single word in the optical literature.

You: The experiment begins when Alice receives another photon whose polarization state is unknown to her, but known to some external preparer.

Us: The experiment begins when Alice receives another photon whose polarization state is unknown to her, but known to a third-party preparer.

Remark: You wanted clarification.

You: She performs a measurement on her two photons – one from the original, entangled pair and the other newly received – and then sends Bob a classical message of only two bits, instructing him how to reproduce that state on his photon.

Us: She performs a measurement on her two photons – one from the original, entangled pair and the other with a state unknown to her – and then sends Bob a classical message of only two bits, instructing him how to reproduce that unknown state on his photon.

Remark: You wanted clarification.

You: Again, the theory has no dynamical description for the "collapse" of the wavefunction.

Us: Furthermore, the theory has no dynamical description for the "collapse" of the wavefunction.

Remark: "Again" makes no sense in this context.

That's pretty much it. Concerning the cartoon, neither of us likes the bottom one you sent us (i.e., the one with the caption, "Well, uh, when a daddy's wave function ..."). I also took a small poll at the lab (five people) and found that the top cartoon (i.e., the one with the caption, "What do you mean ...") was the preferred. And that reflects my preferences too.

Asher, however, is concerned that the cartoon may not be "quantum mechanical" enough and has therefore suggested one from the book "What's so funny about science?" (cartoons by Sidney Harris, from *American Scientist*, reproduced by Kaufmann, Los Altos, CA, 1970 ... 1977). The scene is a mechanic working on a car with the car owner overlooking him. The caption reads, "Actually I started out in quantum mechanics, but somewhere along the way I took a wrong turn."

Asher writes: "If *Physics Today* has a copyright agreement with *American Scientist* (or Kaufmann), this would be the best solution. If this cannot be arranged quickly enough, then let them put 'cause and effect'." This would be agreeable to me, though as I say my own preference is for the first cartoon you supplied. (It seems sufficiently appropriate to me actually, since the decision between one outcome or the other in a quantum measurement has no "cause" describable within physics.) Thus I leave the issue to be between you and Asher as far as that goes. I will abide by either ruling.

Asher's Reply

I am about to leave for a visit to my grandson (and the rest of the family, near Tel-Aviv). He just had his third birthday. I'll print out your letter to Benka and I'll study it later. Thank you for giving a quantum interpretation to the cartoon "cause and effect." Perhaps they can keep the cartoon, but change the caption into something like "Do you mean this resulted from a quantum fluctuation?"

05 February 2000, "Only Trouble Spot"

Let me start with an apology: I apologize for opening a can of worms. It concerns this part of the paper:

If from such a description we can *further* distill a model of a free-standing "reality" independent of our interventions, then so much the better. Classical physics is the ultimate example in that regard. However, there is no logical necessity that this worldview always be obtainable. If the world is such that we can never identify a reality independent of our experimental activity, then we must be prepared for that, too.

Benka wrote:

Your wording is ambiguous, which led to the trouble in the first place. Specifically, you know what "that regard" refers to, but no one else does. To be clear, the passage needs to be longer. Here's an attempt:

> ... Classical physics is the ultimate example of such a description leading (if we so desire) to a model of reality. However, there is no logical necessity for any worldview to be obtainable from a scientific description. Physics is not metaphysics. If the world is such that ...

Please get back to me by Monday.

I don't know about him, but when I read the phrase "in that regard" I immediately back up to the previous sentence and check how the present sentence meshes with it. Despite a serious effort for empathy with him, I can't see his problem. The passage seems crystal clear (and pretty well written) to me. (I like what we've written; I just hate to see it changed.)

But it's too late now: Pandora's box has been opened. You were sure right, when you wrote, "Our formulation was better, but I fear making too many changes in proof, because of the risk of further errors." The question is, how can we please him with the minimal number of changes to our text. Let me propose the following change to the passage (specifically in the second and third sentences).

If from such a description we can *further* distill a model of a free-standing "reality" independent of our interventions, then so much the better. Classical physics, for instance, is the ultimate example of a theory reducible in that way. However, there is no logical necessity that this kind of worldview always be obtainable. If the world is such that we can never identify a reality independent of our experimental activity, then we must be prepared for that, too.

In this formulation, I tried to sidestep two things. 1) The expectation of good literacy on the part of the reader. The second sentence is now more redundant than it needed to be, so he/she won't have to back up to the previous sentence to understand its intent. 2) Benka's apparent misunderstanding of the word "worldview." His attempt of a rewrite seems to indicate that he thinks we don't have one, not simply that ours is different from the kind one finds in classical physics.

For your convenience, let me put the whole paragraph together:

Our purpose here is to explain the internal consistency of an "interpretation without interpretation" for quantum mechanics. Nothing more is needed for using the theory and understanding its nature. To begin, let us examine the role of experiment in science. An experiment is an active intervention into the course of Nature: We set up this or that experiment to see how Nature reacts. We have learned something new when we can distill from the accumulated data a compact description of all that was seen and an indication of which further experiments will corroborate that description. This is what science is about. If from such a description we can *further* distill a model of a free-standing "reality" independent of our interventions, then so much the better. Classical physics, for instance, is the ultimate example of a theory reducible in that way. However, there is no logical necessity that this kind of worldview always be obtainable. If the world is such that we can never identify a reality independent of our experimental activity, then we must be prepared for that, too.

It's a little less literary now, but I think it might fulfil Benka's requirements without compromising my integrity. What a pain to have to play these lawyer games!!

I hope you are having a good day with your grandson. We will likely give Kiki's parents a tour of the city today, showing them the volcano that created these mesas, Oppenheimer's old home, etc.

Presently I am reading a very nice book by Arthur Zajonc – *Catching the Light: The Entwined History of Light and Mind* – in preparation for my interview. I'm quite impressed that an experimenter has such a literary command and extensive knowledge of history! If his personality seems to hold up like his writings, he's bound to be a joy to work with.

05 February 2000, "Also"

I forgot to tell you, beside that complaint of Benka he said the following:

By the way, all the other changes are fine except number 8. You're stuck with "beam splitter."

So if we just get past the hurdle that the last note concerned ...

05 February 2000, to Steve Benka, cc: Asher, "Beaten"

Dear Steve,

After long discussion Asher and I have come to agreement on our one remaining disagreement with you. Let us go with the rewrite you've already proposed, except with the addition of one extra word (which I'll write with in all caps below):

... If, from such a description, we can *further* distill a model of a free-standing "reality" independent of our interventions, then so much the better. Classical physics is the ultimate example of such a model. However, there is no logical necessity for a REALISTIC worldview to always be obtainable. If the world is such that we can never identify a reality independent of our experimental activity, then we must be prepared for that, too.

Let me also convey another message made by Asher:

Thank you for giving a quantum interpretation to the cartoon "cause and effect." Perhaps they can keep the cartoon, but change the caption into something like "Do you mean this resulted from a quantum fluctuation?"

Once again, I will defer to whatever decision you two come to.

Thanks once again for all your attentive help. Now I spend the next two days preparing for my Amherst visit. I will indeed say hello to Bob Hallock for you.

06 February 2000, "Can of Worms"

The only thing my (computer) dictionary says is:

can of worms, Informal. a source of many unpredictable or unexpected problems: Buying a company we know nothing about would be opening up a whole new can of worms. [1965–70]

As you see, they don't trace it back to a biblical origin. But who knows how accurate they are.

08 February 2000, "Copyright Sent"

I just mailed off the copyright agreement for our *PT* article. Also, yesterday (by accident) I talked to Steve Benka on the phone. (I say by accident, because I returned a call of his ... but apparently he had called last week and after our email transactions his questions were no longer relevant.) In any case, he informed me that he changed the caption of the cartoon to be in accord with your suggestion.

Also I received your fax this morning (and your latest email instructions yesterday).

02 May 2000, *"Physics Today* and Way Behind"

Asherism 69: *Shouldn't we have received from Physics Today the comments on our paper, so as to prepare a reply?*

Yes we should have. Do you remember the name of the person who contacted us about this? Maybe one of us should send her a note asking what's up.

I too have been having computer troubles. I got my new laptop last week and it has taken me over five days to set it up properly. (And I've only met partial success.)

Yesterday, I gave my second talk in my series of lectures. It went much too slowly: I tried to pay attention to the ones in the audience who needed more introduction, but that was a mistake. The one high point for everyone, however, was when I exhibited Wigner's 1961 paper where he just missed noticing the no-cloning theorem (and in fact got it wrong). This brought a great smile to Wojciech.

I do still plan to comment on your cats! (I just hope they don't die before I get there.)

10 May 2000, 9:19 AM, "Alive and Well"

Asherism 70: *[9 May 2000:] I was distressed to learn of your forest fire. We occasionally have such problems here. In 1979 the fire was at the bottom of our garden.*

Asherism 71: *[10 May 2000:] Please keep me posted that everything is under control with forest fires, etc. Today (independence day) Aviva and I climbed on our roof to clean it and add some white paint to increase its albedo.*

Asherism 72: *[10 May 2000:] Please write, just to reassure me that you and your family are safe. Richard Jozsa told me that he had seen our paper, and that he disagreed with us.*

Thanks for the concern. Yes, we are still safe ... though the winds have started to howl again. I can hear the slurry bombers flying outside, so I have a feeling they may be bombing very close to us. Unfortunately, I missed the news this morning: I had some very minor surgery yesterday and it has made it difficult for me to sit up. Also Kiki hasn't been able to connect to the emergency update number. We keep our fingers crossed that things won't get worse because of the winds.

I knew that Richard would disagree with us. He has some rather silly notions running around in the back of his mind.

11 May 2000, 11:02 AM, and to others, "Alive but Depressed"

A small note to let you know that Emma, Kiki and I are all OK. We're fairly sure our house and all its contents have been destroyed: we only managed to escape with two suitcases worth. (I even left all my old calculations behind, which struck me as very painful once I realized it.) The whole town is ablaze; I suspect that nothing will be left.

Thank you all for your concern.

Asher's Reply: Displaced Persons

Dear Chris and Kiki,

It must be a terrible feeling to be "displaced persons" but you will overcome that drama. When I was 6 years old, my mother and I were "displaced" from Paris by the advance of German troops (my father was at the front). It took three days of train zigzags to reach the little village where we spent the rest of the war. Five years later, when we returned to Paris, nothing was left of our furniture and other belongings.

In 1956, Aviva and her family were expelled from Egypt (only two suitcases allowed). They had to restart their lives in a new country, speaking an unknown language and with many strange customs. Aviva's parents never fully adjusted to their new life, but she and her brothers have nothing to regret about Egypt.

As painful as things are for you now, you are in your country, there is no war to fear, and you are still young and strong enough to recover and enjoy life. In my office there is a poster with "IF" by Rudyard Kipling. Here are some excerpts:

> If you can keep your head when all around you are losing theirs

> If you can lose ... and start again at your beginnings,
> And never breathe a word about your loss

> Yours is the Earth and everything that's in it

Be strong! We are thinking of you,

Aviva and Asher

18 June 2000, "Absence"

I apologize for my long absence from email. I will be able to look at all you've written (and the original letters) starting Tuesday morning. Thanks for already putting so much time into the project.

11 July 2000, "It's a Boy!"

Many congratulations on the new grandchild! All children are impatient and authoritarian; take it as a compliment that you are so young!

As you should be able to see from this note, I am finally in cyberspace again. The Italian and Greek phone systems were deplorable! I arrived in Munich Sunday, but haven't had a chance to connect until now. Unfortunately the only way I've found a simple means to work completely out of my laptop (given the Compaq and LANL firewalls), is to dial directly to the United States for a connection. So I will likely only be checking and sending mail once a day.

Later this afternoon, I will start the painful task of working on our *Physics Today* replies.

11 July 2000, "End o' Day"

It's getting near the end of the day here, and Kiki and her mom will be returning from Austria soon, at which point I will once again have social demands placed on me.

Today I finally had the courage to look at all the replies to our article and all your related writings – I have procrastinated and procrastinated on this, mostly because I think the fire has made me mentally weak and wanting to shun any criticism. If these letters are the best of the lot sent to *Physics Today*, then that's sad.

I'm quite pleased with much of your drafted reply to the replies. Thank you so for being a better person than me these past weeks. But as you can guess, I disagree with some points (e.g., "physical states" and "it is a set of rules giving the probabilities of macroscopic detection events, FOLLOWING SPECIFIED PREPARATION PROCEDURES.") We'll have to work again to come to some common ground, while still keeping the text snappy and pithy. Tomorrow I will send you a revised draft to see what you think.

Did Ptolemy and Euclid both really reside in Alexandria? Wasn't there quite an amount of historical time between them? And Euclid was Greek, wasn't he?

12 July 2000, "What Muck!"

I think these letters are horrible. What's to be gained from being pitted against this lot? We pored over every word in that article, just to have every deeper idea summarily ignored. I wonder which is more to blame here, the "rigorous in-house review cycle," or instead that every submission was of this quality?

This whole affair, upon reading and thinking more, has turned out to be more complicated than I had wished. What on earth is Sobottka trying to say? I don't think your reply hits the mark, because it seems that ultimately he does *not* want to ascribe an objective reality to the wavefunction.

And Holladay … yuck. Why was he not worried with trying to make some sense of our statements about quantum theory also giving an "effective reality" for some very gross aspects of the world? I.e., those aspects that we can still talk about even without detailed knowledge of any microphysics? If he had done this, maybe he would have answered his own questions. I think we should reinsert some discussion of this into your reply, and make it clear that we are not really instating yet another consideration that we had not in our original article.

Brun and Griffiths: who could have guessed that they wouldn't take this opportunity to advertise their interpretation of quantum mechanics yet again? (I.e., the one Murray Gell-Mann calls the "modern interpretation of quantum mechanics.")

When they talk about "coarse-grainings" what the heck do they think determines such things in a completely objective description of the universe? They can only be expressions of *ignorance* if the theory does not specify them (and it doesn't), and then they are exactly back to the point we started from: saying that the best one can say is about a few degrees of freedom of Nature that do not include the observer. There can be nothing in Nature itself that says, "Thou shalt coarse grain . . . "

Maybe what bugs me about their attempt to inject a free-standing reality into the theory is the impotence of the whole affair. Why can they not appreciate how liberating an idea it is?

Even their reporting of the efficacy of experiment is naïve: "Experimental results guide the development of theory, and are the ultimate arbiters of correctness or incorrectness."

Here's an idea attributed to Einstein by Heisenberg that I love. It captures an extremely important point (in this fuller version that is not quoted so often):

It is quite wrong to try founding a theory on observable magnitudes alone. In reality the very opposite happens. It is the theory which decides what we can observe. You must appreciate that observation is a very complicated process. The phenomenon under observation produces certain events in our measuring apparatus. As a result, further processes take place in the apparatus, which eventually and by complicated paths produce sense impressions and help us to fix the effects in our consciousness. Along this whole path – from the phenomenon to its fixation in our consciousness – we must be able to tell how nature functions, must know the natural laws at least in practical terms, before we can claim to have observed anything at all. Only theory, that is, knowledge of natural laws, enables us to deduce the underlying phenomena from our sense impressions. When we claim that we can observe something new, we ought really to be saying that, although we are about to formulate new natural laws that do not agree with the old ones, we nevertheless assume that the existing laws – covering the whole path from the phenomenon to our consciousness – function in such a way that we can rely upon them and hence speak of "observation."

I am not pushed to the rejection of a free-standing reality in the quantum world out of a predilection for positivism. I am pushed there because that is the overwhelming message quantum THEORY is trying to tell me.

I guess now that the end of the day has arrived again, I will be sending some more concrete changes to the article tomorrow. I hope you will forgive me: this idle time of thinking about things (and running through the whole gamut of angers) seems necessary to congeal my thoughts.

13 July 2000, "Sunny Day Apology"

Today for the first time in about two weeks, Kiki tells me, the sun is shining in Munich. The forecast says that it will disappear again by tomorrow. She has asked if we could go into town, enjoy the sun, window shop, and have lunch at an outside cafe. Despite my obligations, I just couldn't turn her down: I apologize. I want to get these replies off our back too, but it does seem a shame to pass up this opportunity to make Kiki and Emma happy.

I should be able to return to our project this evening or tomorrow with a clean palette. I don't think we'll have any real problem getting things in for a September issue appearance.

14 July 2000, "Lunchtime Report"

I'm working hard on hammering out my changes to our reply: I thought I would give you a quick report before I join Kiki downstairs for lunch. The only *real* trouble I'm having is with Holladay, which I think requires some special care. I hope beyond hope to have you a completed manuscript before the end of the day.

14 July 2000, "Here It Comes"

The end of another day again arrives, but at least this time I am sending a draft of the manuscript with it. There are two versions, one with explanatory footnotes, and one without. My word processor gives the word count in the unfootnoted version as 1520 . . . well below the 1737 that Ms Hanna suggested.

As you will see, the main changes I made in your draft were to the Holladay discussion. I didn't feel that your earlier discussion of him was really getting at the right point. I hope that you find you can agree with some of what I've written.

Then there are further smaller changes throughout. Some of them were put in place to lead up to my other main change: I tried to build a more positive and far-reaching statement into our conclusion. We said exactly that idea throughout (both in our original Opinion and in various places of both this draft and your earlier one), but I thought it should have the ending emphasis. The main reason is that people continually accuse us of pure positivism or pure idealism because of their very sloppy reading of us, and because of their preconceptions (probably based on some childhood trauma). I want to do my best to stop that.

I will be back online tomorrow morning, but there is some chance that Kiki and I will spend much of the day in Munich. In any case, I'm hoping to get the chance to do one more reiteration (if need be) before we fly out early Monday morning.

Being Friday night, I presume you will see Lydia tonight. Please give my warm regards to her and her family, and especially the new little one.

Parsing the Paper

Quote from Draft: We do not claim that contemporary quantum theory is the last word for a description of Nature; it would be hard to imagine how we could have written the last words in our "Opinion" if we did.

Comment: A little slap on the wrist for the empty attention span of Harris and others.

Quote from Draft: Unfortunately, as he hints, . . .

Comment: I added this since he does seem to turn around by the end of his letter.

Quote from Draft: Some people may deplore this situation, but we should make it clear that we were not pushed to the rejection of a free-standing reality in the quantum world out of a predilection for positivism. We were pushed there because that is the overwhelming message quantum theory is trying to tell us.

Comment: These sentences were placed here to make it clear that I, personally, walked into the quantum foundations question as a hard-core realist. I came out of it a new man. That is to say, the cement of my philosophy wasn't set before I got interested in quantum foundations. By the way, the last sentence of this paragraph is throwing a bone to a title of one of David Mermin's recent "Ithaca Interpretation" papers.

Quote from Draft: We do not find any merit in the various alternatives that were proposed to the straightforward interpretation of quantum theory: It is a set of rules giving the probabilities of macroscopic detection events based on any previous experimental information we may have gathered.

Comment: Changed this because, as you know, I am reluctant to say that the only relevant information is preparation information.

Quote from Draft: If that provides a "straitjacket" to our thought, it is only in that it keeps us from contemplating an endless stream of conundrums that come about solely from shunning quantum theory's greatest lesson – that the experimenter cannot be detached from the world he is attempting to describe.

Comment: I really would like to end the paper with a statement that makes the whole affair look positive and hopeful. To say that the experimenter in quantum mechanics is not "detached" is a phrase I like very much. It's one that Pauli used, and now Anton Zeilinger uses a lot.

16 July 2000, "Getting Slow Start"

I'm getting a slow start today: how I can't wait to have my own office at my own home again, with NO visitors to waste my time. (My in-laws are having several house guests this weekend.)

Thanks for sending the revised draft. And MANY thanks for removing that horrible paragraph I wrote on diamonds: immediately upon seeing your comments, I realized that even I didn't agree with what I had written! Still though, I think we have some rough edges to work out on Holladay. I think it is quite important that we tackle his rhetorical question in a head-on way: it (i.e., if the theory is not about reality, then what is it about?)

and the misunderstanding of Brun/Griffiths (i.e., so if there were no people, then there would be no world!) are the same comments we've gotten over and over and over. Beside the written comments we've seen, several have expressed these concerns in person to me (Jozsa, Gisin, Omnès, and Wootters). And we should have a pithy, good answer: in a way, this is a good opportunity for us to flesh that out. There must be a way of expressing it that will make it easy for people to see what we are talking about. Why is it so difficult for them to understand?

I don't think you understood the point of my gambling-house paragraph. I think I will try again on that, because I do think it's an important point, and relevant to the reply on Holladay. Please give me at least one more shot at it. The point – which I will try to re-express more eloquently in the next draft – is that a theory need make NO direct referral to reality in order to be successful. Probability theory is the prime example of that. That is because it is a theory of how to reason in light of the information we have, regardless of the origin of that information. Its success is gauged by the fact of our reasoning as well as we can (by various criteria) in the light of our unpredictable world. Quantum theory is more like that than any other physical theory heretofore. Its formal structure is more a "law of thought" than a "law of physics": the physics is in the information we feed into that formal structure (Hamiltonians, initial conditions, known symmetries, etc.). Nature and our reasoning about it should not be confused.

You took out "straitjacket" because it is a torture device ... but it was not I who put it there in the first place: it was Brun and Griffiths, describing our ideas as working within a straitjacket. I think we should replace it (perhaps making it more explicit that *they* said it first), to help make a frontal attack on their silliness.

I will send you the next iteration by the end of the day today. (With some luck for my sanity, all the house guests ... and perhaps Kiki and her mother ... will all take off for a day trip to somewhere soon!)

16 July 2000, "Blue in the Face"

A couple of more thoughts about Deutsch. [See note to David Mermin, dated 23 July 2000.] Seeing his strong comments does reinforce my resolve for us to be as absolutely clear and as positive as we can be in our reply to critics. But on the other hand, it also reinforces my opinion that to some people we could talk until we are blue in the face, and never make a crack in their preconceived notions. I strongly doubt that Deutsch even read our paper: but *yet* he has an opinion of it. Which of us are the eccentrics, and which of us are the physicists?

More of substance later.

16 July 2000, "A Little Progress"

I think I will go ahead and send a partial draft now. The part in italics I haven't yet touched: it is all the stuff dealing with Holladay's rhetorical question, and, I think, should really be the core of our reply. However, I think I have come to closure on the rest: it seems to me that that is an independent unit, and you can start evaluating your thoughts on it if you wish.

I apologize for not sending you a full draft, but this whole family thing is really getting in my way. If I get a chance to do more this evening (probably not), it will be solely confined to the italicized part of what I'm sending you presently. I really want to shoot for having this whole lot finished before I depart Munich.

In any case, I'll be back tomorrow morning.

Parsing the Paper

Quote from Draft: We do not claim that contemporary quantum theory is the final word for a description of Nature; this should be clear from the last paragraph of our "Opinion" essay.

Comment: How's this version? I didn't think your version was forceful enough (starting with a dependent clause): the guy didn't take more than a moment to read us – probably didn't even read to the end of the article – and he ought to be punished for that.

Quote from Draft: It is these gratuitous additions to quantum theory that should have been the analogs of Ptolemy's epicycles in Harris's story.

Comment: Again, here, I am going for a slightly stronger tone.

Quote from Draft: The mistake Holladay makes in his presumptive question is a common one for those not yet accustomed to quantum lines of thought.

Comment: Presumptive: founded on presumption. Presumption: behavior or language that is boldly arrogant or offensive.

Quote from Draft: He fails to see that we have accepted a distinction that he does not: There is Nature, and then there is our best description of our experimental interventions into it.

Comment: You had changed the wording of this before, but that completely deleted the effect the sentence was supposed to have.

Quote from Draft: Brun and Griffiths call this a "straitjacket," but if it is, it is only in that it keeps us from getting mired in the endless conundrums that arise solely from shunning quantum theory's greatest lesson – that the notion of experiment plays an irreducible role in the world we are trying to describe.

Comment: OK, so now I myself have removed the word "detached" that I like so much. But I'm still shooting for conveying the idea, and would like the paper to end on this note. Does it sound a little better now?

17 July 2000, "A Little Pride"

In the next two notes I will again send two versions of the draft: with and without footnotes.

I must say, I think I am quite proud of the version now! I think it sounds really good, and most importantly conveys some essential thoughts that complement our original "Opinion" exposition. Given so many people's questions, I am quite confident that this was needed.

I hope you will agree with the points ... as I am quite settled on them in my mind now. The discussion of Holladay is now a hybrid of your original and my later versions, and (I think!) greatly improved at getting at the essential points.

We leave very early tomorrow morning for the airport. However, I hope to download email one last time just before then. In any case, I will almost surely check it one last time before dinner this evening (about 8:00 PM Munich time).

It'd be great if we could send our final draft off to Ms Hanna by Wednesday.

17 July 2000, "Great"

I think we've done a fine job. In the next email, I'll send you my latest (only *with* footnotes this time). My changes were only very, very minor changes of English.

Sorry I don't have time to write more: we are starting the long process of a European dinner now.

I will await your last word when I arrive in America; I think we can plan to send everything off to Ms Hanna Wednesday morning.

19 July 2000, "Bigger Change"

I just sent off the final draft to Ms Hanna. You'll note that I modified the final sentence even more than you had suggested: I completely removed the phrase "our getting mired in." I think the sentence flows even better now.

I hope you don't mind me doing that last minute. If you disagree, we can always make a final change at the proof stage.

Now we do a small amount of shopping and then take off for Los Alamos. We had the most horrible experience flying home: delays and delays, so much so that we had to spend the night in Albuquerque last night. And all our checked luggage was lost, one suitcase is still missing (the one with my quantum de Finetti presentation slides!).

21 July 2000, "God, etc."

Asherism 73: *Thank you for sending to Petra a list of interesting problems. I'm afraid some of them are a bit difficult for her at this stage of her studies.*

You can see I've had an undue influence from John Wheeler! I once saw him give an undergraduate a summer research project of "derive quantum mechanics from principles of distinguishability"!! The fellow was floored.

Asherism 74: *I must also say that I am worried by your frequently invoking God in your text. Also, in our last paper, you wrote several times "Nature" which is just a euphemism for the same purpose. As you know, I am a devout atheist, but since we were under the pressure of a deadline, I did not argue with you on this point.*

Yes, you bring up a point that even I have been a little worried about. I would say that I also am a devout atheist. Sometimes, I bring up the imagery of a god creating the world not to be taken seriously, but to emphasize that I am not a pure positivist. I.e., that I believe there is "something" to the universe that is there independently of us. Whether that "something" – I am averse to using the word "reality" because that has no interactive aspect to it, and does not convey the idea that the "something" itself might be malleable and changeable – can be probed with the methods of science, I don't know. But that's not so important to me right now.

I would say that I use the word "nature" for the same reason – to label that which is there independent of us. It is true that capitalizing the word "Nature" denotes it as an object of worship, and in that sense makes it serve as a euphemism for God. In fact I did feel a little uncomfortable in capitalizing it throughout our paper. Looking back on it, I know that I only did that out of peer pressure, i.e., because I have been noticing so many other people do it in print lately.

Shall we decapitalize all the Natures in our paper? I think I would like to do that now myself.

Did you get the fax Ms Hanna sent us?

21 July 2000, "24 Changes"

Well, I count 24 changes to our manuscript. Luckily, only two of them annoy me. Also, they themselves started to decapitalize our Natures: so I'm happy about that. But they missed three, so we'll have to point that out to them.

Now the question is how best can I get these things to you. I did not get a fax of the changes, so I can't fax them to you either. Maybe I will reTEX the changes and send them to you that way.

24

Letters to John Preskill

18 November 1997, "The Radical Eye"

Let me send you some prep work ... while I'm procrastinating from things I really need to be doing. I'll convey the information in my favorite way: quoting/forwarding old emails!

The first, an old – kind of poorly written – note "Dreams of a More Ethereal Quantum," concerns one aspect of the below-the-water stuff: the idea about distilling the MaxEnt-looking part out of quantum theory. It doesn't say much more than I said in the proposal ... but you can't expect too much, it's still just a vague program for a program. That stuff is at the bottom of the note; the early part of the note says a few – even more sketchy – things about Mermin and Quantum Probabilities ... so I thought I would include it for that reason.

The only other readable thing (I think) I've written on this aspect of the below-the-water stuff is on pages 22–26 of my paper with Caves, "Quantum Information: How Much Information in a State Vector" (it's on `quant-ph`). These same pages also contain an extended discussion on probability.

The second note, "Rainy Mondays and Consistent Histories," is a note I wrote Bob Griffiths concerning what I think of his Consistent History stuff. I think it says quite plainly and clearly what I think about probability in quantum mechanics. It's pretty self-contained (i.e., not referring to old conversations, etc.), so you might find it useful reading.

The third, "A Little More Reality," is an example of the excesses that scanners and character-recognition software can lead to! Anyway, it says a bit more about what ontology (or rather what lack of an ontology) I think quantum theory gives us. If you want to see how wacky your postdoc can be in private – I told you I was coming out! – look at the discussion after Asherism 8.

The fourth, "Tyche or Moira?" is again about quantum ontology – it clears up something about the discussion of Asherism 7 in the previous note.

In general, as you'll see if you look at any of this gunk, I'm rarely willing to carry quantum theory so far as to make ontological statements with it – I am more inclined to think that it is a theory about our knowledge. For instance, you might have noticed that I opened up my proposal with "The world we live in is well-described by quantum mechanics." I didn't say, "The world we live in is a quantum mechanical one." (I caught myself in that particular instance ... I came oh so close to compromising my morals!) There must be some ontology – some theory of the "real" – out there, it just doesn't seem to me to be quantum theory. Trying to fit quantum theory to that task seems more like trying to stuff a foot into a glove.

That's about it. Maybe I've said some of these things more clearly in other notes, but these are the most recent ... and easiest to find. I should be around tomorrow (if I can get to sleep tonight) ... as is my way though, you'll probably find my writings more clear than my speech.

21 November 1997, "Closing the Door"

By the way, you know, I believe in reality too. (It's pretty difficult to gulp a world that is completely dependent upon our dreams for its existence.) It's just that the reality I'm groping for is significantly more interactive – or evolutionary – than yours: you have to remember, I grew up in the day of the Graphical Users Interface. You really shouldn't lump all non-Everettistas with the people at *Social Text*. [[I am saying all this with a smile.]]

02 December 1997, "Boojumating"

Thinking about boojumating for the next couple of days has brought my thoughts back to your many-worlds lecture. And I ended up reading a lot of my old correspondence on the subject: in slightly more detail it covers most of the points we discussed the other day, maybe a couple more. Also it shows that I guess in the end I had never really come to grips with accepting the coarse-graining stuff after all.

Anyway, I know that your interest in this has likely long since vanished away. But, just in case there's a little interest left, I'll forward on those old e-mails. Also, I have a hard copy of some of Don Page's objections (in a note sent to Hartle) – if you want to see it, I can bring it around some time.

> But oh, beamish nephew, beware of the day,
> If your Snark be a Boojum! For then
> You will softly and suddenly vanish away,
> And never be met with again!

02 December 1997, "Flying Equations"

I couldn't help but think of the anecdote about John Wheeler's (non)flying equations you told the other day when I came across the following little passage (presumably Biblical in origin):

"I forbade any simulacrum in the temples because the divinity that breathes life into nature cannot be represented."

07 August 1998, "Impressions"

Another cute thing I thought you might like is a quote that ended a letter Adrian Kent sent me this morning. See the connection to one thing we talked about yesterday?

In fragment D 125 ... [Democritus] introduces the intellect in a contest with the senses. The former says "Ostensibly there is colour, ostensibly sweetness, ostensibly bitterness, actually only atoms and the void"; to which the senses retort: "Poor intellect, do you hope to defeat us while from us you borrow your evidence? Your victory is your defeat." You simply can not put it more briefly and clearly.

(Erwin Schrödinger, Nature and the Greeks)

06 April 1999, "Round 1"

I'm finally getting off of my lazy duff and taking care of some of my external duties. In this note I'll fulfill your request that I look at your paper, "Quantum Information and Physics: Some Future Directions."

You wanted citations to Wootters and me for our "visions." For Wootters the best place I know of is his PhD thesis: W. K. Wootters, *The Acquisition of Information from Quantum Measurements*, PhD thesis, The University of Texas at Austin, Austin, TX, 1980. For me, you can cite either my research proposal: C. A. Fuchs, "The Structure of Quantum Information," or my collected emails on the subject: C. A. Fuchs, "Collecting My Thoughts: Some Historical, Foundational, and Forward-Looking Thoughts on the Quantum."

Random comments:

1) Hilbert space is a big place. I think Carl and I were the first people to say it in print and give the idea a quantitative meaning/analysis (though Carl had been saying it in word for some time before that): C. M. Caves and C. A. Fuchs, "Quantum Information: How Much Information in a State Vector?," in *The Dilemma of Einstein, Podolsky and Rosen – 60 Years Later*, edited by A. Mann and M. Revzen, (Annals of The Israel Physical Society **12**), 226–257 (1996). In 1994 after a dinner-time discussion with us, Jeff Kimble popularized it a bit in his quantum-state preparation talks (1-900-SUM-FOCK). One can also find the following phrase quoted at the beginning of Chapter 2 of Volume 1 of Reed and Simon: "Gentlemen: there's lots of room left in Hilbert space." – S. MacLane

2) Holographic universes and other parts of Section III. The other night Charlie Bennett, Herb Bernstein, and I were discussing the foundations of QM, and Charlie was really revealing his distaste for the word "man." (At the root of it, he seems not to realize that the only way Shannon information theory has to define *information* is with respect to ignorance, i.e., probability distributions. Information in universes without ignorant agents ... what could it mean? More importantly, why would one go to the trouble of defining it? Why not stick with a more objective sounding language – like dynamical laws, etc. – and just forget about information?) Anyway, in a moment of

exasperation, I exclaimed, "Why is it that you can trust my observations and thoughts on quantum channels, but not on quantum mechanics itself!?! Why is it that you refuse to give me some benefit of the doubt here?" I suppose I should be happy: his reply was to compare me to Newton and his follies with alchemy. So, OK, I pass on the good will: In my eyes you are like Newton.

01 September 1999, "Flushing Fuchses"

Preskillism 1: *But furthermore, as sympathetic as I am with your remarks about Gleason's theorem, I'm not sure I know how to motivate noncontextuality either. Indeed, as you know, advocates of hidden variables (Bell, Bohm, . . .) have argued that noncontextuality is not well motivated.*

That's quite a good point. Noncontextuality of the *probabilities* bothers me too. (See notes below for whatever they're worth.) This almost always goes as an unemphasized assumption behind Gleason: people just say that a frame function is assumed and let that be that. (At least I haven't seen the term "noncontextuality" used in this way before; the comment by David Mermin indicates that he may not have either.) What you're thinking of when you refer to Bohm and Bell is that they didn't think the notion of a noncontextual *hidden-variable theory* was well motivated. That is, neither of them liked the kind of hidden variable theory that the Kochen–Specker results seem to disallow (but see the Meyer–Kent–Clifton articles of this summer). For a hidden-variable theory to be noncontextual, the truth values of all potential measurement results have to be set before the interaction. For a probability assignment to be noncontextual, the probability of a Hilbert-space projector must be independent of which basis it's associated with (i.e., which measurement interaction it's associated with). Who ordered that? I don't know (but I list my present leaning in the note to Howard below).

Preskillism 2: *Therefore, Deutsch's framework (suitably sharpened) is of potential interest if it enables us to derive the quantum probability rule without directly invoking noncontextuality (as Gleason's argument does). [. . .] So maybe there is something of merit that we can sift from the wreckage of Deutsch's paper . . . Do you agree?*

It could be, though I'm inclined NOT to keep BOTH the state space and the observable space as primitive, when we already know that we can get by without one of those – clearly my personal taste. But if one does accept Insufficient Reason in this context – i.e., that equal amplitudes give equal probabilities – one would still have to be very careful with the remainder of Deutsch's *particular* argument. We were never able to make the rest of it sufficiently rigorous to examine before getting so frustrated as to give up . . . and just write up what we had. (Recall our slight discussion of the principle of substitutability.) Since then, Howard and Jerry have continued to think about it a bit, but I don't think they've been able to come to a sufficient conclusion either.

Safer ground can probably be found in the stuff that Cleve and Braunstein (and Nielsen?) started to write up but never posted. (I talked to Richard about it this summer: he said he believed the result rigorous but not overly compelling and didn't have plans to complete the project.) They explicitly took Insufficient Reason as their starting point. Zurek also has a similar argument that I saw him present in Baltimore (but God knows if there's anything rigorously proved there without loads of side assumptions ... Cleve sure seemed to need them, and I find Cleve very trustworthy). The reference to Zurek is: *Phil. Trans. Roy. Soc. London* **A356** (1998) 1793.

All that said, I have my doubts that your invariance property doesn't already build in noncontextuality almost immediately. The assumptions that Deutsch does use (without invariance) already get to the fact that the value functions can be expressed as expected utilities for some probability distribution. And then from there is it that hard? ... OK, I don't see it immediately: let me think a little harder about this.

07 September 1999, "Two Rabbis in a Bar"

Thanks for the latest note. Carl and I thought a little about your point. We came across nothing great or deep, but let me record some of the equations and thoughts that came out of that discussion (mostly so WE won't forget them).

Basically it looks like there are loads and loads of value functions that satisfy the zero-sum, additivity, and unitary invariance properties. Unitary invariance gives that the value function can only depend upon the amplitudes and utilities (not the basis used in the expansions) and also gives permutation symmetry thereafter:

$$V\left(\sum_i \lambda_i |\phi_i\rangle; \sum_i x_i |\phi_i\rangle\langle\phi_i|\right) = V(\lambda_1, \ldots, \lambda_n; x_1, \ldots, x_n) \tag{24.1}$$

$$= V(\lambda_{\sigma(1)}, \ldots, \lambda_{\sigma(n)}; x_{\sigma(1)}, \ldots, x_{\sigma(n)}), \tag{24.2}$$

where σ is any permutation. Adding to that Deutsch's additivity or Savage's complete transitive ordering we have,

$$V(\lambda_1, \ldots, \lambda_n; x_1, \ldots, x_n) = \sum_i f_i(\lambda_1, \ldots, \lambda_n) x_i \tag{24.3}$$

$$= \sum_i f_i(\lambda_{\sigma(1)}, \ldots, \lambda_{\sigma(n)}) x_{\sigma(i)}. \tag{24.4}$$

If this is going to hold true for all sets of $\{x_i\}$, then we must also have,

$$f_i(\lambda_1, \ldots, \lambda_n) = f_{\sigma(i)}(\lambda_{\sigma(1)}, \ldots, \lambda_{\sigma(n)}). \tag{24.5}$$

Presumably, noncontextuality would force us down to

$$f_i(\lambda_1, \ldots, \lambda_n) = f(\lambda_i). \tag{24.6}$$

Anyway, within this context some of the probability laws we've discussed take the following form:

standard QM rule: $\qquad f_i(\lambda_1, \ldots, \lambda_n) = |\lambda_i|^2,$ (24.7)

BCF²S rule: $\qquad f_i(\lambda_1, \ldots, \lambda_n) = \dfrac{1}{\text{\# of nonzero } \lambda s},$ (24.8)

JP rule (slightly generalized): $f_i(\lambda_1, \ldots, \lambda_n) = \dfrac{g(|\lambda_i|)}{\sum_i g(|\lambda_i|)}$ $\quad$ for any function g. (24.9)

Notice that all of these rules reduce to the standard one when the λs are equal. The point to be made here is that Deutsch's development up to the "pivotal result" does not make a cut between any of these proposed rules. At least up to that stage in the argument, unitary invariance is still pretty weak. So, if unitary invariance does add some extra power it is going to have to be at later stages in his argument, perhaps where he invokes the mysterious principle of substitutability. In particular, we probably ought to retract the statement we wrote you last week, "Indeed you have a point that unitary invariance will pretty much give the whole shebang."

As a point of philosophy, it is quite interesting to see that the unitary invariance of the value functions is distinct from the Gleason assumption of noncontextuality – this may be important even for the strict Bayesian. (You know, we do have significant sympathies with the idea of deriving as much of the quantum probability rule from decision theory as we can.) But, upon thinking about it, I don't understand why you find noncontextuality so unmotivated from the Everett point of view. Can you articulate why you believe that?

I suppose part of my problem in seeing the difficulties you're seeing is that the Everett structure continues to look completely arbitrary to me. If one is going to go already that far in disconnecting the formalism from the world of our experience, why bother with such relatively minor details? Despite your words,

Preskillism 3: *From the Everett viewpoint, then, probability is about the expectations of an agent who is himself part of the system, expectations that are manifested by the behavior of the agent. [. . .] So if we were able to argue that an observer who is part of the quantum system will play a game according to a certain strategy, then perhaps we will have succeeded in "deriving" a probability rule.*

I continue to see no place for probability within the Everett point of view. I just don't know how to say it more forcefully than this: the notion of a decision-making agent is a contradiction in terms within that model of the world. I know you likely read what I wrote Howard the other day even if you didn't comment, but I want the words to gnaw at you:

In the Everett interpretation, we can never ARRANGE to do any experiment, to play any game. There is no sense in which this phrase is meaningful in that interpretation. The Everett world just IS. Experimenters have NO choice to arrange anything in an Everett multiverse; their actions are at best a grand conspiracy set by the initial condition of the universe. They have no will, no possibility to do anything that was not preordained. If an experimenter in the Everett theory perceives himself choosing one game over another, it can be but ILLUSION that he had any choice in the matter.

Charlie Bennett likes to tell the story of two rabbis. One walks into a congregation, throws himself before the ark and exclaims, "Lord, before you I am nothing!" Seeing that, a second rabbi walks in, throws himself violently to the ground; groveling, he screams, "LORD, BEFORE YOU I AM NOTHING!!!" Carried away in the excitement, a common member of the congregation goes to the ark and does the same thing. In shock, the first rabbi looks to the second and says, "Who is HE to think HE's nothing?!?!"

So too, I find myself chuckling, "Who is HE to think that noncontextuality is unmotivated for HIS interpretation of quantum mechanics?!?!" From the point of view I'm trying to construct – that quantum mechanics is about the tradeoff between INFORMATION gain and INFORMATION disturbance – it is truly a problem. For in that model of the world, measurements are invasive creatures and the theory of quantum mechanics is our best attempt to say something in spite of that fact (that truth, reality, noumenon, ding an sich, … whatever you wish to call it). My question is why are measurements invasive but not too invasive? Why are noncontextual hidden variables ruled out, but noncontextual probability assignments endorsed? From this point of view noncontextuality screams out for a deeper explanation. What for the Everettista makes it scream the same way?

08 September 1999, "Smug Note"

Preskillism 4: *"Decision-making agent is a contradiction of terms …":* *I'm not sure I'm ready and willing to debate the issue of free will. The illusion of being able to make decisions (if it is an illusion) is a property of the conscious mind, and can be explained only as such. To the Everettista, this would require a theory of how a conscious observer functions, an observer who is part of a quantum system …*

This is a new tack to me. Since when has not having an explicit "theory of conscious observers" ever slowed down the Everett movement before?? What else should we make of it when Everett says [DeWitt and Graham, p. 65]:

We note that there is no longer any independent system state or observer state, although the two have become correlated in a one–one manner. However, in each *element* of the superposition … the object system state is a particular eigenstate of the observer, and *furthermore the observer-system state describes the observer as definitely perceiving that particular system state.*

How else could we take seriously Figure 2, "A Reversible Simple Mind," in Cooper and van Vechten's paper [D & G, pp. 223–223]. How else could Deutsch in his misconceived thought-experiment [*IJTP*, 1985] say,

After the completion of the measurement, the observer records (in his memory, or in his notebook if necessary) – not the value "N" or "S" of the spin, but only *whether or not he knows this value.* He may write "I, Professor X, F.R.S., hereby certify that at time t''' I have determined whether the value of the North component of the spin of atom-1 is $+\frac{1}{2}\hbar$ or $-\frac{1}{2}\hbar$. At this moment I am contemplating in my own mind one, and only one of those two values. . . ."

How could John Preskill in his much lauded lecture notes even be able to write his Eq. (3.178)??

As advertised, I say all this with complete smugness. The point is, how can all these things be said without a tacit (but nevertheless grand) theory of consciousness working in the background?

How does the Everettista know a priori that conscious states are always associated with ORTHOGONAL quantum states? The brain's a pretty complex system: if we know so little about it, how on earth can we be so confident of capturing its essence in such a glib way? [[Kiki asks me what I'm hungry for. I say, "I don't know, I'm not even hungry." Within a couple of minutes my stomach is roaring. Her information gathering caused a disturbance to my conscious state, you might say. In this respect, my conscious states act a little like nonorthogonal quantum states. Who's to say how far that analogy can or cannot be pushed? I say that tongue in cheek, but my point is firm: I don't think there's a better justification for the standard quantum measurement analysis (in terms of orthogonal observer states) than that the pace was set in the last few pages of von Neumann's book.]] How does the Everettista know a priori that conscious states should be associated with PURE states? [[My body usually runs about 99 degrees F; people who run significantly colder, are usually significantly deader – and I've never caught a dead man thinking. Or to make it sound more relevant, think of the poor girl who was locked in a closet for 13 years while her brain should have been developing speech skills. She was never able to recover: should she be considered as conscious as I? How does the Everettista know a priori that conscious entities do not have to be significantly ENTANGLED with other conscious entities to be conscious in the first place? Again I say that tongue in cheek, but how do you know?]]

Or here's a better one. I feel strongly – you might even say "know" – that I love Kiki. I also feel strongly – you might even say "know" – that I observed a single ion in one of Herbert Walther's ion traps last spring. In the first case, most of us wouldn't presume that a single pure quantum state could be associated with my state of consciousness: "love" is just too loosey-goosey of a notion to imagine being captured in such a simple-minded way. How does the second case differ except out of expedience for the problem at hand?

So the point again: it seems hard to deny that the Everett analysis of measurement doesn't dip into concepts well beyond the domain of standard physical theory. As such I view it as little more than a comforting religion. The point (among others I could give) which the Everettistas have yet to analyze is precisely the point I CONSCIOUSLY leave UNanalyzed: how do experimental outcomes come about and how do they become known?

That is, I take "experiment" as the basic atom out of which we build scientific theories and models of the world. If a model so constructed does NOT cough back up the very possibility of experiment, then I would view it with a leery eye . . . and I suppose that is what I am attempting to get you to do also.

I am a little uncomfortable with the phrase "free will" as you used it above. What I am trying to get at is perhaps better captured by what Hans Primas had to say:

[I]t is a tacit assumption of all engineering and experimental sciences that nature can be manipulated and that the initial conditions required by experiments can be created by interventions using means external to the object under investigation. That is, *we take it for granted that the experimenter has a certain freedom of action which is not accounted for by first principles of physics.* Man's free will implies the ability to carry out actions, it constitutes his essence as an actor. Without this freedom of choice, experiments would be impossible. *The framework of experimental science requires this freedom of action as a constitutive though tacit presupposition.*

The words "free will" are indeed used there, but the more central issue is the one captured in the first sentence. That is about the notion of science and hardly more. If with each succeeding experiment we become more and more confident that quantum mechanics is a good description of the world, how can we at the same time become ever more confident that those same knob-twiddling actions were preordained in the initial condition of the universe?

I keep thinking that if I hit this from enough angles something will eventually shake loose in you. *I want to wake you from your dogmatic slumber* ...

So I'll end with one last act of smugness. You say,

To the Everettista, this would require a theory of how a conscious observer functions, an observer who is part of a quantum system ...

I'll go you one better. In the view I am promoting, I am not only part of a quantum system, I AM a quantum system. You hint that you don't think a Quantum Bayesian can have that. But I do have it: what it means for me to be a quantum system is that I can be analyzed by you or anyone else profitably with the methods of quantum mechanics. If you are lucky enough to have maximal knowledge of me, you would describe what you know via some pure state.

What fear is it that drives you and so many of our colleagues to the excesses (and emptiness) of the Everett worldview?

You want to do quantum cosmology, and you think you can't get by without it? Is that it? Here's how it's done; I'll show you how to do it. You write down a wave function for the universe:

$$| \Psi_{\text{universe}} \rangle.$$

(It *is* that easy.) For some aspects of the universe, this might be an excellent approximation TO WHAT YOU KNOW ABOUT IT. As far as I can see – and I don't believe I am working with blinders on – there are no conceptual difficulties with doing quantum cosmology in just this way.

To think that it would first require a theory of consciousness before we could do quantum cosmology (or any physics at all) in the Bayesian style is just a bunch of malarkey. That's no more true than saying we would need a theory of consciousness before we could open a gambling house in the state of New Mexico. The theory of gambles and decisions seems to work perfectly well in spite of psychology.

You say in your lecture notes:

Preskillism 5: *Those who hold the contrary view (that, even if there is an underlying reality, the state vector only encodes a state of knowledge rather than an underlying reality) tend to believe that the current formulation of quantum theory is not fully satisfactory, that there is a deeper description still awaiting discovery. To me it seems more economical to assume that the wavefunction does describe reality, unless and until you can dissuade me.*

I've now put my best emotion into it, and hopefully enough reasoning to pique your interest. I'm not likely to come back to pester you again if I haven't made a dent by now. But do know this: I do not fit into the category of those who "tend to believe that the current formulation of quantum theory is not fully satisfactory." The only thing that I think is not fully satisfactory is our understanding of why we are compelled to use this theory. When we understand why we cannot obtain more information about a system's behavior than can be captured by a symbol $|\psi\rangle$, we will have learned something profound.

10 September 1999, "Meeker Note"

Thanks for the discussion last night, and thanks for bringing the weakness of my linearity argument to the surface. I got a few things out of the meeting. The biggest one might be this. I was struck more than usual how the Bayesian view seems a bit like beer: that is, in being an acquired taste, or, more accurately, tasting good only after a concerted effort to swallow it more than once. Talking with you helps emphasize how the mode of language Carl and I have gotten ourselves into is so very distinct from the usual language of theoretical physics. This is something we really need to be careful about in our presentations.

Our discussion reminded me of something I once read in an article of Ed Jaynes:

The main suggestion we wish to make is that how we look at basic probability theory has deep implications for the Bohr–Einstein positions. ... Einstein's thinking is always on the ontological level traditional in physics, trying to describe the realities of Nature. Bohr's thinking is always on the epistemological level, describing not reality but only our information about reality. The peculiar flavor of his language arises from the absence of words with any ontological import; the notion of a "real physical situation" was just not present and he gave evasive answers to questions of the form: "What is really happening?"

In some of your formulations of the ways in which we agree, I felt that precisely the same distinction was coming out. The use of the phrase "dynamical laws" is probably a good example. Whenever I used the phrase, I meant something that tells us how to update our predictions. [[A symptom of that is that I find myself questioning whether the dynamical laws (both classical and quantum) are linear simply because Bayes' rule is linear.]] But then I wished to stop there. You, on the other hand, seemed always to use "dynamical laws" in a more ontologized way: those dynamical laws exist in some sense independent of our theory.

I don't find the compulsion to do that. For though I consider myself a realist at heart (for precisely the reason you express in your lecture notes), I don't see physics as an expression of our great knowledge of the world ... but more accurately as an expression of our great ignorance. I see it as an expression of the best stab we can make at things.

Our greatest difference at this stage seems to be a sociological one, or one of temperament or emotion (as you said). I – just stealing the method from Socrates – think we will find our greatest strength by more fully realizing our ignorance and digging for its roots. This is the direction I've been turning toward ever more forcefully since reading the Jozsa, Robb, Wootters paper in 1994 (and choosing my thesis topic soon after that).

When I think of running with the theory (another of your phrases), I mostly think of things like these words of John Wheeler's:

I want you ... to jolt the world of physics into an understanding of the quantum because the quantum surely contains – when unravelled – the most wonderful insight we could ever hope to have on how this world operates, something equivalent in scope and power to the greatest discovery that science has ever yet yielded up: Darwin's Evolution.

You know how Einstein wrote to his friend in 1908, "This quantum business is so incredibly important and difficult that everyone should busy himself with it." ... Expecting something great when two great minds meet who have different outlooks, all of us in this Princeton community expected something great to come out from Bohr and Einstein arguing the great question day after day – the central purpose of Bohr's four-month, spring 1939 visit to Princeton – I, now, looking back on those days, have a terrible conscience because the day-after-day arguing of Bohr was not with Einstein about the quantum but with me about the fission of uranium. How recover, I ask myself over and over, the pent up promise of those long-past days? Today, the physics community is bigger and knows more than it did in 1939, but it lacks the same feeling of **desperate** puzzlement. I want to recapture that feeling for us all, even if it is my last act on Earth.

I cannot shake the feeling that we are lucky to be in a unique time, finally amassing the technical tools to get at what John is asking us to do. Quantum mechanics is about information (in the deepest of senses): run with it, I say – that's the sense.

Another thing that came out of yesterday's discussion is that I became more confident in the force of something I wrote you the other day:

If with each succeeding experiment we become more and more confident that quantum mechanics is a good description of the world, how can we at the same time become ever more confident that those same knob-twiddling actions were preordained in the initial condition of the universe?

Let me try to say it slightly differently. I find it kind of lopsided to accept that experiments help us amass evidence for this or that dynamical theory,[1] and not at the same time accept that they have helped us amass evidence that we are getting ever finer-tuned control of

---

[1] Don't forget, I'm likely to be stating the phrase "dynamical theory" in a more epistemology-soaked way than you are used to.

nature – that finer-tuned control being expressed in our setting the initial conditions of the experiments in any way we please. The force of these statements, I think, finds expression in the general structure we use for all our endeavors in physics: there are dynamical laws, and there are initial conditions. As far as I understand, no one has ever been able to tie the two distinct structures together, as they say, with a "theory of initial conditions." (I suppose I'm willing to bet they never will … but that is certainly a statement of temperament and emotion.)

Where should quantum mechanics go in my eyes? Finding better ways to describe and quantify the "ticklishness" of the world that has been given a technical expression by quantum cryptography and its subsidiaries (for instance I'm thinking of "nonlocality without entanglement" here). To see if there's a good sense in which that same ticklishness can be thought of (in a useful way) as the power behind quantum computing. To get things tidied up from this point of view (or see if it falls) so that the next generation can do something truly astounding. But that's just a feeling in my gut (and probably the spirit of John Wheeler).

12 December 1999, "Freedom"

Preskillism 6: *Free will usually means the ability of conscious beings to influence their own future behavior. Its existence would seem to imply that different physical laws govern conscious systems and inanimate systems. I know of no persuasive evidence to support this viewpoint, and so I am inclined to reject it.*

Can't agree with your second sentence. When you think of physical law, you should say the chant epistemic, epistemic, epistemic. Then you can chime in with Martin Luther King, Jr.: free at last, free at last, thank God, I'm free at last!

Had fun talking to you the other day.

10 October 2000, "Sneak Previews"

Preskillism 7: *In the course of thinking about what to tell my class about Gleason's theorem, I looked again at your paper* quant-ph/9907024, *and pondered again your statement about the noncontextuality assumption:*

This important assumption, which might be called the "nocontextuality" of probabilities, means that the probabilities are consistent with the Hilbert space structure of the observables.

And I realized … I have no idea what is meant by "consistent with the Hilbert space structure of the observables."
What does it mean?

That's a good question. … Because I don't completely know the answer yet. But I think I'm on the tail of the tiger. You'll see what I mean in the next email. That document is a problem set I was writing up for a little meeting Gilles and I organized in Montréal on "Quantum Foundations in the Light of Quantum Information." [See letter to Gilles Brassard, titled "Problem Set Based on Information–Disturbance Foundation Quest," dated 15 May 2000.] It looks a little incomplete because it is. Unfortunately the fire hit just as I was starting to write Problem #2, so I never completely finished the document properly.

The part that's most relevant to your question is Problems #1 and #3 and Proto-Problems #5 and #6. In particular, Problem #1. Have a look at those and then come back to what I'm about to write below.

Since then, we've made progress on a few fronts. With regard to Problem #1, Nurit Baytch and I have now shown that Conjecture 1 holds for at least a set of measure 1 in the parameters. There are a few stubborn holdouts (a discrete set's worth), but I think ultimately we'll fix that. I hope that gives you a better feeling for what the noncontextuality assumption is about: it only meshes well with the orthogonality structure of Hilbert space, not with the linear structure per se or even any other "fixed-angles" structure. Nurit has TEX'd up some of this, but I haven't had a thorough look at her document yet. If you'd like, I can send the details by the end of the weekend.

A more constructive bit of progress has come from Problem #3. I ultimately scrapped the precise wording there because a more natural question is this. Let us think of the most basic elements in quantum measurement theory as the POVMs, not simply the ODOPs (one-dimensional orthogonal projectors). Then a natural generalization of Gleason's problem is to suppose that the probabilities for measurement outcomes are given by a function f from positive operators to the real interval $[0, 1]$ such that $\sum_i f(E_i) = 1$ whenever the E_i form a POVM. This assumption embodies an even stronger kind of "noncontextuality" for quantum probabilities than before. What's neat is that if one makes these assumptions, then one can prove the standard quantum law – i.e., that there must exist a density operator ρ such that $f(E_i) = \text{tr}\rho E_i$ – in a way that mere mortals can understand. The proof is quite simple and memorable (by that I really mean memorizable). This is joint work with Caves, Renes, and another student of Carl's (don't know his last name or even the spelling of his first) ... with actually Renes doing the brunt of the details. A further payoff of this effort is that the proof works in 2-D where the standard Gleason theorem falters. Actually I think that's my favorite feature of the whole thing. (That that might work was my first scientific thought after the fire, so I guess I hold it particularly dear.) Renes has started working on a draft of this: we can send it to you once it's in reasonable form.

Finally, let me say a smidgen about Proto-Problem #5 in conjunction with Proto-Problem #3. I've gathered a very small amount of evidence that the project there might possibly work out. Namely it dawned on me that in all known Kochen–Specker examples all the rays have algebraic-number coordinates. This is a prerequisite for the possibility of a Gleason theorem on that structure. You might sneer that this is an overly technical question, but I think its interpretational import is quite relevant (though I didn't express that so well in my shoddy notes). Let me try to do that idea a little more justice here. The main question is whether most of the structure of quantum mechanics is about what "is" in the world, or instead how it goes ding to our interactions with it. (You know I'm more inclined to the latter.) Well a number of decent scientists (for instance Birkhoff and von Neumann) have thought so much that quantum mechanics is about what "is" that they were willing to rethink what it means for a system to have properties. Out of this arose the idea of a quantum logic: algebraic operations AND, OR, and NOT that satisfy the postulates of orthomodular lattices rather than the postulates of Boolean algebra. Interestingly, B and vN noticed that AND, OR, and NOT can be identified with INTERSECTION, SPAN, and ORTHOGONAL COMPLEMENT for subspaces in a vector space if and only if the (skew-) field of the vector space is one of the three: real numbers, complex numbers, or quaternionic numbers. Rationals won't work; algebraic won't work. So my question to the quantum logicians (like Jeff Bub) is, "Do you hold Gleason's theorem dear?" "If so, then what do you make of the fact that it might work for a structure that is not an orthomodular lattice?" You see, I have no compulsion to believe that QM should only be a tight structure for a theory of "properties" (i.e., a quantum logic). In fact, I would consider it evidence in my favor if Gleason still worked for such a non-lattice-theoretic entity.

If you want updates on any of the other problems in the Montréal problem set, I can send them, but I don't think they're completely relevant to your question.

Hope that helps. I'll try to answer your more logistical email a little later in the week.

25

Letters to Joseph Renes

22 June 2000, "References for Gleason"

Some relevant references. The important ones are Gleason's paper, Cooke *et al.*'s paper, and Pitowsky's paper.

1. R. Cooke, M. Keane, and W. Moran, "An Elementary Proof of Gleason's Theorem," *Math. Proc. Camb. Phil. Soc.* **98**, 117–128 (1981).
2. A. M. Gleason, "Measures on the Closed Subspaces of a Hilbert Space," *J. Math. Mech.* **6**, 885–894 (1957).
3. I. Pitowsky, "George Boole's 'Conditions of Possible Experience' and the Quantum Puzzle," *Brit. J. Phil. Sci.* **45**, 95–125 (1994).
4. I. Pitowsky, "Infinite and Finite Gleason's Theorems and the Logic of Indeterminacy," *J. Math. Phys.* **39**, 218–228 (1998).
5. F. Richman and D. Bridges, "A Constructive Proof of Gleason's Theorem," *J. Func. Anal.* **162**, 287–312 (1999).

11 December 2000, "New Ideas and Old Men"

Thanks for the new ideas. Let me try to make a few comments. I'm in a smoky hotel bar right now; I wonder how this is affecting my laptop's health? I know it's wreaking havoc on mine. I'm still in Vienna, by the way. I leave Tuesday morning. Unfortunately I haven't gotten an ounce written of what I had hoped to.

Renesism 1: *This got me thinking about this stuff, and it seems that in some sense what we're gathering information about, when we make measurements, is the ability to predict future measurements.*

You haven't read my paper with Kurt Jacobs yet, have you? [See `quant-ph/0009101`.] Come on, at least read the Introduction and Footnote 27. Also it might be useful to have a look at the section where we formulate what the notion of information ought to be! (Come on, I didn't write the paper for nothin'.)

Renesism 2: *With this notion it's simple to define the information gained (or lost!) about a hypothetical measurement X when actually measuring Y. It seems first that the structure* *of measurements makes things such that $I_Y(X)$ can be negative, whereas classically this can't be the case (right? knowing more of anything classically means we know more of everything).*

It feels like you're kind of on the right track here, but I think there is a definite problem in grounding this exercise in one observer. That is, if one observer makes a measurement of X, he can always rig the measurement interaction so that he ends up with perfect predictability over Y. The interaction just has to have a conditional unitary so that X-eigenstates are taken to Y-eigenstates.

This is why I think it might be better to ground the problem with respect to two people. Suppose Alice and Bob start with a third system for which they both ascribe a state ρ. Now imagine Alice interacts with it in such a way as to increase her predictability of an X measurement on it, while Bob interacts with it in such a way as to increase his predictability of a Y measurement on it. How high can they make their mutual predictability as a function of ρ, X, and Y?

Why am I so obsessed with always having two players in the game? Because I want to connect all the concerns in quantum mechanics with Bayesianism as much as I can. The issue is: Under what conditions can the various agents in a situation come into better agreement even though they may start with differing or incomplete states of knowledge? That sort of thing is the bread and butter of Bernardo and Smith's book, for instance. But quantum mechanics further adds the issue of noncommutivity to this

So, there will be some limitation to how high the mutual predictability can be. You might find some earlier work on the subject in the papers of Michael Hall (M. J. W. Hall). I'm thinking in particular of a *PRL* of his from about 1994 or 1995. You might ask Carl about it. But I don't think the two player problem is precisely the one Michael addressed.

Renesism 3: *Because I don't know what quantum information is – to me information should be classical, or we should build up the information theory of quantum systems to the point where we make use of von Neumann entropy rather than making quantum information theory an analogous theory to classical information theory.*

I certainly like this statement. In fact I just made a slide for my talk here last week that said:

QM is about Information. $H(X)$. Plain old information in Shannon's sense: uncertainty, lack of predictability.

Renesism 4: *Listening to Bill Wootters's talk yesterday I had a slightly different idea about approaching this – the relationship to entanglement.*

I think that's a good idea too. One gets the feeling that the more entangled two systems are, the more one can predict about measurements on the other system, given only the identity of the measurement. In the ideal case of an EPR pair, one just performs the identical measurement on the system in one's possession.

Unfortunately I can't think of anything more intelligent to say than that.

Let me end by coming back to the point above: noncommutivity. What I really want to know is, "Why noncommutivity?" (That's another way of saying, "Why Hilbert space?") The greatest missing link in my philosophical program is now the answer to that question. How can we make an information–disturbance tradeoff IMPLY the existence of noncommuting states of knowledge? I don't have a clue how to make that happen, but it's got to happen.

It's got to happen. And I've got to go to bed.

05 January 2001, "In Defense of Interactionism"

I don't know what Carl has said, but I like the sound of your "research direction." I doubt any of these references will be overly useful, but I met a guy at the quantum information tutorials in Edinburgh who has spent a lot of time generalizing quantum logics to effect algebras. His name is Roberto Giuntini.

I say I doubt it will be overly useful, because I gathered that, in his heart, he too was looking for the holy grail of most quantum logicians: to find a way of pinning observer-independent properties on the world via algebraic properties of the theory's surface terms (i.e., states, observables, Hamiltonians). In opposition to Carl – I won't say it's heresy – I think that's the wrong way to proceed. I think we'll ultimately find A QUANTUM REALITY, but we'll have to dig deeper than that.

All that said, the reason I like your research direction is because I think understanding the purely algebraic properties of POVMs is an important link nevertheless. I would like to think that quantum mechanics is more about our interface with the world than the world itself ... and that it's at that interface that we'll find our glimpse of a "quantum reality." What is it about the interface that makes our best Bayesian predictions of the quantum mechanical form? That can be an algebraic question just as much as trying to pin naïve properties on the world (regardless of how contrived the logic). In that sense, there is some chance of maybe learning something from the Giuntinis of the world.

30 March 2001, "Web Page Snooping"

Great quote!

> As far as we can discern, the sole purpose of human existence is to kindle a light
> in the darkness of mere being.
>
> (*Carl Jung*)

What's the precise source?

11 April 2001, "Piggybacking Philosophy on Physics"

Renesism 5: *Answering physics questions relative to a static background philosophy will not, I believe, yield what one would want it to – the grand questions are necessarily partly philosophical and thus must be confronted head on.*

I do appreciate the spirit, and I am part of the choir. John Wheeler once said something that made a big impression on me: "Philosophy is too important to be left to the philosophers!"

You can see the implication. If we're going to hope to make real progress we've got to have our feet firmly planted in the practice of physics. Really good physics is the very best of philosophy.

In the words of George Costanza, "Do me a solid buddy."

26

Letters to Mary Beth Ruskai

17 February 2000, "The Quantum View"

What a really nice article! I read it once this evening with dinner (at a place that served grilled chicken – not so easy to scroll through a computer screen under those conditions!). Now, with my jetlag in distinct control – its 3:30 AM for me – let me read it again and make a few comments here and there. (I just arrived in Haifa yesterday for a visit with Asher Peres; I'll be here a little over a week.)

- Why don't you write the following for your requested reference: C. A. Fuchs, "The Structure of Quantum Information," available at [...] [Of course, it is no longer available. Instead, see note to Carl Caves, dated 21 December 1997.]
- I like this view of physics as being like a jig-saw puzzle. Note a typo in this paragraph: you misspelled de Broglie.

Ruskai-ism 1: *As far as I am aware, there is no evidence that either Heisenberg or Schrödinger was motivated by political or social factors.*

Actually there is a pretty strong case that these two guys were indeed quite motivated by cultural factors to find an indeterministic element in physics. And that was long before quantum mechanics took a final form. Two important articles to read in this regard are:

1. P. Forman, "Weimar Culture, Causality, and Quantum Theory, 1918–1927: Adaptation by German Physicists and Mathematicians to a Hostile Intellectual Environment," in *Historical Studies in the Physical Sciences, Vol. 3*, edited by R. McCormmach (U. Pennsylvania Press, Philadelphia, 1971), pp. 1–115.
2. P. A. Hanle, "Indeterminacy Before Heisenberg: The Case of Franz Exner and Erwin Schrödinger," in *Historical Studies in the Physical Sciences, Vol. 10*, edited by R. McCormmach, L. Pyenson, and R. S. Turner (Johns Hopkins U. Press, Baltimore, 1979), pp. 225–269.

Both articles are quite good and do get to the nub of the matter. However, similarly to the point you made in the "suppose-for-the-sake-of-argument" sentence, none of that desire played out in the particular forms for quantum dynamics that these gentlemen proposed.

413

It's probably worth reading and citing these articles, if for no other reason than to show that you are aware of all the issues.

Ruskai-ism 2: *It is often said that mathematics is the language of science. But it would be more accurate to say that mathematics is a family of related languages. The languages of algebra, analysis, geometry, etc. often give very different insights into a problem.*

I like that sentence! Later when you write about gender:

Ruskai-ism 3: *In these situations, I believe that the insights of gender can play a role analogous to that of a mathematical language such as geometry.*

I was reminded that I tend to think similar things about the issue of quantum interpretations. Maybe the different interpretations make no difference in what is and is not possible in quantum phenomena, but they do play a role in directing thought. It seems to me that some interpretations are just dead ends for the inquisitive spirit (I'm thinking of Bohmianism in particular here). Whereas some interpretations – even some that I don't like, like many-worlds – have led to the contemplation of whole new phenomena, and, based on that criterion, are useful regardless of their ultimate efficacy.

- In your quote of Simon, "I speak probability with a marked functional analysis accent: lecturing in Zurich, I couldn't help feeling that I was speaking not hoch probability but only a kind of Schweitzer probability," is there a typo?

Ruskai-ism 4: *Similar remarks apply to the recent attempts by Goldstein, et al. to develop 'Bohmian mechanics', although its advocates assert that it is consistent with standard quantum theory because the Schrödinger equation still holds. However, there is more to quantum theory than the Schrödinger equation, and I am not convinced that no experimental distinction is possible.*

Here, here!!!! Have you read Ghose's paper? Is there anything to it?

Ruskai-ism 5: *Delicate experiments, such as those done by the laboratory groups associated with Aspect, Leggett, and Zeilinger are useful in clarifying some critical aspects of quantum theory. However, to my mind, the most convincing verification of quantum theory is not the individual microscopic experiments, but the fact that no other theory can even come close to explaining so many diverse phenomena (including macroscopic as well as microscopic phenomena); literally, "from atoms to stars." Few jigsaw puzzles fit together so neatly. We are forced to overcome the biases arising from our experience with the familiar macroscopic world of classical mechanics despite the challenge of resolving all questions about the foundations of quantum theory. In the end, quantum theory remains a human construct subject, in principle, to social forces. But it is a theory so remarkable, so different from ordinary experience, that it transcends social and cultural forces.*

Excellent paragraph! We have quite similar views about this. Note one small typo: Leggett should have two t's. More importantly though, Leggett is a theorist: so what laboratory group are you referring to there?

That's about it. Thanks for giving me the opportunity to read through this.

20 February 2000, "Like It Even Better"

Despite your reservations, I enjoyed the added discussion about Bohm. In November, I attended a conference in Naples organized by Dürr and Goldstein, and I was completely turned off to their movement. In my talk, I posed as a challenge for them to draw the Bohmian trajectories associated with any quantum teleportation experiment. And furthermore to tabulate how much work it took to do that versus the extra insight it gives into the phenomenon. Does it give any extra insight? My challenge was summarily dismissed by almost all the participants. The reaction was uniform: "You're missing the point Chris. Bohmians DON'T NEED TO draw the trajectories." Those people got under my skin. When in my second talk, I expressed the point of view that Asher Peres and I take to quantum mechanics (I'll place our *Physics Today* article on the subject in the next mail), they labelled ME as being unscientific!

Anyway, good job. I'll try to dig up those references for you. (It's a little harder with my being in Israel.)

18 April 2000, "Extremal Maps"

I was cleaning up my office today, trying to file some old things, and I came across a paper that might interest you:

S.-H. Kye, "On the convex set of completely positive linear maps in matrix algebra," *Math. Proc. Camb. Phil. Soc.* **122**, 45–54 (1997).

That, however, also prodded my memory that I had once somewhere seen something special about CPMs that leave two points touching the surface of the Bloch sphere. You should probably have a look at:

C.-S. Niu and R. B. Griffiths, "Two Qubit Copying Machine for Quantum Eavesdropping," quant-ph/9810008. (I'm sure the paper's long since appeared; most likely in *PRA*.)

and see if it is indeed relevant to you or has some connection to your work.

19 December 2000, "Myopic Readers"

[NOTE: Beth and I then exchanged some correspondence about the responses she had already received on her article mentioned above. One of those responses referred to my article with Asher Peres in *Physics Today*.]

Ruskai-ism 6: *You are only a rather innocuous footnote.*

See in this connection the return to the attitude of "quantum mechanics works" in the article by Fuchs and Peres ... "What [it] does", these authors write, "is provide an algorithm for computing the consequences of our experimental predictions."

But that is a misquote, and it is exactly the sort of thing that makes me want to keep my ignorant bliss. The actual words were:

Contrary to those desires, quantum theory does *not* describe physical reality. What it does is provide an algorithm for computing *probabilities* for the macroscopic events ("detector clicks") that are the consequences of our experimental interventions.

The word "interventions" is prominent there. It is there *precisely* to refer to the previous paragraph where we write:

To begin, let us examine the role of experiment in science. An experiment is an active intervention into the course of nature: We set up this or that experiment to see how nature reacts.

The word "nature" is a stand-in for all the "stuff" of the world. Why would we use such a word if we didn't believe deep inside that the existence of humankind is only a contingent sort of thing? The main point we were trying to get at is captured by the next few sentences:

We have learned something new when we can distill from the accumulated data a compact description of all that was seen and an indication of which further experiments will corroborate that description. This is what science is about. If, from such a description, we can *further* distill a model of a free-standing "reality" independent of our interventions, then so much the better. Classical physics is the ultimate example of such a model. However, there is no logical necessity for a realistic worldview to always be obtainable. If the world is such that we can never identify a reality independent of our experimental activity, then we must be prepared for that, too.

To paraphrase that for the myopic readers who couldn't see past the large print of our title: the point is, quantum mechanics is much more about our *interface* with the world than the world in and of itself.

The really exciting question to ask, I think, is: What is it about the quantum world that blocks us from going that *further* step? (I.e., the further step of distilling a free-standing reality.) Can we identify the aspect of the theory that demonstrates that in the crispest way? It's here that I think the tools of quantum information theory and cryptography will help us the most. Optimistically, I see us as carrying out that project in our lifetimes. Like I told Jeffrey Bub the other day: [See note to Jeff Bub, dated 10 December 2000.]

Ruskai-ism 7: *On the plus side, I learned that Heisenberg also raised my objection that Bohmian mechanics destroys the symmetry between position and momentum representation.*

I'm sorry, I guess I forgot to tell you that. I knew it too. If she didn't give you the reference, I think I can dig it up.

24 March 2001, "For Other Myopic Readers"

And your note prompted me to reread the rebuttal Asher and I wrote after the appearance of our original *Physics Today* thing. It's pasted below.

It's haughty of me, but I do think that myopia is running rampant in our community ... not only in the reading of our little article, but in the much bigger picture – our hopes for a physical theory. Bohmism is an example of such a dull point of view: If we can just return to the womb of classical physics, everything will be oh so much more warm and comfortable. Yuck!

27

Letters to Rüdiger Schack

28 February 1996, "QM is a Law of Thought"

By the way, did you ever read the paper Carl and I wrote titled "Quantum Information: How Much Information in a State Vector?" In it we talk at some length about this idea of "Quantum Mechanics as a Law of Thought" and other Bayesian sounding things. (We even quote Gertrude Stein?!?!) Of course, all the ideas are loose now, but one day they'll be tight.

13 September 1996, "Silly Latin"

Allow me to continue my tradition of writing you only when I need something translated!!
    I've decided on an epithet for myself. Either:

Quantum Theory is a Law of Thought

or

Quantum Mechanics is a Law of Thought

What do you think? (... and don't tell me I'm plagiarizing Boole!) Anyway, the problem is that I would like to have these translated as closely as possible into Latin. Do you think you're up to the task? I would appreciate it so much.
    I can't wait to see you in Japan; I really miss the old gang.

Rüdiger's First Reply: 19 September 1996

I like "Quantum Mechanics is a Law of Thought" better, but I can't give a good reason for that. I asked my father to translate your epithet into Latin, and he has come up with

Mechanica quantica ex mente orta lex est.

"orta" means something like "emerging from."
    I have a friend who is a Latin specialist. If you want, I can ask him for a second opinion. I am sure my father's translation is correct, but there might be a more elegant and more idiomatic version.
    Looking forward to seeing you in Japan.

Rüdiger's Second Reply: 3 May 2001

[Looking over a preprint version of this samizdat, Rüdiger wrote:]

Concerning your addition at the beginning of the Schack chapter, I told you that "Mechanica quantica ex mente orta lex est" mostly means something trivial like "Quantum mechanics is a law created by mind" or something like that, didn't I? All the experts (three) I asked chickened out in the end. I don't know if you still want to keep the translation, but you should add a footnote that the Latin phrase does NOT capture what you mean. In any case, I am going to make a last attempt to find a better Latin translation.

Rüdiger's Third Reply: 4 May 2001

The big problem for all the Latin specialists I ask (four of them now) is that they misinterpret the English phrase.

Mechanica quantica ex mente orta lex est = qm is a law that has arisen from mind

is not what you want.

Mechanica quantica lex cogitationis est

=

qm is a law of thought

[That is:]

lex = law
cogitatio = [thinking, conception, reflection, reasoning];
 sometimes a particular [thought, idea or intention]
cogitationis = genitive of cogitatio
lex cogitationis = law of thought

You see this is a literal translation, and therefore avoids any narrow interpretation of the English original.[1]

Rüdiger's Fourth Reply: 4 May 2001

Either you capitalize everything or nothing:

Mechanica Quantica Lex Cogitationis Est

MECHANICA QUANTICA LEX COGITATIONIS EST

mechanica quantica lex cogitationis est

---

[1] After much further discussion between Rüdiger, David Mermin, and *the Vatican* (!) – well, actually a professor in Rome who had organized a conference in the Vatican once – it was finally decided that this *really* is the appropriate translation of "Quantum Mechanics is a Law of Thought."

15 October 1996, "Mechanica Quantica"

I'm kind of confused about what you're trying to get at. See my notes below, and try it out on me one more time please. (I realize my notes are pretty muddled in comparison to your one-sentence formulation ... so I hope you have a little patience with me.)

Schackcosm 1: *I have read Hartle's 1968 paper and was quite confused at first. After some thinking, however, I have come to the conclusion that the finite-N part of Hartle's analysis supports my understanding of what makes quantum probabilities special. Here it is in one sentence:*

The knowledge of how to prepare the pure state $|\psi\rangle$ implies the knowledge of how to prepare the product state $|\psi\rangle \otimes |\psi\rangle \otimes \cdots \otimes |\psi\rangle$.

Or:

If one can prepare a pure state, one can prepare any number of independent copies of it.

How is this any different than my being able to prepare N independent (but identical) probability distributions? For instance, what is wrong with saying, "The knowledge of how to prepare a probability distribution $p(x)$ implies the knowledge of how to prepare the product distribution $p(x_1)p(x_2)p(x_3) \cdots$." There are certainly cases where I can do that, and they are purely classical by nature: I could produce 10,000 identical 75/25-weighted coins, flip each of them independently and place them in sealed envelopes (without looking at the outcomes). That creates for me the situation you describe.[2] (Though, granted, it's not doing it for states of maximal knowledge.) Alternatively, in the quantum mechanical setting, I could prepare 10,000 identical mixed states by repeatedly preparing a pure state on a bipartite system and throwing one half of it away.

What is it that makes *pure* quantum states unique in your formulation? And what makes it particularly quantum mechanical in the first place? I'm probably just missing something.

I understand that there are classical situations where I cannot conjure up the situation you describe. For instance, I can give a probability assignment to tomorrow's weather, but I cannot repeat the preparation. But isn't that beside the point?

It sort of seems to me that Carl hit on the crucial point: any "scientific" theory must be capable of dealing with repeatable phenomena. Thus any "scientific" enterprise must be, of its nature, capable of extrapolating from "knowledge of how to prepare" the individual instance to "knowledge of how to prepare" any number of instances.

Schackcosm 2: *I know you have a strong dislike for the word "objective" (Yuck!). Yet it may not be completely misleading to call a quantum pure-state preparation procedure*

---

[2] I didn't understand it at the time, but Rüdiger had hit upon one of the key distinctions between the classical and quantum cases. The two situations are not the same. In the classical case, one is using a nontrivial probability distribution precisely because one is admitting that maximal information is not at hand. Therefore, for just this reason, one would never assign an i.i.d. distribution to the multi-trial space.

"objective." The preparation of a classical system, however, should be called "objective" only if it leads to a completely known state. Otherwise such a preparation always relies on some environment being in an unknown state.

I don't know that I have an opinion on this comment. Does it connect to the previous part of the note?

As a side note, though what about a universal computer built on Toffoli's billiard ball model (for instance). I might start it in a precise configuration that will eventually lead to a proof one way or the other of Riemann's conjecture. I don't know what the answer will be, so – even without environments – I can be in a situation of making probability assignments for the configurations of classical systems. Does this have any impact?

09 August 1997, "Correspondentia"

NOTE: This letter was written with the intent of ultimately being sent to Herb Bernstein. However, the note was *never finished*. During a visit to England soon after the date above, I gave Rüdiger Schack a copy of the unfinished note. This explains my reason for placing the text in this slot.

> "Just as the sexual drive frequently transforms man into a monster, so the elementary category of causality can assume the character of a need, an insatiable craving which overruns everything, and which people will even sacrifice their lives to gratify. It is an indefatigable longing which inflames us ..."
>
> (*Carl Jung, The Zofingia Lectures*)

> "As philosophers we may well find the concept of objective chance troublesome, but that is no excuse to deny its existence, its legitimacy, or its indispensability. If we can't understand it, so much the worse for us."
>
> (*David Lewis, A Subjectivist's Guide to Objective Chance*)

Hey Herb,

It's been a good two weeks since I've talked to you ... how are things? Actually, I was looking for a little philosophical companionship this afternoon; I hope you don't mind my picking on you! For a while I've been wanting to expand on some *silly* thoughts I've been thinking. Maybe you'll be the one to have a sympathetic ear to this stuff. Let me break you in gently, by first repeating two very, very lengthy quotations. I'm sure you'll be able to ferret out the common theme. The first comes from a letter I wrote Greg Comer in 1992 (itself containing lengthy quotations of Pauli). The second is a passage from one of Arthur Fine's articles. (By the way, the latter appears before you with the help of modern technology: a good scanner and a character recognition program.)

---

28 April 1992 (to Greg Comer)

... As per your request, I'll use this as an opportunity to repeat the quote I wrote you the other day. On 29 December 1947, Pauli wrote Fierz:

I'm more and more expecting a further revolutionizing of the basic concepts in physics. In connection with this particularly the manner in which the space–time continuum is currently introduced into it appears to me to be increasingly unsatisfactory. ... Something only really happens when an observation is being made, and in conjunction with which, as Bohr and Stern have finally convinced me, entropy necessarily increases. Between the observations nothing at all happens, only time has, "in the interval," irreversibly progressed on the mathematical papers.

Now I think even more important to me personally is another long quote of Pauli's from a letter to Fierz dated 26 November 1949. For typing practice (ha!), I'll relate this to you in a second. First let me tell you a little about the origin of my interest in this randomness business. In 1986, while writing a paper for a philosophy course I was taking, I used a phrase something like: each individual quantum mechanical measurement outcome is completely undetermined and this indeterminism is NOT due to our ignorance of the true picture of things (as would be the case if a hidden variable theory existed). Well, after the paper was turned in, that simple phrase (or something like it) led me to start wondering: "What could it possibly mean to be 'completely undetermined'?" and "Why is it that if each individual outcome is completely undetermined, nevertheless, in the long run, the outcome statistics eventually settle down to a predictable mean, etc.?" You see I could have easily answered the latter question if there were a hidden variable theory; for then the quantum mechanical probabilities would just represent our ignorance of the actual situation – a situation that would explain one set of frequencies over another. But in the absence of such an external controlling factor how is it that a "completely undetermined" individual outcome also knows something about all the other outcomes that could come about in repeated trials of the same experiment? Doesn't this question hint at a contradiction in terms? It was, in fact, exactly these sorts of questions that eventually led me to identify "the wave function" with a random string rather than sticking with the standard thought of identifying "the wave function" with the "state of the system." Well, anyway, strangely enough, until yesterday I had never seen these questions I raised appear in print. And this is where Pauli's quote comes into the picture.

Bohr's expression "correspondence" served as an aid to me when, then, I was trying to give a name to the positive principle which lies at the basis of quantum mechanics. (*After* putting forward wave mechanics he continues to speak of a "correspondence argument" see *Naturw.* **21**, 245–250, 1933, particularly the passage on page 246, top of the second column.) The statistical behavior of many similar individual systems which have no contact whatsoever with one another ("windowless monads"), without, on the other hand, being causally determined, has, of course, in quantum mechanics been interpreted as the *last* law-governed fact which cannot be further reduced (approximately as was the case for Galileo with respect to uniformly accelerated falling bodies). In my lecture on complementarity, originally published in the journal *Experientia* and now available as an offprint, I thus tried to use the expression "correspondence" in a more general sense than Bohr had, in a way which would specifically characterize the positive side of a quantum mechanical description of nature. It is certainly this *statistical correspondence* which mediates between continuum (wave image) and discontinuum (particle image). (This in a somewhat more general way than the mediation between "quantum theory" and "classical theory" in Bohr's writings). There I did *not* explicitly state that for me the intellectual derivative of the "correspondentia" of the Middle Ages ("correlations") clearly seems to glimmer through in the term "correspondence." In both cases, however, we are concerned with a form of describing nature in terms of laws which transcend normal causality and which are based on some kind of analogy. (This is also the case with Leibniz's prestabilized harmony.)

The single systems of quantum mechanics are "windowless monads" and, nevertheless, the correct fraction can always be found which reacts as calculated (apart, naturally, from the expected statistical fluctuations).

The tertium comparationis of the quantum mechanical case with that of the synchronistic phenomenon is the *mutually tuned behavior of different events* (not bound in a deterministic-causal sense). (It is on this, of course, that the concept of "physical situation" in quantum mechanics rests, not on a direct mutual influence of the objects in question.)

The quantum mechanical situation is naturally not only a degeneration of the more general "synchronicity" (this to be understood as a working hypothesis suggested here), but *also* a "rational generalization" of normal deterministic causality (Bohr). When the fraction is one (instead of between zero and one), that is, of course, a special borderline case, as, indeed, the old deterministic causality seems when observed from the standpoint of quantum mechanics. I have no doubts that the quantum mechanical "statistical correspondence" lies much closer on the side of old determinism than on the side of the synchronicity phenomenon. Observed from the standpoint of this phenomenon quantum mechanics must appear to be a *very weak* generalization of the old causality. And, nevertheless, quantum mechanics seems to me also to have that road sign towards the other direction, towards the one where it is no longer possible to speak of arbitrary reproducibility at all. To me quantum mechanics seems to occupy a kind of intermediate place.

This is my momentary view .... The *success* of the "reasonable belief" – and with it also the possibility of laws of nature – appears to me *always* to rest on an archetypically conditioned coincidence of our expectation (psychologically) with an external natural occurrence (physically). For the abstract arranger there is just *not* any actual difference between "physical–psychological."

Read this about five times and you'll start to see the connections. Long before randomness came around, I studied Leibniz and his monads for just this same reason. The words "synchronicity" and "archetype" in the quote above refer to the concepts used by Jung. You are probably more familiar with this set of concepts than I am – seeing as Joseph Campbell is a fan of Jung.

From Arthur Fine, "Do Correlations Need To Be Explained," in *Philosophical Consequences of Quantum Theory: Reflections on Bell's Theorems*, edited by James T. Sherman and Ernan McMullin (University of Notre Dame Press, Notre Dame, Indiana, 1989).

If we adopt an indeterminist attitude to the outcomes of a single, repeated measurement, we see each outcome as undetermined by any factors whatsoever. Nevertheless, we are comfortable with the idea that, as the measurements go on, the outcomes will satisfy a strict probabilistic law. For instance, they may be half positive and half negative. How does this happen? What makes a long run of positives, for example, get balanced off by the accumulation of nearly the very same number of negatives? If each outcome is really undetermined, how can we get any strict probabilistic order? Such questions can seem acute, deriving their urgency from the apparent necessity to provide an explanation for the strict order of the pattern, and the background indeterminist premise according to which there seems to be nothing available on which to base an explanation. If one accepts the explanationist challenge, then one might be inclined to talk of a "hidden hand" that guides the outcome pattern, or its modern reincarnation as objective, probability-fixing "propensities."

This talk lets us off the hook, and it is instructive to note just how easily this is accomplished. For if propensities were regular explanatory entities, we would be inclined not just to investigate their formal features and conceptual links, but we would make them the object of physical theorizing and experimental investigation as well. However, even among the devotees of propensities, few have been willing to go that far. The reason, I would suggest, is this. Once we accept the premise of indeterminism, we open up the idea that sequences of individually undetermined events can nevertheless display strict probabilistic patterns. When we go on to wed indeterminism to a rich probabilistic theory, like the quantum theory, we expect the theory to fill in the details of under what circumstances particular probabilistic patterns will arise. The state/observable formalism of the quantum theory, as is well known, discharges this expectation admirably. Thus indeterminism opens up a space of possibilities. It makes room for the quantum theory to work. The theory specifies the circumstances under which patterns of outcomes will arise and which particular ones to expect. It simply bypasses the question of how any patterns could arise out of undetermined events, in effect presupposing that this possibility just is among the natural order of things. In this regard, the quantum theory functions exactly like any other, embodying and taking for granted what Stephen Toulmin (1961) has nicely called "ideals of natural order." What then of correlations?

Correlations are just probabilistic patterns between two sequences of events. If we treat the individual events as undetermined and withdraw the burden of explaining why a pattern arises for each of the two sequences, why not adopt the same attitude toward the emerging pattern between the pairs of outcomes, the pattern that constitutes the correlation? Why, from an indeterminist perspective, should the fact that there is a pattern between random sequences require any more explaining than the fact that there is a pattern internal to the sequences themselves?

We have learned that it is not necessary to see a connection linking the random events in a sequence, some influence from one event to another that sustains the overall pattern. Why require a connection linking the pairs of events between the sequences, perhaps some influence that travels from one event in a pair to another (maybe even faster than the speed of light) and sustains the correlation? We have explored part of the answer above. Our experience with correlations that arise in a context in which there generally are outcome-fixing circumstances has led us to expect that where correlations are not coincidental, we will be able to understand how they were generated either via causal influences from one variable to another or by means of a network of common background causal factors. The tangled correlations of the quantum theory, however, cannot be so explained.

The search for "influences" or for common causes is an enterprise external to the quantum theory. It is a project that stands on the outside and asks whether we can supplement the theory in such a way as to satisfy certain a priori demands on explanatory adequacy. Among these demands is that stable correlations require explaining, that there must be some detailed account for how they are built up, or sustained, over time and space. In the face of this demand, the tangled correlations of the quantum theory can seem anomalous, even mysterious. But this demand represents an explanatory ideal rooted outside the quantum theory, one learned and taught in the context of a different kind of physical thinking. It is like the ideal that was passed on in the dynamical tradition from Aristotle to Newton, that motion as such requires explanation. As in the passing of that ideal, we can learn from successful practice that progress in physical thinking may occur precisely when we give up the demand for explanation, and shift to a new conception of the natural order. This is never an easy operation, and it is always accompanied by resistance and some sense of a lost paradise of reason. If we are to be serious about the science that we now have, however, we should step inside and see what ideals it embodies.

The quantum theory takes for granted not only that sequences of individually undetermined events may show strict overall patterns, it also takes for granted that such patterns may arise between the matched events in two such sequences. From the perspective of the quantum theory, this is neither surprising nor puzzling. It is the normal and ordinary state of affairs. This ideal is integral to the indeterminism that one accepts, if one accepts the theory. There was a time when we did not know this, when the question of whether the theory was truly indeterminist at all was alive and subject to real conjecture. Foundational work over the past fifty years, however, has pretty much settled that issue (although, of course, never beyond any doubt). The more recent work related to EPR and the Bell theorem has shown, specifically (although again, not beyond all doubt), that the correlations too are fundamental and irreducible, so that the indeterminist ideal extends to them as well. It is time, I think, to accept the ideals of order required by the theory. It is time to see patterns between sequences as part of the same natural order as patterns internal to the sequences themselves.

A nonessentialist attitude toward explanation can help us make this transition, for it leads us to accept that what requires explanation is a function of the context of inquiry. So when we take quantum theory and its practice as our context, then we expect to look to it to see what must be explained. This leads us to the indeterminist ideal discussed above, and to the "naturalness" of (even distant) correlations. There is a small bonus to reap if we shift our thinking in this direction. For the shift amounts to taking the correlations of the theory as givens not in further need of explanation and using them as the background resources for doing other scientific work. One thing they can do is to help us understand why the theory has correlational gaps. From the very beginning, one wondered about the incompatible observables and why one could not even in principle imagine joint measurements for them. After all, as Schrödinger (I believe) first pointed out, in the EPR situation, one could measure position in one wing and momentum in the other and, via the conservation laws, attribute simultaneous position and momentum in both wings.

The conventional response here has been to point out that only the direct measurements yield values that are predictively useful. Not everyone has been happy with the positivism that seems built into this response. But if we recall the discussion in section 1, then we see that (at least in part) there is a better response at hand. For we have seen how the correlations that the theory does provide actually exclude the possibility that there could be any stable joint distributions for incompatible observables in those states where the correlations are tangled. This shows us that there is no way of augmenting the theory with values for incompatible observables, and distributions for those, that would follow the same lawlike patterns as do the distributions of the theory itself. To put it dramatically, the shadow of the given correlations for compatible observables makes it impossible to grow stable correlations for the incompatibles. There is a sense, then, in which there would be no point in trying to introduce more for incompatible observables than what the theory already provides.

This way of thinking turns the Bell theorem around. Instead of aiming to demonstrate some limitation or anomaly about the theory, this way proceeds in the other direction and helps us understand why the probability structure of the theory is what it is. That understanding comes about when we take a nonessentialist attitude toward explanation, letting the indeterminist ideals of the theory set the explanatory agenda. Such an attitude means taking the theoretical givens seriously, and trusting that they will do good explanatory work. Thus, in the Bell situation, we shift our perspective and use the given quantum correlations (and the simple sort of counting argument rehearsed in section 1) to explain why, even in principle, correlations forbidden by the theory cannot arise. Nonessentialism leads us to engage with our theories seriously, and in detail. In the end, that is how better understanding comes about.

What then of nonlocality, influences, dependencies, passions, and the like, all diagnosed from correlational data? As one good statistician remarked about the similar move from linear regression to causal connection, and as we have seen demonstrated above, "Much less is true."

What do you think of Fine's writing? Isn't it nice and clear? In any case, what I'd like to do in this note is focus mostly on the common thread between the two passages. Namely I want to talk about how quantum phenomena are like "windowless monads." I may come back to Fine's stuff on comparing correlations in EPR-type phenomena to the fundamental assumption of indeterminism, but that won't be the main point of this discussion.

Did you understand my point in the letter to Greg about how one *might* find the existence of a limiting frequency for measurement outcomes in quantum mechanics mysterious? I'm not so sure I said it as clearly as I would have liked to. So perhaps I'll try it one more time ... just to see if I can explain it to *my* satisfaction. (I realize that Fine may have already clarified this to *your* satisfaction with his discussion, but indulge me: I like to beat my potatoes until they're really mushy.)

The point is this. Despite my long-running obsession with and love of quantum mechanics and indeterminism, I have never been completely at ease with the idea of "probabilistic causality." It's there ... and I know that ... and I *like* that ... but I've never felt completely at grips with it. That is to say, I've never been completely at ease with the "hidden hand" that Fine speaks of. The problem seems to be that my mind always tries to slip into the "point of view" of the individual event.

Imagine the following. I have a friend Hideo who is singularly obsessed with preparing his favorite quantum system in the state $|\psi\rangle$ as cleanly and as quickly as he can. His purpose is to make a huge number of copies of $|\psi\rangle$ so that our mutual friend Quentin can test various quantum mechanical statements. In particular, suppose Quentin worries about what he should expect of a measurement corresponding to the projector $|\phi\rangle\langle\phi|$. (Quentin, being just as conscientious as Hideo, is careful to perform his measurements just as cleanly as he can.) So what is it that Quentin can expect? The standard dogma of quantum mechanics gives us two things:

1. The probability of a "yes" answer on a single trial of the measurement is $|\langle\phi|\psi\rangle|^2$.
2. Beyond that probability, the outcome of the measurement is completely unpredictable and, in fact, undetermined.

From a "global" point of view, one might say that there's nothing out of synch about these two statements. This should be especially reasonable given the phrasing I used in item 2 above – namely, "Beyond that probability ... blah, blah, blah."

17 August 1997, "New Babies of Quantum Information"

Very many congratulations on the new baby! I'm glad to hear that both mother and daughter (and presumably the whole family) are doing well. This is quite a family you and Dorothee are building.

It would be nice if there were a generally-agreed-upon definition of "quantum information," but I don't think there is one presently. The problem is the following. A large portion of the community, like the Schumacherites, might say, "Quantum information theory is the subject of quantifying the resources required to transmit intact nonorthogonal quantum states and entanglement through noisy channels." Another portion of the community, like the Holevoites, might say, "Quantum information theory is the subject of quantifying the maximal rate at which classical bits can be sent over quantum mechanical channels." (These, by the way, are the same people that worry about how distinguishable nonorthogonal states are.) Still another portion of the community, like the Caves–Schackites, might say, "Quantum information concerns itself with the various ways in which Hilbert space is big in comparison to classical state spaces." And, I guess I shouldn't forget about the part of the community that worries about the efficient processing of classically defined questions – the quantum computing clique, the Ekertites. They would probably have their own way of saying things.

The truth of the matter is that somehow one needs a definition that encompasses all these ideas, i.e., a reason behind why we all see each other at the same conferences. It can be done ... you just string all the things above together ... but I guess it wouldn't come out too pithily. Good luck. I don't know about a standard reference that emphasizes all the aspects – one should be written. Maybe a good overall reference would be Bennett's *Physics Today* article (last year or the year before last). Other than that I really can't say. Carl and I tried to say something of a summary of "what quantum information is" in our Rosen festschrift, but the things we said weren't nearly all-purpose enough.

29 September 1997, "The Hidden Hand"

Anyway, on a connected subject, I've spent a little time this week reacquainting myself with what has been written on propensities. I'm not yet finished with the project, but maybe I'll go ahead and record some things.

First off, I think it would be worthwhile for you to take a look at the Giere (1973) paper cited in Howson and Urbach (and in the paper Carl and I wrote). If the guy should be damned for not being a subjectivist, he shouldn't be damned for anything too extreme. He is certainly no lover of frequentism. In fact, even now – upon another reading – I think he only differs from us substantially in the argument Carl and I noted (and rejected):

... Having withdrawn earlier objections to Laplace's view, he [Mill] concludes: "Every event is in itself certain, not probable; if we knew all, we should either know positively that it will happen or positively that it will not. But its probability to us means the degree of expectation of its occurrence which we are warranted in entertaining by our present evidence." The role of the assumption of determinism in this argument is clear enough. One will not find the above argument in the writings of Laplace's contemporary heirs, the subjective Bayesians, but they are tacitly committed to a similar position. The most prominent personalists, e.g., Savage and de Finetti, insist that there is only one legitimate concept of probability, that which identifies probability with subjective uncertainty. Once this identification is made, however, one lacks the conceptual apparatus to distinguish uncertainty due to lack of information from uncertainty which no physically possible increase in present knowledge could eliminate. But this is just the distinction between physical determinism and physical indeterminism. Not being able to make the distinction, Bayesians are forced to assume that all uncertainty is due to lack of information, i.e., to assume determinism. Indeed, to admit the possibility of uncertainty not due to lack of information would be to admit the possibility of physical, i.e., nonsubjective, probabilities – an admission personalists refuse to make.

You'll probably remember that Howson and Urbach were pretty rough on the guy. They wrote: "The second objection is more fundamental and seems to be unanswerable. Von Mises's theory may seem stubbornly deficient in empirical content, but the present account is, if anything, even worse. For Giere's single-case propensity theory conveys no information of any sort about observable phenomena, not even, we are told, about relative frequencies in the limit. ... " Well, if that's the view they take of him, then it'll most certainly be the view they take of us too. I say this because Giere takes a stand pretty close to us:

A preliminary point worth emphasizing is that a single-case propensity interpretation provides an extremely natural understanding of the standard formalism for dealing with series of trials. ... Thus, for example, a series of trials is conventionally represented by a sequence of random variables [i.i.d., presumably] ... On my interpretation, the density function gives the propensity distribution on the ith trial. It is as simple as that. There is no need for vague talk about "virtual" infinite sequences, etc. Independence of trials just means that there is no causal connection between the outcome of one trial and the propensity distribution of any other trial. [We would at this point say that maximal information has allowed us identical preparations instead of the no causal connection thing.] ... A central question concerning the relation between single-case propensities and frequencies is whether it is possible to *deduce* values of one from values of the other. The answer, as one would expect since propensities are theoretical, is negative. Consider a series of Bernoullian trials in which the propensity for success is r on each trial. Let f_n be the relative frequency of successes after n trials. In this case the strongest connection one can establish between r and f_n is given by the Bernoulli theorem, namely: [he writes the standard weak law of large numbers], where P measures the propensity of the compound trial which consists of n trials of the original chance setup. It does *not* follow that the sequence of values of f_n has limit r in the ordinary sense [of calculus, that is]. Thus the sequence need not be a [collective] in the sense of von Mises or Reichenbach. Indeed, it is *logically* possible, for example, that $r = 1/2$ and that $f_n = 1$ for any n, although of course the propensity for this outcome approaches zero. The limiting frequency interpretation rules out all such possibilities by convention.

Note his emphasis on "deduce" and "logically". If Howson and Urbach read his paper carefully – and it doesn't look like they did – that should be the only lack of connection between propensities and frequencies that they are criticizing.

Anyway, taking a look at all this I am again struck by how similar it all is. We say "maximal information cannot be completed" where Giere says "propensity" and vice versa (I think). Our main point of departure from Giere is most likely in viewing a distinction between ignorance and objective chance as dangerous (when it comes to mixed-state ensembles for instance). So I think we need to flesh that out more carefully.

I'd like to get hold of Giere's 1976 paper, but it's not in the library here. So it'll probably have to wait until I'm at UCLA Friday – maybe it's in their library.

Let's see, let me say a couple of more things. I also read a paper "Single-Case Probabilities" by David Miller (*Found. Phys.* **21** (1991) 1501–1516. It's not really worth reading, don't worry about it. But I did find one thing interesting in it about the "hidden hand." Miller writes: "A second but not less urgent problem confronted by the propensity interpretation is *the fundamental problem of the theory of chance* – the explanation of objective statistical stability. I shall not discuss this problem here, except to remark that it seems to be well beyond the resources of the subjectivist theory of probability." He cites Popper's Logic of Scientific Discovery (Sec. 49) for the "fundamental problem." Thus the "hidden hand" goes back at least that far! (By the way, it might be fun and/or useful to look at Popper in those pages in a little more detail – I haven't yet – if for no other reason because Popper has had such an inordinate influence on so many.)

The other thing is that I invested a little time in van Fraassen (his book *The Scientific Image*). One thing that I did learn is that the frequentist conception of probability is more distinct from Kolmogorov probability than in just its interpretation. van Fraassen goes to some length to explain that the domain of the "relative frequency function" (that thing which will eventually be labelled "probability" by the frequentists) is not a Borel field. Then he cites de Finetti and Suppes (independently) as having a more complicated argument that shows that it is not even a field. I don't know what all this is worth, but it did seem interesting. Then van Fraassen went on to give his interpretation – "the modal frequency interpretation" – of probability, but I got pretty lost. His main point seems to be that, though a frequency interpretation is useless empirically and can't stand on its own, one can construct a *model* for the world that takes ideas from the old frequentist camp. And that model, he thinks, has some desirable properties. I doubt there's anything to this ... but his name is huge in the philosophy of science since way back in the 70s and he does have a major book on quantum mechanics. So maybe it's at least worth understanding the flavor of his interpretation.

02 October 1997, "Wake Up Call"

I haven't heard from you in a while. You did get my last long letter, right? This morning I think I'll send you a little gift: a passage from J. L. Heilbron's *The Dilemmas of an Upright Man: Max Planck as Spokesman for German Science* (pp. 127–128). I very much enjoyed this because (I think) it led me to the path of thinking about the connections – whether real or not – between "maximal information not being complete" and "introspection."

Although Planck's reconciliation of free will and determinism had been anticipated by others, notably William James, and so was familiar to philosophers, it turns on an argument perhaps too compressed for easy understanding. Planck later set it out clearly and distinctly in his correspondence with Theodor Haering, Professor of Philosophy at the University of Tubingen, whose enlightenment demanded six long letters. First Planck explained that determinism makes sense only when the data on which prediction rests can be obtained without influencing or changing the system under study. "This is the basic presupposition of any sort of scientific knowledge." It does not hold for self-analysis. Hence one's will cannot be a subject of science for one's self, and the question of the will's determination makes no sense to the willing individual. Haering replied that he could not understand why, if scientific determinism holds generally, it does not apply also to "the I"; nor did he perceive why the self must be disturbed in self-examination.

Planck insisted that a disturbance must occur: self-knowledge is a conscious experience; every such experience implies a change of mental state; but self-knowledge has as its object the mental state; therefore self-knowing of the knowing-self is impossible. This was not enough to complete Haering's understanding. It appeared that the philosopher had confused the "phenomenological I ('Me')" with the "actual I ('I')"; the latter can know the former, but not itself. "Therefore," Planck concluded, having reduced his argument, he thought, to the level of his grandchildren, "determinism can never invalidate freedom of the will." The philosopher still had his difficulties: he did not comprehend, he said, how observing an object could change it. Planck replied that observing does not change objects, but subjects; it happens that in the case under investigation, the subject, the knowing I, is identical with the object, the willing I. That did the trick. Haering grasped the argument and Planck the difficulty of discourse with philosophers.

Oh, another paper you may wish to look at: Adrian Kent, "Against Many-Worlds" on the LANL server, `gr-qc/9703089`. Andrew Whitaker brought it to my attention.

06 October 1997, "Germinal Jewel"

Pay no attention to [. . .]; I read the draft tonight while sitting at a coffee shop and am about as pleased as I was two weeks ago when we had our nice conversations. I think there is something of real substance to this "principle of quantum indeterminism" (PQI) – it somehow just pinpoints the right thing. The fact that it gives such a simple, believable justification to the Law of Large Numbers already sets it apart . . . as you show. However, the draft does make me feel like maybe we're closer to the beginning of the road than the end – it'll take me a few days to put my finger on it all. Mainly, don't lose faith in the meantime. It may be just a case of saying things in a slightly different order, and with a little more chutzpah. For instance, I still don't feel that the PQI was stated with sufficient force and in such a way that its necessity will become as pleasing to the reader as we see it. This'll get sorted out; this is the germ of something good.

Let me leave you with a quote from Mermin's 1983 book review of Popper's "Postscript." In it, you can see the germ of *his* present views:

Physicists who have come to take their subject for granted would do well to suppress their aversion to rubbish and read these volumes, both to refresh their awareness of how bizarre their subject has become, and to test their own grasp of its foundations against Popper's view that what is most marvelously intricate and subtle in the behavior of the atomic world is just a mystery and horror to be dispelled by some clear thinking about probability.

07 October 1997, "Two Forgotten Questions"

I forgot to ask you two things last night. The first is, when do you start lecturing? The second is: you cite Percival in the second paragraph for criticism of the assumption that "there exists, in principle, a completely specified preparation procedure for an arbitrary (pure or mixed) quantum state." What is that paper? Can you give me its coordinates?

Last night I did a little more reading on propensity theories – I've been trying to get a handle on what the heck such things are supposed to be. Not much luck so far; with some good fortune, perhaps I can give you a report on my understanding this weekend (along with many further comments on the draft). One thing of note, though, I did reread Feynman's old paper "The Concept of Probability in Quantum Mechanics" (1951). It's not of much use to us, but there is maybe one passage that summarizes his views (as far as they were developed):

I should say, that in spite of the implication of the title of this talk the concept of probability is not altered in quantum mechanics. When I say the probability of a certain outcome of an experiment is p, I mean the conventional thing, that is, if the experiment is repeated many times one expects that the fraction of those which give the outcome in question is roughly p. I will not be at all concerned with analyzing or defining this concept in more detail, for no departure from the concept used in classical statistics is required.

What is changed, and changed radically, is the method of calculating probabilities. . . .

12 October 1997, "The Deep End"

I feel like I've fallen off the deep end: the weekend has now passed and I still have no detailed comments for you. How many apologies will you be able to take before they start seeming meaningless???

Anyway, I'm finding myself doing exactly what I warn the students around me not to do! I can't stop reading and bring myself to action!!! I keep reading about propensities, trying to see what the (careful) propensists think different about their concept than you with your principle of indeterminism. The differences keep seeming smaller to me. Also

I've taken some time to read Jaynes a little better, for instance his Chapter 18 on the A_p distribution.

I guess what I'm trying to do is find a way of expressing your principle of quantum indeterminism in a way that I'm happy with ... so that I might take my own shot at pith. (At least as far as that detail is concerned.) I hope that doesn't take much longer.

On a different note, reading Jaynes has made me think of the following thing. I wonder if we might make use of the example Jaynes does in his two subsections on "Laplace's rule of succession" (pp. 1807–1811). In particular, it would be nice to derive a quantum analog of Eq. (18.22). Namely, where Jaynes' assumes that "the underlying 'causal mechanism' is assumed the same at every trial," we might assume that the same (unknown) quantum state is prepared and the same measurement is performed. Then, putting a uniform (unitarily invariant) distribution on the set of (pure) quantum states, we could go through the same steps as he. Now, the wonderful thing is ... I think ... that nothing whatsoever changes, though it certainly had every right to. This is because of Sýkora's old result that uniform distribution on quantum states corresponds to an induced uniform distribution on the probability simplex. That is, as long as the Hilbert space is complex and not real or quaternionic.

21 November 1997, "Deadbeats for Quantum Reality"

May I ask a favor of you? Would you read the statements below, and tell me whether you think they express the issue fairly accurately. So that someone who hasn't seen MaxEnt before won't be misled, but on the other hand won't be bowled over with details either.

I'll send you the finished product (i.e., the full proposal) in a couple days for your enjoyment.

_____

The year 1957 is significant in physical thought because it marks the penetration of Information Theory into physics in a systematic way – into statistical mechanics in particular. This refers to the *Maximum Entropy* or "MaxEnt" program for statistical mechanics set into motion by E. T. Jaynes. With the tools of Information Theory, one was able for the first time to make a clean separation between the purely *statistical* and the purely *physical* aspects of the subject matter.

Perhaps it would be good to present a mild example of this. Because of MaxEnt, a standard statistical mechanical ensemble, like the canonical ensemble, can finally be seen for what it really is: an expression of the physicist's *state of knowledge* (specified, of course, by the experimental parameters under his control). Though this reveals a subjective aspect for statistical mechanics, the ensemble is not arbitrary: Two physicists working on a single experiment, if true to their states of knowledge, will derive the same distributions for the system's variables. The *structure* of the canonical distribution, with its exponential form, is due purely to the kind of knowledge the experimenter possesses – in this particular case, the expectation value of some observable and nothing else. That is to say, the canonical distribution's form is a theorem of the laws of inference, *not physics*. The physics of the system rests solely in its Hamiltonian and boundary conditions. Thinking that the canonical distribution somehow comes out of physics alone, without reference to what one knows, is to greatly misunderstand the content of statistical mechanics.

14 January 1998, "Practice and Perfection"

Anyway, not that it'll help my standing – it'll probably only hurt – but I offer a little bathroom reading for your contemplation. It takes the form of a letter I wrote David Mermin. It carries the beginning of my new slogan "Bohr was a Bayesian." (In fact a Bayesian good enough to know that Bayes' rule for conditionalizing isn't always valid ... that is, especially when one is confronted with the situation where one's information gathering measurements *necessarily* disturb someone else's predictability.) I hope you enjoy, especially the phraseology that starts up right around Merminition 12.

13 March 1998, "Maximal Information and Schrödinger"

You'll be happy to know that I am jumping into the project today. Wish me sweet luck ... I may need it.

In complement to that, I wonder if you would be interested in looking at some old archival material. I've come across what I think is a remarkable exchange of letters between Pauli, Heisenberg, and Schrödinger in the wake of the EPR paper (June–July 1935). (Steven van Enk has done enough of a translation for me to come to this opinion.) In particular, Schrödinger goes to pains to say things about how in standard quantum theory, pure states correspond to "maximal knowledge." He also seems to stress that, in transferring from one state to another, one gains knowledge but also loses knowledge: the level of knowledge stays maximal, but the particulars of it are about different "things."

Also – though not too connected with our present project – there seem to be some nice remarks by Pauli on how "the separation and combination of systems" should be better understood as a "foundation of quantum theory." And a small (unpublished and never-translated) paper by Heisenberg is included that makes a little more of those remarks.

Would you be willing to read over these and write a summary about their essential content ... especially as it relates to our paper? It'd be even better if you could put a translation onto tape as you did once before for me, but I know that that's asking for a lot of work. Any enthusiasm? The letters are number 412, 413, 414, and 415 in *Wolfgang Pauli: Scientific Correspondence with Bohr, Einstein, Heisenberg a.o. Vol. 2*, edited by Karl von Meyenn (Springer, 1985). If you can't find the book in your library, I could fax you the pages.

Rüdiger's First Reply

The letter exchange is indeed very interesting. I made quite extensive translations, which are a little rough in places. In case you want to cite anything, I'd like to go over it again.

Heisenberg's paper doesn't seem to be relevant for what we are doing: it's all about the impossibility of hidden variables.

Schrödinger is very interesting. He is very hard to translate: he is obviously struggling with the language, trying to formulate deep thoughts in an inadequate language. Pauli, by contrast, is easy to read and easy to translate. That does not mean he is less deep, of course.

I greatly enjoyed Pauli's insults in the first letter.

The most interesting passage is in letter 413. I am not certain though that I do understand it well.

YOU SHOULD LEARN GERMAN!

Letter 412 [Pauli to Heisenberg]
p402

1. Einstein has once more made public remarks on quantum mechanics, in *Phys Rev* of May 15 (with Podolsky and Rosen – no good company btw). As is well known, this always amounts to a catastrophe when it happens. "His conclusion, razor-sharp – that cannot be which should not be" (Morgenstern).

Admittedly I would like to grant him that, should an undergraduate advance objections of this kind, I would consider him quite intelligent and promising. – Because this publication causes a certain danger of public confusion – especially in America – it might be good to send a comment to *Phys Rev*, and I'd like to convince YOU to do it.

Now follows a description of the EPR setup and an Einstein quote. Then, bottom of p403, he says that a pedagogical reply should clarify the difference between

a) The systems 1 and 2 do not interact (= absence of interaction energy)

and

b) The total system is in a state in which subsystems 1 and 2 are independent. (The eigenfunction can be written as a product.)

Definition: This is the case if after a measurement on 2 of an arbitrary quantity F_2 with known result $F_2 = (F_2)_0$ (number) the expectation values of the quantities F_1 stay the same as without the measurement on 2.

Independently of Einstein it seems to me that in a systematic foundation of quantum mechanics one should, to a higher degree, *begin* with the composition and separation of systems, than has been done so far (e.g., by Dirac). ....

One must distinguish different layers of reality: One R, containing all information which can be obtained from measurements on 1 and 2, and one (deducible from R) r, which contains only the information which can be obtained from measurements on 1 alone. One must then show how announcing a measurement result on 2 implies discontinuous changes of r ($r \rightarrow r_A$ or $r \rightarrow r_B$ etc.) (unless the subsystems [Teilchensysteme must be a typo: this means "particle systems", whatever that means. I am sure it should read Teilsysteme, meaning "subsystems"] are independent). And that one is led to contradictions if one tries to explain these changes, without reference to 2, classically or half-classically – e.g., as "hidden properties" of 1.

Elderly Herren like Laue and Einstein are haunted by the idea that qm is correct but incomplete. They think qm can be completed by statements which are not part of qm, without changing the statements which are part of qm. (A theory with such a property I denote – in the logical sense – incomplete. Example: the kinetic gas theory). Maybe you could – in the reply to Einstein – clarify with authority that such a completion of qm is impossible without changing its content.

NB. It is probably only because I recently got an invitation to Princeton for the next winter semester that I have gone to such trouble on these things, which for us are just trivialities. It will be a lot of fun to go there. In any case, I want then to make the Morgenstern motto popular.

The rest of the letter comments on the magnetic moment of the proton.

Letter 413 [Schrödinger to Pauli]
3rd paragraph:

I'd like to know very much if you agree with the following version of the case, behind which I can certainly not go back.

It is well known that there are preparative measurement methods through which one can make a system maximally known, i.e., through which one can transform it into a "pure case." I claim: there are preparative methods for which, after the final isolation of the system being prepared, the experimenter is still free to continue the method either in such a way that a pure case of type A results with certainty, or in such a way that a pure case of type B results with certainty.

So far you agree? Yes? You permit me also to call the pure case a "state" and to say: the experimenter is therefore free, after the system has been isolated, to cause or to prevent the system to end up in a state of type A?

Now, "state" is a word which everybody uses, even Saint PAM [Paul Adrien Maurice], but that doesn't add to its content. One can, however, easily convince oneself that, and in which way, a system really changes when its psi function changes. Any psi function conveys *the same* amount of knowledge. When psi changes, there is never only an increase but always also a loss of knowledge. But knowledge can only be acquired, never lost (except when one goes gaga, which is disregarded here). A loss of knowledge can only occur when the state of affairs has changed. In his sense I say: different psi functions certainly correspond to different states of affairs – or "states." I do not regard this as an illegitimate invoking of the reality dogma. But I'd really like to know what you think about it. And whether you really think that the Einstein case – let's call it thus – doesn't give anything to think about, but is completely clear and simple and self-evident. (This is what everybody said with whom I talked about it for the first time, because they had well learned their Copenhagen credo in unum sanctum. Three days later usually there came the statement: what I said earlier was of course wrong, much too complicated. ...

Letter 414 [Heisenberg to Pauli]
Mentions Bohr's reply to EPR in PR, which claims that EPR is the same as the double slit experiment and therefore nothing new. Then says he is nevertheless tempted to write a paper about the "cut" between system and measurement apparatus. And mentions a paper (or book) by Grete Hermann on the questions, which he likes overall.

Appendix to letter 414:
"Is a deterministic completion of qm possible" (by Heisenberg, unpublished).

The basic argument is that qm predictions are independent of the location of the "cut," but that any such "completion" would not be able to make the same prediction.

Curious, because refuted in a very simple way by Bohmian mechanics.

Letter 415 [Pauli to Schrödinger]

Agrees generally with Schrödinger's view of the Einstein case.

Says that Bohr's reply to EPR contains nothing new, but that he agrees with Bohr that the Einstein case contains nothing but very elementary and direct consequences of the indeterminacy relation.

Then summarizes Bohr's argument.

p420, second paragraph

Now, if one should denote "pure case" by "state"? Already Kramers in his "Lehrbuch der Wellen-mechanik" always puts "physical situation" instead of "state." A pure case of A is a total situation in which the results of certain measurements on A are (to a maximal extent) predictable with certainty. If you call that a "state," I don't object – but it is then indeed so that a change in the state of A – i.e., what can be predicted about A – also in a way different from a direct perturbation of A – i.e., also *after* isolation of A – is within the *free choice* of the experimenter.

3rd para:

in my opinion *there is simply no problem* – and we know this state of affairs even without the Einstein example.

4th para:

I would now like to comment on the general question raised by your handwritten addition – independent of Einstein's example. To what extent is the additional assumption possible that there exist properties, unknown to us today, that distinguish the single realizations of the qm collectives?

I would like to raise this question from the level of the "thou shalt not" to that of a statement that can be decided logically within qm. *Which type of additional assumptions is possible without changing the statistical consequences of qm itself?*

This question is very important, because in the heads of the conservative older gentlemen (Einstein, Laue, etc.) there is an unexterminable wrong analogy with kinetic gas theory. ...

Then he argues that it is impossible to make additional micro-assumptions without changing expectation values – a different approach from Heisenberg's, but refuted by Bohm just the same.

Rüdiger's Second Reply

I missed one interesting part of letter 413 [Schrödinger to Pauli]: the "handwritten addition" on p407:

> I realize that "pure case" (or known ψ function) is a collective as well; and that this collective is created each time through *selection* resulting from reading the (already detached) measurement apparatus. And in both cases because of different choice principles. Thus everything seems intelligible.
>
> But I think that every collective in qm is constituted of completely identical single cases. If one grants me differences between the single cases that constitute the collective, then one admits automatically the incompleteness of the quantum-mechanical

["mechanical] description."[3] Incomplete because I am ordered to regard cases as equal although I know that they are not; because I am forbidden to continue asking although I know that there is something left to ask.

This seems quite relevant to our enterprise.

17 March 1998, "Schnitt"

Schackcosm 3: *Curious, because refuted in a very simple way by Bohmian mechanics.*

Do you think maybe he is relying implicitly in some form or another on locality in his argument? Say, via no interactions between the two sides of the schnitt?

By the way, thanks for the summaries!!!

18 March 1998, "Hi"

Thanks again for doing all the translation work. Yes, I very much wish that I knew German. I can see it happening in the future, but, unfortunately, not immediately.

I was intrigued by Pauli's phrase "layers of reality." What do you think he really meant by that? Did he give any further indication? Perhaps in the no-hidden-variable argument he gave Schrödinger?

Rüdiger's Reply

There is nothing more on this question in the parts I didn't translate. I am intrigued, too. Maybe it is just not a very clear statement. If you replaced "reality" by "description," it would become rather standard.

23 March 1998, "Frightening Depths"

Would you remind me again what you see as the difference between Lewis's "principal principle" and your "principle of quantum indeterminism"? That is, beside the trivial sort of thing that the quantum case applies to an infinite number of random variables (i.e., one for each measurement). It seems to me presently that the statement "maximal information cannot be completed for quantum phenomena" provides the principal principle for the "principal principle."

By the way, did you ever hear from your friend, the Latin expert? One day last week, Hideo snuck into my office, placed a banana on my laptop computer, and changed the

---

[3] Note: Concerning the funny construction ["mechanical] description" Rüdiger later wrote: "It's not a typo. It's exactly like that in the original. I don't know what to make of it."

screensaver to "Mechanica bananica ex mente orta lex est!" So far I've left it without further modification.

23 March 1998, "Bohr's Palate"

> "Between us, we cover all knowledge; he knows all that can be known and I know the rest."
>
> *(Mark Twain, (ruminating on entanglement))*

Schackcosm 4: *Doesn't the principal principle involve two kinds of probability and is therefore unpalatable?*

Indeed it does, but I'm thinking about what similarities and distinctions we can draw between the PP (principal principle) and the PQI (principle of quantum indeterminism) ... don't worry by the way, I don't intend to use these acronyms in the paper: I was just trying to save some writing now ... though I see that I'm defeating that purpose.

The thing I'm thinking is that a Lewisian might see $|\psi\rangle$ as a compact description of all probabilities (for a given physical system) for which the PP is operant.

Schackcosm 5: *"Maximal information cannot be completed for quantum phenomena" entails (does it?) that two copies of a quantum system can be placed into exactly the same state ...*

Not completely, I don't think. One needs something slightly more, perhaps the principle that "knowledge is power" over nature. (Do you know who first said the thing between the quotes? The earliest thing I can find in my quote book is George Eliot.) That is to say, maximal knowledge gives an in-principle reproducibility in experiment (I guess it need not have been so). That maximal information cannot be completed is the quantum mechanical piece of it all.

Schackcosm 6: *Didn't we conclude that statements like the principal principle can be replaced by statements about joint probability distributions?*

True enough. But still the question above.

22 April 1998, "Infinite Patience"

I thought about opening up this note with the sentence, "If there be saints, then by virtue of your infinite patience you are surely one." But I knew that was too sappy and, so, thought better of writing it.

In any case let me thank you. I have spent the whole day navigating the space of your mind, or so it seems. I finally read again, from start to finish (and pretty carefully I might add), your original manuscript, Carl's commentary on it, and finally your commentary on his commentary. I had been avoiding this because I wanted to make some thoughts of my own before doing that – I wanted to sketch an outline for the paper that had a reasonable chance of being independent. I'm surely glad that I did that; but also now I'm surely glad that I read the old stuff again. It again injected me with a warm fuzzy feeling that has so far lasted all day.

Part of that is also surely due to my first real encounter with de Finetti. I've read about half of his 1931 essay "Probabilism" and find that I can hardly put it down (though I had to ... so that I could write this note). It's wonderful. I'm not sure why I wasn't drawn to him so much before ... maybe the time just wasn't ripe. But I'm certainly ready for it now.

Unfortunately I have to close shop in 24 minutes so that we can take Scott Parkins to dinner. So let me just ramble for a couple of minutes.

First, I found the de Finetti article in a special issue of the journal *Erkenntnis* (vol. 31, nos. 2–3, 1989) titled, "Bruno de Finetti's Philosophy of Probability." There look to be several interesting articles in it. So if you haven't encountered this yet, you might be interested in having a look.

Second, I came across a passage that intrigues me in connection to your Principle of Quantum Indeterminism. It is in a book review of Donald Gillies' book *An Objective Theory of Probability*. It states,

His thesis that Kolmogorov's Axioms should be regarded as formulating the general laws of an explanatory theory of random phenomena leads naturally to the realisation that as such they are incomplete and require to be augmented with what Gillies calls an Axiom of Independent Repetitions (p. 90ff). His criticism (pp. 108–118) of the popular thesis that the various limit theorems of mathematical probability supply the connection between theory and experience is to my mind convincing ...

I wonder what that's about? Unfortunately, we don't have a copy of this book here and it'll take me a few days to get it by interlibrary loan (though I'll order it tomorrow). If you get a chance, could you check whether you guys have a copy there? Maybe you're more likely to; I see from his web page that he (Gillies) is at King's College.

23 April 1998, "It Sinks Deeper"

The master de Finetti declares that we should not speak of "unknown" probabilities. For probabilities are not objective entities in their own right. Probability theory is a law of thought. ... ... But quantum mechanics is a law of thought. What does it mean to speak of attempting to clone an unknown quantum state?

I sink deeper into myself.

24 April 1998, "Groping and Moping"

Schackcosm 7: *What does it mean that quantum mechanics is a law of thought?*

That's a good question, the answer of which I keep groping for. I think it means that QM's formal structure or mathematics corresponds to "nothing more" than Bayesian-style reasoning applied to any situation where maximal information is necessarily incomplete. This could be a world in which we must come to intersubjective agreement but yet has the mischievous quality that MY information-gathering questions disturb YOUR predictability. (This appears to be the case in the physical world described by QM.) Or it might just as well be a human mind when it tries to reach some understanding of its future thoughts: the process of introspection.

Schackcosm 8: *Don't take him [de Finetti] too literally here. He understood nothing about quantum mechanics.*

Not to worry, I wasn't. I was just extrapolating based on the thoughts above. This issue of being an "unknown" quantum state has been something that's been bothering me for quite a while. I've tried to invent a better term, but I haven't found one that I am completely satisfied with. If one fails to remember that when one is speaking of an "unknown" quantum state, there is a secret preparator *somewhere* in the background – who in principle could be watching all actions as they unfold – it can lead to a load of trouble.

Schackcosm 9: *If you had a clone of an unknown quantum state, you could not test if it was indeed a clone, could you?*

Indeed, never with infinite precision: just as the flipping of a coin that continually comes up heads can never give – on its own – infinite security that there is actually a bias. Our friend David Mermin keeps asking, "What is the structure of quantum mechanics trying to tell us?" I think it's trying to tell us that it is a law of thought.

27 April 1998, "Fenomeno Aleatorio"

So I ended up not being philosophical at all yesterday – sorry for my hollow promises. I ended up thinking more about making my life more efficient on the computer.

Schackcosm 10: *I don't believe Mermin read it a second time. But I did (not the edited version, though) – and it made me think.* Mechanica quantica ex mente orta lex est. *A bold idea, and I don't think I have ever understood it as clearly as after my last rereading of your notes.*

Thank you for the pleasant compliments. More than is healthy I'm sure, my self-esteem has lately been thriving on such simple personal things. (Rejection letter Number Aleph came in this morning.) If you get a chance, do take a look at the last subsection of the edited version. You'll now find not only Asherisms and Merminitions, but . . .

Schackcosm 11: *Could you remind me of Wheeler's twenty questions?*

The quickest place to look is the notes I wrote on one of Asher Peres' review articles, the discussion after Asherism 8 in particular. (Somewhat connected also is the discussion following Asherism 11.) I believe I sent you those notes, but in case you've discarded them

I'll send them on again. If you really wish, I can look for the clearest statement of it in Wheeler's own writings and give you that reference a little later.

Schackcosm 12: *Your Heisenberg quote "What we learn about is not nature itself, but nature exposed to our methods of questioning" does apply just as well to pre-quantum science. It is a pithy summary of Kant, which gives one the choice of imagining that reality exists independently of any observer.*

True enough. I tried to make some bow to this in my note to Mermin – take a look at the discussion following Merminition 13. (Alternatively do a word search on "noumena.") More to the point though, I do think that Heisenberg (who certainly new of Kant's ideas) was trying to go further than this with his statement . . . though perhaps not as far as I would like him to go. I now think the natural – most conservative, actually – follow-up or rewrite to Kant is in this thing that Herb likes to call "realitty" (i.e., the game-of-twenty-questions ontology). But that hasn't yet been defined well enough to be of great use (in a philosophy of science).

Schackcosm 13: *The existence of a world out there is not a matter of taste any more, not "a mere metaphysical question."*

I don't know, it seems to me to depend upon what lengths one is willing to go to to preserve an "independent reality." They tell me that the Bohm theory does it just fine. The thing that is most important to me – in my public behavior at least – is really which point of view will be most productive for going the next step beyond quantum theory.

Schackcosm 14: *How is Platonism related to your ideas about physical reality? Are mathematical statements like Fermat's last theorem true in an absolute, Platonist sense?*

I've never thought much about the philosophy of mathematics. It seems to me presently though that there's not a lot of difference between Fermat's Last Theorem and my toaster. (See the discussion following Asherism 8 again.) I think I'd likely even say the same thing for mathematical logic (as encoded in Boolean algebra, for instance).

Schackcosm 15: *The Latin phrase may be rather inaccurate – not bad Latin, but not expressing the right thing. I'll make another effort at finding a better translation.*

Oh now this is important! Please do find the best possible translation! (. . . So we'll know for sure that we're in the best of all possible worlds.)

Schackcosm 16: *"Wait, wait, one more last thing: I defer discussion of objective probability once again." Did you ever discuss it with Mermin?*

Indeedy, but you now know everything that I know. I do think what he's trying to do is get at something different than simple frequentism or propensitism. His "objective probability" is founded upon *joint* probability distributions only. So, at least in that simple sense, it must be different from the standard tacks on "objective probability."

Schackcosm 17: *Wouldn't it be marvellous if one could derive Hilbert space from something like the principle of quantum indeterminism?*

Yes, yes, yes, I say! But beyond that, I'm sort of empty of thought right now.

OK, to work finally on THE project. I want to finish reading the de Finetti volume today.

30 April 1998, "One Stone Unturned"

The point of the last note? Nothing really, I just like lists ... you'll see what I mean. The next note I'm sending contains (much of) the bibliography for the paper. Please check it over when you get a chance and tell me what I've missed ... certainly no hurry on this. One thing I know that I have missed is de Finetti's books. Unfortunately I haven't been able to get hold of them yet: it's a shame, because I'm (mentally) ready for them now. Maybe I should buy them ... they're just so darned expensive.

Believe it or not, I do believe that I will cite each and every one of the items you're about to see. There's a lot of emphasis on the objectivist/propensity stuff. Reason? Basically because subjectivists – the few that there are in this game – will feel pretty happy with the argument. However each and every objectivist – take Ian Percival as an example – will be thinking something like, "Well, you clearly haven't taken the time to understand my version of the story!" Oh yeah?

Now the challenge: To make the paper longer than the reference list!!!

1. L. E. Ballentine, "Can the Statistical Postulate of Quantum Theory Be Derived? – A Critique of the Many-Universes Interpretation," *Found. Phys.* **3**, 229–240 (1973).
2. F. C. Benenson, "Randomness and the Frequency Definition of Probability," *Synthese* **36**, 207–233 (1977).
3. P. Benioff, "Possible Strengthening of the Interpretative Rules of Quantum Mechanics," *Phys. Rev.* D **7**, 3603–3609 (1973).
4. P. A. Benioff, "Finite and Infinite Measurement Sequences in Quantum Mechanics and Randomness: The Everett Interpretation," *J. Math. Phys.* **18**, 2289–2295 (1977).
5. P. Benioff, "A Note on the Everett Interpretation of Quantum Mechanics," *Found. Phys.* **8**, 709–720 (1978).
6. P. Benioff, "On the Correct Definition of Randomness," in *PSA 1978: Proceedings of the 1978 Biennial Meeting of the Philosophy of Science Association, Vol. 2*, edited by P. D. Asquith and I. Hacking (Philosophy of Science Association, East Lansing, Michigan, 1981), pp. 63–78.
7. G. Boole, *An Investigation of the Laws of Thought*, (Dover, New York, 1958).
8. R. B. Braithwaite, "On Unknown Probabilities," in *Observation and Interpretation: A Symposium of Philosophers and Physicists*, edited by S. Körner (Academic Press, New York, 1957), pp. 3–11.
9. S. L. Braunstein and C. M. Caves, "Wringing Out Better Bell Inequalities," *Ann. Phys.* **202**, 22–56 (1990).
10. J. Bub, "Popper's Propensity Interpretation of Probability and Quantum Mechanics," in *Induction, Probability, and Confirmation*, Minnesota Studies in the Philosophy of

Science, Vol. VI, edited by G. Maxwell and R. M. Anderson, Jr. (U. of Minnesota Press, Minneapolis, 1975), pp. 416–429.

11. M. Bunge, "Possibility and Probability," in *Foundations of Probability Theory, Statistical Inference, and Statistical Theories of Science, Vol. III*, edited by W. L. Harper and C. A. Hooker (D. Reidel, Dordrecht, 1976), pp. 17–33.

12. A. Cassinello and J. L. Sánchez-Gómez, "On the Probabilisitic Postulate of Quantum Mechanics," *Found. Phys.* **26**, 1357–1374 (1996).

13. C. M. Caves, "Information and Entropy," *Phys. Rev.* E **47**, 4010–4017 (1993).

14. C. M. Caves and C. A. Fuchs, "Quantum Information: How Much Information in a State Vector?," in *The Dilemma of Einstein, Podolsky and Rosen – 60 Years Later*, edited by A. Mann and M. Revzen, *Ann. Israel Phys. Soc.* **12**, 226–257 (1996).

15. R. Cooke, M. Keane, and W. Moran, "An Elementary Proof of Gleason's Theorem," *Math. Proc. Camb. Phil. Soc.* **98**, 117–128 (1981).

16. B. de Finetti, "Probabilism," *Erkenntnis* **31**, 169–223 (1989).

17. D. Deutsch, "Quantum Theory as a Universal Physical Theory," *Int. J. Theor. Phys.* **24**, 1–41 (1985).

18. B. S. DeWitt, "The Many-Universes Interpretation of Quantum Mechanics," in *Proceedings of the International School of Physics "Enrico Fermi" Course IL: Foundations of Quantum Mechanics*, edited by B. d'Espagnat (Academic Press, New York, 1971), pp. 211–262.

19. J. Earman, *Bayes or Bust?: A Critical Examination of Bayesian Confirmation Theory*, (MIT Press, Cambridge, MA, 1992).

20. H. Everett, III, "'Relative State' Formulation of Quantum Mechanics," *Rev. Mod. Phys.* **29**, 454–462 (1957).

21. H. Everett, III, "The Theory of the Universal Wave Function," in *The Many-Worlds Interpretation of Quantum Mechanics*, edited by B. S. DeWitt and N. Graham (Princeton U. Press, Princeton, 1973), pp. 3–140.

22. E. Farhi, J. Goldstone, and S. Gutmann, "How Probability Arises in Quantum Mechanics," *Ann. Phys.* **192**, 368–382 (1989).

23. J. H. Fetzer, "Statistical Probabilities: Single Case Propensities vs. Long-Run Frequencies," in *Developments in the Methodology of Social Science*, edited by W. Leinfellner and E. Köhler (D. Reidel, Dordrecht, 1974), pp. 387–397.

24. J. H. Fetzer, "Probability and Objectivity in Deterministic and Indeterministic Situations," *Synthese* **57**, 367–386 (1983).

25. R. P. Feynman, "The Concept of Probability in Quantum Mechanics," in *Proceedings of the Second Berkeley Symposium on Mathematical Statistics and Probability*, edited by J. Neyman (U. of California Press, Berkeley, 1951), pp. 533–541.

26. A. Fine, "Do Correlations Need To Be Explained?," in *Philosophical Consequences of Quantum Theory: Reflections on Bell's Theorem*, edited by J. T. Cushing and E. McMullin (U. of Notre Dame Press, Notre Dame, 1989), pp. 175–194.

27. D. Finkelstein, "The Logic of Quantum Mechanics," *Trans. N.Y. Acad. Sci.* **25**, 621–635 (1963).

28. M.-C. Galavotti, "Comments on Patrick Suppes 'Propensity Interpretations of Probability,'" *Erkenntnis* **26**, 359–368 (1987).

29. M.-C. Galavotti, "Anti-Realism in the Philosophy of Probability: Bruno de Finetti's Subjectivism," *Erkenntnis* **31**, 239–261 (1989).

30. A. J. M. Garrett, "Making Sense of Quantum Mechanics: Why You Should Believe in Hidden Variables," in *Maximum Entropy and Bayesian Methods (Paris, France, 1992)*, edited by A. Mohammad-Djafari and G. Demoment (Kluwer, Dordrecht, 1993), pp. 79–83.

31. R. N. Giere, "Objective Single-Case Probabilities and the Foundations of Statistics," in *Logic, Methodology and Philosophy of Science IV*, edited by P. Suppes, L. Henkin, A. Jojo, and Gr. C. Moisil (North-Holland, Amsterdam, 1973), pp. 467–483.

32. R. N. Giere, "Review of D. H. Mellor, *The Matter of Chance*," *Ratio* **15**, 149–155 (1973).

33. R. N. Giere, "A Laplacean Formal Semantics for Single-Case Propensities," *J. Phil. Logic* **5**, 321–353 (1976).

34. R. N. Giere, "Propensity and Necessity," *Synthese* **40**, 439–451 (1979).

35. D. A. Gillies, *An Objective Theory of Probability*, (Methuen, London, 1973), Chaps. 4–5.

36. N. Gisin, "Propensities and the State-Property Structure of Classical and Quantum Systems," *J. Math. Phys.* **25**, 2260–2265 (1984).

37. A. M. Gleason, "Measures on the Closed Subspaces of a Hilbert Space," *J. Math. Mech.* **6**, 885–894 (1957).

38. N. Graham, "The Everett Interpretation of Quantum Mechanics," Ph.D. thesis, University of North Carolina, Chapel Hill, NC (1970).

39. N. Graham, "The Measurement of Relative Frequency," in *The Many-Worlds Interpretation of Quantum Mechanics*, edited by B. S. DeWitt and N. Graham (Princeton U. Press, Princeton, 1973), pp. 229–253.

40. S. Gutmann, "Using Classical Probability to Guarantee Properties of Infinite Quantum Sequences," *Phys. Rev.* A **52**, 3560–3562 (1995).

41. J. B. Hartle, "Quantum Mechanics of Individual Systems," *Am. J. Phys.* **36**, 704–712 (1968).

42. G. Hellman, "Randomness and Reality," in *PSA 1978: Proceedings of the 1978 Biennial Meeting of the Philosophy of Science Association, Vol. 2*, edited by P. D. Asquith and I. Hacking (Philosophy of Science Association, East Lansing, Michigan, 1981), pp. 79–97.

43. R. I. G. Hughes, *The Structure and Interpretation of Quantum Mechanics*, (Harvard U. Press, Cambridge, MA, 1989), Chap. 8.

44. P. Humphreys, "Why Propensities Cannot Be Probabilities," *Phil. Rev.* **94**, 557–570 (1985).

45. E. T. Jaynes, "Clearing Up Mysteries – The Original Goal," in *Maximum Entropy and Bayesian Methods (Cambridge, England, 1988)*, edited by J. Skilling (Kluwer, Dordrecht, 1989), pp. 1–27.

46. E. T. Jaynes, "Probability in Quantum Theory," in *Complexity, Entropy and the Physics of Information*, edited by W. H. Zurek (Addison-Wesley, Redwood City, CA, 1990), pp. 381–403.

47. E. T. Jaynes, *Probability Theory: The Logic of Science*, book in preparation. Incomplete preliminary draft available at ftp://bayes.wustl.edu/.

48. R. C. Jeffrey, *The Logic of Decision*, (McGraw-Hill, New York, 1965).

49. R. C. Jeffrey, "Mises Redux," in *Basic Problems in Methodology and Linguistics*, edited by R. E. Butts and J. Hintikka (D. Reidel, Dordrecht, 1977), pp. 213–222.

50. R. Jeffrey, "Reading *Probabilismo*," *Erkenntnis* **31**, 225–237 (1989).

51. A. Kent, "Against Many-Worlds Interpretations," *Int. J. Mod. Phys.* **A5**, 1745–1762 (1990).
52. H. E. Kyburg, "Propensities and Probabilities," *Brit. J. Phil. Sci.* **25**, 358–375 (1974).
53. H. E. Kyburg, Jr., "Chance," *J. Phil. Logic* **5**, 355–393 (1976).
54. I. Levi, "Probability Exists (but just barely)!" in *Logic, Methodology and Philosophy of Science VII*, edited by R. Barcan Marcus, G. J. W. Dorn, and P. Weingartner (North-Holland, Amsterdam, 1986), pp. 367–385.
55. D. Lewis, "A Subjectivist's Guide to Objective Chance," in *Studies in Inductive Logic and Probability, Vol. II*, edited by R. C. Jeffrey (U. of California Press, Berkeley, 1980), pp. 263–293.
56. T. J. Loredo, "From Laplace to Supernova SN 1987A: Bayesian Inference in Astrophysics" in *Maximum Entropy and Bayesian Methods (Dartmouth, U.S.A., 1989)*, edited by P. F. Fougère (Kluwer, Dordrecht, 1990), pp. 81–142.
57. D. H. Mellor, *The Matter of Chance*, (Cambridge University Press, Cambridge, 1971).
58. N. D. Mermin, "The Ithaca Interpretation of Quantum Mechanics," *Pramana* **51**, 549–565 (1998).
59. N. D. Mermin, "What Is Quantum Mechanics Trying To Tell Us?," *Am. J. Phys.* **66**, 753–767 (1998).
60. D. Miller, "Single-Case Probabilities," *Found. Phys.* **21**, 1501–1516 (1991).
61. P. Mittelstaedt, "The Objectification in the Measuring Process and the Many Worlds Interpretation," in *Symposium on the Foundations of Modern Physics*, edited by P. Lahti and P. Mittelstaedt (World Scientific, Singapore, 1991), pp. 261–279.
62. P. Mittelstaedt, "Is Quantum Mechanics a Probabilistic Theory?," in *Potentiality, Entanglement and Passion-at-a-Distance*, edited by R. S. Cohen, M. Horne, and J. Stachel (Kluwer, Dordrecht, 1997), pp. 159–175.
63. M. Mugur-Schachter, "The Quantum Mechanical Hilbert Space Formalism and the Quantum Mechanical Probability Space of the Outcomes of Measurement," in *Foundations of Quantum Mechanics and Ordered Linear Spaces*, edited by A. Hartkämper and H. Neumann (Springer-Verlag, Berlin, 1974), Lecture Notes in Physics, No. 29, pp. 288–308.
64. W. Ochs, "On the Strong Law of Large Numbers in Quantum Probability Theory," *J. Phil. Logic* **6**, 473–480 (1977).
65. W. Ochs, "Concepts of Convergence for a Quantum Law of Large Numbers," *Rep. Math. Phys.* **17**, 127–143 (1980).
66. D. N. Page, private communication to J. B. Hartle, 28 October 1996.
67. A. Peres, "What is a State Vector?," *Am. J. Phys.* **52**, 644–650 (1984).
68. A. Peres, *Quantum Theory: Concepts and Methods*, (Kluwer, Dordrecht, 1993).
69. I. Pitowsky, "From George Boole to John Bell – The Origins of Bell's Inequality," in *Bell's Theorem, Quantum Theory and Conceptions of the Universe*, edited by M. Kafatos (Kluwer, Dordrecht, 1989), pp. 37–49.
70. I. Pitowsky, "George Boole's 'Conditions of Possible Experience' and the Quantum Puzzle," *Brit. J. Phil. Sci.* **45**, 95–125 (1994).
71. I. Pitowsky, "Infinite and Finite Gleason's Theorems and the Logic of Indeterminacy," to appear in *J. Math. Phys.* (1998).

72. K. R. Popper, "The Propensity Interpretation of the Calculus of Probability and the Quantum Theory," in *Observation and Interpretation: A Symposium of Philosophers and Physicists*, edited by S. Körner (Academic Press, New York, 1957), pp. 65–70.

73. K. R. Popper, *Quantum Theory and the Schism in Physics*, (Hutchinson, London, 1982).

74. J. Preskill, "Chapter 3. Foundations of Quantum Theory II: Measurement and Evolution," in *Lecture Notes for Physics 229: Advanced Mathematical Methods of Physics*, available on the World Wide Web at http://www.theory.caltech.edu/people/preskill/ph229/.

75. H. Reichenbach, *The Theory of Probability: An Inquiry into the Logical and Mathematical Foundations of the Calculus of Probability*, (U. of California Press, Berkeley, 1949).

76. C. Rovelli, "Relational Quantum Mechanics," *Int. J. Theor. Phys.* **35**, 1637–1678 (1996).

77. W. C. Salmon, "Propensities: A Discussion Review," *Erkenntnis* **14**, 183–216.

78. L. J. Savage, *The Foundations of Statistics*, (Dover, New York, 1972).

79. C. Schneider, "Two Interpretations of Objective Probabilities," *Philosph. Naturalis* **31**, 107–131 (1994).

80. B. Skyrms, *Causal Necessity: A Pragmatic Investigation of the Necessity of Laws*, (Yale U. Press, New Haven, 1980), Chap. IA.

81. B. Skyrms, "Statistical Laws and Personal Propensities," in *PSA 1978: Proceedings of the 1978 Biennial Meeting of the Philosophy of Science Association, Vol. 2*, edited by P. D. Asquith and I. Hacking (Philosophy of Science Association, East Lansing, Michigan, 1981), pp. 551–562.

82. B. Skyrms, *Pragmatics and Empiricism*, (Yale U. Press, New Haven, 1984).

83. B. Skyrms, "Probability and Causation," *J. Econometrics* **39**, 53–68 (1988).

84. S. Spielman, "Physical Probability and Bayesian Statistics," *Synthese* **36**, 235–269 (1977).

85. W. Spohn, "A Brief Remark on the Problem of Interpreting Probability Objectively," *Erkenntnis* **26**, 329–334 (1987).

86. E. J. Squires, "On an Alleged 'Proof' of the Quantum Probability Law," *Phys. Lett.* A **145**, 67–68 (1990).

87. M. Strauss "Two concepts of probability in physics," in *Logic, Methodology and Philosophy of Science IV*, edited by P. Suppes, L. Henkin, A. Jojo, and Gr. C. Moisil (North-Holland, Amsterdam, 1973), pp. 603–615.

88. P. Suppes, "New Foundations for Objective Probability: Axioms for Propensities," in *Logic, Methodology and Philosophy of Science IV*, edited by P. Suppes, L. Henkin, A. Jojo, and Gr. C. Moisil (North-Holland, Amsterdam, 1973), pp. 515–529.

89. P. Suppes, "Some Further Remarks on Propensity: Reply to Maria-Carla Galavotti," *Erkenntnis* **26**, 369–376 (1987).

90. P. Suppes, "The Transcendental Character of Determinism," in *Midwest Studies in Philosophy, Volume XVII: Philosophy of Science*, edited by P. A. French, T. E. Uehling, Jr., and H. K. Wettstein (U. of Notre Dame Press, Notre Dame, 1993), pp. 242–257.

91. S. Sýkora, "Quantum Theory and the Bayesian Inference Problems," *J. Stat. Phys.* **11**, 17–27 (1974).

92. B. C. van Fraassen, *Laws and Symmetry*, (Clarendon Press, Oxford, 1989).

93. B. C. van Fraassen, *Quantum Mechanics: An Empiricist View*, (Clarendon Press, Oxford, 1991).

94. R. von Mises, *Probability, Statistics and Truth*, (Dover, New York, 1981).

95. J. von Neumann, *Mathematical Foundations of Quantum Mechanics*, translated by R. T. Beyer, (Princeton U. Press, Princeton, 1955), p. 298, fn. 156.

96. J. von Neumann, "Quantum Logics (Strict- and Probability-Logics)," in *John von Neumann: Collected Works, Vol. IV*, edited by A. H. Taub (Macmillan, New York, 1962), pp. 195–197.

97. J. von Plato, "de Finetti's Earliest Works on the Foundations of Probability," *Erkenntnis* **31**, 263–282 (1989).

98. C. F. von Weizsäcker, "Probability and Quantum Mechanics," *Brit. J. Phil. Sci.* **24**, 321–337 (1994).

99. E. P. Wigner, "On Hidden Variables and Quantum Mechanical Probabilities," *Am. J. Phys.* **38**, 1005–1009 (1970).

24 May 1998, "Nature Says No"

Well, nature apparently doesn't like our tinkering with these anti-realist ideas so much. Despite my attempt at optimism the other day, I've hardly had a productive moment since the last time I wrote you. I had to stay at home both Thursday and Friday. Friday I became feverish. This morning my ears started filling with fluid, feeling as if they were going to burst. So I finally went to see a doctor: now my system is stuffed full of clarithromycin, codeine, acetaminophen, phenylephrine HCl, and God only knows what else. I hope these drugs will have some effect soon – I will soon go crazy otherwise.

In some of my delusional moments I have been thinking about your mathematical Platonism question. The more I think about it, the more I really dislike the idea. A world with such a stable, untouchable substrate strikes me as a horrible, still-born world.

31 August 1998, "Comfort in Numbers"

I found something today that I think you might be interested in: an essay by Heisenberg titled "Quantum Mechanics and Kantian Philosophy" in his book *Physics and Beyond*. He makes some points about Kant's a priori that I heard you making last week. Take a look if you get a chance.

21 January 1999, "Stupendous Literature Man"

Apparently, the phrase "The world is presented to us only once" and slight variants thereof are popular ones. For I have now seen them used to mean three different things. The quotes below speak for themselves. In the original outline of Fuji, however, I had only meant the version that Richard Jeffrey (and Ernst Mach) are talking about.

Mara Beller, "The Sokal Hoax: At Whom are We Laughing?," *Physics Today*, September 1998, pp. 29–34:

In fact, Einstein was no "naive realist," despite such caricaturing of his stand by the Copenhagen orthodoxy. He ridiculed the "correspondence" view of reality that many scientists accept uncritically. Einstein fully realized that the world is not presented to us twice – first as it is, and second, as it is theoretically described – so we can compare our theoretical "copy" with the "real thing." The world is given to us only once – through our best scientific theories.

Erwin Schrödinger, *Mind and Matter* (1958), p. 51:

It is the same elements that go to compose my mind and the world. This situation is the same for every mind and its world, in spite of the unfathomable abundance of "cross-references" between them. The world is give to me only once, not one existing and one perceived. Subject and object are only one. The barrier between them cannot be said to have broken down as a result of recent experience in the physical sciences, for this barrier does not exist.

Erwin Schrödinger, *Mind and Matter* (1958), p. 63:

Nay, may we call a world that nobody contemplates even that? [...] But a world, existing for many millions of years without any mind being aware of it, contemplating it, is it anything at all? Has it existed? For do not let us forget: to say, as we did, that the becoming of the world is reflected in a conscious mind is but a cliché, a phrase, a metaphor that has become familiar to us. The world is given but once. Nothing is reflected. The original and the mirror-image are identical. The world extended in space and time is but our representation.

Erwin Schrödinger, *Nature and the Greeks* (1954):

Plato, Aristotle, and Epicurus emphasize the import of being astonished. And this is not trivial when it refers to general questions about the world as whole; for, indeed, it is given us only once, we have no other one to compare it with.

Albert Einstein, "Elementare Überlegungen zur Interpretation der Grundlagen der QuantenMechanik," in M. Born, Scientific Papers Presented to Max Born (Hafner Publishing, New York, 1953), pp. 33–40:

Nature as a whole can only be viewed as an individual system, existing only once, and not as a collection of systems.

Carl's further attribution of the translation of this to H. Freistadt is apparently incorrect; or at least I couldn't find this phrase in that article – it is just a review article of Bohmianism. Carl, where did you actually see this line quoted? Was it in Cushing's book or something?

Now for the *pièce de resistance*. I'm not sure I agree with all of the larger context that Jeffrey adds to the quote – I think it still imagines that the observer is "detached" in the way of classical physics (and you guys know I hate that) – but here it is, the thing I was originally thinking of.

Richard Jeffrey, "de Finetti's Radical Probabilism," in *Probabilità e Induzione – Induction and Probability*, edited by P. Monari and D. Cocchi, Biblioteca di Statistica, CLUEB, Bologna, pp. 263–275:

de Finetti's probabilism is "radical" in the sense of going all the way down to the roots: he sees probabilities as ultimate forms of judgement which need not be based on deeper all-or-none knowledge. ...

In science de Finetti expected determinism to give way to probabilism. ...

[Physical arguments are not alone decisive.] Still less are philosophical arguments alone decisive. But de Finetti's philosophical view does see determinism as a state of mind masquerading as a state of nature, and sees causality as a fancied magical projection into nature of our own patterns of expectation. Beneath the mask of determinism is a state of mind – certainty – that is intelligible enough, but (de Finetti argues) generally inappropriate. This state of mind needs to be replaced by a more appropriate, probabilistic one. That's one prong in his attack on determinism.

The other corresponds to Mach's aphorism,[4] "Nature is only there once." In indecision or ignorance we envision alternative courses of events, each of which is governed by the laws of nature, i.e., each is logically consistent with certain sentences in universal form which (let us suppose) correctly describe and predict the one actual course of events. Causal laws are magical projections into nature of such confident descriptions and expectations. But it is over the whole mental array of *possibili* courses of nature that we distribute judgmental probabilities, giving no special treatment to the unknown course that corresponds to reality – much as we might like to.

25 January 1999, "Oh Hibernia"

I think my most accurate way of describing this time after Emma's birth can be summed up with the word "hibernation." On the one hand, it feels like we get no sleep at all; while on the other, it feels like we do nothing but sleep! I hope that sympathy you offered me last week is still holding steady. I've hardly been working at all, I must say.

I have read all the Galavotti papers again. They are marvelous. Have you read all of them? I'll append the whole list below.

Another paper that I think should be of *great* interest to us is Schrödinger's "The Present Situation in Quantum Mechanics," translated by J. D. Trimmer, *Proc. Am. Philos. Soc.* **124**, 323–338 (1980). You can also find it in the Wheeler–Zurek volume. The important piece in it is his emphasis (and eloquent explanation) that a pure state corresponds to "maximum information" even though that information is incomplete. Sound familiar? What he sees as the real cutting point between classical and quantum, however, is that maximal information about the whole generally does not determine maximal information about the parts. This, he tries to state, is the essence of entanglement. Note, in particular, that he does not draw the line where we usually do in our discussions: namely, at the fact that maximal information

---

[4] "In der Natur gibt es keine Ursache und keine Wirkung. Die Natur ist nur einmal da" *Die Mechanik in ihrer Entwicklung*, Leipzig, 1883, ch. 4, §4 (Die Oekonomie der Wissenschaft), par. 3; p. 455.

is complete classically but not complete quantum mechanically. I think this may be an important point. And I think it is perfectly in keeping with his pre-quantum prejudices: Schrödinger believed in physical *indeterminism* well before quantum mechanics ever came upon the scene. (See for instance Paul Forman's discussion of Schrödinger's and Exner's indeterminism in his huge article.) Thus – I think – for him, *classical* physical description already included cases where maximal information could not be made complete ... so of course that concept alone couldn't be used to make the cut between classical and quantum. I wish he had said these things outright in his article (and not leave me to so much detective work), but unfortunately he didn't.

By the way, I doubt Hibernia (i.e., Ireland) has anything to do with the word hibernate, but I thought it would help make a good title.

1. M. C. Galavotti, "Comments on Patrick Suppes 'Propensity Interpretations of Probability'" *Erkenntnis* **26**, 359–368 (1987).
2. M. C. Galavotti, "Anti-Realism in the Philosophy of Probability: Bruno de Finetti's Subjectivism," *Erkenntnis* **31**, 239–261 (1989).
3. M. C. Galavotti, "The Notion of Subjective Probability in the Work of Ramsey and de Finetti," *Theoria* **LVII**, 239–259 (1991).
4. M. C. Galavotti, "Operationism, Probability and Quantum Mechanics," *Found. Sci.* **1**, 99–118 (1995).
5. M. C. Galavotti, "F. P. Ramsey and the Notion of 'Chance'," in *The British Tradition in 20th Century Philosophy: Proceedings of the 17th International Wittgenstein Symposium, 14–21 August 1994, Kirchberg am Wechsel (Austria)*, edited J. Hintikka and K. Puhl (Hölder-Pichler-Tempsky, Vienna, 1995), pp. 330–340.
6. M. C. Galavotti, "Probabilism and Beyond," *Erkenntnis* **45**, 253–265 (1996).

25 January 1999, and to Carl Caves, "Oh Hernia"

Just when you think things are getting better and better ... then you stay up most all night. Last night was not a good night. Carl sometimes says I should thank him for the hernia I got back in '94 and the good life it brought. Thank you Carl.

Schackcosm 18: *I reread that section of Forman's article. Do you have the complete reference? My copy of it does not give the source.*

The reference is:

P. Forman, "Weimar Culture, Causality, and Quantum Theory, 1918–1927: Adaptation by German Physicists and Mathematicians to a Hostile Intellectual Environment," in *Historical Studies in the Physical Sciences*, Vol. 3, edited by R. McCormmach (U. of Pennsylvania Press, Philadelphia, 1971), pp. 1–115.

A more focused study on Schrödinger's thoughts on indeterminism is:

P. A. Hanle, "Indeterminacy before Heisenberg: The Case of Franz Exner and Erwin Schrödinger," in *Historical Studies in the Physical Sciences*, Vol. 10, edited by

R. McCormmach, L. Pyenson, and R. S. Turner, (Johns Hopkins U. Press, Baltimore, MD, 1979), pp. 225–269.

This reference has little to do with Schrödinger, but while we're on the Forman thesis and I have the paper out in front of me:

P. Forman, "*Kausalität, Anschaulichkeit* and *Individualität*, or How Cultural Values Prescribed the Character and the Lessons Ascribed to Quantum Mechanics," ... oops I don't have the full reference on this one. I'll have to get back to you on that one.

01 August 1999, and to Carl Caves, "Recording the Vague"

I've got a few moments (hours?) to kill while waiting in Heathrow – we just discovered that Kiki's passport expired a month ago – so I thought I'd use this time to record some rather vague thoughts.

The first of my troubles has to do with the usual way of justifying that quantum time evolutions correspond to trace-preserving completely positive maps. Preserving positivity and trace, I have no problem with: we can justify this straight out with Gleason's theorem. But where does the strong assumption of linearity come from? Where does the strong assumption of *complete* positivity come from? The question I want to ask is whether we can give Bayesian motivations for either of these assumptions?

Think first about linearity. The motivation for it given by Kraus (and the other operationalists) is that quantum states correspond to equivalence classes of preparations. So, if we don't know the preparation completely, we will know just as little about the time evolved version of it as we did about the original: the time evolved state will be a convex combination of the individual time evolved states. Another way of saying this is that the set of states forms a convex structure, and the time evolutions complement that convex structure quite nicely. This is their motivation for linearity.

But from our point of view, a quantum state need not necessarily correspond to a preparation – to use Carl's phrase, it is a subjective entity. Quantum states are states of knowledge regardless as to how that knowledge actually came about. So the first thing we might ask is, why should the class of subjective entities possess a convex structure? I think we've started to find a partial answer to that in the quantum de Finetti theorem. Can we, by twist or turn, answer the question of linearity in a similar fashion?

The point here is that it is only the de Finetti theorem (both classically and quantumly) that allows us to make sense of probability distributions over subjective entities. If linearity is going to come into quantum mechanics via purely Bayesian means, then it seems to me that it is going to have to come in at just this point.

But – today at least – I'm having a hard time making this idea concrete. Perhaps we could play with the following sort of ideas. Assume outright that time evolutions must map density operators to density operators, but make no a priori assumptions about linearity, etc. Say that a time evolution is individual to a system (and identical on identical systems), if whenever we have an exchangeable sequence of density operators, the time evolved versions also form an exchangeable sequence – that seems like a perfectly good notion of individual time evolution to me. Now use the quantum de Finetti theorem for both the initial and final states. Is there some argument we can use to ensure that probabilities in both de Finetti representations should be the same? Perhaps by invoking some principle about the difficulty of learning? But then, even if we could get that far, where do we take the argument from there? Just something to think about. Clearly there's a long way to go, but I have a hard time shaking the feeling that there might just be something here that a little clear thought can conquer.

Perhaps one more way to say it all is that maybe just maybe linearity comes straight out of Bayes' rule as applied to learning.

Next point: *complete* positivity, who ever said that entangled states had to exist and that they must be preserved? My trouble is mainly this. The usual argument for complete positivity relies on extending the system (to an arbitrarily large one) and making the assumption that a "superagent" trying to describe the extended system must also have his valid states of knowledge (density operators) evolve to valid states of knowledge. But why should we have to invoke such a superagent within our point of view? Why should we ever have to look outside the given system? Why would we, who are concerned with a fixed set of n-level quantum systems, ever worry about the point of view of someone outside our little world? (For all we know, there is no one outside our little world?) Clearly we *should* care if we believe in the existence of the Church of the Larger Hilbert Space ... if we were lovers of the many-worlds point of view. But we're not, so why should *we* bother?

What I'm trying to get at is this. Surely there is a characterization of the completely positive maps that is intrinsic to the given system. The question is, what in our program will get at that characterization in a natural way? This may be a little lame, but the only thing I can think of right now is something to do with a Dutch book. If there were a bookie who didn't believe in the complete positivity of quantum evolutions, could we find a series of bets that would lead him to a sure loss? Maybe. Any ideas?

OK, that's my two vague thoughts about time evolution. I look forward to any reactions you might have.

Finally, let me – for the benefit of Carl – turn to something that I discussed very briefly with Rüdiger while still in Cambridge. What is the firm ground behind the Bayesian point of view of QM? What is the real stuff underneath it all? Carl and I both have had tendencies to think that it is the Hamiltonians and interactions between systems. But that may be troublesome. Just as we can ask "What is an unknown quantum state?" we should also ask "What is an unknown Hamiltonian or interaction?" The Kraus representation of a completely positive map is not unique, nor is a unitary representation in the Church of the Larger Hilbert Space. If Carl finds the nonuniqueness of the density operator decomposition as the most damning thing for an objective interpretation of quantum probabilities, then it seems that he should find the same pretty damning for the objective interpretation of quantum time evolutions. Where we go now with this, I'm not sure ... but as above, it may be worth thinking about.

That's it for now.

18 August 1999, and to Carl Caves, "A Hated Phrase"

I'm coming ever more to hating phrases like the following:

However, if we assume that macroscopic objects can be described by quantum theory, then the orthodox view predicts that under certain conditions, objects such as the pointers on the dials of an apparatus do not have the property of being at a well defined location, or more dramatically, that a cat does not have the property of being dead or alive. This is clearly in conflict with our observations.

Since when has such a property as "live" versus "dead" been such a clear cut thing? The breaking point between live and dead in the years before cardiopulmonary resuscitation was certainly very different from what it was afterward. With better and better technology, one finds that death is not so irreversible. So, what is it that makes "live" versus "dead" such an all-fired objective property? And maybe most importantly, what point of principle distinguishes this situation from our ability to manipulate superpositions?

I swiped the phrase above from Robert Spekkens' webpage; he's one of the modal interpretation guys I met last week. (He and his advisor Sipe, by the way, along with a student of Leslie Ballentine were quite interested in the Q de Finetti theorem.[5] They were the only ones beside D'Ariano who spent a significant amount of time trying to understand its statement thoroughly and the method of proving it.)

Just thinking out loud.

29 August 1999, and to Carl Caves, "Penrose Tiles"

I want you ... to jolt the world of physics into an understanding of the quantum because the quantum surely contains – when unraveled – the most wonderful insight we could ever hope to have on how this world operates, something equivalent in scope and power to the greatest discovery that science has ever yet yielded up: Darwin's Evolution.

(John Archibald Wheeler, 13 March 1998)

Quantum states are states of knowledge, we say, not states of nature. Embracing this is surely the surest path to John's vision. But, Roger Penrose and Richard Jozsa counter with the opposite opinion. Here's their argument. From *The Emperor's New Mind*, pages 268–269:

Despite the fact that we are normally only provided with probabilities for the outcome of an experiment, there seems to be something *objective* about a quantum-mechanical state. It is often asserted that the state-vector is merely a convenient description of "our knowledge" concerning a physical system – or, perhaps, that the state-vector does not really describe a single system but merely provides probability information about an "ensemble" of a large number of similarly prepared systems. Such sentiments strike me as unreasonably timid concerning what quantum mechanics has

---

[5] I later learned that the student's name was Joseph Emerson. He is now a professor of applied mathematics at the University of Waterloo.

to tell us about the *actuality* of the physical world. [Actually, in Richard Jozsa's version of this sentence he substituted the word "cowardly" for "timid."]

Some of this caution, or doubt, concerning the "physical reality" of state-vectors appears to spring from the fact that what is physically measurable is strictly limited, according to theory. Let us consider an electron's state of spin, as described above. Suppose that the spin-state happens to be $|\alpha\rangle$, but we do not know this; that is, we do not know the *direction* α in which the electron is supposed to be spinning. Can we determine this direction by measurement? No, we cannot. The best that we can do is extract 'one bit' of information – that is, the answer to a single yes/no question. We may select some direction β in space and measure the electron's spin in that direction. We get either the answer **YES** or **NO**, but thereafter, we have lost the information about the original direction of spin. With a **YES** answer we know that the state is *now* proportional to $|\beta\rangle$, and with a **NO** answer we know that the state is *now* in the direction opposite to β. In neither case does this tell us the direction α of the state *before* measurement, but merely gives some probability information about α.

On the other hand, there would seem to be something completely *objective* about the direction α itself, in which the electron "happened to be spinning" before the measurement was made. [Footnote: This objectivity is a feature of our taking the standard quantum-mechanical formalism seriously. In a *non*-standard viewpoint, the system might actually "know," ahead of time, the result that it would give to *any* measurement. This could leave us with a *different*, apparently objective, picture of physical reality.] For we *might* have chosen to measure the electron's spin in the direction α – and the electron has to be prepared to give the answer **YES** with *certainty*, if we happened to have guessed right in this way! Somehow, the "information" that the electron must actually give this answer is stored in the electron's state of spin.

It seems to me that we must make a distinction between what is "objective" and what is "measurable" in discussing the question of physical reality, according to quantum mechanics. The state-vector of a system is, indeed, *not measurable*, in the sense that one cannot ascertain, by experiments performed on the system, precisely (up to proportionality) what that state is; but the state-vector *does* seem to be (again up to proportionality) a completely *objective* property of the system, being completely characterized by the results that it must give to experiments that one *might* perform. In the case of a single spin one-half particle, such as an electron, this objectivity is not unreasonable because it merely asserts that there is *some* direction in which the electron's spin is precisely defined, even though we may not know what that direction is. (However, we shall be seeing later that this "objective" picture is much stranger with more complicated systems – even for a system which consists merely of a *pair* of spin one-half particles.)

But need the electron's spin have any physically defined state *at all* before it is measured? In many cases it will *not* have, because it cannot be considered as a quantum system on its own; instead, the quantum state must generally be taken as describing an electron inextricably entangled with a large number of other particles. In particular circumstances, however, the electron (at least as regards its spin) *can* be considered on its own. In such circumstances, such as when its spin has actually previously been measured in some (perhaps unknown) direction and then the electron has remained undisturbed for a while, the electron *does* have a perfectly objectively defined direction of spin, according to standard quantum theory.

And, almost identically, from *Shadows of the Mind*, pages 312–315:

There are many versions of the viewpoint according to which the state vector $|\psi\rangle$ is *not* regarded as providing an actual picture of a quantum-level physical reality. Instead, $|\psi\rangle$ would be taken to serve only as a calculational device, useful merely for calculating probabilities, or as an expression of the experimenter's "state of knowledge" concerning a physical system. ... Sometimes, it is even argued that $|\psi\rangle$ *cannot* describe a quantum-level reality since it makes no sense at all to talk of a "reality" for our world at that level, reality consisting only of the results of "measurements". ...

Yet, if we accept that there must be a reality of some kind that holds at the quantum level, we may still have doubts that this reality can be accurately described by a state vector $|\psi\rangle$. There are various arguments that people raise as objections to the "reality" of $|\psi\rangle$. ... [He lists a few arguments here.]

All this is true, yet it remains hard to take the opposite position either: that the state vector $|\psi\rangle$ is somehow physically "unreal," perhaps encapsulating merely the sum total of "our knowledge" about a physical system. This, I find very hard to accept, particularly since there appears to be something very subjective about such a role of "knowledge." *Whose* knowledge, after all, is being referred to here? Certainly not mine. I have very little actual knowledge of the individual state vectors that are relevant to the detailed behaviour of all the objects that surround me. Yet they carry on with their precisely organized actions, totally oblivious to whatever might be "known" about the state vector, or to whoever might know it. Do different experimenters, with different knowledge about a physical system, use different state vectors to describe that system? Not in any significant way; they might only do so if these differences were about features of the experiment that would be inessential to the outcome.

One of the most powerful reasons for rejecting such a subjective viewpoint concerning the reality of $|\psi\rangle$ comes from the fact that whatever $|\psi\rangle$ might be, there is always – in principle, at least – a *primitive measurement* whose **YES** space consists of the Hilbert-space ray determined by $|\psi\rangle$. The point is that the physical state $|\psi\rangle$ (determined by the ray of complex multiples of $|\psi\rangle$) is *uniquely* determined by the fact that the outcome **YES**, for this state, is *certain*. No other physical state has this property. For any other state, there would merely be some probability, short of certainty, that the outcome will be **YES**, and an outcome of **NO** might occur. Thus, although there is no measurement which will tell us what $|\psi\rangle$ actually *is*, the physical state $|\psi\rangle$ is uniquely determined by what it asserts must be the result of a measurement that *might* be performed on it. This is a matter of counterfactuals again, but we have seen how important counterfactual issues actually are to the expectations of quantum theory.

To put the point a little more forcefully, imagine that a quantum system has been set up in a known state, say $|\phi\rangle$, and it is computed that after a time t the state will have evolved, under the action of **U**, into another state $|\psi\rangle$. For example, $|\phi\rangle$ might represent the state "spin up" ($|\phi\rangle = |\uparrow\rangle$) of an atom of spin $\frac{1}{2}$, and we can suppose that it has been put in that state by the action of some previous measurement. Let us assume that our atom has a magnetic moment aligned with its spin (i.e., it is a little magnet pointing to the spin direction). When the atom is placed in a magnetic field, the spin direction will precess in a well-defined way, that can be accurately computed as the action of **U**, to give some new state, say $|\psi\rangle = |\rightarrow\rangle$, after a time t. Is this computed state to be taken seriously as part of physical reality? It is hard to see how this can be denied. For $|\psi\rangle$ has to be prepared for the possibility that we *might* choose to measure it with the primitive measurement referred to above, namely that whose **YES** space consists precisely of the multiples of $|\psi\rangle$. Here, this is the spin measurement in the direction $\rightarrow$. The system has to know to give the answer **YES**, with *certainty* for that measurement, whereas *no* spin state of the atom *other* than $|\phi\rangle = |\rightarrow\rangle$ could guarantee this.

How do we move past this? I am coming more and more to the opinion that the solution lies in discarding the term "measurement" for labelling what it is that occurs in the "quantum measurement process." The problem is that the word "measurement" conveys too much of a confusing sense that the purpose of the measurement outcome is to reveal something that *is* (or at least *was*) "there" independently of our having looked. I think a much better terminology – one that may start to crack this problem – can be built from the terminology used in L. J. Savage's book.

Decisions, Acts, and Consequences. Let me introduce this in a rather long-winded way by quoting some long passages that have had a significant influence on me. The first is a piece from Doug Bilodeau's paper "Why Quantum Mechanics is Hard to Understand," (not yet published and apparently not going to be, quant-ph/9812050): [See note to Asher Peres, titled "The Deep Intervention," dated 26 December 1998.]

The second selection is an old article that I will quote in full. It is by Markus Fierz: "Does a physical theory comprehend an 'objective, real, single process'?" from *Observation and Interpretation in the Philosophy of Physics*, edited by S. Körner (Dover, NY, 1957), pp. 93–96.

"Does a physical theory comprehend an 'objective, real, single process'?
by M. Fierz

The so-called 'orthodox' interpretation of quantum theory has been criticized in recent years from different sides, by Einstein, Schrödinger, and de Broglie. These critics think that the ideas about quantum theory put forward by Bohr and others are much too far away from those points of view that have been successful and productive in the development of our science for the last three hundred years. They claim, against the 'orthodox' or 'Copenhagen' school, that wave-mechanics is an incomplete theory, even inside its field of application, because this theory does not comprehend really an 'objective and real single physical process.'

Now everybody acknowledges the great success of wave-mechanics in explaining quantitatively a multitude of physical phenomena. And nobody denies that our theory, as it stands, is incomplete, as there is no mathematically irreproachable relativistic generalization. Just because of this the question arises, in what direction we should search for a better relativistic theory. It seems to us that our 'heterodox' colleagues think it might be helpful to look for a theory which can be interpreted more akin to classical mechanics or field theory. Surely it is not their aim to translate only the physical contents of quantum theory as it stands, into a language different from the one of Bohr.

I think, indeed, that the ideas developed by Bohr during the growth of quantum theory will not lose their leading character. I further expect that the new features characterizing Bohr's way of thinking will even be more dominant in a new and better relativistic quantum theory. Such a theory will lead to physical ideas even more different from those of old.

Only the future can decide if I am right or wrong. At the moment there is no possibility of proof, as no such a theory has been discovered. But we can try to justify our point of view by plausible arguments. To do so, we may ask: does classical theory describe really an 'objective and real single process'? Our critics would answer this question in the positive – at least I think so, as they are in quest of a 'realistic' theory.

If one takes 'real' as opposed to 'ideal' – against this one could argue with good reasons indeed – one might say: a physical theory never describes real events. It treats always idealized systems, as only these are capable of mathematical treatment. One may, however, claim the idealization to be a simplification only, not referring to any essential features of the system under consideration. But the distinction between the essential and the inessential already contains a theoretical element; and theory itself seems to me something ideal. I don't want to pursue further the questions following from this consideration. This would lead us to rather general problems, not directly linked to our special question. This question is: does classical physics describe or comprehend single physical events? It does indeed, but only in so far as these can be looked at as an example, contained in a class or ensemble of similar events. The judgement that such events are similar and form a class is naturally always a theoretical one. And by this judgement some features of a given event must be taken as inessential. To these so-called inessential features now belong those which make our event a single and unique one, showing up here and now. So, in some sense, physical theory never comprehends a real single process.

To make this statement clearer, I refer to Kepler's theory of the solar system, a theory belonging to a pre-physical state of science. Kepler was convinced that the sun is the centre of the universe. He held the universe to be a singular harmonic structure, a cosmos, contained in the sphere of the fixed stars. The infinite space of Giordano Bruno, containing an infinity of worlds, seemed to him to be a horrible exile. His own theory was meant to explain the one harmonic universe. He wanted to understand why there are just 6 planets, circling at well defined distances around the sun. By constructing, in a most ingenious way, the spheres of the planets with the help of the 5 regular or 'platonic' bodies, he thought he had reached his goal. The order in which he had to arrange the bodies seemed to him to be mathematically unique, and as there are just five of them, he claimed to have explained the uniqueness of our world. This still remarkable theory is very beautiful, but we no longer believe in it.

In Newton's theory, which we think to be right, the solar system is just one example taken from the innumerable systems of planets created by God in infinite space. Newton had no ability, nor did he feel any need, to explain the number 6 of our planets – and there are really more than 6.

So Newton's theory just does not achieve what Kepler held to be of utmost importance for a theory of the solar system. It foretells neither the number of the planets nor their distances from the sun. There is also no longer any essential difference between planets and comets, a difference which was very marked in all old theories of the solar system.

The strength of Newton's theory is not to explain the unique structure of this system, but to comprehend in general terms the movement of any system of planets, independently of the number and the bulk of the celestial bodies contained in it.

Classical mechanics does not give equations for a definite movement of the masses in a system, but for all their possible movements. The real movement taking place depends on initial conditions, corresponding to the constants of integration of differential equations. These initial conditions have to be known experimentally. From the theoretical point of view they are arbitrary. If there were a law stating only one initial condition to be possible, the theory would lose much of its sense, as 'almost all' of its statements would refer to something impossible. There would be no reason either to assume the only path possible to be embedded in a family of curves, fulfilling certain differential equations of second order. I think the theoretical arbitrariness of the initial conditions, the fact that they have to be given experimentally, is an essential feature of classical mechanics. It shows this theory to be one belonging to an experimental science, where the experimenter has the freedom to interfere with a system and may form an initial state corresponding to his aims.

Against this, one may point out that there are systems, as the system of planets, where we cannot change the initial state. That's true. But the laws ruling over the movements of the planets are the same as those ruling over the fall of a stone on earth. These we can explore experimentally, and we can verify them with different examples. One is the solar system, in which Jupiter with his moons can be looked at as a second small and independent one. Similarly we are able to understand the processes in the interior of the sun, where we can't do any experiments. They can be explored with the help of terrestrial experiments, because these processes are not peculiar to the sun. By this again the sun becomes but an example of a system, where such processes take place. It seems to me, that all this was clear to Newton, when he wrote his second *regula philosophandi*:

> Ideoque effectuum naturalium eiusdem generis eaedem sunt causae: descensus lapidum in Europa et America, lucis in igne culinari et in Sole, reflexionis lucis in terra et in planetis.

From our point of view, the idea of Laplace that the whole world is nothing but a huge mechanical machine seems rather queer. In this picture there is no room for an experimenter as all initial conditions are given for ever. This is quite contrary to Newton's outlook, who even in his cosmological speculations assumed God to be the great experimenter, who changes from time to time, according to his purpose, the state of the world. In Laplace's view, we cannot take the whole world as an example of a possible world, as by definition there doesn't exist any other one. The situation becomes quite different indeed if we base our cosmology on the assumption that the world is homogeneous in the large scale. If this is done, every region big enough can be looked at as representative of the whole world. A given region, for example the one we are able to survey, can be taken as an example for any region. In such a theory everything distinguishing one region from another must be looked at as accidental, and so in some sense as inessential. Thus such a theory needs to be a statistical one.

We may sum up all this with the statement that physical laws refer to reproducible phenomena only. Correspondingly, an experiment is meaningful only if it is reproducible. Although physics is not bound to experimentally reproducible events, it can only treat phenomena where every single event can be taken as representative of an ensemble or class of similar events.

Now in some sense every real event is something single and unique and happens never again. But this feature of reality is outside physical theories. As this is true for classical theory, we should not be astonished, if it is more so in quantum theory. This is a statistical theory, and as such very well fitted to treat reality, in so far it consists of reproducible phenomena. Therefore it is a logical and natural development. The dominant role the experimenter or observer plays here, if we interpret the theory as Bohr does, clarifies the general feature of physical theories. To me this seems definite progress.

If somebody wants to construct a theory, comprehending the unreproducible reality of a single event, and this seems to me the only other consequent alternative, he should not only discard Bohr's ideas on what a physical theory can be, but even those of Newton. He should look for a theory which in its whole character is more akin to Kepler's ideas – and even Kepler's theory, as he himself understood it, is highly platonic!

I do not believe that in such a way the actual problems of relativistic quantum theory come nearer to a solution."

The point of both stories is that it is our "acts" and their "consequences" that are the ground upon which scientific theories are built. Only with hindsight, it seems to me, can one *in certain circumstances* distill the idea of a "measurement" of a property intrinsic to the system acted upon. Our acts and their consequences are the iron posts of scientific reality; the intrinsic properties they "reveal" is just so much papier maché we place in between (when we can).

The wonderful lesson of quantum mechanics is that it brings these points to life. When confronted with a quantum system, we can *decide* to act upon it in one way or the other in any of a number of nonequivalent ways. The set of acts available to us is the set of positive operator valued measures: each POVM corresponds to an *act* we might perform on the system. The set of possible *consequences* associated with an act corresponds to the separate operators within the act's affiliated POVM.

Savage, in his development, goes further to tip his hat to "states of the world." These, when combined with the acts, lead to the consequences. But it doesn't seem to me that this notion is overly needed to get at a system much like the one he develops: what role do the "actual" states of the world play in our decisions? None, it seems. (Rüdiger, by the way, has been wanting to push this point of view for quite some time in the context of many worlds. But there it seems doomed to failure to me: the multiverse is not only a world void of states; it is a world void of decisions, void of acts, and certainly void of consequences.)

So we are left with the point of view that quantum measurement is about decisions, acts, and consequences. What is a *quantum system*? It is nothing more than a conduit from acts to consequences. It takes the place of the "states of the world" in the Savage system, but it is a black box that need not be cracked (and appears not to be crackable). A *quantum state* is an "if-then." It is the compendium of probabilities that summarizes our beliefs about the consequences of our acts.

Let me return to Penrose. The point is that from this view the quantum state is specifically *not* a property intrinsic to the quantum system. It is a statement about what we can expect of the consequences to our actions. If you want to call it a "property," then it seems to me at best it is a *relational* property.

I don't think it is unsafe to liken this situation to the cryptographic problem of remote coin tossing. Alice places either bit of her choosing, 0 or 1, in a locked safe and sends it to Bob. Bob, after receiving the safe, publicly announces whatever bit he has in mind. Alice then relinquishes the key to her safe. The outcome of the official "coin toss" is the XOR of the two previous bits. How would Penrose interpret this process? Is the outcome of the coin toss an *objective* property of the bit in Alice's safe? Would he answer **YES**? I don't think he would, and I am left wondering why he would do otherwise for the quantum state.

Decisions. Acts. Consequences.

22 September 1999, "Andrei"

My present opinion – the one that has really evolved since we last met – is that there is probably no choice in the classical equations of motion. That is, once one fixes that the subject of discourse is about the evolution of probability distributions over a phase space. Then to say that one has maximal knowledge about the evolution is to say that the distinguishability of any "unknown probabilities" neither increases nor decreases. One then has that the overlaps between any two phase space distributions must be constant. And presumably – but this is a research program, not a definitive statement – an analogue of

Wigner's (quantum) theorem on symmetries will step in to save the day. The only "flows" with that property (connected to the identity) are "unitary" flows. And these correspond to the standard Hamiltonian dynamics that we all know. That's the idea anyway. (This, by the way, has evolved from my starting to think of classical physics as well as quantum physics as a "law of thought." We have become accustomed to thinking of classical physics as more "ontological" than its quantum counterpart ... which we have started think of as more "epistemological." But I now think that was just wrong-headed. Kant was a pretty smart man.)

17 June 2000, "Quantum Pedophiles"

I'm reading this article of Jaynes ("Predictive Statistical Mechanics") that I wrote you about and I'm realizing that actually I had never seen it before. This is the first time I'm reading it, and it (or at least the first seven pages of it) is wonderful!

The astonishing thing is that the program he pushes for in this paper is significantly similar to the one I've been pushing for: an information–disturbance foundation for quantum mechanics. The main difference is that he hasn't yet reconciled himself to the idea that "maximal information cannot be made complete (even conceptually) because of the strength of the disturbance."

Remember what Bennett wrote me:

Probably to one uncorrupted by many worlds, your idea seems perfectly sensible and beautiful. My second thought is the realization that you encountered Jaynes in your formative years, at around the mental age I was when I encountered whatever makes me loyal to my Church. In other words, in both cases, we encountered what might be described as a pedophilic idea, an idea so seductive as to be dangerous to our youthful selves and make us dangerous to others when we grew up, or were thought by others to have grown up.

Until today I hadn't realized how true that might be.

03 September 2000, "One Sentence?"

What could I do to entice you to carefully translate just one sentence for me? It's below, you can't miss it. The passage is from: C. P. Enz, "Quantum Theory in the Light of Modern Experiments," in *Advances in Scientific Philosophy: Essays in Honour of Paul Weingartner on the Occasion of the 60th Anniversary of his Birthday*, edited by G. Schurz and G. J. W. Dorn (Rodopi, Amsterdam, 1991), pp. 191–201.

I hope to have conveyed with the above discussion of experiments the Schopenhaueresque feeling for the *freedom* of the quantum objects for potentially *being* both, particle and field, located and moving, in phase and numbered, ..., before facing the *necessity* of being determined by the *action* of the measurement. Indeed, one of the new ideas Schopenhauer proclaimed is this: "Ich hingegen sage: jedes Wesen, ohne Ausnahme, *wirkt* mit strenger *Nothwendigkeit*, dasselbe aber *existiert* und ist was es ist, vermöge seiner *Freiheit*." But this new thinking became possible only through an ever finer subdivision of the objects of science or, to let Pauli speak once more: "Science is a systematic refinement of the concepts of everyday life revealing a deeper and [...] not directly visible reality behind the everyday reality of the colored, noisy things."

Rüdiger's Reply

My pleasure:

I, however, say: every being, without exception, *acts* with strict *necessity*, but *exists* and is what it is, by virtue of its *freedom.*

05 April 2001, "Waking in the Hundred Acre Wood"

Thanks for the vote of confidence. I needed that first thing this morning. (But I guess insecure people always need that, no matter what the timing.)

I'll fax you the figures as soon as I get in to the office this morning.

Thinking of von Neumann measurements as more special than other POVMs carries a certain anti-Bayesian danger. Not least of all by laying more credence on the Church of the Larger Hilbert Space point of view than I prefer to do: it says there is something special about extending the Hilbert space; that we have a necessary reason to do it, even if only to understand measurement. (One can write incomplete sentences in email without my wrath.) But also, I am reluctant to take the popular expo of POVMs as a necessary thing because of this way of viewing quantum-state change as a generalization of Bayes' rule (that I've been promoting). You can read about it in my NATO paper when it's finished (hopefully by the end of the month), or we can talk about it in Sweden.

From still another front, I try to make the point sharp in my talks: I put up a slide that has two columns. On the left-hand side there is a list of various properties for von Neumann measurements. On the right-hand side, there is an almost identical list of properties for POVMs. The only difference is that it is MINUS the orthonormality condition required of a von Neumann measurement. Then I have an arrow pointing up to the orthonormality condition on the von Neumann side with this caption: "Does the addition of this one extra assumption really make the process of measurement any less mysterious?" I follow that by saying, "I imagine myself teaching quantum mechanics for the first time and taking a vote with the unconditioned students. Which set of postulates for measurement would you prefer? [Pause.] They'd only look at me with blank faces. And that's the point! It'd make no difference to them, and it should make no difference to us."

Damn, now I spoiled my talk in Sweden!

28

Letters to Robert Schumann

22 February 2000, "Your Project"

(Please remind me of our conversation in Italy. I remember your name, but I can't quite picture you.)

Your proposal sounds interesting: it wasn't too bold. Why don't we carry on just as you suggested. I'd be happy to supervise a project of yours (or perhaps collaborate) ... as long as the work concerns something that interests me. Presently most of my interest is in quantifying different aspects of the trade-off between information and disturbance in quantum mechanics. This includes deriving various properties for some information functions that have little coverage in the existing literature so far. If we could find something that pleases us both in that general area, I think it'd be great. (Presently I'm not too interested in tackling entanglement questions: I can't see the forest for the trees in that subject.)

About collaborations, I only have one hard and fast rule at the outset: that is that author ordering be alphabetical. I place my form letter concerning this issue below. If you feel that this would hurt you, we can agree to keep MY role in the project in a strictly supervisory capacity.

Let me know what you think: I'm always within email contact. (I'm in Haifa, Israel for the week. Next week I will be back in New Mexico. Then a week following that I will be in Japan.)

23 February 2000, "OK"

Now that you've prodded my memory, I remember you quite well. I'm sorry that I had forgotten.

I'm glad to hear that you wouldn't mind thinking about information vs. disturbance issues. My ultimate hope is to use them for a better grounding of quantum foundations, but there are plenty enough practical things (like quantum cryptography) that they are connected to that the wider community will still have respect for the work.

Two possible projects come to my mind presently. One has to do with extending our knowledge of a quantity that Bill Wootters calls the subentropy. This is because I want to use it as a quantification of "how much" one knows, when one possesses a density operator. From there I plan to use it to explore how much "information" *I* lose about a system when *you* learn something about it (via a measurement). I already have preliminary results showing that this will be a quite interesting investigation. The practical implication of this question is that it can also tell us something about controlling quantum systems (a kind of quantum control theory).

The other project has to do with quantifying how "quantum" a set of quantum states are with respect to each other. In particular I quantify it by a game related to quantum cryptography: how good is the best intercept–resend strategy? (This is closely connected to something I call the "minimal disturbance at maximal information.") What is interesting about the measure I propose is that it *appears* that for a Hilbert space of dimension d, one can always find a set of d^2 states that is just as quantum (with respect to each other) as they can possibly be. I.e., one does not have to make use of the full Hilbert space of quantum states to see maximal quantumness. I have just a load of open questions about this new information function.

In the case of the first project, the two things to read are:

1. R. Jozsa, D. Robb, and W. K. Wootters, "Lower bound for accessible information in quantum mechanics," *Physical Review A*, **49**, pp. 668–677, 1994.
2. W. K. Wootters, "Random quantum states," *Foundations of Physics*, **20**(11), pp. 1365–1378, 1990.

Especially the first of these papers. What I mean by "reading" a paper is working through every step (with pencil and paper), making sure that you understand exactly what they are doing. I estimate that it will take about a week (or possibly two) of real work to get through these papers in that kind of way. As I recall, that is how long it took me, and I was at the same academic level as you.

In the case of the second project, the only thing I have in print so far is: C. A. Fuchs, "Just Two Nonorthogonal Quantum States," quant-ph/9810032. This unfortunately is a sort of review paper with no detailed calculations. Also the direction I'm going in now is to generalize that concept to n-state ensembles. I do, however, have 23 pages of notes and calculations prepared already on this project. I think they're pretty readable already. If you will send me a mailing address I will have them shipped express to you as soon as I get back to Los Alamos early next week.

I have picked both projects so that only a minimal amount of reading is required. (That is because in both cases almost no literature exists.) So it is just a case of understanding a minimal amount of stuff, and then jumping into the research waters yourself. Once you understand both these areas, I'll send you a list of open questions, ideas and suggestions. Then we can sift through that and see where to go next. I'd be more than happy to roll up my sleeves and contribute to either project as you develop it.

Also could you send me a fax number? I may be able to fax you a subset of the notes from here. Finally, can you read and write LaTeX? When I collaborate long distance, I most always write my notes in LaTeX so that formulas can be displayed easily.

I'm happy to hear that you've gotten Bill Wootters thesis. In my opinion there's no deeper thinker in our field than Bill. (And I think that thesis was really on the right track: it would be wonderful to get quantum mechanics out of a maximization principle. I think the only thing Bill was missing is that it will have to do with both information AND disturbance.) By the way, to get to know me a little better philosophically – and the direction I'll be sending you, but in a more technical way – you might *skim* some of the stuff I've posted at my website.

20 March 2000, "Ontology"

I see you found my letter "Fuchsian Genesis" but somehow you didn't absorb the main point of it. I do want ontology, just as much as the next guy. I just don't want a cheap ontology. I think quantum mechanics is giving us the grand opportunity to lift the veil a bit, to see how alive and changeable the world really is. Maybe look over that letter once again, but also have a look at:

Letter to Landauer, 6 July 1998
Letter to Landauer, 28 July 1998
Letter to Bill Wootters, 2 July 1999.

The project I intend to have you working on, as I see it, is one of quantum ontology: refining the tools (in your case the subentropy function) that will help us lift the veil even further. I want to understand the subentropy's properties better to see what they can better tell us about the tradeoff between information gain and information loss by various observers in a measurement situation. It is an epistemological tool that will help us say something about an ontological property of the world: the world is sensitive to our touch.

You were right that I do want a principle. But I don't think that precludes "inferential power." Just the opposite. The way I see it is that once we pin down the principle behind quantum mechanics, then the floodgates to new science will open before us. I like to liken our situation with quantum mechanics to that of Lorentzian mechanics five or so years before Einstein. The Lorentz transformations (as listed on the back cover of Taylor and Wheeler's book, for instance) existed – that's why they're called the Lorentz transformations. But the principle they captured was a mystery. When Einstein finally saw a principle – that spacetime is a manifold – everything changed. This was not because it immediately gave much more inferential power, but because it laid the groundwork for the much greater realization to come, the general theory of relativity. Einstein needed to see that spacetime was a manifold first before he could even imagine asking whether it was a curved manifold. We too need to see that quantum mechanics hints that the world is a _____ before we can contemplate asking whether it is a _____ _____. Whatever this great principle is, it's likely to be right here in front of us just waiting that we fix our attention upon it. When we break the mental barrier and see it, the world will change.

By the end of the week, I will send you a concrete project that I think you should go through the motions of. It'll be based on a paper that Kurt Jacobs and I are writing presently. It is to explicitly work out a relation for the gain and loss of subentropy for a two-level system. It is the sort of thing that I know can be done – I could do it myself if I had sufficient stamina – and so, at the very least, will make you feel that you've done something real. (I once had a quite satisfying summer digging post holes for a gas pipeline company. There was no question that at the end of the day I had done something.) Also, I will collect the book by Aczél and Daróczy before departing for Scotland and start to think more deeply while I am there whether we can give some classical sense to the subentropy.

See further poetry below. [See note to Greg Comer, titled "Rabbits on the Moon," dated 20 February 2000.]

20 March 2000, "Interventions"

You might also have a look at the article Asher Peres and I wrote for *Physics Today* (March issue). It comes out sounding slightly more negative than I had hoped for – between fighting with Asher and fighting with the editors over the appropriate choice of words, I finally just gave up. But the message is there between the lines if you look closely. I say, give it up if what you're looking for is a free-standing reality independent of our interventions: we're part of the world and we're learning that we have to take that into account in a deep way.

By the way, I never adequately said anything about Jaynes. Charlie was right, he was quite an influence on me. I'm pretty Bayesian to the core now when it comes to probability theory. Where Jaynes and I differ though is in what we think the ignorance can be about when one writes down a probability function. Jaynes explicitly thought that probability could only concern the truth value of a hypothesis. This led him to think that there must be some kind of hidden variables underneath quantum mechanics. I, on the other hand, am willing to attach probabilities to the consequences of our interventions into the course of Nature. Those consequences have no truth value before the intervention (and maybe not even after it). Just because one has ignorance, it does not follow that it can always be relieved (as Jaynes had hoped).

20 March 2000, "The Blank"

Schumannism 1: *Essentially, talking about disturbance reminds me of Bohr's rhetoric, and indeed your position sounds like his because you don't have the _____ in which quantum theory takes place.*

OK, I'll fill the blank:

nou·me·non ('nüme,nän) *noun*
plural **nou·me·na** (-ne)
Philosophy.

1. An object that can be intuited only by the intellect and not perceived by the senses.
2. An object independent of intellectual intuition of it or of sensuous perception of it. Also called thing-in-itself.
3. In the philosophy of Kant, an object, such as the soul, that cannot be known through perception, although its existence can be demonstrated.

The noumenon, whatever it is, is malleable, changeable, utterly sensitive to the touch. So much so, that the only grasp we can get on it necessarily involves information, unadulterated agent-centered information. (Information is, by definition, not an ontological entity.) That's the direction I want to go anyway, at least until someone shakes some better sense into my head.

It's indeed probably better to focus on a little mathematics today. Do you completely understand the derivation of Eq. 3.273 in my thesis? Also the manipulations between Eqs. 3.284 and 3.288? You'll have to be able to do manipulations like that for the project I'll propose to you by the end of this week.

25 August 2000, "Start of the Day"

Schumannism 2: *My general feeling is to agree with the Bayesian view because of its consistency – although it leaves a hollow feeling in someone who, deep down, is a realist. (Can we use a subjective interpretation in understanding the physical universe?)*

Deep down, I am a realist, though one of a strange kind of sort (more like Schopenhauer). The thing is not to confuse probability and reality; the two things are different. Bayesians are pretty good about this. But most of the time they carry that forward in too naive a fashion.

Anyway, I say this more eloquently below. [See note to David Mermin, titled "Great Garrett Quote," dated 23 July 2000.] Today, I spend the whole day with your thesis.

29

Letters to Abner Shimony

16 December 1999, "Evolutionary Laws Paper"

I obtained this email address for you from Asher Peres. I have seen in an advertisement that you have written an article titled "Can the Fundamental Laws of Nature Be the Results of Evolution?" I'm quite interested in that subject. If possible, I would like to obtain a copy of your paper. You can either send a hard copy to my address below, or simply email a file of it (I can read TEX, LATEX, PostScript, MS Word).

[NOTE: The updated reference to that paper is A. Shimony, "Can the Fundamental Laws of Nature Be the Results of Evolution?," in *From Physics to Philosophy*, edited by J. Butterfield and C. Pagonis (Cambridge University Press, Cambridge, 2000), pp. 208–223.]

20 January 2001, "Nonontology"

I'm writing to invite you to a meeting and to remind you of a challenge you once presented (in the hope that the latter might help entice you to come). All the information about the meeting is further below. That's the easy part.

About the challenge, I refer to something you wrote in 1978: [See note to Jon Waskan, dated 11 May 1999.]

Well, opposing you, I do find a nonontological interpretation of quantum mechanics quite appealing. On the other hand, I think you were absolutely on the mark with your statement, "I do not believe that a fully worked out and coherent formulation of a nonontological interpretation ... exists in the literature." Moreover, I don't think much has been done to fill the gaps in the 22 years since you wrote that. BUT, as David Mermin will tell you, I am a mad optimist, and I believe in my heart that we're just on the verge of getting there. The missing link all this time has been that we didn't have the proper tools for EVEN asking the proper questions. That, however, I believe is changing with the advent of quantum information theory and quantum computing. And thus my involvement in this meeting and the particular list of invitees you'll find below. (Also in a separate note, I'll forward the invitation I wrote to Henry Folse. I think this will give you a better feeling about what I'm up to.)

I do hope you can join us. I want to crack this problem you posed so long ago, and I think your attendance would be just wonderful in that regard.

04 April 2001, "Using Shannon to Build on Bohr"

Shimonyism 1: *I am puzzled by your remark that you would read the paper that I sent you. What paper was that?*

I was referring to the paper you recommended in this passage:

Shimonyism 2: *You have a good memory and an even better temperament, to continue to be optimistic about your nonontological program. I hope to discuss the matter with you, in Sweden or elsewhere, but I suggest strongly that before we meet you read something relevant that I've written since 1978: "Reality, Causality, and Closing the Circle", in vol. 1 of my collection of essays entitled Search for a Naturalistic World View (Cambridge University Press, 1993).*

You later sent it to me in the mail. I have absorbed most of it, but will certainly give it another read before June.

Shimonyism 3: *I have no plans to write a paper especially for the conference at present, though I should be happy to send old papers that might be of interest to you and other participants.*

I had only meant that we would try to find time to discuss the paper above during the meeting. If you have any other papers that are appropriate, please send them to me and I'll see to it that they make it to onto the display table with the other preprints/reprints. That's what I really meant by being a virtual participant.

The main thing I want to explore the prospects for is captured in the original note I wrote you concerning Växjö. I'll place that note below so you can read it again. I was especially intrigued by your remarks about a *similarity* between the programs of Kant and Bohr. In my own mind, I'm even more intrigued by the possibility of a *similarity* between the programs of Schopenhauer and Bohr ... and that's what I want to flesh out. It seems to me that quantum mechanics does give us a little knowledge of the thing-in-itself. But that thing-in-itself is not the state vector, nor is it some set of hidden variables: Rather it is that *property* of quantum systems which keeps us from ever knowing more of how they will react to measurement situations than can be captured by a state vector. It is the thing that shunts us away from the potential to even theorize a reasonable hidden-variable theory. That property I view as something active, interesting, and good. Giving some mathematical substance to it is what I'm about, and precisely where I think the tools of quantum information will be most handy. As I see it, most of quantum mechanics – but not all – is about the optimal processing of subjective information (not so unlike Bayesian probability) in light of that fundamental situation (imposed by the world).

I hope that helps you in the selection of papers you might send.

30

Letters to Jon Waskan

24 April 1999, "Thinking Kantian Thoughts?"

The other day I wrote the little essay below [cf. "Fuchsian Genesis" addressed to Greg Comer, dated 22 April 1999] for myself, and then searched up a reason to send it to my friend Greg Comer. It would have been better to send it to you I think. It strikes me as sounding a little Kantian, though with an extra twist: namely, that we are even further removed from the "ding an sich" than he had imagined. On my view, we know that the quantum formalism CANNOT correspond to reality (whatever it may be). My understanding of Kant is that he would have said that *classical* physics may or may not be a reflection of the real thing, but we can never know. It does form part of the categories of understanding and is therefore necessary to make sense of the world; but correspondence to reality is a different issue – in fact, one that cannot be decided. In the case of the quantum world, I think it is (reasonably) safe to assume that the terms in quantum theory cannot be mapped to any sort of "ding an sich." That's why it strikes me as a still stronger form of Kantianism.

Am I way off track here? Any thoughts?

Jon's Reply

> Chris said: My understanding of Kant is that he would have said that *classical* physics may or may not be a reflection of the real thing, but we can never know.

This is what Kant *should* have said. In fact, he said that the noumenal realm does not have any of the properties (time, space, causation) attributable to the known world of objects and events. I have often wondered about this claim myself, and I have never been able to come up with a convincing reason to prefer Kant's view to yours.

> Chris said: It does form part of the categories of understanding and is therefore necessary to make sense of the world; but correspondence to reality is a different issue – in fact, one that cannot be decided. In the case of the quantum world, I think it is (reasonably) safe to assume that the terms in quantum theory cannot be mapped to any sort of "ding an sich." That's why it strikes me as a still stronger form of Kantianism.

This part I'm not so sure about. With the disclaimer that Kant's project was to give a sense of necessity to Newton's laws (they were not just inductive generalizations a la Hume): When we do science, according to Kant, we are examining a world of objects and events conditioned by the faculties of understanding. This does not always require immediate perception (Critique A225/B272). It only requires that these objects have some causal influence on us (again, we're in the phenomenal realm).

Kant was a realist about theoretical entities. E.g., though we cannot see it, he thought there were good reasons (e.g., hot springs and volcanoes at various points around the planet) for believing in a "central fire." He also wrote, "From the perception of the attracted iron filings we know of the existence of a magnetic matter ... Our knowledge of the existence of things reaches, then, as far as perception and its advance according to empirical laws can extend" (A226/B273). While there remains the possibility that the noumenal realm conforms to the same laws as those governing thought, science has no bearing on this issue according to Kant. The big question is this: What would Kant say about the world of very small theoretical objects? Even subatomic objects should conform to the rules of the understanding that he outlined. When mere observation starts altering the object of inquiry something seems to be amiss.

Maybe he has three options:

1. Kant can say that we just don't know the mechanisms, so it all looks mysterious (And how many times has *that* happened in the history of science?).
2. Kant can revise his system and say that there were more rules of thought conditioning heaven and earth than were dreamt of in his philosophy (highly unlikely that Kant would admit this since his analysis of the faculties was both exhaustive and without possibility of error; this may be close to what you're after though).
3. He can admit he was wrong and that when we do science we really are investigating the noumenal realm. At the macro level, the laws of thought and the laws of nature are in conformity (it sure makes good evolutionary sense that this would be the case). In the case of the really big and really small, however, all bets are off. The best we can do is gain some predictive leverage within the constraints imposed by our own representational system. That would explain the proliferation of metaphors rooted in Newclidian (hey, I like that) spacetime. Space is curved, malleable, and what-not.

Kant, I imagine, would have to go with (1).

11 May 1999, "Nails and Heads"

You hit the nail on the head with the second of your three options:

Waskanism 1: *Kant can revise his system and say that there were more rules of thought conditioning heaven and earth than were dreamt of in his philosophy (highly unlikely that Kant would admit this since his analysis of the faculties was both exhaustive and without possibility of error; this may be close to what you're after though).*

That is to say, I think this is indeed the sort of thing I'm after. (But – warning! – I don't care so much about what Kant himself would have thought of the program. I'm just encouraged to know that it is along the same lines as the one he set out with respect to classical physics.)

I read a passage in an old paper by Abner Shimony the other day [A. Shimony, "Metaphysical Problems in the Foundations of Quantum Mechanics," *International Philosophical Quarterly* **18**, 3–17 (1978)], and it struck me that it was a partial elaboration of (and answer to) the point you made. I'll scan in the passage for both our sakes:

I shall discuss the latter of these two options first – the abandonment of a realistic interpretation of quantum mechanics of physical systems – since it is the more familiar in the literature and indeed seems to have been the view of Niels Bohr, and with some variations, of various other advocates of the Copenhagen interpretation of quantum mechanics. Bohr's former assistant, Petersen, quotes Bohr as saying, "There is no quantum world. There is only an abstract quantum physical description. It is wrong to think that the task of physics is to find out how nature *is*. Physics concerns what we can say about nature." Petersen says elsewhere that the radical way in which Bohr broke with tradition was to abandon an *ontological* mode of thinking. I am not fully confident that this is the most accurate exegesis of Bohr's rather cryptic philosophical statements, but it does seem to me to be a way in which one can deny the strong version of EPR's reality criterion. It is possible, in my opinion, that we shall be forced to accept this nonontological philosophy as a last resort, but for several reasons I find it very unappealing.

(3) Most important, I do not believe that a fully worked out and coherent formulation of a nonontological interpretation of quantum mechanics exists in the literature, in writings of Bohr or of any one else. We have a standard of comparison, for there is one great and quite fully worked out nonontological philosophy which antedates quantum mechanics: namely, the transcendental philosophy of Kant. It is, of course, central to Kant's philosophy that we have no knowledge of things in themselves, but only knowledge of phenomena. But then Kant undertakes the obligation of explaining why our knowledge has the structure that it does have, if that structure cannot be attributed to the things-in-themselves: the structure is imposed by the knowing subject, the faculties of intuition supplying the forms of space and time and the faculty of understanding supplying the categories. I am completely unconvinced by Kant's explanation of the structure of our knowledge, but I do admire him for his sense of responsibility in undertaking to work out a detailed epistemology, instead of elliptically stating a program. What, according to Bohr, plays the role of Kant's transcendental self in establishing the structure of our physical knowledge? It seems not to be the knowing self since Bohr explicitly denies that he is any kind of idealist. If it is the character of experimental apparatus, then a nest of troublesome problems is opened: is the apparatus to be considered as "real" in an unequivocal sense, so that the position is not nonontological at all, but rather is a kind of ontological commitment to macrophysicalism? And is it not strange that the macrophysical apparatus has to be described in microscopic terms, and specifically quantum mechanically, in order to understand how it works – which would be a peculiar kind of macrophysicalism? In short, I am gladly willing to say that Bohr is a great phenomenologist of scientific research – that he has said some profound things about experimental arrangements – but I do not see that he has been able *successfully* to bypass the question of characterizing the intrinsic states of physical systems.

In opposition to Mr. Shimony, I do find a nonontological interpretation of quantum theory quite appealing. However, I have to agree with him that no one has worked out the details of such a thing yet. (But then he gets way off track again by talking about experimental apparatuses.)

There's so much great work to be done!

23 August 1999, "The Point"

I'm sitting here reading another article about the similarities between Bohr's thought and Kantianism [C. Chevalley, "Philosophy and the Birth of Quantum Theory"]. And I found a couple of sentences that seemed relevant to my discussions with you – ones that I think should be kept in mind whenever we tackle this subject again. Let me record them so that I'll have them in my files.

If we change our standpoint and look back at the genesis of quantum theory as if it belonged to the history of philosophy, we see that the destructuration of the theoretical language of classical theories had as its counterpart a destructuration of the philosophical language of Kant's theory of knowledge. This internal breakdown took support from a tradition that is little visible now but can be identified, and it left almost nothing intact except Kant's fundamental stance that we are finite beings and hence cannot ever compare our representations to essences.

Jon's Reply

I see your Chevalley and raise you a Hegel:

> Kant's so-called critique of the cognitive faculties, Fichte's [doctrine that] consciousness cannot be transcended nor become transcendent, Jacobi's refusal to undertake anything impossible for Reason, all amount to nothing but the absolute restriction of Reason to the form of finitude ...
>
> *(Hegel, Faith and Knowledge, p. 64)*

The philosophical tradition "that is little visible now" is probably that of Jacobi and Fichte.

31

Letters to Bill Wootters

31 October 1997, and to Asher Peres, "Premonition?"

I was awakened this morning just in the middle of a complicated dream. I only remember very fuzzy fragments, but there is one in particular that I thought you two would enjoy hearing about. Someone walked up and handed me a newspaper clipping. The headline read, "Bill Wootters and World Renowned Teleporter Asher Peres Attain New Heights." I didn't see the article, but the headline was accompanied by a picture of you two both wearing lederhosen and Alpine hats, smiling at the camera!!

21 July 1998, "Quantum Giggles"

I'm sitting at a little sidewalk cafe in Benasque, Spain thinking of you and all your efforts to find a compelling structure underneath quantum mechanics.

It's strange, but for me this has turned out to be a very foundation-oriented conference. Carl Caves and I have been working to construct a quantum version of de Finetti's theorem in classical probability theory. de Finetti's theorem gives the "subjectivist" a way to interpret an "unknown probability" within his framework: probability distributions over the probability simplex are nothing more than shorthand for a certain kind of probability assignment (called "exchangeable") over a large multi-trial space. What is an unknown quantum state? If you take quantum states to be states of knowledge, as we do, then an unknown quantum state is a troublesome concept. We think that it can be fixed up in roughly the same way – one just has to identify the right notion of an "exchangeable density operator." We've got a weak result in that respect, but are still searching for a stronger one.

On top of all that, Adrian Kent showed up and so I've been having more philosophical fun than I should. Even Richard Jozsa has jumped into the debate. Richard thinks quantum states are much realer than I think is safe. So the discussions have been lively at the least ... and scathing at the most!

I don't know that I've ever heard you take a strong stance on what you think a quantum state represents. What is a quantum state? I know you once mentioned that you think the many worlds interpretation is a clean solution, but that you like a mystery. What's the best indication your mystery gives to the meaning of a quantum state?

On another strand, I reread your little paper "Is Time Asymmetry Logically Prior to Quantum Mechanics?" yesterday. That was fun: I do like to think that time asymmetry is prior to it all – for quite a while I've believed that the flow of time is one of those iron posts that John Wheeler speaks of. The reversible equations of motion that we use in our predictions are just the papier-mâché of theory we put in between. I don't suppose you've made any progress in the direction defined by that paper? If you've ever had any more thoughts on it, I'd love to hear them.

Let me now give you a bit of an introduction to the silly stuff I'm gonna send you. I think it is pretty darned hard to uphold the idea that a quantum state somehow corresponds to an objective state of affairs out there in nature. That is not to say, however, that I think quantum mechanics is completely divorced from some kind of ontological statement. There are lots of structures within quantum mechanics: POVMs, tensor products, entanglement, unitary evolutions, etc., etc. To fixate on taking the quantum state itself as the sole possible referent to reality in the theory and to struggle hard to interpret that weird reality, strikes me as really misguided. That's what these scattered notes are about.

30 August 1998, "Oh So Close"

For historical interest, let me tell you about something I ran across the other the other day in case you haven't seen it. It's an essay by Eugene Wigner titled "The Probability of the Existence of a Self-Reproducing Unit" (found in *The Logic of Personal Knowledge: Essays in Honor of Michael Polanyi*, Routledge and Kegan Paul, London, 1961). What's so interesting about it is that it is concerned with the possibility of "cloning" in the biological sense and it identifies that question precisely with "cloning" in the quantum mechanical sense!! Somehow though – amazingly! – he never quite got to a crisp statement of the no-cloning theorem. Instead he takes as given an interaction S between the "organism" and the "nutrient" and asks if there are any solutions $|v\rangle$ (state of organism), $|w\rangle$ (initial state of nutrient), and $|r\rangle$ (final state of nutrient) to the cloning equation

$$S|v\rangle|w\rangle = |v\rangle|v\rangle|r\rangle.$$

He concludes that for a random S, the probability of there being a solution is vanishingly small.

What's even more surprising is that he really did get to the brink of the problem with cloning. He just didn't see the conclusion. For he says, "There must be many states $|v\rangle$ all of which represent a living organism. . . . Let us denote the n vectors which represent living organisms by $|v^k\rangle$ . . . . Then every linear combination of the $|v^k\rangle$ will also represent a living state." He apparently didn't see that this just can't be done at all for any S!!! Instead, he says that only Ss "tailored so as to permit reproduction" will work.

In light of this, I find it now even more impressive that you fellows were able to get to the nub of what unitarity is: no cloning, no information without disturbance. Wigner had just the right sort of question in front of himself for ferreting this out and somehow missed it.

On another matter, I took your story of your conversation with your son about moving his hand with his will as a good excuse to read and think about some of Schrödinger's old writings again. (I always get a little overly philosophical on Sundays.) They're just marvelous.

16 September 1998, "The Aphorism"

Last night, I saw another episode of the show on PBS that I told you about: the name is "Oliver Sacks: The Mind Traveler." This one was nice, I think, in that it really brought even closer to home your aphorism about the dog and its understanding of the universe. The topic was a genetic disease called William's Syndrome (due to part of the 7th chromosome being missing). What's really intriguing about this is that this "defect" seems to enhance people's verbal and social skills ... making them extraordinarily pleasant to be around and extraordinarily receptive to other people's feelings, facial expressions, etc. Moreover, it seems even to enhance natural musical ability! Many of the afflicted have perfect pitch. In the tradeoff however, one looses a strong sense of spatial relations and even very basic mathematical skills (as in counting accurately). The example in the show that most clearly exhibited your aphorism was in the asking of a six year old, socially very adept girl – almost adult-like actually – to rebuild a pattern in blocks that was just built in front of her and left for her examination: it was just a simple cross, like this +. After something of a struggle, the best she could come up with was a T ... and then she didn't even perceive that there was a difference between the two patterns. Really very intriguing.

14 November 1998, "Magic Eight Ball"

By the way, have I shared the imagery of the Magic Eight Ball with you before? It has its roots in a debate/discussion with Michał Horodecki about what it is that is really sent through quantum channels anyway. We were talking about teleportation in particular. Michał said, "Something certainly travels from Alice to Bob." I said, "If you ask me, the only thing that is teleported is what a Victor has the right to say about what's in Bob's hands. Quantum systems are more like empty boxes out of which answers pop when we ask questions; they don't actually contain the state vectors we ascribe them." Anyway, I returned from Spain and told Kiki about this business – I caught her in a rare moment when she actually wanted to talk about this(!) – and she said, "Oh you mean quantum systems are really Magic Eight Balls?" I loved it! Especially since the classical point of view is that the world is made of little billiard balls. The last vestige of Democritus's atomism is the eight ball ... the magic eight ball!

17 November 1998, "Supplement on Undergrads: In Case It's Useful"

Undergraduate Research and Quantum Information

When I was an undergraduate at The University of Texas I had the opportunity to be associated briefly with the research group of John Archibald Wheeler. Two things about Wheeler's style made a great impression on me. First, he viewed research for both graduates and undergraduates in precisely the same light: it's a frying pan and you've just got to jump into it! It got him results, and it trained a generation of excellent theoretical physicists. The second thing came in a question–answer session at the end of a seminar. Someone in the audience asked, "Do you see a difference between the students at Princeton and the students at UT?" His answer was just as clean and as simple as it should have been, "Yes I do; the students at Princeton *know* they're smart. Next question."

If you want to know my philosophy of how to advise research, then you need go no further than the paragraph above. Great discoveries are waiting to be made at all levels of science. And if there ever was a frying pan to jump into for the undergraduate, it is quantum information. Some of the greatest discoveries of our field have been very literally at the level of a third-year undergraduate quantum mechanics course. There is not a student who has studied Vol. III of *The Feynman Lectures on Physics* who could not have discovered quantum teleportation for himself. There is not a student at that level who could not have discovered the idea behind quantum cryptography. Wonderfully, these are not isolated incidents: there is a sense in which they define what the field is about. The field is about looking at quantum mechanics in a new way and wringing everything we can from it. The only tool a student really needs for a start in quantum information is to *know* that he's smart.

The bulk of present-day research in quantum information is truly an interdisciplinary effort. Take quantum mechanics, computer science, information theory, and linear algebra, put them in a bowl and mix. Because the field is in its infancy, the use of undergraduate-level ingredients from each of those disciplines is far from exhausted. There is just so much fun work to be done; one cannot help but be thankful for the army of eager, questioning undergraduates that will teach their professors so much.

02 January 1999, "The Holidays"

A little while ago I opened up a nice holiday greeting from Asher that contained a photo of you, Peter Shor, Richard Jozsa, and me. I wrote Asher, "Who would guess that these giants would get so close to me as to be captured in a single photograph?!?" Anyway, I hope you are doing well and are happy for the holidays.

My romantic side tends to think that there's something quite important about 1999 and indeed about 2000. I keep coming back to the talk John Wheeler gave in Santa Fe in 1994. He showed a slide of a coin with Planck's face on it and said, "In 1900 Planck discovered the quantum. The end of the century is drawing near; we only have six years left to understand why the quantum is here. Wouldn't that be a tribute?!" Maybe I'm foolish, but I really don't think it's out of our reach.

24 January 1999, "Throwing Away Old Scraps"

I was just cleaning up my old office (i.e., Emma's new room) and ran across a note I had made about a paper of Araki's. The paper is H. Araki, "On a Characterization of the State Space of Quantum Mechanics," *Comm. Math. Phys* **75**, 1–24 (1980). In it (near the end as I recall), he makes the same observation that you do in your paper "Local Accessibility of Quantum Information." The reals are too big, the quaternions are too small; complex spaces are just right. There, now that that information is part of the community of communicators, I can safely throw away my old scrap.

24 May 1999, "The Consequence of Indeterminism"

Yesterday, I made the decision to go with LANL over Amherst this year. Today, with the decision set in stone, I find myself feeling rather empty. The wonderful thing about indeterminism is the openness it gives our futures. But the awful thing about indeterminism is the openness it gives our futures.

Thanks a million for the thoughts you sent. The second set was especially inspiring. Connected to one comment you made, I couldn't help but think of two small stories about Max Planck that I had scanned into my computer a few months ago. I'll place them below; you'll see the connection.

Planck traced the discovery of his vocation to the teaching of an instructor at the gymnasium, Hermann Müller, who awakened an interest, which became a passion, to "investigate the harmony that reigns between the strictness of mathematics and the multitude of natural laws." In 1878, at the age of twenty, Planck chose thermodynamics as the subject of his doctoral dissertation, which he wrote in four months. He recalled that his professor at the University of Munich, Philipp von Jolly, had counseled against a career in physics on the ground that the discovery of the principles of thermodynamics had completed the structure of theoretical physics. That had not dissuaded Planck, who had his compulsion and also an objective far removed from the principal ambition of today's physicists. He had no wish to make discoveries, he told Jolly, but only to understand and perhaps to deepen the foundations already set.

(J. L. Heilbron)

Many kinds of men devote themselves to Science, and not all for the sake of Science herself. There are some who come into her temple because it offers them the opportunity to display their particular talents. To this class of men science is a kind of sport in the practice of which they exult, just as an athlete exults in the exercise of his muscular prowess. There is another class of men who come into the temple to make an offering of their brain pulp in the hope of securing a profitable return. These men are scientists only by the chance of some circumstance which offered itself when making a choice of career. If the attending circumstance had been different they might have become politicians or captains of business. Should an angel of God descend and drive from the Temple of Science all those who belong to the categories I have mentioned, I fear the temple would be nearly emptied. But a few worshippers would still remain – some from former times and some from ours. To these latter belongs our Planck. And that is why we love him.

(A. Einstein)

02 July 1999, "Small Piece of WV"

Last night after our conversation, on my 40-minute walk home, I kept thinking about "what's missing?" from my two little pieces below. I think I can put it in a slogan: "The world can be moved."

There is a reason we are stuck with a physics that is the ability to win a bet: if the world can be moved, you can't ask for much more.

That sentence will make more sense if you read the stuff below. [See note to Paul Benioff dated 10 June 1999.]

13 July 1999, "Weyling Away My Time"

Here's that Weyl quote concerning indeterminism that I was talking about tonight. I snatched it from Paul Forman's article "Weimar Culture, Causality, and Quantum Theory, 1918-1927: Adaptation by German Physicists and Mathematicians to a Hostile Intellectual Environment."

Finally and above all, it is the essence of the continuum that it cannot be grasped as a rigid existing thing, but only as something which is in the act of an inwardly directed unending process of becoming .... In a given continuum, of course, this process of becoming can have reached only a certain point, i.e., the quantitative relations in an intuitively given piece S of the world [regarded as a four-dimensional continuum of events] are merely approximate, determinable only with a certain latitude, not merely in consequence of the limited precision of my sense organs and measuring instruments, but because they are in themselves afflicted with a sort of vagueness .... And only "at the end of all time," so to speak, ... would the unending process of becoming S be completed, and S sustain in itself that degree of definiteness which mathematical physics postulates as its ideal .... Thus the rigid pressure of natural causality relaxes, and there remains, without prejudice to the validity of natural laws, room for autonomous decisions, causally absolutely independent of one another, whose locus I consider to be the elementary quanta of matter. These "decisions" are what is actually real in the world.

23 July 1999, "Autumn Leaves"

Boy you really took me back in time tonight – both to old thoughts about the generation of reality and to memories of a sad time. Below is an old attempt to capture some of those thoughts ... strangely enough, both the kinds you brought back tonight. (The source is from a letter to my friend Greg Comer dated 19 September 1992.) Looking back on it, I wonder why I stopped thinking about the ideas there.

I truly had a good time tonight. Thanks for sharing all those thoughts with me! There is just so much to this world, isn't there?

Greetings from Chapel Hill – once the seat of some of the deepest theoretical investigations around revealing the mutual underlying structure of spacetime and quantum theory. Or should I have said "underlying structurelessness?" So how goes it buddy? Nothing new on this end – just as the greeting hints. Just a lonely old man living out his days in the most mindless fashion possible. Remember the good old days when my notes were filled with nothing but philosophy – good old metaphysics. Remember the days when there was nothing else on my mind?

Waking up the other morning, I discovered part of the origin of the dream I wrote you about [9 July 1992] where you had entered five paintings in an art show … one painting of which wispily revealed the words "Where Randomness Persists." For some reason I woke up, put on my glasses, but remained in bed – just staring at the ceiling and walls. On one patch of wall next to my chest-of-drawers I noted some thumbtacks still remaining from Mary's move. The five thumbtacks were in just the formation I wrote you about! They used to hold up five small impressionistic paintings (postcards) all by Monet I think. This contrasts with the dream; for there all the paintings were surreal. It's funny, though, in my first days here at Chapel Hill – when I was first really focusing on this randomness business – I was seriously starting to consider an impressionistic picture for the world's structure. I would walk down the "Quantum Gravity Trail" and look at the falling autumn leaves. Here or there I would see a pattern that would faintly remind me of something – maybe a pot or a shovel or a kite. And I would think that that sort of process was the origin of structure just as in quantum mechanics. But then later it dawned on me that, for instance, there was nothing intrinsically kite-like in the set of leaves that I was picturing as a kite. With another glance I might see that set of leaves as depicting something more akin to an arrow. The structure emerging from the leaves was imposed by me … within bounds, of course, but nevertheless imposed by me. More of essence here I realized was the fact that I need never impose a structure when looking at some random pattern, but if I ever did, that very action would bring into existence another clear-cut requirement. If I were to ever say "Oh, that's an arrow pointing that way," I could never simultaneously say "But it's also an arrow pointing the other way." Everything I could ever say of the leaves was necessarily constrained to be consistent. This is where the old idea of "randomness with consistency" comes from. And where this idea that "consistency" is an adequate (and perhaps preferred) replacement for "realism" as a foundation for metaphysics. And hence this is how I started to think of the world as an impressionistic painting. Up close one can see whatever one wants in the painting, but from a view far away one is almost assuredly constrained to seeing one unique image. (I remember a note from you – after your first visit to Paris – saying that you've had similar thoughts.) Now for some reason, in the dream, impressionism was replaced with surrealism. I can only wonder if the phrase "Where Randomness Persists" dually references the five-painting pattern that used to house a set of impressionistic works and my decaying but unsettled relationship with Mary. And I am led to ponder more deeply the reason for surrealism over impressionism. You know, I think I eventually discarded part of the autumn leaves analogy – it just didn't take into account the crucial aspect of quantum theory that the leaves are falling because somebody's shaking the tree. The surrealist painter creates his own world – not one completely constrained by something already given. Whether this fits into anything, I'm not sure.

Bill's Reply

I too really enjoyed our conversation that night.

Your 1992 letter, with its invocation of the consistency constraint, is wonderful and also happened to remind me of something I once wrote in a journal (though I think what you said makes more sense than my journal entry). I just now dug it up. It was pretty far out stuff, mostly about how there is only one consciousness, which manifests itself as different people in pretty much the same way as

I can focus my attention at one moment on one thing and at another moment on something else and yet still be a single conscious being. "Now suppose that there is in fact only one consciousness, the results of whose focusing its attention are consciousnesses as we know them. When it focuses on a certain thing, it is me; when it focuses on something else, it is someone else, or it is me at a later time, etc. Thus, my own focusing of my attention is not only an analogy but also a special case of this." At some point in this metaphysical prose I suggested that, though inconsistency was somehow necessary, the world we experience might be a world of minimum inconsistency: "[One could] make a mathematical model of focusing [in the above sense], and define 'degree of inconsistency' and try to minimize it. Local minima will correspond to stable consciousnesses." That's from January, 1973.

02 August 1999, "Consistent Inconsistencies"

What a wonderful idea that the world in totality may not be consistent, but that local minima in its inconsistency might be interesting little islands. January 1973! Don't you think it's high time we started doing something scientific with these ideas!?!? (Ok, ok, so your PhD was an example in that direction. But there's still so much further to go!)

20 August 1999, "Echoes"

Once upon a time, you wrote me this:

Woottersism 1: *What I believe about the meaning of a quantum state is mostly negative: I don't think it can be taken literally as an element of reality; neither is it merely an expression of what we know. I think it somehow straddles the fence between objective and subjective, just as information does. Information is embodied physically, as Landauer points out, but at the same time, information is not information unless it is information for somebody. The same physical object can hold different amounts of information for different people. (E.g., for someone who already knows the state of the object, it holds no information.) I think a quantum state likewise, though maybe in a more profound way, lies in the region between objective and subjective, or between ontology and epistemology.*

In going through some Pauli writings again yesterday, I thought I heard a faint echo. From a letter to Markus Fierz, 12 August 1948:

"The layman usually thinks that when he says 'reality' he is speaking of something obviously known; while it seems to me just to be the most important and exceedingly difficult task of our time to work for the establishment of a new idea of reality. This is also what I mean when I always emphasize that science and religion *must* have something in common. (I do *not* mean 'religion within physics' nor also 'physics within religion' – since both would be one-sided – but inclusion of both into a whole.) What appears to me to contain the new idea of reality I would like to tentatively call: the *idea of the reality of the symbol*. A symbol is, on the one hand, a product of the efforts of man and, on the other hand, a sign for an objective order in the universe of which man is only part. It owns something of the old notion of God and also something of the old notion of object."

And again, from a letter to Fierz dated 26 October 1954:

Sought: an *"intermediate realm"* (between matter and spirit, mental or respectively "neutral")

If you have a little time lying about, I would really appreciate it – to the level of a beer, if not a whole meal or an evening of child sitting – if you would tackle (at least in a meager way) the question I posed after you wrote what you did above.

Challenge: What do you see as the cleanest, simplest argument indicating that the quantum state has some objective (ontological) content? Can you articulate that? What aspect of the quantum state lies on the objective side of the straddle?

Is your argument different than what Roger Penrose writes in *The Emperor's New Mind* (pages 268 and 269) or, almost identically, in *Shadows of the Mind* (pages 312 through 315)? (If you don't have a copy of that, I have some of the passages scanned in and I can send them to you.) This issue is a bit important to me for the paper that Carl, Rüdiger and I are writing on the quantum de Finetti theorem; but it's also quite important for the large scheme of things. If your reasoning is different from Penrose's, I would like to understand it.

Bill's Reply

I confess that I haven't taken the time to think as hard about this as I should have, or if I did think hard about it when you first posed the challenge, the thoughts I had then are not immediately accessible to me now. But here is a response, for what it's worth.

Consider Penrose's spin-1/2 particle (of pp. 268–9 in ENM) in a pure state. Now, not everyone will necessarily regard it as being in a pure state. It may be in a mixed state for you and in a pure state for me. But if it is in a pure state for me, then I will find that everyone I talk with for whom the particle is in a pure state will agree on *which* pure state it's in. In that sense the pure state is an objective property. It's just as objective as the orientation of the desk that I'm sitting at. Now, whether or not that makes it part of *reality* is something one can still argue about. (In the many-worlds theory neither the direction of spin nor the orientation of the desk is an element of reality.) I don't want to claim that the direction of spin is "pure reality." Like everything, its existence requires perceivers. But the fact that everyone I talk to agrees on the state is the aspect that lies on the "objective side."

30 September 1999, "Your Dog"

Physics is mathematical not because we know so much about the physical world, but because we know so little: it is only the mathematical properties we can discover.

(Bertrand Russell)

12 February 2000, "Feels Right"

Still another thing to come out of the process is that I've been reading Arthur Zajonc's book *Catching the Light: The Entwined History of Light and Mind*. It's great! I think you'd enjoy it too. Especially part of the discussion on some of Rudolf Steiner's ideas:

[T]he physical world is the fruit of the moral world. Pure hearts truly will illumine future worlds. Or, if we harbor darkness within, then a dark world will be its lawful consequence in the distant future. We are cocreators of the world not only through the deeds of our hands but, in even greater measure, through the spiritual impulses we foster inwardly.

13 February 2000, "Zteiner"

Woottersism 2: *I like the Steiner quote.*

Actually its a Zajonc quote, from the section on Steiner. (Sorry to bug you with another email: I just couldn't leave a misimpression.)

32

Letters to Anton Zeilinger

06 January 1999, "The (Un-)Detached Observer"

Yesterday I ran across your paper "On the Interpretation and Philosophical Foundation of Quantum Mechanics" posted on the web, and was very much impressed with its depth. Most importantly, however, I am sad to say, it reminded me that I still owe you some papers.

I have remedied that now by sending you a pile of things that I have written on the relation between information gain and quantum state disturbance. These are relevant to your request above, and, I think, even more relevant to the thoughts in your paper. I am in great agreement with the idea that there may be a deeper foundation to quantum mechanics, one along the lines that Pauli and Wheeler attempted to sketch. Quantum mechanics, it seems to me, is an expression of the best we can say in a world where the observer is not detached (to use Pauli's phrase). Recognizing and appreciating that lack of detachedness is the most exacting tool we have for peering into the greater future of quantum theory. This I think is the ultimate role of quantum information theory. (For the opposite point of view, however, you may be interested in having a look at John Preskill's paper "Quantum Information: Its Future Impact on Physics" at http://theory.caltech.edu/~preskill/.)

I hope you will enjoy the things I have sent you; it is always my attempt to be as clear as I possibly can be when I write. If you have written anything more on the point of view you espouse in the paper above, I would very much appreciate learning its coordinates.

Anton's Reply

Thank you very much for your e-mail. I look forward to receiving the package of your papers. I fully agree with the basic spirit of what you said. We have to learn a little more about the role of us in the Universe, which certainly cannot be the role of a "detached observer."

Since I wrote the paper you mentioned I have kept thinking about these problems and I feel that I made a significant step forward (one always loves one's own children best!) in having been able to identify a foundational principle for quantum mechanics. A copy is in the mail to you.

The basic point in all these considerations is that the old idea in physics of a world "out there" which we just observe and the consequent role of physics to just describe what already "is" out there is too naïve. In essence, what can be said about Nature has a constitutive contribution on what can be "real." I would be very interested to hear your responses on my paper.

11 July 2000, "Need Reference"

It was great to finally meet you the other day. Could you do me a favor and give me the proper page numbers (both beginning and end) for the article below? I'm preparing a little article titled "Resource Material for a Paulian/Wheelerish Conception of Nature" and I want it to be as complete as possible.

A. Zeilinger, "On the Interpretation and Philosophical Foundation of Quantum Mechanics," in *Vastakohtien todellisuus, Juhlakirja Professori K. V. Laurikaisen 80-vuotispäivänä*, edited by U. Ketvel, A. Airola, R. Hämäläinen, S. Laurema, S. Liukkonen, K. Rainio, and J. Rastas (Helsinki University Press, Helsinki, 1996). Also archived at http://www.ap.univie.ac.at/users/Anton.Zeilinger/philosop.html.

[The page numbers were 167–178.]

20 July 2000, "Our Mutual Interest"

I have just informed the ESI secretary that I would like to come to Vienna for the dates November 27 through December 17. I would mostly like to have that time to think, and reason, and have some time to interact with you on our mutual interest (namely, getting down to the essence of the undetached observer in QM).

I have had a chance to look at your *FP* paper that you suggested, and also a paper by you and Brukner. I sympathize with the direction of your thought, but I am reluctant to think that that specific course goes deep enough for my tastes. So we should have plenty to discuss.

In the meantime, if you would like to become more familiar with my ideas, please go to my website listed below. The long file titled "Notes on a Paulian Idea" should be particularly interesting for you. (If you're worried that it might be a waste of your time, get a referee report from David Mermin first.) I plan to go public with a longer version of that file soon, on quant-ph, just before my move.

Which brings me to some other good news: I have now obtained a permanent research position at Bell Labs. We will be building a strong group in quantum information, and my personal bent will be in turning it toward foundational studies in quantum cryptography. I'm quite serious about getting at this "undetached observer" business in a more technical way – that provides my real motivation. I should be starting there in September or October.

33

Other Letters

11 April 1996, to Michiel van Lambalgen, "Thesis"

About myself (in case you are wondering) I am a physicist. My research specialty is "quantum information theory," a field that concerns a hodgepodge of things including quantum cryptography, quantum computing, and statistical inference problems having to do with quantum mechanical systems. You may have heard of some of these things through Paul Vitanyi or André Berthiaume (Vitanyi's postdoc). I am presently a postdoc working for Gilles Brassard and Claude Crépeau in Montréal; starting October, I will have a three year position at Caltech.

I was once interested in the mathematics of randomness because, though I am Bayesian through and through for all other uses of probability, I believed that probabilities for quantum mechanical measurement outcomes were something different ... in fact something more akin to the frequentist conception. Thus I put a lot of effort into studying von Mises, Church, Kolmogorov, Martin-Löf, Chaitin, etc. (I didn't find your papers until I had pretty much abandoned this belief, though I'm not sure that I am completely over it!) I had hoped that there might be some mathematical connection between the structures found in quantum theory (vector spaces, positive-operator valued measures, etc.) and the structures required to formalize the notion of random sequences ... at least that was my motivation.

I thought about you again the other day because – while in search of another paper – I ran across a citation of your thesis. (The paper is "A Limiting Frequency Approach to Probability Based on the Weak Law of Large Numbers" by R. E. Neapolitan, *Philosophy of Science* **59** (1992) 389.) Also there have been some claims lately by Sidney Coleman – I'm sure you've never heard of him, but he's a pretty well-known physicist – that he has *derived* that strings of repeated quantum mechanical measurements on identical preparations give rise to strings random in the sense of Martin-Löf. I'm pretty confident that this is humbug ... because it is based on some earlier work by Hartle claiming to do such a thing for frequencies. And I know that that stuff, in the end, turned out to be a misinterpretation of something that was nothing more than a Law of Large Numbers argument (with a bit of a twist).

485

(By the way, there were some efforts in the early 1970s by Paul Benioff to explicitly introduce "randomness" as a postulate of quantum mechanics. Are you aware of these? The papers are pretty steeped in the jargon of symbolic logic and should be right up your alley. However, I don't think anything useful ever came of them.)

In any case, I have now started thinking about these things again. I look forward to seeing your thesis and latest things ... and, if you are interested, continuing this conversation.

15 April 1996, to Michiel van Lambalgen, "Getting Started"

Just a few replies with my coffee to get the morning started.

You say,

> No work on randomness that I am aware of differentiates between classical and quantum mechanics; but if you have an idea how this could be done, I would be most grateful for a few hints...

I wish I had an idea! As I say, I am quite Bayes- (i.e., subjective probability-) oriented in all matters except possibly quantum phenomena. I've never been able to buy it that an infinite repetition of tosses of a classical coin cannot give rise to all heads, HHHHH... *However* I am sometimes – though now *very* rarely – inclined to believe that cannot be the case for quantum coins ... for just the sorts of reasons von Mises used mistakenly for classical ones. Thus, in my mind, all the work on defining randomness was tacitly about truly indeterministic phenomena – of which quantum measurement outcomes are the paradigmatic (and ONLY as far as I am concerned) example – whether the authors intended it that way or not. This is why I would have said that "no work on randomness [...] differentiates between classical and quantum mechanics." It is a tautology. And this is why I had hoped to see some similarity in the structure of the two theories. However, who knows, perhaps after reading your thesis I'll come away with a different opinion.

Of course, compounding the problems of such an idea (i.e., relating quantum and randomness) is the fact that there is no unique notion of randomness within mathematics (as your papers were the first to really teach me). But as I've said, I'm already disposed to leaving these ideas behind.

Now I much more strongly believe that even the probabilities in quantum theory are of the Bayesian cast, quantifying degrees of belief based on prior information. One must dig deeper for the indeterminism ... which is most certainly there whether one adopts a subjective view of probability or not. I think the point worth hanging on is this. In going from classical physics to quantum physics, it is the nature of the alternatives in the probability assignments that changes, not the notion of probability itself. In this respect, maybe I do have something for you to think about, because I find myself using language somewhat similar to the things I see in your writings.

(Oops, my coffee's already run out; so much for a short note!) There is a pretty clear sense in which the outcomes of measurements on classical systems can be said to be "already there" before any measurement is performed; any surprisal on the experimenter's part upon seeing the outcome can be taken to be due to his ignorance of the "actual" situation. However, for quantum systems, the measurement outcomes "are not already there" independent of the measurement (much as I've seen you contend for random sequences). The surprisal on the part of the experimenter is again due to ignorance, but not an ignorance that could have been removed beforehand as in the classical situation. I like to paraphrase this situation with the following words. "Maximal" information in classical physics is complete in the sense that, once it is fixed, all possible measurement outcomes can be predicted perfectly; maximal information in quantum physics is *not* complete (in this sense) and cannot be completed. Because, again, the outcomes "are not already there" before measurement. Is there some real resemblance between these two uses of "not already there"??

There is a new idea that intrigues me: that perhaps, with very few side conditions, quantum theory is the unique expression for how to use probabilities when "maximal information is not complete." That is to say, this situation so restricts the use of probability assignments that, in the end, what is left can be identified with the standard structure of a quantum theory (i.e., probabilities being generated by vectors in a linear space and positive-operator valued measures). However, I don't have much of a clue about how to tackle this (still too vague) idea.

If any of this has piqued your interest, I can send you a paper by Carl Caves and myself where we lay these ideas out in a little more detail.

19 July 1996, to Sam Braunstein, "The Prior"

While in Torino, you really got me interested in the old Cox Box question again. I noticed in this version of the book that Jaynes makes some points about how there are still quite a few questions about how to set priors when you don't even know how many outcomes there are to a given experiment, i.e., you don't even know the cardinality of your sample space. That, it seems to me, has something of the flavor of quantum mechanics ... where you have an extra freedom not even imagined in classical probability. The states of knowledge are now quantum states instead of probability distributions; and one reason for this is that the sample space is not fixed – any POVM corresponds to a valid question of the system. The number of outcomes of the experiment can be as small as two or, instead, as large as you want.

However I don't think there's anything interesting to be gained from *simply* trying to redo the Coxian "plausibility" argument but with complex numbers. It seems to me that it'll more necessarily be something along the lines of: "When you ask me, 'Where do all the quantum mechanical outcomes come from?' I must reply, "There is no where there." (with apologies to Stein again!) That is to say, my favorite "happy" thought is that when we know how to properly take into account the piece of prior information that "there is no where there" concerning the origin of quantum mechanical measurement outcomes, then we will be left with "plausibility spaces" that are so restricted as to be isomorphic to Hilbert spaces. But that's just thinking my fantasies out loud.

22 January 1997, to Harald Weinfurter, "Random Recall"

I see that I had a somewhat poor recall of what von Neumann actually said about the generation of random numbers. Nevertheless, they're still kind of interesting. The reference is an article titled "Various Techniques Used in Connection with Random Digits," and it is found in The Collected Works of von Neumann, p. 768 (I don't have the particular volume written down).

The first quote is the following: "... Two quantitatively different methods for the production of random digits have been discussed: physical processes and arithmetical processes. ... There are nuclear accidents, for example, which are *the ideal of randomness* [my emphasis], and up to a certain accuracy you can count on them." Then he goes on to discuss technical problems about collecting random numbers in this way.

A little later he comes to pseudo-random numbers, i.e., ones generated by an algorithm, about which he says the following: "Any one who considers arithmetical methods of producing random digits is, of course, in a state of sin. For, as has been pointed out several times, there is no such thing as a random number – there are only methods to produce random numbers, and a strict arithmetic procedure of course is not such a method."

I hope you can find a use for these nice quotes.

29 July 1997, to Ken Alder, "Rosenthal Thesis"

Actually I have read the books by Kuhn and Heilbron that you suggest. My interest is much more in Paul Forman's ideas; I've read, I believe, three of his papers ... the main one being the one you cite. The thing that caught my eye about the Rosenthal thesis is that it apparently provides some commentary on Forman. So, yes, I would still like to obtain a copy of it: if nothing else, perhaps the secondary references will be useful to me. Also there's the basic curiosity of seeing what someone else thinks of Forman – I haven't seen too much commentary on his ideas up to now.

My main interest along these lines concerns scientists' thoughts on "indeterminism" before the birth of quantum mechanics. Of course, I am also interested in this issue of their social and cultural conditioning, but that is somewhat secondary for me. On top of the things mentioned so far, I've taken some time to study C. S. Peirce, Pais's comments on the air about Copenhagen surrounding the Bohr–Kramers–Slater paper, and also a long paper on Schrödinger and Exner (I forget the author's name). If you have any further suggestions on this subject, I would certainly appreciate it – my literature search, I'm sure, has been far from exhaustive. I profess nothing more than dilettantism in this!

Also, if you can find it, I would like the reference to Garber – I don't believe that I've heard of her before. And, finally, thanks for the tip about Kojevnikov; I will look him up soon ... I didn't realize that we had any historians around here!

PS. If I can find it, I'll dig up a letter I wrote a few years ago that may amuse you: the subject is a (playful) application of Forman's ideas to my research in the early 1990s.

(It'll be written in LaTeX, but it should still be readable with a little effort if you don't have the means to compile it.) Maybe, it'll also help you see what interests me.

18 February 1992 (from a letter to Greg Comer)

So anyway, as promised, I intend to make this note cheerier than the last. In fact I think I'll do this by describing something that delighted me to no end the other day. (And in the process maybe we'll both learn a little more about the "proper" way to implement the "Law without Law program." At any rate, I'll get some quotes into my computer that I've been wanting to get there.) You see, lately I've become quite interested in the way the LWL idea has started to infiltrate popular culture. Examples of it can now be found everywhere. I'll show you a few in just a couple of seconds. Why am I thinking about this? I'm not really sure, but probably because I don't think I'm up to any real physics right now. Nevertheless, I should say that, to some extent, I suppose that I've always considered myself a "child of the times" (*??zeitkinder??*). For instance, whenever predicting social behaviors, my usual justification is simply, "I'm an average guy and I figure he'd do what I'd do." (The word "do" here not necessarily to be taken in Charlie's usual sense!) But this of course leads me to wonder whether my randomness obsession is nothing more than a product of my cultural milieu (that's a $5.00 word) or, alternatively, evidence that I'm part of the fuel for this fire. As of right now I'm not sure which of these is the proper point of view to take or whether it really matters at all. What these considerations do lead to, though, is the following story and set of thoughts.

I suppose I should start with the examples I alluded to. At the most superficial level recall the postcard in my office that I found in the Haight-Ashbury district of San Francisco – a scene of Einstein observing the hand of God rolling a pair of dice with the Crab Nebula in the background. Or similarly the graffito I saw on a restroom wall at the Hole-In-The-Wall in Austin signed by "The Random Man." (Speaking of which, I came very close to writing below this the formal expression for a random string and signing it "The REAL Random Man.") Stretching it – I know – but consider some of the lines from the movie *Slacker* (a film about the fringe culture in Austin). For instance, the line of the Dostoevski Wannabe, "Who's ever written a great work about the immense effort required in order not to create?" Or that of the Old Anarchist, "And remember the passion for destruction is also a creative passion." Or the Disgruntled Grad Student, "Every action is a positive action even if it has a negative result." And, in particular, the soliloquy of Having A Breakthrough Day:

> It's like I've had a total recalibration of my mind, you know. I mean it's like I've been banging my head against this nineteenth century thought type ... ah, what ... thought mode, constructs, human construct. Well the wall doesn't exist; it's not there, you know. I mean they tell you to look to the light at the end of the tunnel – well there is no tunnel. There is no structure. The underlying order is chaos. Man. I mean everything's in one big ball of fluctuating matter – constant state of change. You know. I mean it's like across that great quantum divide is this new consciousness. And you know I don't know what that's gonna be like, but I know that we're all part of it. I mean it's new physics. You can't look at something without changing it. You know, anything. I mean man that's like almost beyond my imagination. Just like that butterfly flappin' it's wings in Galveston and somewhere down the road a pace it's gonna create a monsoon in China. You know.

A mouth full, huh? But I should also add that the same character followed this with the later remark, "That's all right, time doesn't exist." (By the way, the definition of a slacker is a member of "a new generation of young people, primarily centered around college campuses, that rejects the values of the generation before them.")

So what do you think? A valiant attempt of the literate layman? I don't know; I don't know. I do know, though, that something about all this (or at least the more extensive quotes) gives me the wrong feel. I wish I could articulate it a little better at this point, but I don't think I can. It's just something about the turning of phrase in these passages (and the countless others that I haven't been lucky enough to get down) that leads me to believe their authors see nature's indeterminism as happening within some "background." But it seems that the true indeterminism – quantum mechanical indeterminism – is the farthest thing it could possibly be from that. The quantum mechanical indeterminism doesn't come about from an indiscriminate swerve in the path of an atom; it comes from the point of contact between the theory and the world – the measurement. And the laws of nature aren't the accumulated effect of a multitude of lawless fireflies flashing against a black screen. The laws of nature arise out of some sort of Gödelian web: only pattern/law/guess confirming bodies (like people) can perform quantum mechanical measurements even though they themselves can in turn be treated quantum mechanically by a similar body. Furthermore, you've already seen my guess that the "throw of a die" isn't quite the correct model for quantum mechanical indeterminism: the repeated throw of a die can create a nonrandom string.

"On with the story!" you say. "Tell me how this delighted you to no end." Well ... actually all that didn't, but what it reminded me of did. And that was Jammer's contention that, "...certain philosophical ideas of the late nineteenth century not only prepared the intellectual climate for, but contributed decisively to, the formation of the new conceptions of the modern quantum theory." The excitement started when this memory led me to two papers that I had long ago copied but never bothered to read. Their subject: "indeterminism in German physics before quantum theory." Let me just start with a list of some physicists/philosophers who were indeterminists (or at least toyed with the notion) well before quantum mechanics was around: Maxwell (!!), C. S. Peirce, Larmor, Boltzmann, Exner (Schrödinger's graduate advisor), Schrödinger himself, von Mises, Schottky, Nernst, Reichenbach, and Weyl. What's very interesting is that almost exactly across the board when these people spoke of indeterminism they were speaking of an indeterminism for "little" objective events (or for the collisions of bodies) within spacetime! (Perhaps not so different from the pop culture we see now.) The one notable exception was Weyl. And this is exactly where the delight comes in. Recall that once upon a time a certain brash young man wrote to you that:

> If you turn to ... Wheeler's contribution "Law without Law" [you will find] an illustration of a 3-D letter **R** with the following caption: "What we call reality, symbolized by the letter R in the diagram, consists of an elaborate papier-maché construction of imagination and theory fitted in between a few iron posts of observation." From various considerations ... I see no alternative but to view quantum mechanical measurements as Wheeler's "few iron posts of observation" and spacetime as at least part of his "elaborate papier-maché construction of imagination and theory fitted in between."

Compare and contrast this (and the bits and pieces of our discussions that you can recall) with the wonderful set of words by Weyl that I found in the paper, "Weimar Culture, Causality, and Quantum

Theory, 1918–1927: Adaptation by German Physicists and Mathematicians to a Hostile Intellectual Environment" by Paul Forman.

> Finally and above all, it is the essence of the continuum that it cannot be grasped as a rigid existing thing, but only as something which is in the act of an inwardly directed unending process of becoming .... In a given continuum, of course, this process of becoming can have reached only a certain point, i.e., the quantitative relations in an intuitively given piece S of the world [regarded as a four-dimensional continuum of events] are merely approximate, determinable only with a certain latitude, not merely in consequence of the limited precision of my sense organs and measuring instruments, but because they are in themselves afflicted with a sort of vagueness .... And only "at the end of all time," so to speak, ... would the unending process of becoming S be completed, and S sustain in itself that degree of definiteness which mathematical physics postulates as its ideal .... Thus the rigid pressure of natural causality relaxes, and there remains, without prejudice to the validity of natural laws, room for autonomous decisions, causally absolutely independent of one another, whose locus I consider to be the elementary quanta of matter. These "decisions" are what is actually real in the world.

Perhaps there's not all that much to compare – Weyl, I think, sees the continuum as being in a state of formation by a physical process rather than as a thought construct and he certainly views the fundamental "decisions" in a more objective light than allowed by quantum mechanics. But nevertheless there is a certain similarity between these ideas. I guess it's just kind of nice to know that even if you can't keep good company with your contemporary colleagues, there's at least somebody from the past with which you can.

16 October 1997, to Lucien Hardy, "Wispy Words"

I was rummaging through the library today, and I found this little thing that reminded me of you. It is a transcript of a panel discussion at the Symposium on the Foundations of Modern Physics 1994 in Helsinki. The point of interest is something that Zeilinger said:

... we don't know why events happen, as expressed by Bell. Let me explain a little bit what I mean by that. By quantum phenomenon we mean the whole unity from preparation via evolution and propagation to detection. Then there is an uncontrollable element somewhere in this chain. It can be called the reduction of the wave packet. Or it can be in the many worlds interpretation the unexplainability of the fact that I find my consciousness in one given universe and not in the others. In a Bohm interpretation it can be the fact that I cannot control the initial condition. As an experimentalist I would say that there is some uncontrollable element from the following point of view. When doing a Stern–Gerlach experiment, for example, with an x-polarized spin, I cannot predict that this spin will go up, this one will go down, etc. There is something beyond my control. My personal feeling is that we have found for the first time in physics that *there are things which happen without sufficient reason*. This, I think, is a very profound discovery. I don't know whether there is a way to understand this or not. I feel there might be a way to understand why the world is so strange but we have not understood that yet. In my opinion this is so, because we really don't know what information is. We don't know what it means to collect information about the world. There is some world out there. In the words of Professor Laurikainen, in a very specific sense we have created the whole universe. But in some sense it exists without us. We have to understand therefore what it means to collect information about something which is not as much structured as we think.

Is that roughly what you remember? They are wispy words indeed!

10 October 1997, to Eli Yablonovitch, "Entanglement and Correlation"

Thanks for being such a good participant at my talk last week. I really enjoyed having you there … I must have because I woke up the other day thinking about one of the points you brought up. I just want to address it quickly while it's still on my mind.

If I can paraphrase, I think you said at some point something like "entanglement is another word for correlation." And then we had a slight discussion of things to do with Bell inequalities. Anyway, what I'd like to do is emphasize again the distinction between "classical correlation" and "entanglement" – this is something that people have only started to focus on in an applied way in the last five years. (One thing that prompts me to this is something I overheard at a meeting in England three weeks ago: this annoying fellow named […] – have you ever met him? – said far more graphically than you, "As far as I'm concerned, 'entanglement' is just baby talk for 'correlation'." So I'm taking my frustrations with him out on you!! Aren't you lucky?)

The main point is that "classical correlation" ultimately boils down to … or can be thought of as … simple probabilistic correlation between things objectively existent on the two players' sides. "Alice and Bob previously entered a lottery for which they were the only two players. The lottery commission sent them an announcement of the winnings. They haven't opened their envelopes yet, but the message in one envelope says that one is the winner and the message in the other says that one is the loser." Entanglement, on the other hand, expresses the *potential* for correlation. "Alice and Bob will eventually perform some measurements on their EPR pairs and the outcomes of their measurements *will be* correlated. However, before their measurements are performed, there are no objectively existent variables that are correlated in the sense of the last example. Different measurements can and will lead to different correlations." The distinction is important conceptually and also useful for technical applications. In a certain sense entanglement is a kind of "all-purpose correlation" just waiting to be baked into something real. (You should think of it as a modern version of the "Martha White's All-Purpose Flour" that Tennessee Ernie Ford used to advertise.)

One of the tasks of quantum information theory is to invent ways that such all-purpose correlation can be more useful than plain old classical correlation. There are now several: Bennett and Wiesner's "superdense coding," quantum teleportation, quantum cryptography, entanglement-enhanced classical communication, Richard Cleve's communication complexity games, better control of frequency standards, etc., etc. That's the motivation.

Because of this, one of the hot research topics in this field is to actually quantify the amount of classical correlation *and* entanglement in a bipartite density operator. One would like to know the exact ways the quantities trade off each other, etc.

16 October 1997, to Glenn Starkman, "Michelson Lecture Proposal"

A few weeks ago I wrote a wispy, philosophical-sounding little letter to an old friend. I think I can give you no better introduction to what I would like to say in more detail in the Michelson Lectures than to quote it directly.

I'm nearing the end of the flight and feeling a little philosophical. I hope you'll let me entertain you for a while. Lately I've been thinking about the airy nothings of quantum mechanics again. It's been a long time since I've done that to any extent – it's sort of refreshing.

Indeterminism and entanglement. The first is an old friend, that you know. The second, though, every day takes my heart a little more. In a certain way, indeterminism couldn't live without entanglement: the EPR argument would have triumphed over indeterminism if entanglement hadn't *also* led to a necessary violation of Bell inequalities. I believe in the ultimate indeterminism of quantum mechanical measurement outcomes just because of the experimental confirmation of Bell inequality violations and the experimental confirmation of Special Relativity. I've said this to you before (probably three years ago), but now it's starting to weigh on my mind more heavily. If I want to understand quantum indeterminism, then I must also understand entanglement: the argument goes in just that order.

Luckily for me, I think, the field of Quantum Information is especially suited to that purpose. Viewing entanglement as a new resource is the main thing on everybody's mind. In fact, I'm starting to feel that the situation we're in can be likened to the beginning of thermodynamics. What is heat, energy, work? No one knew at the outset; some thought them fluids, some thought them vital forces much like the soul, and so on. However, one thing did become clear eventually: no informed judgment on that fundamental question stood a chance until there existed a quantitative theory of thermodynamics. Without that, one could have never come across the mechanical theory of heat and the corollary of atomism that it led to.

So what is this thing called entanglement? What is its use? That we're just starting to understand. If I had to put it in a phrase right now, I would say it is "all-purpose correlation." Alice and Bob come to me and say, "Give us a little correlation, something that we can both have and no one else can possess. We think we're going to need it pretty badly tomorrow." I say, "Sure, no problem, just tell me which variables you'll be needing correlated and I'll do the trick for you." They say, "Sorry, we don't know which ones we'll need correlated yet. A lot of that will depend upon what we actually encounter tomorrow."

In the classical world, Alice and Bob would have been out of luck. But because the world is quantum, I actually can do something for them. I can give them a little "all-purpose correlation." And it turns out that that stuff can be really useful for several tasks. (In fact, we're finding ever more uses all the time.)

Thus, in a certain way, I'm starting to be impressed that "entanglement" shares a strong analogy to "energy." Both fulfill similar roles in our engineering endeavors: they are "all-purpose" essences that can be used for various beneficial tasks. Once we understand that in real depth, I think we'll finally put a dent in this question of "How come the quantum."

The "tasks" this letter refers to include quantum cryptography, quantum state teleportation, error correction for quantum computers, entanglement-enhanced classical communication over noisy channels, better control of frequency standards, and the list goes on. All these things sum together to form the new and growing fields of Quantum Information Theory and Quantum Computation.[1]

I propose for the *colloquium* associated with the Michelson Lectures to give an introductory talk on some of these new applications of quantum entanglement. Chief in the list will be quantum state teleportation and a clean, pretty example from Communication Complexity Theory due to Cleve and Buhrman.[2] These examples and the technical details associated therewith will form the backbone of the talk. However, along the way, I plan to weave connections all the way from the very fundamental (the Einstein–Podolsky–Rosen argument and the Bell inequalities) to the very applied (the power of quantum computing to factor large integers). The hope is that it will come out as a fairly entertaining and useful mix. The level of presentation should be accessible to anyone who has had a basic undergraduate quantum mechanics course and has some knowledge of simple Dirac notation. The title of the talk could be, "What Can You Do With Quantum Entanglement?"

I propose to explain, for the *three technical talks* of the Michelson Lectures, detailed aspects of Quantum Information Theory more closely associated with my own research. The three talks will build on each other, each expanding upon one of the "tasks" mentioned above.

Lecture 1: *Optimal Quantum Measurements and the Distinguishability of Quantum States.*

Arbitrarily large amounts of "classical" information can be dumped into a single finite quantum system – for instance, the spin of a simple spin–1/2 particle – just by preparing it in one of many nonorthogonal quantum states. The amount of information so dumped corresponds to the logarithm of the number of possible preparations. On the other hand, in general very little of that information can be retrieved reliably. This is because there are no physical means for distinguishing nonorthogonal quantum states. This principle, closely related to the "no-cloning theorem" of quantum mechanics, is the engine that powers quantum cryptography. In this lecture I will build the background required for quantifying just how much information can be retrieved in a situation like this. Topics will include generalized quantum measurements, the distinguishability of mixed-state density operators, and various information theoretic measures associated with that question.

---

[1] For your reference, two recent introductory/expository articles on the subject can be found in the pages of *Physics Today*: pages 19–21 of the October 1997 issue, and pages 24–30 of the October 1995 issue. Web links to several introductory expositions and resources on the subject can be found in John Preskill's "Physics 229" homepage: http://www.theory.caltech.edu/people/preskill/ph229/.

[2] See "Substituting Quantum Entanglement for Communication" by Richard Cleve and Harry Buhrman at http://xxx.lanl.gov/abs/quant-ph/9704026.

Lecture 2: *Sending Classical Information on Noisy Quantum Mechanical Channels.*
People like to put "classical" information – like the stories in today's newspaper – into quantum systems for a very simple reason: to get it from one place to another. Since the world is quantum mechanical, this, in the last analysis, is exactly what one *always* does in transmitting information. So, what is the highest rate at which one can hope to transmit information on a quantum system? That is, what is the largest number of bits per transmission that one can hope to obtain? The answer to this question is not yet completely known, but great progress toward a general conclusion has been found recently. This lecture will focus on reporting the most general results known so far; this will require integral use of the measures introduced in the first lecture. With that, one can start to explore at last two other questions that reveal rather surprising aspects of the world. Can a channel's noise ever be evaded more effectively by first entangling the separate transmissions before sending them on their way? Can it ever help to encode classical information in nonorthogonal quantum states? Simple examples of both effects will be demonstrated.

Lecture 3: *Quantifying Entanglement.*
The name of the game in Classical Information Theory is to make the correlation between sender and receiver as high as possible. This is what communication is. A corresponding game in Quantum Information Theory is to try to make the amount of "all-purpose correlation" or entanglement between sender and receiver as high as possible. Such a thing is useful, for instance, when one needs to port the program on one quantum computer onto another. In the classical case, Shannon's "mutual information" quantifies the question. What is the corresponding quantity in Quantum Information Theory? How does one quantify entanglement? This is the subject of this lecture, which focuses on the two most interesting ways of carrying this project out. The first has to do with how much "pure" entanglement one can hope to squeeze out of a noisy transmission. The second has to do with reversing the process. Both of these quantities are interesting in their own right, and go quite some way toward showing how different "classical information" is from "quantum information."

All three lectures will assume the same level of expertise on the part of the audience that the colloquium does, though they will likely be presented in a more intensive way. Audience participation will be encouraged (and even begged for if it's not forthcoming on its own!). In all cases, the goal will be on getting the essential ideas across to the audience – to draw pictures in their minds that won't vanish upon leaving the auditorium.

My most personal stake in giving these lectures is to try to convey some of the excitement in this new field, to spread the word. I firmly believe the field holds forth great potential not only in its technical application but for our deeper understanding of the world in its entirety. It is just a question of planting the seed in enough fresh young minds. I deem that the Michelson Lectures will provide a good vehicle for just this purpose and I thank you for your consideration.

19 January 1998, to Daniel Kevles, "Herzfeld History"

I'm wondering if you can help me track down the identity of my academic great-great-great-grandfather. My interest in this came originally from my winning the Michelson award at Case Western Reserve U. Since John Wheeler is my academic great-grandfather and had once described his PhD advisor as "the great optical physicist . . ." in front of me, I was hoping there was some chance that his advisor's advisor was Michelson. (I would have liked to use this to start off my lectures at CWRU.) Anyway that turned out not to be the case: Wheeler's advisor, I have since found out, was Karl F. Herzfeld (whom you write a bit about in your book *The Physicists*). I have been able to find out that Herzfeld received his doctorate in Vienna in 1914 (both from *Sources for History of Quantum Physics* and *American Men of Science*), but I have not been able to trace his advisor. Could it have possibly been Boltzmann?? Do you know of any other source by which I might ferret out this information?

PS. One tidbit that I find really interesting is that Mehra and Rechenberg have traced the usage of the phrase "quantum mechanics" back to at least a 1921 paper by Herzfeld!

19 January 1998, to Gary Herling, "Herzfeld"

Thanks for sending on your friend's suggestions. It would have indeed been great if Herzfeld had been Boltzmann's student! I have been able to ascertain that Herzfeld got his doctorate in Vienna in 1914 – this, I think, rules out Boltzmann in the lineage. Boltzmann committed suicide in 1906. Herzfeld focused mostly on spectroscopy and statistical mechanics. He had at least Wheeler and Heitler as students. However I still don't know who Herzfeld's advisor was.

In any case, my search hasn't been fruitless! The most interesting tidbit I've picked up is this: Herzfeld may actually have been the first person to introduce the term "quantum mechanics" into the literature!! Born thought that he had done it himself, but his first usage of it was in a paper of 1924. Lorentz preceded that in 1923 with "the mechanics of quanta." However, Herzfeld had a paper of 1921 titled, "Quantum Mechanics of Atomic Models." (All this according to a footnote in Mehra and Rechenberg, vol 4.)

22 January 1998, to Alan Hájek, "Bohr was a Bayesian?"

I was quite intrigued the other day to find (from Sam Braunstein) that you have some interest in the subject of "objective chance." I do too, in connection with quantum mechanics . . . though in a somewhat negative way. I would very much like to talk to you in detail about these things. A small part of my research program in quantum information theory is to give some substance to the slogan "Bohr was a Bayesian" (. . . though a Bayesian good enough to know that Bayes' rule for conditionalizing isn't always valid . . . especially when one is confronted with a world where one's information gathering measurements *necessarily* disturb someone else's predictability.) Or more importantly, historical facts aside, to give some substance to the idea!

14 March 1998, to Ryszard Horodecki, "Jaynes' Principle"

A warm thank you to you and Michał [Horodecki] for sending me a copy of your upcoming paper on Jaynes' Principle and quantum-state compression. I gave it a quick read with much interest this morning. Indeed I am very sympathetic with efforts such as this. For a long time now it has been my pet "idea for an idea" to use quantum information theory in aid of finding a foundation for quantum theory itself that has something of the flavor of Jaynes' efforts in classical statistical mechanics. As I wrote Asher Peres once:

My own pet "idea for an idea" (as Wheeler would say) is that the quantum theoretical description – along with its Hilbert-space structure – is forced upon us when the information we have about a physical situation/system is of a certain variety. In this way, quantum theory is not so different from Ed Jaynes' molding of classical statistical mechanics and the maximum entropy principle. It just so happens in the latter case that the prior information is in the form of an expectation value for some physical quantity that can be assumed objective and independent of measurement: this prior information along with a small few desiderata lead uniquely to the canonical probability assignment. In the case of uniquely quantum problems, we start with a different kind of prior information. The *hope* is that, upon pinpointing the nature of that information (and again a few small desiderata), one would see that a "wave-function assignment" is the natural (and unique) way of summarizing what we know and what we can predict.

The extra piece of knowledge that we seem to have concerning quantum phenomena is that the probability assignments we make will be ultimate ... and, in particular, non-updateable. I.e., if we make a formal assumption of "no hidden variables" and use a decent principle of unbiased inference – as Jaynes' principle seems to be for the case where one is given an expectation value and nothing more – will we end up with a structure formally identical to quantum mechanics? I wish I knew. Of course, the principle, whatever it may turn out to be, will not be the MaxEnt principle itself: the emphasis is rather on the "flavor" of the MaxEnt principle. Anyway, I say slightly more about this idea on pages 23–26 of my paper quant-ph/9601025 with Carl Caves. And I hope to say still a little more in an upcoming paper with Caves and Schack about Bayesianism and the quantum probability law.

05 May 1998, to Ronald Giere, "Quantum Propensity & Quote"

I am a physicist trying to make sense of whether "the" notion of propensity is of any use for understanding quantum theory. (In fact, some colleagues and I are preparing a manuscript on the subject presently.) Thus I have come across several of your old papers from the 1970s on propensity and objective chance. (I might add that I think these are the *very best* that I've seen on the subject so far.)

Perhaps you can help me out in some of my bibliographic troubles for the project? Most importantly, in your paper "A Laplacean Formal Semantics for Single-Case Propensities" (*J. Phil. Logic*, 1976), you attribute a nice little quote to Albert Einstein:

I can, if worst comes to worst, still realize that God may have created a world in which there are no natural laws. In short, a chaos. But that there should be statistical laws with definite solutions, i.e., laws which compel God to throw the dice in each individual case, I find highly disagreeable.

But then you say, "I have been unable to find the exact reference for this one."

Have you been able to find the source of the quote in the intervening years? If not, what was your original source in the first place? Is this "quote" actually your own paraphrase of a conversation that you had with someone? If so, who? I would like to quote it myself and want to make the reference as complete as possible.

Besides that, though – but less importantly for my immediate manuscript – can you point me to any significant critiques/endorsements of your theory of propensity?

05 May 1998, to Ronald Giere, "Propensity Stuff"

Many thanks for trying to rethink your steps on the Einstein quote. As far as I recall, there's no mention of any such thing in Fine's book. (I read it pretty carefully a couple of years ago, and, in any case, a quote like this would have caught my attention ... given my interests. Also there's one technical point in that the oldest essay in the volume didn't appear until 1976, the same year as your Laplacean-semantics paper – so it's not very likely you could have read the full book then.)

The closest that I've found to it is from a letter, Einstein to Solovine, dated 30 March 1952:

You find it strange that I consider the comprehensibility of the world (to the extent that we are authorized to speak of such a comprehensibility) as a miracle or an eternal mystery. Well, a priori one should expect a chaotic world which cannot be grasped by the mind in any way. One could (yes one should) expect the world to be subjected to law only to the extent that we order it through our intelligence. Ordering of this kind would be like the alphabetical ordering of the words of a language. By contrast, the kind of order created by Newton's theory of gravitation, for instance, is wholly different. Even if the axioms of the theory are proposed by man, the success of such a project presupposes a high degree of ordering of the objective world, and this could not be expected a priori. That is the "miracle" which is being constantly reinforced as our knowledge expands.

But this really doesn't have the same intent as your quote. If you ultimately think of where you found it, please let me know at any time.

Thanks also for the remarks on the old propensity theory. I was aware of Humphreys' critique. In case you would like the appropriate references, at least two of them are:

1. P. W. Humphreys, "Is 'Physical Randomness' Just Indeterminism in Disguise?," in *PSA 1978: Proceedings of the 1978 Biennial Meeting of the Philosophy of Science Association*, Vol. 2, edited by P. D. Asquith and I. Hacking (Philosophy of Science Association, East Lansing, Michigan, 1981), pp. 63–78.
2. P. Humphreys, "Why Propensities Cannot Be Probabilities," *Phil. Rev.* **94**, 557–570 (1985).

A related argument was proposed by Peter Milne:

3. P. Milne, "Can There Be a Realist Single-Case Interpretation of Probability?," *Erkenntnis* **25**, 129–132 (1986).

I myself am striving to formulate a relatively personalist account of quantum probabilities – that is one with a notion of "objective chance" that is no more reified than Ramsey's 1928 version of the concept. In searching through the propensity/chance literature, I've mostly been trying to see that all my bases are covered.

10 August 1998, to Richard Beyler, "Your Thesis"

Last night I read your article "Targeting the Organism" (*Isis*, 1996) and enjoyed it very much. Lately I've taken a dilettante interest in the particular thoughts of Pauli and Jordan on the foundations of quantum mechanics, and your article has been helpful in that regard. (I am a physicist at Caltech, by the way, not a historian.) My interest mostly comes from a "fixation on the issue of determinism" (as you describe Jordan), which lays at the foundation of my line of research (quantum information theory/quantum cryptography) more than any other aspect of quantum mechanics. Anyway, in this regard, I am quite interested in reading your Ph.D. thesis. I'm writing this note, mostly, to ask how I might obtain a copy. If you have an offprint of it lying around that you could mail to me that would be great! Alternative to that, though, if have a Microsoft Word, TEX/LATEX, or PostScript file of it, then you could email it here and I could print it myself.

16 August 1998, to Adán Cabello, "Mermin and His Correlata"

Last week, I was fortunate enough to receive a sneak preview of your paper "Quantum Correlations as Local Elements of Reality" from my (visiting) office-mate Asher Peres. It is a nice paper and very well written. In particular, I was very pleased to discover the existence of another level-headed Copenhagenist out there! Given the number of physicists trying to turn away from the Copenhagen interpretation, we are in grave need of as many eloquent supporters for our point of view as possible.

However, unfortunately, I think your article will make no affect on David Mermin. I know because I have tried exactly the same argument as yours on him before (once in person last December, then once in some email correspondence). He rejected it then; I suspect he'll reject it now. In a separate mail, I'll send you an excerpt from my "Quantum Correspondence File" with the appropriate passages: the most relevant piece of it is in the section titled Merminition 11.

The problem hinges on what David means by the word "correlation." His definition is fairly strict: from his perspective, the correlation between particles 1 and 4 does *not* change as a result of the situation you describe. For what he means by "correlation" is the set of all joint probability distributions for "measurement outcomes" (on systems 1 and 4 separately) that can be derived from the *reduced density operator* of that joint system. Because the reduced density operator on the 1–4 system does not depend on whether Experiment 1 or 2 is performed – in fact, it doesn't depend at all on actions taken on the 2–3 system – the "correlation" between systems 1 and 4 does not change either. For David, it doesn't matter that the experiments you describe allow the experimenter to update *his* quantum state for the 1–4 system. Moreover, it doesn't matter that the experimenter can toggle between different amounts of *entanglement* for those systems. It doesn't seem to matter to him that with the receipt of the classical information gathered from your experimenter, Alice and Bob at systems 1 and 4 can "activate" different amounts of useful entanglement. (For the whole spectrum of things that can be done in between these extremes for a situation more general than the one you describe, you might be interested in taking a look at our paper "Entanglement of Assistance" quant-ph/9803033.)

You may not like it that he is ultimately inconsistent in his use of the term quantum state – i.e., not specifying with what/whom it is in respect to – but that is his usage, and there's not much we can do about that. You also may not like his ignoring entanglement as a reasonable notion of (potential) correlation – I didn't – but, again, it's his prerogative to define his terms as he wishes. In essence, David says this himself in Section VIII of his paper "What Is Quantum Mechanics Trying to Tell Us?".

In summary, though I liked your exposition, much of the force of your argument is lost explicitly because in your Assumption (b) David is only talking about the reduced density operator ... not the correlations due to the quantum states determined from measurements on other systems.

Thank you again for the opportunity – unbeknownst to you – for a little thought!

19 August 1998, to Richard Cleve, "The Word Entanglement"

The two best sources I know of on the history of the original EPR argument and the subsequent development of the idea of entanglement are: the article, Max Jammer, "The EPR Problem In Its Historical Development," in *Symposium on the Foundations of Modern Physics: 50 Years of the Einstein–Podolsky–Rosen Gedankenexperiment*, edited by P. Lahti and P. Mittelstaedt (World Scientific, Singapore, 1985), pp. 129–149, and the book, *The Shaky Game: Einstein Realism and the Quantum Theory*, by Arthur Fine (U. Chicago Press, 1986).

Indeed Schrödinger was the first person to use the word "entanglement" ... and he meant just exactly what we mean in the modern sense. The first appearance in print is with the words,

When two systems, of which we know the states by their respective representatives, enter into temporary physical interaction due to known forces between them, and when after a time of mutual influence the systems separate again, then they can no longer be described in the same way as before, viz. by endowing each of them with a representative of its own. I would not call that *one* but rather *the* characteristic trait of quantum mechanics, the one that enforces its entire departure from classical lines of thought. By the interaction the two representatives (or ψ-functions) have become entangled.

This comes from his article "Discussion of Probability Relations Between Separated Systems," *Proceedings of the Cambridge Philosophical Society* **32**, pp. 555–563, 1935 (submitted 14 August, read 28 October). He makes it clear that he is thinking of the modern concept of entanglement in Section 2 of the article, where he writes,

Let x and y stand for all the coordinates of the first and second systems respectively and $\Psi(x, y)$ for the normalized representative of the state of the composed system, when the two have separated again, after the interaction has taken place. What constitutes the entanglement is that Ψ is not a product of a function of x and a function of y.

According to Fine, Schrödinger was already saying these things to Einstein in a June 7 letter,

Schrödinger continues by focusing on certain mathematical aspects of the EPR example. These have to do with the expansion of the state function of a composite system into a bilinear series of functions defined only on the component systems. He points out that the composite EPR case, after the interaction has effectively ceased, is a very exceptional one in this regard, for there all the coefficients of the bilinear expansion are identical. ...

The original EPR paper appeared May 15 in *Physical Review*.

Schrödinger continues with a much more detailed set of thoughts on entanglement in his paper "The Present Situation in Quantum Mechanics," reprinted (and translated from German) in the book edited by Wheeler and Zurek "Quantum Theory and Measurement." It made its first appearance in *Die Naturwissenschaften* **23**, pp. 807–812, 824–828, 844–849 (1935). It's worth reading if you've got the time.

What is intriguing about the Jammer article is that it makes it clear that Einstein had a rudimentary notion of entanglement in his head as early as 9 July 1931. This is exhibited in a letter from Ehrenfest to Bohr on that date explaining a conversation he had had with Einstein. Jammer wrote:

... what Einstein had in mind is confirmed by a letter which Ehrenfest wrote to Bohr on July 9, 1931. As Ehrenfest reports, Einstein uses the photon-box no longer to disprove the uncertainty relation but "for a totally different purpose." For the machine, which Einstein constructs, emits a projectile; well after this projectile has left, a questioner can ask the machinist, by free choice, to predict by examining the machine alone *either* what value a quantity A *or* what value an even conjugate quantity B would have if measured on the projectile. "The interesting point," continued Ehrenfest, "is that the projectile, while flying around isolated on its own, must be able of satisfying totally different non-commutative predictions without knowing as yet which of these predictions will be made ..."

01 October 1998, to Richard Beyler, "More Pauli"

It would be wonderful if you could convert your dissertation into MS Word format. Your effort will certainly benefit more than just me: we write these words so that people will read them. If you are going to do the conversion, I will hold off contacting University Microfilms for a while – just let me know when/if you've given up and what I should do.

Quantum cryptography concerns the business of disseminating a "secret key" among communicators who may want to communicate privately at some point in the future. Once the secret key is in their hands, and they have verified that no one else knows it, then they can rest assured that their encryption scheme will be completely secure. The engine that powers it is the particular form of indeterminism provided by quantum mechanics: it gives a method by which the complete secrecy of the key can be verified. This is something that cannot be done with classical physics. I've dug up for you an old list of popular articles that Gilles Brassard put together in 1994; I'll paste it below.

Often in these articles, you will see it expressed that the source of quantum cryptography's efficacy is the "Heisenberg uncertainty principle." But, mostly I think that comes from a lack of a very deep understanding of quantum mechanics on these writers' parts. Also it comes from taking the easy way out – i.e., attaching a quick explanation in terms the writers think the readers will feel comfortable with. If you ask me, the real source of the efficacy is what Pauli so often called "the lack of the detached observer" in the quantum world. Thus a little of my interest in Pauli. If you take a real interest in this, I'll provide you with some technical references later.

About Pauli's thought, my main sources are the following:

1. W. Pauli, *Writings on Physics and Philosophy* (Springer-Verlag, 1994).
2. K. V. Laurikainen, *Beyond the Atom: The Philosophical Thought of Wolfgang Pauli* (Springer-Verlag, 1988).
3. K. V. Laurikainen and C. Montonen, eds., *Symposia on the Foundations of Modern Physics 1992: The Copenhagen Interpretation and Wolfgang Pauli* (World Scientific, 1993).

Laurikainen can be a bit annoying in his repetition and in his hero worship, but still his articles are good (if meager) sources of Pauli's more private thoughts.

I haven't run across any comparative studies of Pauli and Jordan outside of the references I found in your paper. If you know of any newer material on these fellows' thoughts, please let me know.

13 October 1998, to Itamar Pitowsky, "Rational Gleason"

Well I'm finally giving your paper "Infinite and Finite Gleason's Theorems ..." the frontal attack it deserves. I've worked my way up to the top of page 224 – where the proof of the regular version of Gleason's theorem ends – and have been having a great time. (That in spite of the myriad typos infused throughout the manuscript, AARGH!) In fact I'm giving a talk on it at this week's "quantum information journal club" that we have here. If you'd ask me for one thing that I wish you had included, I would have to say a motivational sermon or two for the graphs G_1 and G_2. Who ordered them? What line of reasoning led to considering those graphs in particular? You haven't written a longer, more expository, version of the paper where you do this, have you?

Let me run one speculative idea past you; tell me what you think. One of the great things I'm going to learn – when I get all the way through your paper(!) – is the interesting restrictions placed on states even when they're defined on finite sets of rays. My question, then, is this: How far do you think one has to go before recovering the whole standard probability rule? For instance, how much is the continuum really needed for this problem? Might one be able to derive a full version of Gleason's theorem for Hilbert spaces over the *rational* complex field?

Itamar's Reply

I don't think the theorem will remain true if you stick with a dense set such as the rationals. Years ago I've shown a very bizarre example to the contrary. I "constructed" an example of a three dimensional function f with values 0 and 1 with the following property: If x is ANY direction then $f(x) + f(y) + f(z) = 1$ for all directions y, z orthogonal to x (and to each other) *except countably many*. The proof is based on the continuum hypothesis (!!!) and is not very difficult. The set on which $f(x) + f(y) + f(z) = 1$ is nonmeasurable and big and dense. The proof appears in *Phil. of Science* around 1985. (I'm in Canada right now so I don't have the reference, I can send you a copy when I'm back in a couple of weeks.) There may be dense subsets on which Gleason's theorem is true but they have to have a special structure.

19 October 1998, to John B. Kennedy, "Tensor Products"

Actually my question is pretty simple ... nothing to get too intrigued about. I read your paper "On the empirical foundations of the quantum no-signalling proofs" (*Phil. Sci.*, 1995) and enjoyed it quite a bit. I think you're really on the mark with it!

At the end of the Introduction you say, "One direction for future research would be to canvass such systems [i.e., nonstandard axiom systems] for clean justifications of the tensor product formalism." To what extent have you carried that out? Have you found any interesting tidbits (or even uninteresting ones, for that matter) in that direction? Alternatively, or even better, have you written any follow-up papers to this one that I might get hold of? Has there been any discussion in the literature about your paper? Those are my main questions.

One little point about some of the language you use. In your summary you write, "Tensor Product Space: assumes the impossibility of wholly new states in the combined system … in short, no gain or loss of states." I think that's a little bit of a dangerous expression. In a certain sense, we get loads more states when we combine systems. There are all the product states (the ones isolated to each system alone), but then on top of that we have all the entangled states! The product states form a set of measure zero! Moreover just look at the dimensionality change. When one combines systems of dimensions x and y, the result is a system with dimension xy. These remarks, I'm sure, are trivial to you … and that leads me to think you mean something else. But I'm not completely sure what that is.

One thing that might interest you in relation to the tensor product combo rule, is that there is a certain sense in which it is only well behaved for complex Hilbert spaces. It is not such a nice thing for real or quaternionic spaces. Suppose that one means by a "possible state" any density operator (Hermitian, positive semi-definite, trace 1) on the Hilbert spaces of a given dimension. Then, for instance, if we take the tensor product of two real Hilbert spaces, one of dimension x and one of dimension y, the set of operators constructible from (real) superpositions of tensor product operators will not turn out to be equivalent to the set of "possible states." The set of possible states is strictly larger: for it consists of all density operators constructible on a space of xy dimensions. The two things are not the same for real Hilbert spaces (though they are for complex spaces). Another way to say it is the following. For separate isolated systems described by complex Hilbert space quantum mechanics, we can completely reconstruct the overall state (if we have many copies) from local measurements only. We need only know the measurement outcomes and the correlations between them. For real Hilbert space quantum mechanics, local measurements and the correlations between the outcomes are not enough for state reconstruction. You can read about this in:

W. K. Wootters, "Local Accessibility of Quantum States," in *Complexity, Entropy and the Physics of Information*, edited by W. H. Zurek (Addison-Wesley, Redwood City, California, 1990), pp. 29–46.

In general, I'm just interested sorting out the physical essence of the tensor producting assumption and how that fits into a Copenhagenistic framework of understanding quantum mechanics.

28 October 1998, to Charles Seife, "Yeeks!"

Let me first tell you my view of what defines "quantum teleportation." It's an affair that involves at least three participants: Alice and Bob, who run the business, and Victor, their customer. Victor has some quantum system that he knows something about; the description of what he knows is called a "quantum state." It's that description that he want's transferred from the physical system he has, onto a system in the possession of Bob. Perhaps Victor is sitting in New York, and he needs a physical system in Paris to be describable in some particular way: for instance, he needs to know that there is at least one photon in Paris for which he can ascribe a polarization angle equal to a binary encoding of his *secret* middle name, Rumpelstiltskin. When teleportation is complete, it is that description that has been teleported. People often say that it is an unknown quantum state that is teleported, but what that means is that it is unknown to Alice and Bob. There always has to be a Victor in the background, or there wouldn't be a hell of a lot of use in running this business. It's much like AT&T: they're more than willing to transport secret messages for you, without ever snooping, as long as you pay your bill. No customers, no business. This is what teleportation strives for. But there's a little more to it. Alice and Bob can't just be a courier service like Federal Express; they're not allowed to physically transport the actual system Victor gives them to its destination. They are only allowed to physically transport some "side information" that they generate themselves. I present this as if it's a restriction, but actually it's an advantage – they advertise it. This method of getting the quantum state from one place to the other is what makes their business special. It means that they don't have to worry about stormy weather screwing up Victor's package while in transit. And on top of that, Bob might even be subcontracting for Victor in that he promises not to reveal his location to Alice. These things can be done as long as Alice and Bob have previously gotten together and shared a little "entanglement." This is the physical resource their company relies on to stay in business. Intel would be out of business without silicon for their chip; AT&T would be out of business without copper for their wire. Alice and Bob would be (should be!) out of business without entanglement.

03 November 1998, to Daniel Gottesman, "The Bill Wootters Fan-Club Page"

Well, I've finally gotten off my duff and written the reviews I promised you: you're right it is a pretty enjoyable process – I hope it won't be too long before I do some more! As you'll see below, I can be a pretty wordy fellow.

Review of:

William K. Wootters, *The Acquisition of Information from Quantum Measurements*, Ph.D. Thesis, The University of Texas at Austin, 1980.

Lately I've been finding in my work in quantum information that I'm starting to feel a little more like an engineer than a physicist. *Transactions on Information Theory* is after all published by the IEEE. I find it becomes easier and easier to forget the original goal of my education – to understand nature itself ... regardless of how we might make use of that knowledge. I fall asleep thinking, "What can I do that will impress IBM or AT&T?" There's got to be more to our work than this!

In that regard, this little book is a bright star of hope. For his Ph.D. work, Wootters had the courage to ask where quantum mechanics itself might come from. And more interesting than that, he saw that the answer might have something to do with information theory!

This work sets out, for the most part, to derive the standard quantum probability rule from a variational principle, one that involves information. For instance, could it be the case that Nature is so constructed that the following game tells us something deep?

Someone hands you many, many identical copies of an otherwise completely unknown (pure) quantum state; on each shot, you perform the same quantum measurement and tabulate the outcome. The goal of the game is to attempt to identify the unknown state from the statistical information that so accumulates. Since the information is statistical, only some of it is useful for the task. But, how much? Could it be the case that if the probability rule for the measurement outcomes were anything other than the one we know and love, then the rate at which useful information accumulates would not be optimal?

Well, it is ... ALMOST. I say, "almost" because it is true for quantum mechanics with real Hilbert spaces, not for quantum mechanics with complex Hilbert spaces. Aarrgh! Unfortunately, things turn out to be not so clean for complex spaces, though perhaps not all hope is lost (Wootters has a chapter devoted to this issue). Some crucial connection between quantum mechanics and information theory is still missing. There's still work to be done!

This book is a delight to read, and surely inspirational for getting as far as it did. I recommend it to all in our field for a little reading just before sleep: we should all dare to dream.

Review of:
Eugene P. Wigner, "The Probability of the Existence of a Self-Reproducing Unit,"
in *The Logic of Personal Knowledge: Essays Presented to Michael Polanyi on his Seventieth Birthday* (Routledge & Kegan Paul, London, 1961), pp. 231–238.

The no-cloning theorem first discussed by Wootters and Zurek and (independently) by Dieks is now understood to be a significant part of the foundation of quantum information theory. But have you ever explained it to another physicist and received a reaction of the form, "Is that it? That's the big deal everyone is talking about?" I have. And it's no wonder: the issue of no-cloning boils down to almost an immediate consequence of unitarity – inner products cannot decrease. In fact, Wigner's famous theorem on symmetries even shows that the group of time-continuous, inner-product preserving maps on Hilbert space is strictly equivalent to the unitary group. Therefore, it comes as quite a shock to see that Wigner himself just missed the no-cloning theorem! In this paper, Wigner tackles the question, "How probable is life?" He does this by identifying the issue of self-reproduction with the existence of the types of map required for the cloning of quantum states. He doesn't tackle the question of cloning for a completely *unknown* quantum state head on, but instead analyses the "fraction" of unitary operators on a tensor-product Hilbert space that can lead to a cloning transformation for at least some states. Nevertheless, he states quite clearly that an arbitrary linear superposition of clonable states ought also to be clonable. But this, of course, cannot be.

I think this paper is quite interesting from the historical point of view of our field: it gives us an appreciation of the beauty and simplicity of that little theorem in a way that simply learning about it cannot provide. It gets at the heart of something deep in very present physical terms, terms that even a great mind like Wigner's missed.

Review of:

William K. Wootters, "Entanglement of Formation of an Arbitrary State of Two Qubits" *Physical Review Letters* **80**, No. 10, pp. 2245–2248 (1998).

This paper is a linear-algebraic tour de force. It proves the correctness of a surprisingly weird-looking closed expression for the entanglement of formation of a general mixed state on two qubits. The expression itself had been conjectured in an earlier paper by Hill and Wootters. For those of you who hope to find a surprise or two in the "information functions" of quantum information – something, say, beside the same old, same old von Neumann entropy – this paper is a winner. More importantly for our field, however, are the nontrivial mathematical techniques this paper introduces for issues such as this. These techniques should have a wider range of application, for instance, in obtaining exact expressions for other entanglement measures.

Review of:

Lane P. Hughston, Richard Jozsa, and William K. Wootters, "A Complete Classification of Quantum Ensembles Having a Given Density Matrix," *Physics Letters* A **183**, pp. 14–18 (1993).

Abner Shimony likes to say that entanglement gives rise to "passion at a distance." He does this because when Alice performs a measurement on A of an entangled system AB, *something* changes for B, BUT that change cannot be used for the purpose of communication with a Bob at B. If you ask me, this is language that is just asking for trouble; it is language that is poised to confuse a generation of new physicists. Something indeed does change for B, what Alice can *say* of it. But it is nothing more than that; to think that it is truly a physical change with respect to B alone – especially one that is so contrived as to not lead to communicability – is to open up a sink hole. We are dealing here with changes of states of knowledge.

Within this context, it is quite reasonable and quite interesting to ask how many different ways Alice's knowledge can change. Depending upon which measurement Alice wishes to perform on A, there will be any of a number of different state assignments for B that follow from that. What are they, and what are their probabilities? This is the main question addressed in this little paper. It has a very clean answer: a state assignment can be created for B from a measurement on A if and only if that state falls within a pure-state decomposition of the marginal density operator of B. Moreover, the probabilities of Alice's measurement outcomes correspond precisely to the probabilities of the ultimate state assignments.

This result, it turns out, is not of purely academic interest. It has had a wide range of application in several more applied problems in quantum information: it is crucial for the proof that no quantum bit commitment schemes exist, it plays a crucial role in proving the exact expression for the entanglement of formation for two qubits, and it is crucial for defining the notion of the entanglement of assistance. This theorem is one that all quantum information theorists should have incorporated into their tool bag.

Can more be said? Actually, it turns out that this result even has some historical significance. For, unknown to the authors above, the same question was raised and even partially answered by Erwin Schrödinger in a 1936 paper! [E. Schrödinger, "Probability Relations between Separated Systems," *Proc. Cam. Phil. Soc.* **32**, 446–452 (1936).]

Review of:
Richard Jozsa, Daniel Robb, and William K. Wootters,
"Lower Bound for Accessible Information in Quantum Mechanics,"
Physical Review A **49**, pp. 668–677 (1994).

There is a sense in which much of quantum information theory reduces to only a small subset of classical information theory. For ultimately there is always classical information going into a problem (in terms of quantum-state assignments) and classical information coming out of a problem (in terms of the results of measurement outcomes). In between there is just a very restricted class of transition probabilities between the input and output alphabets – a class that is absolutely minuscule in terms of the ones contemplated by classical information theory.

However, there is a much more important sense in which quantum information is something new, something sui generis. Within it, we can free ourselves from explicitly having to consider the ultimate input and output: transmitting quantum states for the sake of simply transmitting quantum states has a certain beauty of its own. For, at the very least, it gives rise to some miraculous information quantities that stand out in their own right. It is hard to shake the feeling that these quantities will not ultimately be just as important as Shannon's original quantification of information.

The quantity of information that arises in this paper, the subentropy, though it is ultimately motivated by a problem of classical information transmission, is exactly of this flavor. Who would imagine that by replacing a trace with a determinant in a rather abstruse expression of the Shannon entropy one would find something of any interest? Moreover that that quantity would be the answer to a well-defined, physically motivated problem? No one I know, except maybe these three guys.

The subentropy is a strange beast; here is how it arises. Consider the problem of extracting the most information possible about the identity of an unknown (pure) quantum state. What "unknown" means in this context is that a set of states along with a listing of their a priori probabilities – an ensemble – is given; it is just the actual identity of the state in front of the observer that is not known. The maximal information that can be extracted via a measurement is called the accessible information. The subentropy of the density operator of the given ensemble turns out to be a lower bound on that information. Moreover, the subentropy is the best lower bound that depends only on the density operator – for a given density operator, there is always an ensemble for which the accessible information equals the subentropy. What is really most interesting is how the subentropy can be so similar to the von Neumann entropy at the same time as being so different from it. For instance, the von Neumann entropy is bounded above by the logarithm of the dimension of the Hilbert space; the subentropy, on the other hand, is bounded above by a more complicated function of the dimension with in turn is ultimately bounded above by one minus Euler's constant.

What other uses will we find for this intriguing little quantity? That remains to be seen. But this paper is one to keep in the back of one's mind.

27 November 1998, to Tim Ralph, "Quantum Teleportation: A QND View?"

We have read with interest your comment "Quantum Teleportation: An Operational View." Thank you for giving us that opportunity. But our learning from you in this case is a double-edged sword. This is because your writings show that you pretty much missed the boat by analyzing teleportation through the language of conventional QND. No doubt you did this because you were most familiar with that language, but that really only gets in the way for this problem. The problem is that the questions for which QND was invented are too remote from the question at hand to be an efficient tool for its analysis. Quantum information is something new. As such, its operational analysis requires tools of a different sort than have arisen in the old classical-communication and parameter-estimation scenarios you are familiar with. We think your comment proves this point.

In this note, we plan to make this just as crystal clear as we can.

In quantum teleportation, the task set before Alice and Bob is to get a *quantum state* – one that is *unknown* to them – from one side of the laboratory bench to the other. It is assumed at the outset that the only resources available for this transmission are a classical channel and some previously shared entanglement. (By the term "classical channel" we mean, strictly speaking, a channel that allows only the transmission of some set of orthogonal quantum states, not arbitrary superpositions of them.) The question of the operational verification of teleportation becomes this: how can we know when Alice and Bob had to have used some of their entanglement in the aid of their transmission? If there are certain transmission characteristics that they could not possibly have met without the use of their previously shared entanglement, then that will be the sure signature. Meeting the minimal such signature is what we call "unconditional quantum teleportation." On these last three sentences it seems we all can agree.

Where our differences appear to hinge is in the italicized words that start the last paragraph: *quantum state* and *unknown*. Let us go at them one at a time.

What does it mean to get a quantum state from one side of the bench to the other with minimal distortion? We chose to gauge this by the "fidelity" between the input and output. To be precise, this is defined in the following way. If the input state – as known by some third party Victor – is $|\psi_{in}\rangle$ and the output is (generally a mixed-state density operator) ρ_{out}, then the fidelity is given by

$$F = \langle\psi_{in}|\rho_{out}|\psi_{in}\rangle. \tag{33.1}$$

This measure has the nice property that it equals 1 if and only if $\rho_{\text{out}} = |\psi_{\text{in}}\rangle\langle\psi_{\text{in}}|$. Moreover it equals 0 if and only if the input and output states can be distinguished with certainty by *some* quantum measurement. The thing that is really important about our particular measure of "fidelity" is these last two properties. It captures in a single, simple and convenient package the extent to which *all possible* measurement statistics that can be produced by the output state – if a measurement were performed on it(!) – will match what the input would have specified.

For instance, take *any* observable (generally a positive operator-valued measure or POVM) $\{E_\alpha\}$ with measurement outcomes α. (What you call "symmetric detection" is an example of this where $E_\alpha = \frac{1}{\pi}|\alpha\rangle\langle\alpha|$ and the $|\alpha\rangle$ are coherent states, but it could be any observable whatsoever: photon counting, measuring a single quadrature, or what have you.) If that observable were performed on the input system, it would give a probability density for the outcomes α given by

$$P_{\text{in}}(\alpha) = \langle\psi_{\text{in}}|E_\alpha|\psi_{\text{in}}\rangle. \tag{33.2}$$

If the same observable were performed on the output system, it would have instead given a probability density of

$$P_{\text{out}}(\alpha) = \text{tr}(\rho_{\text{out}}E_\alpha). \tag{33.3}$$

One can gauge the similarity of these two densities by their statistical overlap:

$$\text{overlap} = \int \sqrt{P_{\text{in}}(\alpha)P_{\text{out}}(\alpha)}\, d\alpha. \tag{33.4}$$

It turns out that regardless of what observable we are talking about,

$$\text{overlap}^2 \geq \langle\psi_{\text{in}}|\rho_{\text{out}}|\psi_{\text{in}}\rangle. \tag{33.5}$$

Moreover there exists an observable that will actually give precise equality in this expression.[3] So, it is in this sense that the fidelity captures a fact about all possible measurements.

What now about the measures of transmission quality that you would have your readers consider? Well they are certainly good measures of something that has been of great interest in QND, but they hardly capture the essence of good teleportation. In fact there is a sense in which they only capture an infinitesimal bit of it: the upshot is that you are gauging teleportation quality by the performance of *just two* of the possible observables in Eq. (33.4). From our point of view that is an immediate handicap for your criterion. And this remains, regardless of how it may be used in combination with any other ingredients. For true-quality teleportation, one should transfer *every* aspect of the unknown quantum state from one side of the bench to the other (at least to within some tolerance). With this point, it is easy to see the trouble spot in your criterion. The problem is that two state

---

[3] For the original proof of this inequality and an expression of the observable that gives it equality, see Fuchs and Caves, "Mathematical Techniques for Quantum Communication Theory," *Open Sys. and Info. Dyn.* **3**, 345–356 (1995). Or you can find it in the more readily available paper: Barnum, Caves, Fuchs, Jozsa, and Schumacher, "Noncommuting Mixed States Cannot Be Broadcast," *Phys. Rev. Lett.* **76**, 2818–2821 (1996).

vectors can be completely orthogonal – and therefore just as different as they can possibly be – and still give rise to the same x statistics *and* the same p statistics.

For an easy example of this, just consider the two state vectors $|\psi_+\rangle$ and $|\psi_-\rangle$ whose representations in x-space are

$$\psi_\pm(x) = \left(\frac{2a}{\pi}\right)^{1/4} \exp\left((-a \pm ib)x^2\right), \tag{33.6}$$

for $a, b \geq 0$.[4] In k-space representation, these state vectors look like

$$\tilde{\psi}_\pm(k) = \left(\frac{a}{2\pi}\right)^{1/4} \sqrt{\frac{a \pm ib}{a^2 + b^2}} \exp\left(\frac{-a \mp ib}{4(a^2 + b^2)} k^2\right). \tag{33.7}$$

Clearly neither x measurements nor p measurements can distinguish these two states; for in both observables both wave functions differ from each other only by a local phase function. However, if we look at the overlap between the two states we find:

$$\langle \psi_- | \psi_+ \rangle = \sqrt{\frac{a(a + ib)}{a^2 + b^2}}. \tag{33.8}$$

By taking $b \to \infty$, we can make these two states just as orthogonal as we please. Suppose now that $|\psi_+\rangle$ were Victor's input into the teleportation process, and – by whatever means – $|\psi_-\rangle$ turned out to be the output. By your criterion, this would be perfect teleportation. But it certainly isn't so!

Of course it is clear that you're trying to get at something else. But the example above, contrived though it is, already shows a certain weakness in your criterion. It is true, as you point out, that a *cheating* Alice cannot simultaneously get hold of two conjugate observables. This is just as true for these two state vectors as for any others – the uncertainty principle is a fact of life. But so what of it? What does it necessarily have to do with the ability to detect a cheating Alice and therefore define an operational criterion for experimental teleportation?

Just to give you a far-fetched example, let us make use of the (almost orthogonal) states $|\psi_+\rangle$ and $|\psi_-\rangle$ once again. Consider the case where Alice and Bob are privy to the fact that Victor wishes to teleport these and only these states. At any shot, they know that Alice will be given one of these states; they just don't know which one. Then, clearly, they need use no entanglement whatsoever to get the quantum states across the bench. A cheating Alice need only perform some measurement $\mathcal{O}$ with eigenstates very close to $|\psi_+\rangle$ and $|\psi_-\rangle$ and send that classical information to Bob over the classical channel. Bob simply uses that information to resynthesize the appropriate state at his end. No entanglement has been used, and yet your criterion doesn't catch them in the cheat.

Well, of course, our criterion of fidelity in Eq. (33.1) doesn't catch them either. But this example will help us define the second problem in your comment much more sharply. The issue is this: What does it mean to say that Alice and Bob are given an *unknown* quantum

---

[4] This nice example was provided by Jason McKeever, one of our new graduate students at Caltech.

state? If you dig deep, you'll see that there's about only one useful meaning that can be given to this phrase. It means that Alice and Bob know that Victor will be drawing his states $|\psi_{in}\rangle$ from some fixed, given set S. Alice and Bob's lack of knowledge about which state Victor will actually draw is necessarily quantified by some probability ascription $P(|\psi_{in}\rangle)$. That's what an unknown quantum state means in the most general setting.

The last example above forces the following statement upon us:

All useful criteria for "unconditional quantum teleportation" must be anchored in whatever S and $P(|\psi_{in}\rangle)$ are given. A criterion is senseless if the states to which it is to be applied are not mentioned explicitly.

If S consists of orthogonal states, then no criterion whatsoever (short of watching Alice and Bob's every move) will ever be able to draw a distinction between true teleportation and the sole use of the classical side channel. Things only become interesting when the set S consists of two or more *nonorthogonal* quantum states.

With that caveat, one can start to make a little sense out of why you might have been led to the particular criterion that you and Lam proposed. From our first days as physics majors we are taught a mantra: Measurement causes disturbance. But then almost immediately all our instructors botch it by saying things like, "Measuring x necessarily disturbs p and vice versa." What could it possibly mean to say this when we know from the observed Bell inequality violations that not both x and p can exist with simultaneously definite values? How can something be disturbed when it's not even there? Fortunately quantum information and quantum cryptography in particular provide us a way out of that historical conundrum. When one is given an unknown state drawn from a set of nonorthogonal states, it is the gathering of information about the state's identity that causes a necessary disturbance. But that disturbance has nothing a priori to do with conjugate *classical* variables. Instead the disturbance is to the state itself.[5] Herein lies the key to a good teleportation criterion.

We, in particular, used Eq. (33.1) to define a notion of average fidelity between input and output:

$$F_{av} = \int_S P(|\psi_{in}\rangle) \, F(|\psi_{in}\rangle, \rho_{out}) \, d|\psi_{in}\rangle. \qquad (33.9)$$

Depending upon the precise states in S and the given form of $P(|\psi_{in}\rangle)$ there will be any number of different maximum values that F_{av} can take on in the presence of a cheating Alice and Bob (i.e., an Alice and Bob who only use their classical channel and make no use of their entanglement). Only one thing is for sure: if S contains some distinct nonorthogonal states, a cheating Alice and Bob can never achieve $F_{av} = 1$.

Let us drive this home before returning to a comment about your criterion. Suppose $S = \{|\psi_0\rangle\}$ consists of just one state. Then a cheating Alice and Bob can certainly achieve

---

[5] If you would like a lengthier discussion of this issue, have a look at: C. Fuchs and A. Peres, "Quantum State Disturbance vs. Information Gain: Uncertainty Relations for Quantum Information," *Phys. Rev.* A **53**, 2038–2045 (1996) or C. Fuchs, "Information Gain vs. State Disturbance in Quantum Theory," *Fort. der Physik* **46**, 535–565 (1998).

$F_{av} = 1$. Alice needs perform no measurement nor send any classical information in order to do this: Bob only has to synthesize the single fixed state for his output at the correct times. One way of saying this is that our teleportation criterion is so ridiculously nonstringent that it is trivial to achieve.

So let us consider a more stringent criterion. Take $S = \{|\psi_0\rangle, |\psi_1\rangle\}$ to consist of just two nonorthogonal states (with inner product x). Suppose the two states are deemed to occur with equal probability. Then it can be shown[6] that the best thing for the cheating duo to do is this. Alice should measure an operator whose orthogonal eigenvectors symmetrically bestride $|\psi_0\rangle$ and $|\psi_1\rangle$. She then sends that information to Bob. He synthesizes two states $|\tilde{\psi}_0\rangle$ and $|\tilde{\psi}_1\rangle$ depending upon the outcomes, being careful to tweak them so that the overlap between input and output is as high as it can possibly be on average. (In this case that means that Bob's states will have a slightly larger inner product than the original two states.) When all that is said and done, the best fidelity that can be achieved in this way is

$$F_{av} = \frac{1}{2}\left(1 + \sqrt{1 - x^2 + x^4}\right). \tag{33.10}$$

Even in worst case $\left(\text{when } x = 1/\sqrt{2}\right)$, this fidelity is always relatively high – it is always above 0.933.

What does this mean? Again it means that the criterion of unconditional teleportation is awfully weak. It is so easy for Alice and Bob to cheat when they know that their input is only one of two possibilities that the average fidelity can just about be made to approach one with hardly any effort.

As we pile more and more states into S it becomes ever harder for Alice and Bob to cheat. But even then things can still be nontrivial for finite-dimensional Hilbert spaces. Consider the case where S consists of every normalized vector in a Hilbert space of dimension d and assume S equipped with the uniform probability distribution. Then it turns out that the maximum value F_{av} can take on is[7]

$$F_{av} = \frac{2}{d+1}. \tag{33.11}$$

For the case of a single qubit, as in the original teleportation scheme of Bennett and company, Alice and Bob would *only* have to achieve a fidelity of 2/3 before they could claim unconditional quantum teleportation for Victor. BUT, again, that is only if Victor can be sure that Alice and Bob know absolutely nothing about which state he will input other than the dimension of the Hilbert space it lives in.

This last case tells us something very nice about the teleportation of continuous-variable quantum states. For this essentially corresponds to the limit $d \to \infty$ above. If Victor can be sure that Alice and Bob know nothing whatsoever about the quantum states he intends to teleport, then on average the best fidelity they can achieve in cheating is strictly zero!

[6] Fuchs and Peres, op. cit.
[7] H. Barnum, *Quantum Information*, Ph.D. thesis, University of New Mexico, 1998.

In this case, seeing any nonzero fidelity whatsoever in the laboratory would signify that unconditional quantum teleportation had been achieved.

To use this last criterion for our experiment here at Caltech, however, would have been a little ridiculous. For any reasonable Alice and Bob that had wanted to cheat would know that it is pretty difficult for the Victor using their services to create nonclassical states for his light. Therefore we explicitly made the assumption for our case that S contains the coherent states $|\alpha\rangle$ with a gaussian distribution describing the probability density on that set (it matters not where the mean of the gaussian is placed). It turns out that in the limit the variance of the gaussian approaches infinity, i.e., the distribution of states becomes ever more uniform, the *very best possible* average fidelity that can be achieved by a cheating Alice and Bob is

$$F_{av} = \frac{1}{2}. \tag{33.12}$$

Or to say it one more time, if Alice and Bob know that Victor will be handing off a coherent state but they have no clue which one it will be, then the best fidelity they can get upon cheating is $1/2$. "Very best possible" means that one must consider all measurements allowed within quantum mechanics for Alice's part and all possible translations of that classical information into quantum-state synthesis for Bob's part.

17 April 1999, to Salman Habib, "Tightivity/Quantum Sensitivity"

Well it's a little tight, but I believe I said everything I've ever wanted to say ... on a single page! (Shows how little I have to say.)

Quantum Sensitivity and Control

Our world is a quantum mechanical one. What should we make of that? In a way, the answer to this question was once less positive than it is today. For though quantum theory is a tool of unprecedented success in understanding the phenomena about us, the usual intellectual lesson derived from it is one of limitations. This is seen in almost any introductory lecture on the Heisenberg uncertainty principle: it is presented as if the world had short-changed us. "The task of physics is to sober up to this and make the best of it."

But that is a lesson of the past. Recent years have seen the start of a significantly more positive, almost intoxicating, attitude about the basic role of quantum mechanics. This is evidenced no more abundantly than in the field of quantum information and computation. Its point of departure is not to ask what limits quantum mechanics places upon us, but instead what novel, productive things we can do in the quantum world that we could not have done otherwise. In what ways is the quantum world fantastically better than the classical one?

Perhaps the two most striking examples are quantum key distribution and Shor's quantum factoring algorithm. In the first, quantum mechanics allows communicators to transmit cryptographic keys in a way that (hidden) eavesdropping on them can be excluded out of hand. This is impossible in the classical world because no connection exists between the information gatherable about a physical system and the disturbances induced upon it in the process. In the second, one finds that algorithms designed for computers built of unabashedly quantum components can factor large integers exponentially faster than anything designed for classical computers. To give a quick instance, consider a 600-digit number that is known to be the product of two (secret) primes. The number of computational steps required of a classical computer to crack it into its two components is roughly 10^{34}. In contrast, the corresponding number of steps for a quantum computer is only 10^{11}. Quantum computing gives 23 orders of magnitude greater efficiency here!

These two examples are the most outstanding of the class, and there is well-founded hope that they are the tip of a technological iceberg. There is no doubt, however, that they are the tip of a physical iceberg. What is being uncovered in a *rigorous* way for the first time since quantum theory's inception is a sense in which the quantum world is wildly more "sensitive to the touch" than the classical world: irritate it in the right way and the result is a pearl. One has to wonder if this itself is not the very essence of quantum mechanics.

Exploring three well-defined aspects of this sensitivity is the underpinning of my research program. Information, disturbance, entanglement. Let me say a word about each.

Information. People encode information in the states of quantum systems for a reason: to get it from one place to another. Since the world is quantum mechanical, this is, in the last analysis, what is always done for information transfer. Once this is recognized forthright, hosts of issues arise that have no analog in classical information theory. Key among these are ways of coding information into *nonorthogonal* quantum states so that it *cannot* be retrieved with complete reliability. What are the ultimate limits on that retrievability? This question defines a field of research in itself. Why bother? It turns out that for noisy channels it is better to take a hit in retrievability at the outset before transmission – the information becomes more resilient to noise in the long run. Understanding the "capacity" of noisy channels is the greatest of open problems. Solving it would put quantum information theory where the classical theory was in 1948.

Disturbance. Nonorthogonal states are not only remarkable in that *some* of the information they encode is lost forever, but in that they are magical canaries for the theft of *any* of the remaining. Measurements on nonorthogonal states cause detectable disturbances. This is the engine that powers quantum cryptography – an engine only now beginning to be understood. What is the ultimate tradeoff between information gain and state disturbance for nonorthogonal states? What is the ultimate secrecy capacity – the fraction of secret bits per transmission – that can be achieved with such alphabets? These are technical questions that will find solid answers in the near term. They are also the beginning of a deep set of foundational questions. Is there a sense in which the structure of quantum mechanics is optimal for these numbers? Much work remains to be done.

Entanglement. In 1935 Schrödinger coined the term and said of it, "I would not call that *one* but rather *the* characteristic trait of quantum mechanics." It is the stuff Einstein–Podolsky–Rosen pairs and Bell inequalities are made of. On the one hand, entanglement is the very embodiment of the information–disturbance principle – the quantum sensitivity to touch. On the other, it is a phenomenon *sui generis*. Most distinctly, it can be viewed as a physical resource in its own right, lending itself to information applications of every sort. The greatest issue is how to quantify entanglement and, further, to distill how such measures connect with the previous two categories. Progress along these lines is inevitable.

The T-8 group at LANL offers a unique set of resources for the continuance and expansion of the research just described. Information, disturbance, and entanglement all enter in crucial ways into the theory of quantum control systems. Environmental perturbations become entangled with the system to be controlled. To stave that off, control systems must gather information about the perturbation.

But then the system is perturbed further in the process. Ultimately one must understand the interplay and interrelation of all these factors to make the best of them – to irritate the quantum world in just the right way. Working on this project would be a grand adventure. Thank you for your consideration.

08 June 1999, to Todd Brun, "Information Theoretic Entanglement"

Brunism 1: *As you might have guessed, I have a question. I've recently been toying with ideas of quantum information in consistent histories, with the vague intent of trying to understand what it is. Most recently, I've been looking at entanglement.*

A definition of entanglement that occurs fairly naturally in CH is the following (translated into standard QM terms):

"Entanglement is quantified by the difference between the minimum missing information in a general measurement and the minimum missing information in a set of local measurements (suitably defined)."

What a coincidence that you should ask that just now! Within the last few days I (along with a student here, Bob Gingrich, and Michael Nielsen) have been readdressing just that question. Unfortunately things don't look so well. As you know, "the minimum missing information in a general measurement" is identical to the von Neumann entropy of the density operator under consideration (so long as the general measurements are restricted to be rank-one POVMs). So one might consider comparing $S(\rho_{AB})$ to $S(\rho_A)$ and/or $S(\rho_B)$ to get at a measure of entanglement. Indeed a good start is to think about $S(\rho_{AB}) - S(\rho_A)$. If the state ρ_{AB} is separable (i.e., violates *no* Bell inequalities, or equivalently can be written as a convex combination of product states), then that quantity is always positive. However, for a lot of entangled states it goes negative. Unfortunately it does not go negative for all entangled states.

Nicolas Cerf and Chris Adami have put in a lot of time talking about the quantity

$$S(\rho_A \otimes \rho_B) - S(\rho_{AB}) = S(\rho_A) + S(\rho_B) - S(\rho_{AB}),$$

and calling that a "measure of entanglement" (at least in their early papers). But almost all of that work is casuistry as far as I am concerned: this quantity too makes no unique break between separable and nonseparable. It ranges all over the map for both kinds of states.

You gave yourself some good leeway by saying "the minimum missing information in a set of local measurements (suitably defined) ... (Only one party? Two uncorrelated? Two correlated? Any product state?)." The point of the remark above is that two-party uncorrelated seems no good. Two-party correlated may be better, but how to get a clean characterization of all such things? The difficulty with this is what our nine-state eight-author paper is about (quant-ph/9804053). POVMs consisting of product states? The problem with this is that I don't know what meaning can be given to such a construction (again because of the considerations in the nine-state paper).

The reason I've become interested in this question again is mostly because I'd like to give another forward thrust to an old idea of mine. It is that "entanglement" itself is a shorthand for a more fundamental situation: namely that information gathering (about what someone else knows) and disturbance (to what they can predict) go hand in hand in the quantum world. The theory of quantum mechanics is our best attempt at synthesizing what can be said in light of that situation. "I know something about the system in front of you, how it will react (statistically) when I prod it this way or that. And I know something about you – how you will react statistically when I prod you this way or that – and I even know that you will try to learn something about what I know of the system. BUT after you have carried that project through, I know that I can say less of the system than I could have before. And indeed I can say less of you than I could have before." That is what entanglement IS ABOUT ... or at least this is what is at trial in my mind. (I put a little more meat on these bones in the letters to David Mermin dated 15 February 1998, 04 September 1998, and 08 September 1998 ... just in case you're not immediately seized with the desire to call this rubbish.)

In this connection, it is my present hope that one can use this as a background to give some substance to (and, more importantly, extend) Schrödinger's 1935 remark:

This is the reason that knowledge of the individual systems can *decline* to the scantiest, even to zero, while that of the combined remains continually maximal. Best possible knowledge of a whole does *not* include best possible knowledge of its parts – and that is what keeps coming back to haunt us.

Let me give you a couple of pointers for where to do a little reading about entanglement (from an information theoretic point of view):

1. R. Horodecki and M. Horodecki, "Information-Theoretic Aspects of Inseparability of Mixed States," *Phys. Rev.* A **54**, 1838–1843 (1996).
2. M. Horodecki, P. Horodecki, and R. Horodecki, "Separability of Mixed States: Necessary and Sufficient Conditions," *Phys. Lett.* A **223**, 1–8 (1996).

The second of these papers is also on quant-ph; the first appears not to be.

I hope that's a little helpful. If you have any good ideas or thoughts, please keep me informed!

03 August 1999, to Alan Fowler, "Landauer Memorial"

I apologize for not having replied to your invitation to Rolf's memorial meeting until now, but as it turns out today was my first day back at the office after being away for six weeks.

Very unfortunately, I must decline your warm invitation; my schedule is already booked for the days of Sept. 13 and 14. However, I would still like to make a contribution to the meeting if in some way the opportunity arises. Below, I paste a part of one of my email exchanges with Rolf. It stands out in my mind because of the sheer sweetness with which he won me over ... just before telling me that I didn't understand him at all. In return, I spanked him a bit by pointing out that he was fighting a battle with someone who was not me. Rolf's final response was short but gracious and, I think, indicative of his personality as a whole. [See correspondence with Rolf Landauer collected herein.]

He was a good man, and I will miss him.

18 August 1999, to Rob Spekkens, "Modal Stuff"

Please do send me a copy of your paper as soon as you get it written. I want to devote more attention to understanding what I like/dislike about the modal point of view. I haven't thought about it that much before.

A few things crop up in my mind that I want to record now.

1) Are the probabilities in the decomposition objective or subjective? If subjective, from whose point of view? If objective, then define explicitly what an objective probability might mean.
2) Why is it that the universal wave function in your theory plays the role of a compendium of possible worlds (along with their associated probabilities), but the other wave functions – the ones cropping up in your preferred decomposition – correspond to actual properties? Why would two things with such different ontological status be described by mathematical objects with indistinguishable mathematical structures? (They're both given by vectors in a vector space.)
3) Why a decomposition into d orthogonal components? What physical criterion singles out that number? I know you do it because you want an easy rule to get to the probabilities. But what dismisses, for instance, always decomposing the state into d^2 components (all of which when turned into projectors are linearly independent in the operator sense or some other such rule) and using a more contrived method for distilling probabilities? What compels you to making the assumptions you do? Why such a blatantly minor variation on Everett?
4) Why minimize the Shannon entropy per se, other than the herd effect? Why not any of the Rényi entropies? Why not the Bayesian probability of error for inferring the initial state via knowledge of the final state? (This of course would require you to look at the universal wave function at two different times when ferreting out your decompositions ... and would thus be harder. But if you're trying to get at a certain stability property – the world sort of hops about the least – this to me seems to be better motivated than raw Shannon entropy.)

If you write me back, I'm sure to think about and mull over your answers. But there's no guarantee that I'll write back until I get the itch again.

23 August 1999, to Sylvan Schweber, "Living with Law without Law"

I have just had the most wonderful time reading your two articles:

1. S. S. Schweber, "Physics, Community and the Crisis in Physical Theory," in *Physics, Philosophy, and the Scientific Community*, edited by K. Gavroglu, J. Stachel, and M. W. Wartofsky (Kluwer, Dordrecht, 1995), pages 125–152.
2. S. S. Schweber, "The Metaphysics of Science at the End of a Heroic Age," in *Experimental Metaphysics: Quantum Mechanical Studies for Abner Shimony, Vol. 1*, edited by R. S. Cohen, M. Horne, and J. Stachel (Kluwer, Dordrecht, 1997), pages 171–198.

and starting a search on many of your references cited therein.

What I would like to know is if you have written anything further expanding on the theme in these two articles? If you have, I would appreciate learning of their coordinates or perhaps receiving reprints/preprints.

I am a physicist at Caltech working in quantum information theory and quantum cryptography. I have had a long-standing interest in John Wheeler's ideas on "law without law" – that is how your articles attracted my attention. (I'm giving a plenary talk on quantum information at the New England Section Fall APS meeting this November; perhaps I'll meet you there and we can talk about all this much more?)

24 August 1999, to Pawel Wocjan, "Quantum Theory from QKD"

I've finally found a few moments to think about some of the things I discussed with you. Namely, that of all theories with a trade-off between information and disturbance, quantum theory might just be the optimal one. I did a search through my files and found three papers that might be of relevance for such a project:

1. I. Csiszár and J. Körner, "Broadcast Channels with Confidential Messages," *IEEE Trans. Inf. Theory* **IT-24** (1978) 339–348.
2. R. Ahlswede and I. Csiszár, "Common Randomness in Information Theory and Cryptography – Part I: Secret Sharing," *IEEE Trans. Inf. Theory* **39** (1993) 1121–1132.

and, maybe most importantly,

3. U. Maurer, "Secret Key Agreement by Public Discussion from Common Information," *IEEE Trans. Inf. Theory* **39** (1993) 733–742.

If these papers exist, then probably even more relevant things can be found in the literature. But these may be a good starting point for some thought. (And they may form a good base for a literature search for some more up-to-date things.)

If you have any ideas or thoughts, I would love to hear them.

24 August 1999, to Dominik Janzing, "Bopp Stuff"

About Bopp, if it is Friedrich (Fritz) Arnold Bopp that you are speaking of, then I have found a little bit about him. Max Jammer, in his book *The Philosophy of Quantum Mechanics*, has a few pages devoted to him. He writes, "Bopp's theory, of which only a very brief outline could be presented, seems to have found, in spite of its logical and philosophical consistency, only a very limited amount of interest among physicists and philosophers." Almost all the references are in German except one: F. Bopp, "The Principles of the Statistical Equations of Motion in Quantum Theory," in *Observation and Interpretation in the Philosophy of Physics*, edited by S. Körner (Dover, NY, 1957), pp. 189–196. As it turns out, I have that volume at home; so I will have a look at the paper. Jammer also cites the following paper: W. M. Machado and W. Schützer, "Bopp's Formulation of Quantum

Mechanics and the Einstein–Podolsky–Rosen Paradox," *Anais da Academia Brasileira de Ciencias* **35** (1963) 27–35. I would like to get my hands on that article, but it may be pretty hard to.

07 January 2000, to Jeremy Butterfield, "Page Numbers for Shimony"

Could you do me a favor? Shimony sent me a copy of his paper "Can the Fundamental Laws of Nature Be the Results of Evolution?" appearing in your book *From Physics to Philosophy*: I would like to enter the paper into my database, but I don't have the page numbers. Could you please give me both the beginning and ending page numbers for the article?

In case the book hasn't made it to press yet, here's one typo that you may have overlooked. Shimony's spell checker must not have known the word "observership," for the word "ownership" consistently appears in its place (there are at least two instances). For instance, Wheeler (1977) is listed as "Genesis and Ownership" instead of "Genesis and Observership."

07 February 2000, to Eugen Merzbacher, "Missing Copenhagen"

Aha! You ask me a new question, but you don't answer my old one? Perhaps I should try to extract some leverage with this . . . ! Oh, I suppose I'll just place it below to prod your memory and leave it to your conscience.

You said,

Fortunately, your piece leaves a number of issues untouched or unresolved. For a long time I have thought about writing an essay on what is meant by the term "physical system," including some history of that slippery concept. I'll probably never do it, but your opinion piece has brought this matter back to my mind.

and I replied with

I would be interested in hearing your thoughts about the "slippery concept of physical system." One of the things that fascinates me now is our freedom to decompose a large Hilbert space into any tensor product decomposition we wish. We label the components in that decomposition to be "systems," but that is imposed by us, not Nature. In what way do you mean "slippery?"

Thanks for sending me the information about the play *Copenhagen*. I had tried to get to London to see it this summer when I was in Cambridge for six weeks, but things never panned out. My colleague Rüdiger Schack saw it and said that it's marvelous.

27 July 2000, to Robert Garisto, "The Zing of Empty Space"

Thanks for the note. I'm still thinking about your question, which now strikes me as a truly important one.

Q: What is real about a quantum system?

A: That "sensitivity to the touch," that "zing!" that keeps us from ever having more information about it than can be encoded in a quantum state.

That's where I left it. But you asked, "Where are these systems? Where do they live?" Could this say anything about spacetime that we may have missed before?

Good question!

14 November 2000, to Steven van Enk, "Always One Theory Behind"

If (by a miracle) you end up reading any of the thick document I gave you yesterday *and* (by chance) you find any typos in it, I'd appreciate it if you would note them and let me know.

I've been thinking about your "philosophical" question all morning. (I'm in the Chicago airport right now.) And, it dawned on me that maybe the following is a better language than Bohr's. As you might remember, I'm already quite inclined to reexpress the process of quantum "measurement" in a language borrowed from the economists. Let me review that briefly before getting back to some Bohrish things. (I'll place an older, long account far below.) The keywords are DECISIONS, ACTS, and CONSEQUENCES. The experimentalist DECIDES to perform this or that ACTION upon a quantum system. Such actions will generally have some unpredictable CONSEQUENCES and it is those consequences that we more standardly call a measurement outcome.

Within this wording, your question becomes this: Must our acts and their consequences always be expressed in the language of classical physics? Here's where I think I've made some headway. A better way, I think, to express the situation is that we have no choice but to use a language that is "one theory behind." In this, I have been greatly influenced by a passage (attributed to Einstein) from one of Heisenberg's books. You can find the passage in "The Activating Observer," but let me repeat it here.

W. Heisenberg, *Physics and Beyond: Encounters and Conversations*, translated by A. J. Pomerans (Harper & Row, New York, 1971), pp. 63–64. Heisenberg reports Einstein as having once said the following to him:

> It is quite wrong to try founding a theory on observable magnitudes alone. In reality the very opposite happens. It is the theory which decides what we can observe. You must appreciate that observation is a very complicated process. The phenomenon under observation produces certain events in our measuring apparatus. As a result, further processes take place in the apparatus, which eventually and by complicated paths produce sense impressions and help us to fix the effects in our consciousness. Along this whole path – from the phenomenon to its fixation in our consciousness – we must be able to tell how nature functions, must know the natural laws at least in practical terms, before we can claim to have observed anything at all. Only theory, that is, knowledge of natural laws, enables us to deduce the underlying phenomena from our sense impressions. When we claim that we can observe something new, we ought really

to be saying that, although we are about to formulate new natural laws that do not agree with the old ones, we nevertheless assume that the existing laws – covering the whole path from the phenomenon to our consciousness – function in such a way that we can rely upon them and hence speak of "observation."

Now, Einstein probably thought the situation of having to express "observations" in a language "one theory behind" was to be a provisional thing: At the completion of the new theory's construction, one would just do away with it. But maybe that was never the case, and Bohr was onto that train of thought. (Or maybe he wasn't, and it's just me instead.)

Anyway, that's my little thought for the morning. I'll mail this to you when I get to Caltech.

14 January 2001, to Rob Pike, "Wispy Sunday"

I'm listening to Dinah Washington this Sunday morning. And that always makes me feel wispy. But part of it is coming from this session I'm organizing for the Sweden meeting.

I was impressed with some things you said at cookie time Friday. I think they show we have some overlap in our thoughts about quantum information. In particular, your remark about how computer scientists work, perhaps, in an overly restrictive framework by thinking quantum computing is captured (exclusively?) by the idea of an exponentiation of computational paths struck me as on the mark. In that connection, I think you might enjoy reading Andy Steane's paper, "A Quantum Computer Needs Only One Universe," quant-ph/0003084. (Steane, if you don't know him, independently invented the idea of quantum error correction at the same time as Peter Shor.)

I too keep thinking that entanglement and exponentiation are more a surface phenomenon of a deeper kind of "zing!" that quantum systems have. The two old notes below put that more in the context of our business. And they seem suitably wispy to fit my Sunday mood: so I thought I'd forward them to you. When I rediscovered the remark I made to Mermin about Bell Labs at the end of one of them, I thought it was so apropos! [See note to Gilles Brassard, titled "Memories of Montréal," dated 17 June 2000, and note to David Mermin, titled "Zing!," dated 20 July 2000.]

18 January 2001, to Max Tegmark, "100 Years of the Quantum"

[AUTHOR'S NOTE: Given the influence of John Wheeler on my passion for the Paulian idea, it seems only appropriate that this volume should include a bibliography of John's post-Everett-enthusiasm quantum writings. (By the adjective post-Everett-enthusiasm, I refer to the period in Wheeler's published writings from roughly 1971 through to the end of 2000.) Max Tegmark's posting of quant-ph/0101077 on 17 January 2001 presented the right circumstances for me to place my personal bibliography from that period into the medium of email. Five references and two quotes have been added since the email's original delivery.]

1. J. A. Wheeler, "Transcending the Law of Conservation of Leptons," in *Atti del Convegno Internazionale sul Tema: The Astrophysical Aspects of the Weak Interaction* (Cortona "Il Palazzone," 10-12 Giugno 1970), Accademia Nationale die Lincei, Quaderno N. **157** (1971), pp. 133–164.
2. J. A. Wheeler, "From Relativity to Mutability," *The Physicist's Conception of Nature*, edited by J. Mehra (D. Reidel, Dordrecht, 1973), pp. 202–247.
3. C. W. Misner, K. S. Thorne, and J. A. Wheeler, "Beyond the End of Time," Chap. 44 in *Gravitation*, (W. H. Freeman, San Francisco, CA, 1973), pp. 1196–1217.
4. J. A. Wheeler, "From Mendeléev's Atom to the Collapsing Star," in *Philosophical Foundations of Science*, edited by R. J. Seeger and R. S. Cohen (Reidel, Dordrecht, 1974), pp. 275–301.
5. J. A. Wheeler, "The Universe as Home for Man," *Am. Sci.* **62**, 683–691 (1974). This is an abridged, early version of Ref. [8].
6. J. A. Wheeler, "From Magnetic Collapse to Gravitational Collapse: Levels of Understanding Magnetism," in *Role of Magnetic Fields in Physics and Astrophysics*, Ann. NY Acad. Sci. **257**, edited by V. Canuto (NY Academy of Sciences, NY, 1975), pp. 189–221.
7. J. A. Wheeler, "Another Big Bang?" *Am. Sci.* **63**, 138 (1975). This is a letter in reply to a reader's comments on Ref. [5].
8. J. A. Wheeler, "The Universe as Home for Man," *The Nature of Scientific Discovery: A Symposium Commemorating the 500th Anniversary of the Birth of Nicolaus Copernicus*, edited by O. Gingerich (Smithsonian Institution Press, City of Washington, 1975), pp. 261–296, discussion pp. 575–587.
9. C. M. Patton and J. A. Wheeler, "Is Physics Legislated by Cosmogony?," in *Quantum Gravity: An Oxford Symposium*, edited by C. J. Isham, R. Penrose, and D. W. Sciama (Clarendon Press, Oxford, 1975), pp. 538–605.
10. J. A. Wheeler, "Include the Observer in the Wave Function?," Fundamenta Scientiae: Seminaire sur les Fondements des Sciences (Strasbourg) **25**, 9–35 (1976).
11. J. A. Wheeler, "Genesis and Observership," in *Foundational Problems in the Special Sciences: Part Two of the Proceedings of the Fifth International Congress of Logic, Methodology and Philosophy of Science, London, Canada – 1975*, edited by R. E. Butts and J. Hintikka (D. Riedel, Dordrecht, 1977), pp. 3–33.
12. J. A. Wheeler, "Include the Observer in the Wave Function?," *Quantum Mechanics, a Half Century Later: Papers of a Colloquium on Fifty Years of Quantum Mechanics, Held at the University Louis Pasteur, Strasbourg, May 2–4, 1974*, edited by J. Leite Lopes and M. Paty (D. Reidel, Dordrecht, 1977), pp. 1–18. This is a reprint of Ref. [10].
13. C. M. Patton and J. A. Wheeler, "Is Physics Legislated by Cosmogony?," in *Encyclopedia of Ignorance: Everything You Ever Wanted to Know about the Unknown*, edited by R. Duncan and M. Weston-Smith (Pergamon, Oxford, 1977), pp. 19–35. This is an abridged version of Ref. [9].
14. J. A. Wheeler, "The 'Past' and the 'Delayed-Choice' Double-Slit Experiment," *Mathematical Foundations of Quantum Theory*, edited by A. R. Marlow (Academic Press, New York, 1978), pp. 9–48.
15. J. A. Wheeler, "Parapsychology – A Correction," *Science* **205**, 144 (1979). This correction refers to something stated in Wheeler's talk at ?? (reported later in corrected form in Ref. [24]).
16. J. A. Wheeler, "Frontiers of Time," in *Problems in the Foundations of Physics, Proceedings of the International School of Physics "Enrico Fermi," Course LXXII*,

edited by G. Toraldo di Francia (North-Holland, Amsterdam, 1979),
pp. 395–492.

17. J. A. Wheeler, "The Quantum and the Universe," in *Relativity, Quanta, and Cosmology in the Development of the Scientific Thought of Albert Einstein, Vol. II*, edited by F. de Finis (Johnson Reprint Corp., New York, 1979), pp. 807–825.

18. J. A. Wheeler, "The Superluminal," New York Review of Books, 27 September 1979, p. 68.

19. J. A. Wheeler, "Collapse and Quantum as Lock and Key," *Bull. Am. Phys. Soc.*, Series II **24**, 652–653 (1979).

20. J. A. Wheeler, "Beyond the Black Hole," in *Some Strangeness in the Proportion: A Centennial Symposium to Celebrate the Achievements of Albert Einstein*, edited by H. Woolf (Addison-Wesley, Reading, MA, 1980), pp. 341–375, discussion pp. 381–386.

21. J. A. Wheeler, "Pregeometry: Motivations and Prospects," in *Quantum Theory and Gravitation: Proceedings of a Symposium Held at Loyola University, New Orleans, 23–26 May 1979*, edited by A. R. Marlow (Academic Press, New York, 1980), pp. 1–11.

22. J. A. Wheeler, "Law without Law," *Structure in Science and Art*, edited by P. Medawar and J. Shelley (Elsevier, Amsterdam, 1980), pp. 132–154.

23. J. A. Wheeler, "Delayed-Choice Experiments and the Bohr–Einstein Dialogue," in *American Philosophical Society and the Royal Society: Papers Read at a Meeting, 5 June 1980*, (American Philosophical Society, Philadelphia, 1980), pp. 9–40.

24. J. A. Wheeler, "Not Consciousness but the Distinction Between the Probe and the Probed as Central to the Elemental Quantum Act of Observation," in *The Role of Consciousness in the Physical World*, edited by R. G. Jahn (Westview Press, Boulder, CO, 1981), pp. 87–111.

25. J. A. Wheeler, "The Participatory Universe," *Science81* **2**(5), 66–67 (1981).

26. J. A. Wheeler, "The Elementary Quantum Act as Higgledy-Piggledy Building Mechanism," in *Quantum Theory and the Structures of Time and Space: Papers Presented at a Conference Held in Tutzing, July, 1980*, edited by L. Castell and C. F. von Weizsäcker (Carl Hanser, Munich, 1981), pp. ??–??.

27. J. A. Wheeler, "The Computer and the Universe," *Int. J. Theo. Phys.* **21**, 557–572 (1982).

28. J. A. Wheeler, "Particles and Geometry," in *Unified Theories of Elementary Particles: Critical Assessment and Prospects*, Lecture Notes in Physics **160**, edited by P. Breitenlohner and H. P. Dürr (Springer-Verlag, Berlin, 1982), pp. 189–217.

29. J. A. Wheeler, "Bohr, Einstein, and the Strange Lesson of the Quantum," in *Mind in Nature: Nobel Conference XVII, Gustavus Adolphus College, St. Peter, Minnesota*, edited by R. Q. Elvee (Harper & Row, San Francisco, CA, 1982), pp. 1–23, discussion pp. 23–30, 88–89, 112–113, and 148–149.

30. J. A. Wheeler, "Physics and Austerity: Law without Law," University of Texas preprint, 1–87 (1982).

31. J. A. Wheeler, "Black Holes and New Physics," Discovery: Research and Scholarship at the University of Texas at Austin **7**(2), 4–7 (Winter 1982).

32. J. A. Wheeler, "On Recognizing 'Law without Law': Oersted Medal Response at the Joint APS-AAPT Meeting, New York, 25 January 1983," *Am. J. Phys.* **51**, 398–404 (1983).

33. J. A. Wheeler, "Elementary Quantum Phenomenon as Building Unit," in *Quantum Optics, Experimental Gravity, and Measurement Theory*, edited by P. Meystre and M. O. Scully (Plenum Press, New York, 1983), pp. 141–143.

34. J. A. Wheeler, "Law without Law," in *Quantum Theory and Measurement*, edited by J. A. Wheeler and W. H. Zurek (Princeton University Press, Princeton, 1983), pp. 182–213.

35. J. A. Wheeler, "Guest Editorial: The Universe as Home for Man," in *The Dynamic Universe: An Introduction to Astronomy*, edited by T. P. Snow (West Pub. Co., St. Paul, Minnesota, 1983), pp. 108–109. This is an excerpt from Ref. [8].

36. J. A. Wheeler, "Physics and Austerity," in *Krisis*, Vol. 1, No. 2, edited by I. Masculescu (Klinckscieck, Paris, 1983), pp. 671–675.

37. J. A. Wheeler, "Jenseits aller Zeitlichkeit," in *Die Zeit: Schrifter der Carl Friedrich von Seiemens-Stiftung*, Vol. **6**, edited by A. Peisl and A. Mohler (Oldenbourg, Munchen, 1983), pp. 17–34.

38. R. S. Armour, Jr. and J. A. Wheeler, "Physicist's Version of Traveling Salesman Problem: Statistical Analysis," *Am. J. Phys.* **51**, 405–406 (1983).

39. J. A. Wheeler, "Quantum Gravity: The Question of Measurement," in *Quantum Theory of Gravity: Essays in Honor of the 60th Birthday of Bryce S. DeWitt*, edited by S. M. Christensen (Adam Hilger, Bristol, 1984), pp. 224–233.

40. J. A. Wheeler, "Die Experimente der verzögerten Entscheidung und der Dialog zwischen Bohr und Einstein," in *Moderne Naturphilosophie*, edited by B. Kanitscheider (Königshausen and Neumann, Würzburg, 1984), pp. 203–222. This is a slightly abridged, German version of Ref. [23].

41. J. A. Wheeler, "Bits, Quanta, Meaning," in *Problems in Theoretical Physics*, edited by A. Giovannini, F. Mancini, and M. Marinaro (University of Salerno Press, Salerno, 1984), pp. 121–141.

42. J. A. Wheeler, "Bits, Quanta, Meaning," in *Theoretical Physics Meeting: Atti del Convegno, Amalfi, 6–7 Maggio 1983*, (Edizioni Scientifiche Italiene, Naples, 1984), pp. 121–134. This should be a reprint of Ref. [41].

43. W. A. Miller and J. A. Wheeler, "Delayed-Choice Experiments and Bohr's Elementary Quantum Phenomenon," in *Proceedings of the International Symposium, Foundations of Quantum Mechanics in the Light of New Technology*, edited by S. Kamefuchi, H. Ezawa, Y. Murayama, M. Namiki, S. Nomura, Y. Ohnuki, and T. Yajima (Physical Society of Japan, Tokyo, 1984), pp. 140–152.

44. J. A. Wheeler, "Bohr's 'Phenomenon' and 'Law without Law'," in *Chaotic Behavior in Quantum Systems: Theory and Applications*, edited by G. Casati (Plenum Press, New York, 1985), pp. 363–378.

45. J. A. Wheeler, "Niels Bohr, the Man," *Phys. Today* **38**(10), 66–72 (1985).

46. J. A. Wheeler, "Delayed-Choice Experiments and the Bohr–Einstein Dialogue," in *Niels Bohr: A Profile*, edited by A. N. Mitra, L. S. Kothari, V. Singh, and S. K. Trehan (Indian National Science Academy, New Delhi, 1985), pp. 139–168. This is a reprint of Ref. [23].

47. J. A. Wheeler, "Hermann Weyl and the Unity of Knowledge," *Am. Sci.* **74**, 366–375 (1986).

48. J. A. Wheeler, "Niels Bohr: The Man and his Legacy," in *The Lesson of Quantum Theory*, edited by J. de Boer, E. Dal, and O. Ulfbeck (Elsevier, Amsterdam, 1986), pp. 355–367.

49. J. A. Wheeler, " 'Physics as Meaning Circuit': Three Problems," in *Frontiers of Nonequilibrium Statistical Physics*, edited by G. T. Moore and M. O. Scully (Plenum Press, New York, 1986), pp. 25–32.

50. J. A. Wheeler, "Foreword," in J. D. Barrow and F. J. Tipler, *The Anthropic Cosmological Principle*, (Oxford University Press, Oxford, 1986), pp. vii–ix.

51. J. A. Wheeler, "How Come the Quantum" in *New Techniques and Ideas in Quantum Measurement Theory*, edited by D. M. Greenberger, *Ann. New York Acad. Sci.* **480**, 304–316 (1987).

52. J. A. Wheeler, "Foreword" in H. Weyl, *The Continuum: A Critical Examination of the Foundation of Analysis*, translated by S. Pollard and T. Bole (Thomas Jefferson University Press, Kirksville, MO, 1987), pp. ix–xiii. This is an excerpt from Ref. [47].

53. J. A. Wheeler, "World as System Self-Synthesized by Quantum Networking," *IBM J. Res. Develop.* **32**, 4–15 (1988).

54. J. A. Wheeler, "World as System Self-Synthesized by Quantum Networking," in *Probability in the Sciences*, edited by E. Agazzi (Kluwer, Dordrecht, 1988), pp. 103–129. This is a reprint of Ref. [53].

55. J. A. Wheeler, "Hermann Weyl and the Unity of Knowledge," in *Exact Sciences and their Philosophical Foundations*, edited by W. Deppert *et al.* (Lang, Frankfurt am Main, 1988), pp. 366–375. This is an expanded version of Ref. [47].

56. J. A. Wheeler, "Bits, Quanta, Meaning," in *Festschrift in Honour of Eduardo R. Caianiello*, edited by A. Giovannini, F. Mancini, M. Marinaro, and A. Rimini (World Scientific, Singapore, 1989), pp. 133–154. This should be a reprint of Ref. [41].

57. J. A. Wheeler, "Information, Physics, Quantum: the Search for Links," in *Proceedings of the 3rd International Symposium on Foundations of Quantum Mechanics in the Light of New Technology*, edited by S. Kobayashi, H. Ezawa, Y. Murayama, and S. Nomura (Physical Society of Japan, Tokyo, 1990), pp. 354–368.

58. J. A. Wheeler, "Recent Thinking about the Nature of the Physical World: It from Bit," in *Frontiers in Cosmic Physics: Symposium in Memory of Serge Alexander Korff*, Ann. NY Acad. Sci. **655**, edited by R. B. Mendell and A. I. Mincer (NY Academy of Sciences, NY, 1992), pp. 349–364.

59. J. A. Wheeler, *At Home in the Universe*, (American Institute of Physics Publishing, New York, 1992).

60. J. A. Wheeler, "Time Today," in *Physical Origins of Time Asymmetry*, edited by J. J. Halliwell, J. Pérez-Mercader, and W. H. Zurek (Cambridge University Press, Cambridge, 1994), pp. 1–29.

61. J. A. Wheeler (with K. W. Ford), *Geons, Black Holes, and Quantum Foam: A Life in Physics*, (W. W. Norton, New York, 1998).

62. J. A. Wheeler, letter to Carroll Alley, 13 March 1998.

I want you and Einstein to jolt the world of physics into an understanding of the quantum because the quantum surely contains – when unravelled – the most wonderful insight we could ever hope to have on how this world operates, something equivalent in scope and power to the greatest discovery that science has ever yet yielded up: Darwin's Evolution.

You know how Einstein wrote to his friend in 1908, "This quantum business is so incredibly important and difficult that everyone should busy himself with it." … Expecting something great when two great minds meet who have different outlooks, all of us in this Princeton community expected something great to come out from Bohr and Einstein arguing the great question day after day – the central purpose of Bohr's four-month, spring 1939 visit to Princeton – I, now, looking back on those days, have a terrible conscience because the day-after-day arguing of Bohr was not with Einstein about the quantum but with me about the fission of uranium. How recover, I ask myself over and over, the pent up promise of those long-past days? Today, the physics community is bigger and knows more than it did in 1939, but it lacks the same feeling of **desperate** puzzlement. I want to recapture that feeling for us all, even if it is my last act on Earth.

63. J. A. Wheeler, "The Eye and the U," letter dated 19 March 1998 to Abner Shimony, with carbon copies to Arthur Wightman, Peter Mayer, Demetrios Christodoulou, Larry Thomas, Elliot Lieb, Charles Misner, Frank Wilczek, Kip Thorne, Ben Schumacher, William Wootters, Jim Hartle, Edwin Taylor, Ken Ford, Freeman Dyson, Peter Cziffra, and Arkady Plotnitsky.

You have known me long enough and well enough to believe me when I say that I would willingly give up an arm or a leg to understand "How Come the Quantum?" or "How Come Existence?" – or give up both to understand both. As you know, nobody is taking a question seriously unless he educates himself enough about it to hypothesize an answer to it. My tentative answer is stated nowhere more compactly than in Act IV, Scene 1, of Shakespear's *The Tempest*, "We are such stuff as dreams are made on" – and symbolized in the Eye and the U at the right:

64. J. A. Wheeler, "Information, Physics, Quantum: The Search for Links," in *Feynman and Computation: Exploring the Limits of Computers*, edited by A. J. G. Hey (Perseus Books, Reading, MA, 1999), pp. 309–336.
65. J. A. Wheeler, " 'A Practical Tool,' But Puzzling, Too," New York Times, 12 December 2000.

What is the greatest mystery in physics today? Different physicists have different answers. My candidate for greatest mystery is a question now a century old, "How come the quantum?"

What is this thing, the "quantum"? It's a bundle of energy, an indivisible unit that can be sliced no more. Max Planck showed us a hundred years ago that light is emitted not in a smooth, steady flow, but in quanta. Then physicists found quantum jumps of energy, the quantum of electric charge and more. In the small-scale world, everything is lumpy.

And more than just lumpy. When events are examined closely enough, uncertainty prevails; cause and effect become disconnected. Change occurs in little explosions in which matter is created and destroyed, in which chance guides what happens, in which waves are particles and particles are waves.

Despite all this uncertainty, quantum physics is both a practical tool and the basis of our understanding of much of the physical world. It has explained the structure of atoms and molecules, the thermonuclear burning that lights the stars, the behavior of semiconductors and superconductors, the radioactivity that heats the earth, and the comings and goings of particles from neutrinos to quarks.

Successful, yes, but mysterious, too. Balancing the glory of quantum achievements, we have the shame of not knowing "how come." Why does the quantum exist?

My mentor, the Danish physicist Niels Bohr, made his peace with the quantum. His "Copenhagen interpretation" promulgated in 1927 bridged the gap between the strangeness of the quantum world and the ordinariness of the world around us. It is the act of measurement, said Bohr, that transforms the indefiniteness of quantum events into the definiteness of everyday experience. And what one can measure, he said, is necessarily limited. According to his principle of complementarity, you can look at something in one way or in another way, but not in both ways at once. It may be, as one French physicist put it, "the fog from the north," but the Copenhagen interpretation remains the best interpretation of the quantum that we have.

20 January 2001, to John Smolin, "Shannon Meets Bohr"

As I told you on the phone, it's happening to me again: I've gotten another opportunity to gather my friends in an exotic place to thrash out the idea of "Quantum Foundations in the Light of Quantum Information." This time I've been asked to organize a session at a larger conference in Växjö, Sweden titled "Quantum Theory: Reconsideration of Foundations." The main organizer is Andrei Khrennikov, and he is planning to have 40–50 people attending, with further sessions on Bohmian mechanics, GRW mechanics, and other issues in quantum foundations (Bell inequalities, Kochen–Specker theorems, etc.).

In flavor and constitution, I plan to make our session much like the meeting Gilles Brassard and I held in Montréal last spring. There the theme was organized around the strong feeling that we'll find the greatest things and technologies to come out of quantum mechanics when we finally grasp the parts of it that make us feel the most uncomfortable. The theory is begging us to ask something new and profound of nature.

This time the list of confirmed attendees (all invitees so far) includes: Gilles Brassard, Jeffrey Bub, Carl Caves, Lucien Hardy, Richard Jozsa, Pekka Lahti, David Mermin, Asher Peres, Itamar Pitowsky, John Preskill, Rüdiger Schack, and Ben Schumacher. Depending upon how the money holds out, I might be able to call up even a few more in the community.

I think we'll have a double-edged opportunity at this meeting, so I'm looking quite forward to it. First, we'll get a chance to continue the lines several of us started in Montréal – at that time, the attendees were Bennett, Bernstein, Brassard, Hayden, Jozsa, Mermin, Schack, Schumacher, and Wootters. But second, we may be able to play the role of educators to a larger community interested in dabbling in these same issues. That is, we will have an opportunity to get them to think of quantum information as a tool for exploring quantum foundations.

The dates for the meeting are June 17 to 22. We think this choice will help make for quite a pleasant time: Midsummer's eve is June 21, and Mermin has pointed out that picking wild strawberries in the midnight twilight can be huge fun. Furthermore, there are significant local festivities planned around that time of year.

The information I'd like to get from you (as soon as possible!) is:

1) Can you confirm positively that you would like to come?
   and
2) What kind of financial resources will it take for us to get you there?
   a) Full expenses paid?
   b) Local expenses paid?
   c) Nothing?

Concerning question 2), you can count this letter as an invitation, so feel free to answer any of the three options a), b), or c). However, the more money I can get the invitees to throw into the pot (if they have it available), the more interesting people I can get to Sweden to keep us entertained. I've been told that I can have 3–4 invitees all expenses paid and 5 invitees with local expenses paid. You can see that I'm trying to stretch my limits! The way I've been able to do this is that, indeed, a significant number of attendees have *volunteered* to pay their nonlocal expenses, i.e., their travel expenses. (In this group are Caves, Hardy, Mermin, and Preskill. Jozsa and Schack have also committed to contribute in the case that it will make for a more interesting meeting. I think your presence would do that!)

I really hope you can attend; I think we'll have a lot of fun. Please let me know your thoughts as soon as you can! (Like, today or tomorrow maybe?) The sooner Khrennikov and I can get this preliminary information (dates in particular), the sooner we can start throwing stones at all the other problems.

26 January 2001, to Jeremy Butterfield, "Worlds in the Everett Interpretation"

Anyone who knows me knows that I am rather down on attempts to interpret quantum mechanics along Everett-like lines. I think the most funny and telling statement of this in the present context is that, whereas Mr. Wallace speaks of "Everettians," I often speak of "Everettistas." Thus, I am almost surprised that you sent me this paper to referee.

My difficulties come not so much from thinking that an Everett-like interpretation is inherently inconsistent or that parallel worlds tax the imagination too much. It's more that this line of thought strikes me, at best, as a complete dead end in the physical sense. At worst, I fear it requires us to tack on even more ad hoc structures to quantum theory than we already have. (Here, I'm thinking of a preferred basis for the Hilbert space and a preferred tensor-producting of it into various factors.) For these reasons, among umpteen others, I have always been inclined to an epistemic interpretation of the quantum state. Doing this has helped me (personally) to focus the issue to asking, "What is this *property* of the quantum world – i.e., reality – that keeps us from ever knowing more of it than can be captured by the quantum state?" To that extent, I consider myself something of a realist who – just as David Deutsch – takes the wavefunction absolutely seriously. BUT absolutely seriously as a state of knowledge, not a state of nature. I do well believe we will one day shake a notion of reality from the existing theory (without adding hidden variables, etc.), but that reality won't be the most naïve surface term floating to the top (i.e., the quantum state). When we have it, we'll really have something; there'll be no turning back. Physics won't be at an end, but at a beginning. For then, and only then, will we be able to recognize how we might extend the theory to something bigger and better than quantum mechanics itself.

All that said, you're going to be surprised by my evaluation of this paper. Without hesitation, I recommend you publish it! Of all the things I've read on Everett-like interpretations over the years, this paper has struck me as the most reasonable of the lot. This is a nice paper. It changed none of my views, but it caused me deep pause for thought. What better honor can one have from a sparring partner?

I am not in a position to judge whether the paper is a significant step forward over the many *recent* papers it cites – I've never read any of them – but I do know this much: I have walked away from several talks of the "most modern" treatment of Everett and not had a clue what was being talked about. So, even if my evaluation (as an outsider) of the paper's technical merit leaves something to be desired, I think Mr. Wallace's paper clearly serves a purpose within our community. The analogies between the Everett-like structure he proposes and the spacetime of general relativity are indeed intriguing and worthy of thought. For me personally – whatever their ultimate merit – they give a new stone on which to hone the arguments for an epistemic interpretation of the quantum state.

Let me just make a few minor comments to round out this report.

I very much enjoyed the discussion:

So there are no theory-neutral observations: rather, there is an existing theory in terms of which our observations are automatically interpreted, and which we must take as our starting point when interpreting the theories of physics.

I've seen a discussion like this before, and it probably should be cited. Here are two references from my personal archive:

1. W. Heisenberg, *Physics and Beyond: Encounters and Conversations*, translated by A. J. Pomerans (Harper & Row, New York, 1971), pp. 63–64.
2. M. Jammer, "The Experiment in Classical and in Quantum Physics," in *Proceedings of the International Symposium, Foundations of Quantum Mechanics in the Light of New Technology*, edited by S. Kamefuchi, H. Ezawa, Y. Murayama, M. Namiki, S. Nomura, Y. Ohnuki, and T. Yajima (Physical Society of Japan, Tokyo, 1984), pp. 265–276.

Section 10 (the big table). This may have been my favorite part of the paper. But I would do this: put the relativity column on the left and the Everett column on the right. Somehow, I found it much easier to read in that manner. It just seemed more natural to recognize a feature in relativity first, and then look for the analogous feature in Everett.

27 January 2001, to David Wallace, "The Test Particle"

I presume by now you've seen my thoughts on your Everett article. I meant it all; it's a very nice paper. I'd like to encourage you to place the article on the quant-ph archive. I'd like to refer some of my friends to it to get a debate going. Having the paper easily accessible will help get that off the ground.

One piece of analogy (or disanalogy) that you didn't explore very much is the geodesic. Is there an analog to the geodesic in your Everettian system? One thing that strikes me is that you might start seeing a conceptual divergence here. It's not clear to me that one can concoct a good notion of "test particle" for this game: even adding the smallest system possible to an existing multiverse (I hate that term) doubles its Hilbert space dimension (acting as if the Hilbert space is finite to begin with). There is no such thing as a small system maintaining a negligible perturbation to the larger system.

27 February 2001, to Växjö Meeting participants, "Claude Elwood Shannon"

To those friends I contacted for the "Shannon meets Bohr" session in Växjö:

We at Bell Labs were all deeply moved yesterday to learn of Claude Shannon's death over the weekend. Like for many Alzheimer's patients, however, we knew that it may have been a blessing in disguise. You can read more about Shannon's life in the New York Times obituary section today: http://www.nytimes.com/pages/national/text/index.html#obits

Some weeks ago, I found myself using the following words for a recommendation letter I was asked to write:

As it turns out, today January 16, we had a dedication ceremony at Bell Labs for a bronze bust of Claude Shannon, the founder of classical information theory. That struck me as symbolic. Since joining the laboratory, I have been asking myself over and over what role I might play in furthering the legacy of Shannon.

Those words hold just as true for me for this midsummer's meeting. Let us use information theory as a sword and finally defeat this mystery of the quantum.

Postpartum

16 May 2000, "History Reduced to Eight Pages"

The past exists only insofar as it is recorded in the present.
(John Archibald Wheeler)

Dear friends and family who have written or
called with concern over the Los Alamos fires,

I apologize at the outset for sending a mass mailing like this. All of you hold a unique place in my mind, but time is scarce at the moment.

Sometime Wednesday or Thursday last week, the Cerro Grande fire invaded our home in Los Alamos and took essentially all that was in it. The only exceptions were our car, a couple of suitcases of clothes and memories, and most importantly, our lives. Kiki, Emma, and I are all physically well. So are our "boys," the golden retrievers Albert and Wizzy. The last few days we have received shelter and comfort in Albuquerque in the home of Carl Caves and Karen Kahn. The boys have been in the foster care of Chris Hains (where they likely view themselves as having a resort vacation).

We saw the remains of our home Sunday: all that was left solid was the chimney. The thought of it makes us quite sad at times. I count as my greatest loss all the "intellectual property" that never made it or could not make it into my computer – my thousands of pages of meticulous calculation, my sad poems from sad times and my pre-electronic correspondence, my collection of over 700 physics, math, and philosophy books. The evidence of my academic life has been reduced to the eight pages of my vita. Kiki counts as her greatest loss her many family heirlooms, her red and her yellow "biker" jackets, her antique cookbooks, the quilts and cookie cutters of her grandmother, some forgotten jewelry of great sentiment, and the large library of children's books she had built up for Emma.

In the grand scheme of things, all of this will of course fade. Our daughter is a great source of strength and purpose. The responsibility of rebuilding a good life for her sake, if for nothing else, is at the top of our minds.

With this loss, as with most events in my life, I have tried to look for a lesson of deeper significance. It won't come as a surprise to most of you that my thoughts have turned toward quantum mechanics. What I have always found wonderful in the theory is its indication that our world is more malleable than was thought in classical times. With our experimental interventions into Nature, we have the opportunity to shape it in unforeseen and perhaps wonderful ways. How, now that so many of my records are gone, I hope there is a grain of truth in John Wheeler's words. For then, the past would be every bit as open as the future. I could see that as a kind of strange recompense in this time of trouble. But in this, just as for all things, only time will tell.

Thoughts of all,

Chris

Index of Names

Printed in the United States
by Baker & Taylor Publisher Services

Printed in the United States
by Baker & Taylor Publisher Services